IORA International Conference on Operations Research (IORA-ICOR 2017)

Competing in the Era of Analytics

IOP Conference Series: Materials Science and Engineering
Volume 332

Tangerang Selatan, Indonesia
12 October 2017

Editors:

Toni Bakhtiar
Agnes Puspitasari Sudarmo
Henry Kurniawati

ISBN: 978-1-5108-6112-1
ISSN: 1757-8981

Printed from e-media with permission by:

Curran Associates, Inc.
57 Morehouse Lane
Red Hook, NY 12571

Some format issues inherent in the e-media version may also appear in this print version.

Printed by Curran Associates, Inc. (2018)

For permission requests, please contact the Institute of Physics at the address below.

Institute of Physics
Dirac House, Temple Back
Bristol BS1 6BE UK

Phone: 44 1 17 929 7481
Fax: 44 1 17 920 0979

techtracking@iop.org

Additional copies of this publication are available from:

Curran Associates, Inc.
57 Morehouse Lane
Red Hook, NY 12571 USA
Phone: 845-758-0400
Fax: 845-758-2633
Email: curran@proceedings.com
Web: www.proceedings.com

Table of contents

Volume 332

Indonesian Operations Research Association - International Conference on Operations Research 2017

12 October 2017, Tangerang Selatan, Indonesia

Accepted papers received: 9 March 2018
Published online: 5 April 2018

IOP Conf. Series: Materials Science and Engineering **332** (2018) 011001 doi:10.1088/1757-899X/332/1/011001

IOP Conference Series:

Materials Science and Engineering

Editors:
Toni Bakhtiar (*Bogor Agricultural University*)
Agnes Puspitasari Sudarmo (*Universitas Terbuka*)
Heny Kurniawati (*Universitas Terbuka*)

UNIVERSITAS TERBUKA

Organized by: Faculty of Mathematics and Natural Sciences, Universitas Terbuka
Supported by: Indonesian Operations Research Association, Department of Mathematics Bogor Agricultural University, Department of Mathematics Universitas Padjadjaran, Department of Computer Science Universitas Pakuan, Faculty of Engineering Universitas Katolik Indonesia Atma Jaya, Department of Management Universitas Indonesia

IOP Conf. Series: Materials Science and Engineering **332** (2018) 011001 doi:10.1088/1757-899X/332/1/011001

THE CONFERENCE
IORA International Conference on Operations Research 2017

Date: 12th October 2017 (Thursday), 08.00 – 17.00
Venue: Universitas Terbuka Convention Center (UTCC)
Jl. Cabe Raya, Pondok Cabe, Pamulang,
Tangerang Selatan 15418, Indonesia

In the spirit to promote decisions based analytics through OR/MS, the theme of the conference is

"Competing in the Era of Analytics."

The primary objectives of the conference are:

1. to facilitate interaction between OR/MS researchers and academicians in discussing current challenges that need to be addressed as well as highlighting new developments of methods, algorithms, and tools in the field,
2. to provide OR/MS researchers, academicians and practitioners an appropriate platform for sharing experiences, communication and networking with other experts within the nation and from around the world in maximizing the contribution of OR/MS for sustainable growth, promoting of a knowledge-based economy, and utilizing the limited resources.

IOP Conf. Series: Materials Science and Engineering **332** (2018) 011001 doi:10.1088/1757-899X/332/1/011001

FOREWORD
IORA International Conference on Operations Research 2017

Conference Chair:
Dr. Agnes Puspitasari Sudarmo, *Universitas Terbuka, Indonesia*

It is well-known that the use of data in decisions making is not a new idea. But the field of business analytics that was born in the mid-1950s, with the advent of analytical tools that could digest a bulky quantity of information and perceive patterns in it far more quickly than the unassisted human mind ever could. Taking full advantage of big data's potential means companies must comprehend analytics into their strategic vision and utilize it to provide better and faster decisions, i.e., promote decisions based on analytics rather than instinct, while in other side, volume of data continues to double every three years as information surges in from digital platforms. Thus, analytical capability helps decision makers look beyond their own perspective in discerning real pattern and expecting opportunity.

Operation research as well as management science (OR/MS) has had an impressive contribution on improving the efficiency of numerous organizations around the world by offering a best solution. In the process, OR/MS has made a significant support to increasing the productivity of the economies of various countries. In this era of data-driven analytics, OR/MS is an ultimate tool for technical professionals who want to acquire the knowledge and skills required to incorporate analytics to solve real business problems.

This second conference, IORA International Conference on Operations Research 2017, is held in conjunction with Universitas Terbuka National Seminar on Mathematics, Sciences, and Technology 2017. The conference and seminar initiate to bring together OR/MS researchers, academicians and practitioners, whose collective work has sustained continuing OR/MS contribution to decision-making in many fields of application. It can be considered as good platforms for the OR/MS community, particularly in Indonesia, to meet each other and to exchange ideas. Thank you!

■

IOP Conf. Series: Materials Science and Engineering **332** (2018) 011001 doi:10.1088/1757-899X/332/1/011001

IORA ICOR 2017

WELCOME REMARKS
IORA International Conference on Operations Research 2017

From the President
Indonesian Operations Research Association (IORA)
Prof. Sudradjat Supian

Drawn extensively from the divisions of mathematics and science, operations research (OR) applies cutting-edge statistical analysis and mathematical modeling to address a number of conflicting interests in inventory planning and scheduling, production planning, transportation, financial and revenue management and risk management as well as to improve decision-making mechanism. Yet, the importance of analytics inclusion into managerial decision making has grown significantly in the recent years. Massive amounts of data are now available for many organizations and businesses to be analyzed to support decision making process. How will big data fundamentally change what we do in OR? Analytics—the scientific process of transforming data into insight for making better decisions is now our key point.

For this conference we choose the following theme for our stand of work: "Competing in the Era of Analytics." It is our conviction that OR can make significant contribution to this emerging situation and challenging domain of research. It seems that the practice of big data analytics would fall entirely in the field of OR. By this conference we aim to promote the increase in the use of OR as a practical tool for problems in many aspects of data analysis. The ability to analyze large and complicated problems with operations research techniques is expected to suggest better decisions.

Establishment of Indonesian Operations Research Association (IORA) in 2014 is evidently intended to reinforce the above mentioned initiative. We hope IORA can be considered as good platforms for OR researchers, academicians and professionals in Indonesia to meet each other, exchange ideas and strengthen their collaboration.

Welcome to Tangerang Selatan, Indonesia, and welcome to IORA-ICOR 2017.

■

IOP Conf. Series: Materials Science and Engineering **332** (2018) 011001 doi:10.1088/1757-899X/332/1/011001

COMMITTEE

IORA International Conference on Operations Research 2017

Conference Chair: Dr. Agnes Puspitasari Sudarmo, *Universitas Terbuka*

Program Chair: Prof. Toni Bakhtiar, *Bogor Agricultural University*

Advisory Committee:

Prof. Ojat Darojat, *Universitas Terbuka*
Prof. Tian Belawati, *Universitas Terbuka*
Dr. Sri Harijati, *Universitas Terbuka*
Prof. Sudradjat Supian, *Universitas Padjadjaran*
Dr. Amril Aman, *Bogor Agricultural University*
Prof. Hadi Sutanto, *Atma Jaya Catholic University*
Prof. Fatma Susilawati Mohamad, *Universiti Sultan Zainal Abidin, Malasiya*
Prof. Abby Tan Chee Hong, *Universiti Brunei Darusalam*
Prof. Peerayuth Charnsethikul, *Kasetsart University, Thailand*
Prof. Abdul Talib bin Bon, *Universiti Tun Hussein Onn, Malaysia*
Prof. Soewarto Hardhienata, *Universitas Pakuan*
Prof. Nguyen Phi Trung, *Ho Chi Minh City University of Technology and Education, Vietnam*
Prof. Mustafa Mamat, *Universiti Sultan Zainal Abidin, Malaysia*
Prof. Edy Soewono, *Institute of Technology Bandung*

Scientific Committee:

Dr. Admi Syarif, *Lampung University*
Prof. Djati Kerami, *Universitas Indonesia*
Prof. Asep K. Supriatna, *Universitas Padjadjaran*
Dr. Subchan, *Kalimantan Institute of Technology*
Prof. Ilias Mamat, *Quest International University Perak, Malaysia*
Prof. Ismail bin Mohd., *Univeristi Malaysia Perlis, Malaysia*

Board of Reviewers:

Prof. Toni Bakhtiar, *Bogor Agricultural University*
Prof. Soewarto Hardhienata, Universitas Pakuan
Prof. Sudradjat Supian, *Universitas Padjadjaran*
Prof. Hadi Sutanto, *Atma Jaya Catholic University*

IOP Conf. Series: Materials Science and Engineering **332** (2018) 011001 doi:10.1088/1757-899X/332/1/011001

Dr. Ema Carnia, *Universitas Padjadjaran*
Dr. Herlina Napitupulu, *Universitas Padjadjaran*
Dr. Jaharuddin, *Bogor Agricultural University*
Dr. Nursanti Anggriani, *Universitas Padjadjaran*
Dr. Ratih Dyah Kusumawati, *Universitas Indonesia*
Dr. Sukono, *Universitas Padjadjaran*
Dr. Subiyanto, *Universitas Padjadjaran*

External Affairs and Publication Committee:

Heny Kurniawati, *Universitas Terbuka*
Dina Mustafa, *Universitas Terbuka*
Pramono Sidi, *Universitas Terbuka*
Sitta Alief Farihati, *Universitas Terbuka*
Agung Prajuhana Putra, *Universitas Pakuan*

Finance Committee:

Sri Kurniati Handayani, *Universitas Terbuka*
Ema Kurnia, *Universitas Pakuan*
Eman Lesmana, *Universitas Padjadjaran*
Eneng Tita Tosida, *Universitas Pakuan*

Event Management Committee:

Tengku Eduard Sinar, Universitas Terbuka
Fajar Delli Wihartiko, *Universitas Pakuan*
Farida Hanum, *Bogor Agricultural University*
Prapto Tri Supriyo, *Bogor Agricultural University*

IOP Conf. Series: Materials Science and Engineering **332** (2018) 011002 doi:10.1088/1757-899X/332/1/011002

Peer review statement

All papers published in this volume of *IOP Conference Series: Materials Science and Engineering* have been peer reviewed through processes administered by the proceedings Editors. Reviews were conducted by expert referees to the professional and scientific standards expected of a proceedings journal published by IOP Publishing.

IOP Conf. Series: Materials Science and Engineering **332** (2018) 012001 doi:10.1088/1757-899X/332/1/012001

Trends in currency's return

A Tan[1], M Shahrill[2], S Daud[3] and E Leung[4]

[1]Faculty of Science, Universiti Brunei Darussalam, Jalan Tungku Link, BE1410, Brunei Darussalam.
[2]Sultan Hassanal Bolkiah Institute of Education, Universiti Brunei Darussalam, Jalan Tungku Link, BE1410, Brunei Darussalam.
[3]UBD School of Business and Economics, Universiti Brunei Darussalam, Jalan Tungku Link, BE1410, Brunei Darussalam.
[4]Faculty of Science, Universiti Brunei Darussalam, Jalan Tungku Link, BE1410, Brunei Darussalam.

Corresponding author: abby.tan@ubd.edu.bn

Abstract. The purpose of this paper is to show that short-ranged dependence prevailed for Singapore-Malaysia exchange. Although, it is perceived that there is some evidence of long-ranged dependence [1,2], it is still unclear whether Singapore-Malaysia exchange indeed exhibits long-ranged dependence. For this paper, we focus on the currency rate for a sixteen-year period ranging from September 2002 to September 2017. We estimate the Hurst parameter using the famous rescaled R/S statistics technique. From our analysis, we showed that the Hurst parameter is approximately 0.5 which indicates short-ranged dependence. This short memory property is further validated by performing a one-tailed z-test whose alternative hypothesis is that the Hurst parameter is not 0.5 at 1% significance level. We conclude that the alternative hypothesis is rejected. The existence of short memory proves that the behaviour of the exchange rate is unpredictable, supporting the efficient market hypothesis, which states that not only is price movement completely random but also tomorrow's prices are predicted by all the information in today's prices.

1. Introduction

Over the last ten years or more, there has been overwhelming evidence [3,4,5] stressing that returns of assets are strongly correlated. More importantly, as the time lags between the observations increase, its autocorrelation function decays at a rate that is too slow to be considered as an exponential rate. In fact, for a very large time lag between observations, a power law decay best describes the decay of its autocorrelation function. This phenomenon is well-known as the long memory effect. However, there is considerably less studies on the properties of the foreign exchange market.

This paper is divided into several sections. In the next section, we introduce short process and explain evidence of its existence. We then give examples of existing short memory process. In the penultimate section, we present numerical results of the R&S statistics and deduce the Hurst parameter from our result.

2. Short memory process and its application

A stochastic process can be categorized as having either short or long memory [6,7]. Define a covariance stationary process, X_t , that has covariance, $\lambda_s = \text{cov}(X_t, X_{t-s})$. If the sum of covariance function is finite, the process is said to have short memory. The covariance of the short memory process decays at a rate best described by exponential decay ie e^{-s} where time lag $s \gg 1$. It should be noted that the exponential rate at which the short memory process decays is much more rapid than the power law decay of the long memory process. For a short memory process, its variance $\mathrm{var}(X_t)$ grows linearly according to $\mathrm{var}(X_t) \sim t$.

In finance, ARCH [8] is the first short memory process used to model the short-ranged dependence of the volatility process although the use of short memory process in finance can be traced back to Bachelier [9] who assumes that change in the asset price is modelled by a simple random process with short memory properties. In 1973, Fischer Black and Myron Scholes [10] derived the first universally accepted option pricing model which was primarily based on the asset price following a Geometric Brownian motion model which is another short memory process.

3. Evidence of short memory in financial market

There is considerable evidence [11,12,13] suggesting that asset price and foreign exchange return exhibit short memory properties. This implies the unpredictability of foreign exchange market in the longer run. It is widely accepted that the foreign exchange market is more unpredictable than before. This can be partly attributed to larger trading volumes in foreign exchange trading which might sometimes render central bank intervention less effective. Moreover, there are numerous unpredictable factors that determine the strength of the currency such as inflation, interest rate, government policies, political events etc. Such developments are occurring with increasing frequency which probably explain the unpredictability of the market.

4. Summary of short memory model

The autoregressive conditional heteroscedasticity (ARCH) family of models has been widely used in the financial time series [8,14]. It does not capture the long memory property. Other short memory process includes ARMA, AR model [15,16] etc.

In continuous time setting, there exists a family of continuous random processes [17] which captures both the short and long memory property known as Fractional Brownian motion, B_H where H is the Hurst parameter. For the case of short memory process, H takes value of 0.5 and this particular Fractional Brownian motion is also known as Standard Brownian motion [18]. Define a standard Brownian motion as a random process $B = \{B_t : t \in [0,\infty)\}$ which has the basic properties as follows

1. $B_t = 0$ with probability 1
2. B_t has stationary and independent increments
3. B_t is normally distributed with mean 0 and variance t for each $t \in (0,\infty)$
4. $\frac{dB_t}{dt} = \eta(t)$ where $\eta(t)$ is the Gaussian White noise

There are various methods to estimate the Hurst parameter H namely the R/S statistic, the correlogram method, variance plot etc [16].

5. Rescaled R&S methodology

We used the rescaled range method to deduce the long memory properties of the Singapore-Malaysia currency exchange. In total, we have 4090 observations spanning 16 years. This method depends on ranges to be analysed. Size of each range is $\frac{1}{2}, \frac{1}{4}, \frac{1}{8}, \frac{1}{16}$ and $\frac{1}{32}$ of the entire data series.

Subsequently, we calculated the mean for each range. For instance, the range with 2 intervals will require two means to be deduced. In total, there will be 1+2+4+8+16+32=63 means to be deduced.

The formula for mean,m, is $m = \frac{1}{n}\sum_{i=1}^{i=n} x_i$, where x_i is the observation. In the next step, we create a new series of deviation for each range by subtracting its corresponding mean from the range as follows:

$y_t = x_t - m$ where

y_t is the new time series adjusted from deviation from mean. Having created 63 new series, we will work out the running total of each range's deviation from its mean as follows

$$y = \sum_{i=1}^{n} y_i$$

We will the calculate the widest difference between the largest and smallest running total for each series denoted as R_i which is given as follows

$$R_i = \max(y_1, y_2, \dots, y_n) - \min(y_1, y_2, \dots, y_n)$$

We also calculated the standard deviation, S, of each range and subsequently deduce the $\log(R/S)$ for each range. For range with more than one interval, we averaged out the values of $\log(R/S)$. To deduce the Hurst parameter, we plot the graphs of the averaged $\log(R/S)$ against its corresponding log intervals. The gradient of the graph will actually give us the Hurst parameter.

6. Analysis

Table of log(interval) and $\log(R/S)$ is shown below (Table 1). Graph of $\log(R/S)$ against log(interval) are plotted in Figure 1. We plot best-fit straight line and deduce that the gradient of the straight line is 0.46327 thus proving that the currency return does not exhibit any long memory properties.

Table 1. log(interval) and $\log(R/S)$

log(interval)	$\log(R/S)$
3.61161	1.71437
3.31059	1.37076
3.00945	1.19455
2.70842	1.15628
2.40654	1.05241
2.10721	0.93695

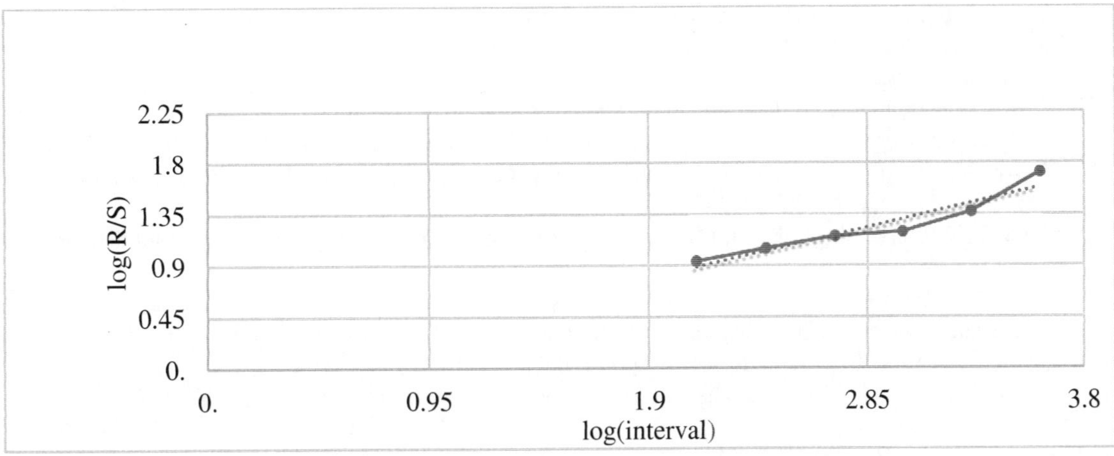

Figure 1. Plot of log(*R/S*) against log(interval)

Other properties of this plot are as follows:
Slope coefficient: 0.463265047
Intercept coefficient: -0.086907537
Standard error of slope: 0.077309364
Standard error of intercept: 0.224569686
R-squared: 0.899770228
Standard error of regression: 0.097347598
F-test overall: 35.90830186
Degrees of freedom (n-k): 4
Regression SS: 0.34028699
Residual SS: 0.037906219

Lastly, we want to show that the Hurst parameter is not different from 0.5 by performing a statistical test. We found that the scatter plot is roughly linear and the residual plot does not exhibit any pattern and has equal variance. We conduct a two-tailed *t*-test on the Hurst parameter at 1% significance level with 4 degrees of freedom. Define null and alternative hypothesis as follows:
Null hypothesis, H_0: $H = 0.5$
Alternative hypothesis, H_1: $H \neq 0.5$

The test statistics, t is given by $t = \frac{0.463265047-0.5}{0.077309364.} = -0.47$ which is greater than the *t*-critical value of -4.604. Thus, we do not reject the null hypothesis.

7. Conclusion

It is proven that the return of Singapore-Malaysia exchange rate does not exhibit long ranged-dependence. This result is consistent with [11,12,13] which strongly suggests that the exchange rate exhibits an unpredictable pattern and will revert to its long-term mean.

References

[1] Soofi Abdol S, Shouyang W, and Yuqin Z 2006 Testing for long memory in the Asian foreign exchange rates *Journal of Systems Science and Complexity* **19.2** 182-190
[2] Cheung Y-W 1993 Long memory in foreign-exchange rates *Journal of Business & Economic Statistics* **11.1** 93-101
[3] Ding Z, Granger C W J, and Engle R F 1993 A long memory property of stock market returns and a new model *Journal of empirical finance* **1.1** 83-106
[4] Henry O T 2002 Long memory in stock returns: some international evidence *Applied Financial Economics* **12.10** 725-729
[5] Arouri M E H, et al. 2012 Long memory and structural breaks in modeling the return and volatility dynamics of precious metals *The Quarterly Review of Economics and Finance* **52.2** 207-218.
[6] Wang F, et al 2006 Scaling and memory of intraday volatility return intervals in stock markets *Physical Review E* **73.2** 026117
[7] Chow K V, et al. "Long-term and short-term price memory in the stock market." Economics Letters 49.3 (1995): 287-293.
[8] Engle R F 1982 Autoregressive conditional heteroscedasticity with estimates of the variance of United Kingdom inflation *Econometrica: Journal of the Econometric Society* 987-1007
[9] Bachelier L 1900 *Théorie de la spéculation* (Gauthier-Villars)
[10] Black F and Scholes M 1973 The pricing of options and corporate liabilities *Journal of political economy* **81.3** 637-654
[11] Eitelman P and Vitanza J 2008 *A non-random walk revisited: Short-and long-term memory in asset prices*
[12] Limam I 2003 Is long memory a property of thin stock markets? International evidence using Arab

countries *Review of Middle East Economics and Finance* **1.3** 251-266
[13] Barkoulas J T, et al. 2000 Persistent dependence in foreign exchange rates? A reexamination *Working Papers in Economics* **252**
[14] Duan J-C 1997 Augmented GARCH (p, q) process and its diffusion limit *Journal of Econometrics* **79.1** 97-127
[15] Weiss A A 1984 ARMA models with ARCH errors *Journal of time series analysis* **5.2** 129-143
[16] Taylor S J 2008 *Modelling financial time series* (World Scientific)
[17] Mandelbrot B B and Ness J W V 1968 Fractional Brownian motions, fractional noises and applications *SIAM review* **10.4** 422-437
[18] Revuz D and Yor M 2013 Continuous martingales and Brownian motion *Springer Science & Business Media* **293**
[19] Granger CWJ and Marmol F 1998 *The correlogram of a long memory process plus a simple noise*

A spatial model for a stream networks of Citarik River with the environmental variables: potential of hydrogen (PH) and temperature

A Bachrudin[1], N B Mohamed[2], S Supian[3], Sukono[4*], Y Hidayat[5]

[1,5]Department of Statistics, Faculty of Mathematics and Natuaral Sciences, Universitas Padjadjaran, INDONESIA

[2]School of Informatics and Applied Mathematics, Universiti Malaysia Terengganu, MALAYSIA

[3,4]Department of Mathematics, Faculty of Mathematics and Natuaral Sciences, Universitas Padjadjaran, INDONESIA

*Corresponding author e-mail: sukono@unpad.ac.id

Abstract. Application of existing geostatistical theory of stream networks provides a number of interesting and challenging problems. Most of statistical tools in the traditional geostatistics have been based on a Euclidean distance such as autocovariance functions, but for stream data is not permissible since it deals with a stream distance. To overcome this autocovariance developed a model based on the distance the flow with using convolution kernel approach (moving average construction). Spatial model for a stream networks is widely used to monitor environmental on a river networks. In a case study of a river in province of West Java, the objective of this paper is to analyze a capability of a predictive on two environmental variables, potential of hydrogen (PH) and temperature using ordinary kriging. Several the empirical results show: (1) The best fit of autocovariance functions for temperature and potential hydrogen (ph) of Citarik River is linear which also yields the smallest root mean squared prediction error (RMSPE), (2) the spatial correlation values between the locations on upstream and on downstream of Citarik river exhibit decreasingly

Keywords: Stream networks, spatial dependence, autocovariance function, kernel convolution, kriging.

1. Introduction

A Classical geostatistics is widely used in environmental science and ecology which are normally applied the statistical instruments such as variogram, semivariogram, covariance functions and almost all of the statistical instruments are based upon a Euclidean distance framework (e.g. see Cressie, [2], Recently, there was a new class of geostatistics that is in the basis of non-Euclidean distance [9], Curriero, [4; 5], Rathbun, [12], and Higdon, [7], Barry & Ver Hoef, [1]). Curriero [4], pointed out that conditions of a metric are not sufficient proof of the validity of distance to yield positive definite functions. Such distances cannot be used without proof in covariance and semivariogram models. Covariance $\mathrm{cov}(d(s_i, s_j))$ calculated using metric $d(s_i, s_j)$ must satisfy the non-negative definiteness property. Briefly, a metric distance that must meet the geometric properties and covariance function based on the metric distance must satisfy the non-negative definiteness property.

Ver Hoef et al.[14] and Cressie et al. [3] developed a geostatistcs for non-Euclidean distance which are applied for a stream networks. The non-Euclidean distance is called as a stream or hydrological distance. They built several autocovariance functions which were permissible for stream distance met the both properties. Using moving-average construction (kernel convolution), they extended a spatial model for stream networks. Ver Hoef et al. [14] developed the spatial model based on both stream distance and flow direction, namely tail-down and tail-up and this paper will discuss tail-up model. A process of this model moves from downstream to upstream which is the opposite of flow direction. In the tail-up model, it is possible that flow volume from one upstream segment is larger than the flow volume for the other upstream segment, then it might want to weight according to flow volume. The weighting processes can be used a flow volume, a stream order, or a proxy variable such as a watershed area, or other measures.

The objective of this paper is to analyze a capability of a predictive on two environmental variables, potential of hydrogen (ph) and temperature using ordinary kriging.

2. A Spatial model for a stream networks

As has mentioned previously that Ver Hoef et al. [14] developed a new class of geostatistics using a non-Euclidean distance for a stream networks, and this distance is well known as a stream or a hydrological distance. However, for studies related to hydrology, using Euclidean distance is not appropriate. Since the movement of materials in the water only occurs within stream networks, it is more appropriate to use distance defined only along the steam. Stream distance between any two points is obtained by summing up the length of segments which connect them [8]. The different stream and Euclidean distance is described in Figure 1. One problem arises is that how to develop a permissible autocovariace function using stream distance. An answer of this problem is described as follows.

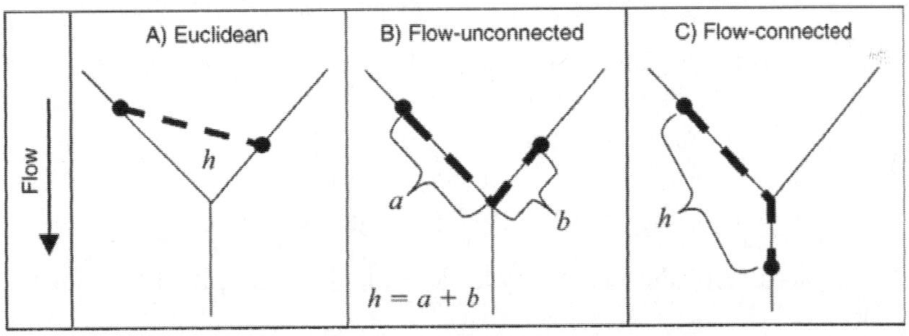

Figure 1. Stream and Euclidean distance

Barry and Ver Hoef [1] showed that large class of autocovariance functions can be developed by creating random variables as the integration of a mowing average function over white noise random process:

$$Z(s) = \int_{-\infty}^{\infty} g(x-s|\boldsymbol{\theta})W(x)dx \tag{1}$$

where $W(x)$ the white noise is random process and $g\left(x-s|\boldsymbol{\theta}\right)$ is the moving average function. Based on the moving average construction, the autocovariance of $Z(s)$ and $Z(s+h)$ is given as follows (Ver Hoef et al. [14]):

$$\begin{aligned} C(h|\boldsymbol{\theta}) &= Cov(Z(s), Z(s+h)) \\ &= \int_{-\infty}^{\infty} g(x-s|\boldsymbol{\theta})g\left(x|\boldsymbol{\theta}\right) \end{aligned} \tag{2}$$

Equation (2) is a permissible autocovariance for stream networks and h is a stream distance. Proof of Equation (2) can be found in appendix. From Equation (2) can be obtained a variety of valid covariance functions which is given in Table 1 as follows:

Table 1. Permissible autocovariance functions

Name	**Autocovariance function**
Linear with sill	$C(h\|\boldsymbol{\theta}) = \begin{cases} \theta_0 + \theta_1 & h = 0 \\ \theta_1\left(1 - \dfrac{h}{\theta_2}\right) & 0 < h < \theta_2 \\ 0 & h \geq \theta_2 \end{cases}$
Spherical	$C(h\|\boldsymbol{\theta}) = \begin{cases} \theta_0 + \theta_1 & h = 0 \\ \theta_1\left(1 - \dfrac{h}{2\theta_2} + \dfrac{1}{2}\left(\dfrac{h}{\theta_2}\right)^3\right) & 0 < h < \theta_2 \end{cases}$
Mariah	$C(h\|\boldsymbol{\theta}) = \begin{cases} \theta_0 & h = 0 \\ \theta_1 \ln\left(\dfrac{h/\theta_2 + 1}{h/\theta_2}\right) & 0 < h \end{cases}$
Exponential	$C(h\|\boldsymbol{\theta}) = \theta_1 \exp\left(-h/\theta_2\right), h > 0$

Equation (2) and the autocovariance functions in Table are described in detail in Table 1. In a steam networks, there will be a finite number of stream segments, and it gives index them arbitrarily with $i = 1, 2, 3$ as Figure 2. One of stream networks that is called as a tail up model, the moving average function begins from downstream (segment $i = 3$ in Figure 2) , then the moving average into two parts, with one part going up segment $i = 2$, and the other going up segment $i = 3$. If it uses no weighting at all, so variance of the random variable is $\int_{-\infty}^{\infty}\left(g\left(x|\boldsymbol{\theta}\right)\right)^2 dx + v_j^2$ from Figure 2 for r_3 would be greater than for r_2. When the moving average spilt into parts, then the variance of each parts could be a different. To keep the variances are constant, sum of each weight segment must be one. For example, let ω_1 and ω_2 be a weight for segment $i = 2$ and $i = 3$ in Figure 2, so to keep the variances are constant that must be $\omega_1 + \omega_2 = 1$. This constraint that stream networks is to keep a stationary process.

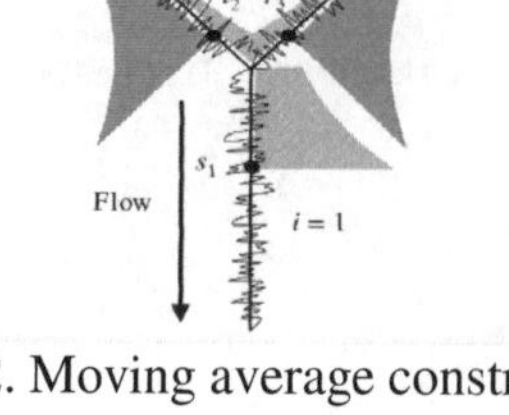

Figure 2. Moving average constructions

Autocovariance function in Equation (2) incorporates a weighting ω of a segment of stream networks is given as follows:

$$C\left(s_i,t_j\middle|\boldsymbol{\theta}\right)=\begin{cases}0 & \text{if s and t are not flow-connected,}\\ C(0)+v_j^2 & \text{if s = t,}\\ \prod_{k\in B_{s_i,t_j}}\sqrt{\omega_k}\,C\left(d(s_i,t_j\right) & \text{otherwise}\end{cases} \tag{3}$$

where v_j^2 is a discontinuity at $h=0$ which is the nugget effect in geostatistical terminology, $d\left(s_i,t_j\right)$ is the steam networks, $C\left(d\left(s_i,t_j\right)\right)$ can be found in Table 1, and B_{s_i,t_j} is the set of stream segments between two locations, including the segment for the upstream location, but excluding the segment for the downstream location. If the weighting is applied in a river, so it can incorporate flow volume. In the absence of any flow volume characteristics, it can simply weight each split in the moving average function by $\frac{1}{2}\sqrt{2}$ [14].

The ultimate objective of this paper is to measure a predictive ability for stream networks which is measured by root mean square prediction error (RMSPE):

$$RMSPE=\sqrt{\frac{\sum_{i=1}^{n}(z_i-\hat{z}_i)}{n}} \tag{4}$$

It is estimated by leave-one-out cross-validation after estimating the covariance parameters. Where $\hat{z}_i$ is the ith site kriged using all the other $n-1$ sites.

A kriging method is normally applied in geostatistics to predict $z(s)$ at point s_0 using a model:

$$Z(s_0)=\sum_{i=1}^{n}\lambda_i Z\left(s_i\right) \tag{5}$$

Using the unbiased constraint $\sum_{i=1}^{n}\lambda_i=1$, and minimizing variance of predictor error, λ_i can be computed the following expression as follows:

$$\boldsymbol{\lambda}^t=\left(\lambda_1,\ldots,\lambda_n\right)=\left(\mathbf{c}+\mathbf{1}\frac{\left(1-\mathbf{1}^t\Sigma^{-1}(\boldsymbol{\theta})\mathbf{c}\right)}{\mathbf{1}^t\Sigma^{-1}(\boldsymbol{\theta})\mathbf{1}}\right)\Sigma^{-1}(\boldsymbol{\theta}) \tag{6}$$

where $\boldsymbol{\Sigma}(\boldsymbol{\theta})$ a covariance matrix is whose elements is in Equation (3), $\mathbf{c}=\left(C(s_0-s_1),\ldots,C(s_0-s_n)\right)^t$. The interpolation $\hat{z}(s_0)$ gives a best linear unbiased predictor [2].

3. Methodology

Many applications of this model are widely used on a river (e.g. see Gareta et al. [6], Peterson et al. [10], [11]),. A case study of this paper is the basin of the Citarik River which is a small watershed of the Citarik River. This river is located in Bandung regency. A study area is described by Figure 3. The collected dataset has two environmental variables (ph and temperature) sampled at 25 sites. These variables are potential of hydrogen (PH) and temperature. Addition to these variables, to measure a spatial dependence is a volume discharge.

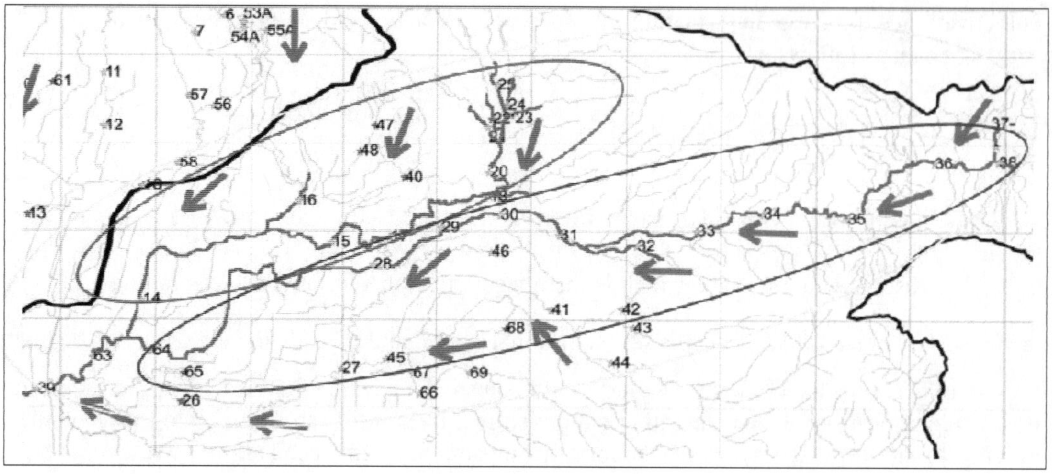

Figure 3. Study area

After computing a matrix of stream distance, then compute a matrix of covariance $\mathbf{\Sigma}(\boldsymbol{\theta})$ whose elements are in Equation (3) for each autocovariance function is in Table 1. Maximum likelihood is applied to Estimate of the parameters $\boldsymbol{\theta} = (\theta_0, \theta_1, \theta_2)^t$ by assuming that $\mathbf{Z(s)} = (Z(s_1), \ldots, Z(s_n))^t$ is normally multivariate distributed, or $\mathbf{Z(s)} \sim N(\mathbf{0}, \mathbf{\Sigma}(\boldsymbol{\theta}))$. Finally, compute RMSPE using Equation (4). Its computation is used R programming and a mapping of the prediction results is employed GIS software.

4. Empirical Results

Figure 4 shows the histogram of potential hydrogen (PH) and temperature indicated a crude distribution of their variables. The histograms plays an important role in a diagnostic of the probability distribution. Besides the histogram, some descriptive statistics of the two variables are computed to see some statistical measures shown in Table 2.

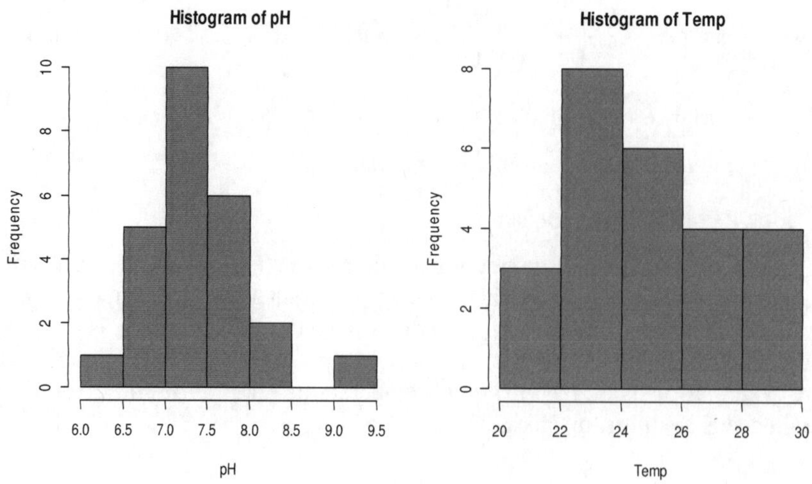

Figure 4. Histogram of PH and temperature

Table 2 Descriptive statistics for PH and temperatures

PH		Temperature	
Mean	7.44	Mean	25.04
Standard Error	0.135277	Standard Error	0.567098
Median	7.3	Median	25
Standard Deviation	0.676387	Standard Deviation	2.835489
Kurtosis	2.473181	Kurtosis	-0.77294
Skewness	1.106285	Skewness	0.003662
Range	3.2	Range	10
Minimum	6.3	Minimum	20
Maximum	9.5	Maximum	30
Sum	186	Sum	626
Count	25	Count	25

Following Figure 4 and Table 2 show that the histogram shape of PH variable tends leptokurtic distribution and has heavier tails distribution supported by the estimate kurtosis is a positive value. In addition, this distribution exhibits a non-symmetric shape. For temperature variable shows that the distribution of data is fairly symmetric since the estimate of skewness is almost zero. Because of the negative kurtosis value is less than zero, then the distribution is light tails and is called a platykurtic.

Table 3. Model selection of autocovariance functions using aic statistic and the parameter estimates of the selected model

Environmental Variables	AIC Statistic			The Estimated Parameters		
	Exponential	Spherical	Linear	θ_0	θ_1	θ_2
Temperature	111.93	124.01	102.30	0	256.96	1194.36
PH	195.47	233.25	198.92	3.89	23.81	19999.96

Table 3.shows the least values of AIC statistics are respectively linear and Exponential auto covariance function for the variables of temperature and PH, but the values of AIC statistic for PH are not significantly different, so it is concluded the best fit of auto covariance for both of the variables is linear. For Linear that the estimated parameters are $\hat{\theta}_1 = 256.96$, and $\hat{\theta}_2 = 1194.36$ and for ph that the estimated parameters are $\hat{\theta}_0 = 3.89$ $\hat{\theta}_1 = 23.81$, and $\hat{\theta}_2 = 19999.96$. Of this result indicates that dependency of ph is longer than temperature to the quality of water along on the Citarik River, but the degree of dependence is higher.

The Table 4 and 5 indicate a spatial correlation between the locations of the Citarik River. The first column In Table 3 shows that the spatial correlation between the Location 37 and the locations 33, 36, 35, 34, 31, 30, 29, 28, 64, 63, and 69 are 0.234, 0.211, 0.192, 0.179, 0.079, 0.075, 0.071, 0.067, 0.05, 0.023, and 023, respectively. The values of the spatial correlations equal to zero in Table 3 and 4 which indicate unconnected flows between the locations. Following this result can be concluded that spatial correlation values between the locations on upstream and on downstream of Citarik river exhibit decreasingly, as well as spatial correlation in Table 4. The spatial correlation of temperature is generally higher than PH.

Table 4. The Spatial Matrix of PH

Titik	37	38	36	35	34	33	32	31	30	29	28	64	24	25	23	21	22	20	18	17	15	16	14	63	39
37	1																								
38	0	1																							
36	0.234	0.24	1																						
35	0.211	0.21	0.45	1																					
34	0.192	0.19	0.41	0.46	1																				
33	0.179	0.18	0.38	0.43	0.5	1																			
32	0	0	0	0	0	0	1																		
31	0.079	0.08	0.34	0.38	0.4	0.44	0.23	1																	
30	0.075	0.08	0.16	0.18	0.2	0.21	0.22	0.5	1																
29	0.071	0.07	0.15	0.17	0.2	0.2	0.21	0.4	0.5	1															
28	0.067	0.07	0.14	0.16	0.2	0.19	0.2	0.4	0.4	0.47	1														
64	0.05	0.05	0.11	0.12	0.1	0.14	0.15	0.3	0.3	0.35	0.37	1													
24	0	0	0	0	0	0	0	0	0	0	0	0	1												
25	0	0	0	0	0	0	0	0	0	0	0	0	0	1											
23	0	0	0	0	0	0	0	0	0	0	0	0	0	0	1										
21	0	0	0	0	0	0	0	0	0	0	0	0	0	0	0	1									
22	0	0	0	0	0	0	0	0	0	0	0	0	0	0	0	0	1								
20	0	0	0	0	0	0	0	0	0	0	0	0	0.06	0.06	0.12	0.12	0.12	1							
18	0	0	0	0	0	0	0	0	0	0	0	0	0	0	0	0	0	0	1						
17	0	0	0	0	0	0	0	0	0	0	0	0	0.02	0.02	0.05	0.05	0.05	0.21	0.22	1					
15	0	0	0	0	0	0	0	0	0	0	0	0	0.02	0.02	0.05	0.05	0.05	0.193	0.2	0.46	1				
16	0	0	0	0	0	0	0	0	0	0	0	0	0	0	0	0	0	0	0	0	0	1			
14	0	0	0	0	0	0	0	0	0	0	0	0	0.01	0.01	0.02	0.02	0.02	0.079	0.08	0.19	0.21	0.2	1		
63	0.023	0.02	0.05	0.05	0.1	0.06	0.07	0.1	0.2	0.16	0.17	0.23	0	0	0.01	0.01	0.01	0.036	0.04	0.09	0.09	0.1	0.23	1	
39	0.023	0.02	0.05	0.05	0.1	0.06	0.07	0.1	0.2	0.16	0.17	0.23	0	0	0.01	0.01	0.01	0.034	0.04	0.08	0.09	0.1	0.21	0.5	1

Table 5. The spatial matrix of temperature

Titik	37	38	36	35	34	33	32	31	30	29	28	64	24	25	23	21	22	20	18	17	15	16	14	63	39
37	1.00																								
38	0.00	1.00																							
36	0.23	0.23	1.00																						
35	0.20	0.20	0.43	1.00																					
34	0.17	0.18	0.38	0.44	1.00																				
33	0.16	0.16	0.38	0.40	0.46	1.00																			
32	0.00	0.00	0.00	0.00	0.00	0.00	1.00																		
31	0.07	0.07	0.30	0.34	0.39	0.43	0.23	1.00																	
30	0.06	0.06	0.14	0.16	0.18	0.20	0.21	0.46	1.00																
29	0.06	0.06	0.13	0.15	0.17	0.18	0.20	0.43	0.47	1.00															
28	0.05	0.05	0.12	0.14	0.15	0.17	0.18	0.39	0.43	0.46	1.00														
64	0.04	0.04	0.08	0.09	0.10	0.11	0.12	0.27	0.29	0.31	0.34	1.00													
24	0.00	0.00	0.00	0.00	0.00	0.00	0.00	0.00	0.00	0.00	0.00	0.00	1.00												
25	0.00	0.00	0.00	0.00	0.00	0.00	0.00	0.00	0.00	0.00	0.00	0.00	0.00	1.00											
23	0.00	0.00	0.00	0.00	0.00	0.00	0.00	0.00	0.00	0.00	0.00	0.00	0.00	0.00	1.00										
21	0.00	0.00	0.00	0.00	0.00	0.00	0.00	0.00	0.00	0.00	0.00	0.00	0.00	0.00	0.00	1.00									
22	0.00	0.00	0.00	0.00	0.00	0.00	0.00	0.00	0.00	0.00	0.00	0.00	0.00	0.00	0.00	0.00	1.00								
20	0.00	0.00	0.00	0.00	0.00	0.00	0.00	0.00	0.00	0.00	0.00	0.00	0.06	0.06	0.12	0.12	0.12	1.00							
18	0.00	0.00	0.00	0.00	0.00	0.00	0.00	0.00	0.00	0.00	0.00	0.00	0.00	0.00	0.00	0.00	0.00	0.00	1.00						
17	0.00	0.00	0.00	0.00	0.00	0.00	0.00	0.00	0.00	0.00	0.00	0.00	0.02	0.02	0.05	0.05	0.05	0.20	0.21	1.00					
15	0.00	0.00	0.00	0.00	0.00	0.00	0.00	0.00	0.00	0.00	0.00	0.00	0.02	0.02	0.04	0.04	0.04	0.18	0.19	0.44	1.00				
16	0.00	0.00	0.00	0.00	0.00	0.00	0.00	0.00	0.00	0.00	0.00	0.00	0.00	0.00	0.00	0.00	0.00	0.00	0.00	0.00	0.00	1.00			
14	0.00	0.00	0.00	0.00	0.00	0.00	0.00	0.00	0.00	0.00	0.00	0.00	0.01	0.01	0.02	0.02	0.02	0.07	0.07	0.17	0.19	0.20	1.00		
63	0.02	0.02	0.04	0.04	0.05	0.05	0.05	0.12	0.13	0.14	0.15	0.22	0.00	0.00	0.01	0.01	0.01	0.03	0.08	0.08	0.09	0.07	0.22	1.00	
39	0.02	0.02	0.04	0.04	0.05	0.05	0.05	0.12	0.13	0.14	0.15	0.22	0.00	0.00	0.01	0.01	0.01	0.03	0.08	0.07	0.08	0.08	0.20	0.46	1.00

Based on the Table 6 shows that the smallest value of RMSPE for ph and temperature for linear, respectively that the results are consistent to using AIC statistic.

Table 6. Performances of models and predictions using RMSPE

Autocovariace Function	**RMSPE**	
	PH	**Temperature**
Linear	0.62	0.66
Spherical	0.70	0.69
Exponential	0.73	1.89

It is usually in forecasting analysis to compare the predicted result with the actual or real data shown in Figures 5 and 6. For temperature, based on Figure 5 shows that there is a positive correlation between actual and prediction whose determination coefficient is 57%. For PH, Figure 6 shows that there is weakly correlated and determination coefficient equals only 16%. Comparing the predicted result and real data in the Citarik River is shown in Figures 7 and 8. Figures 7 and 8 shows that prediction values of PH and temperature make more "alkali" and become a high temperature, respectively. These mappings are appropriate to analyzing by means of quantitative results. Anyway, the predictions are almost the same result as {6) and [10] that means RMPE for PH is less than 1, and for temperature is less than 3, respectively.

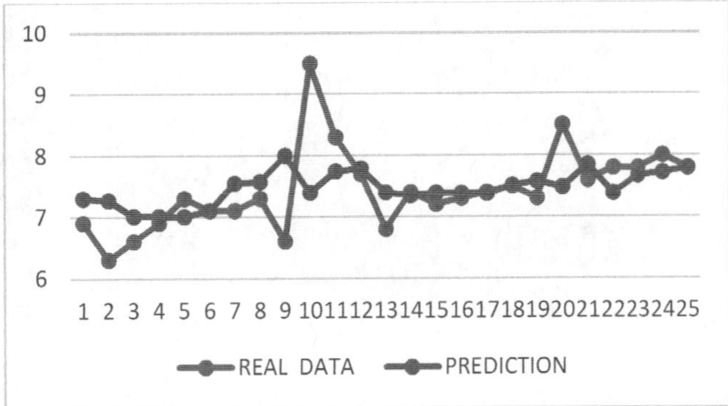

Figure 5. Plot actual and prediction for temperature

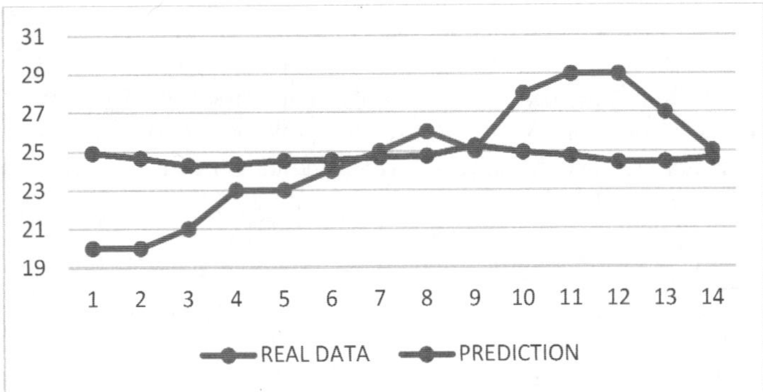

Figure 6. Plot actual and prediction for PH

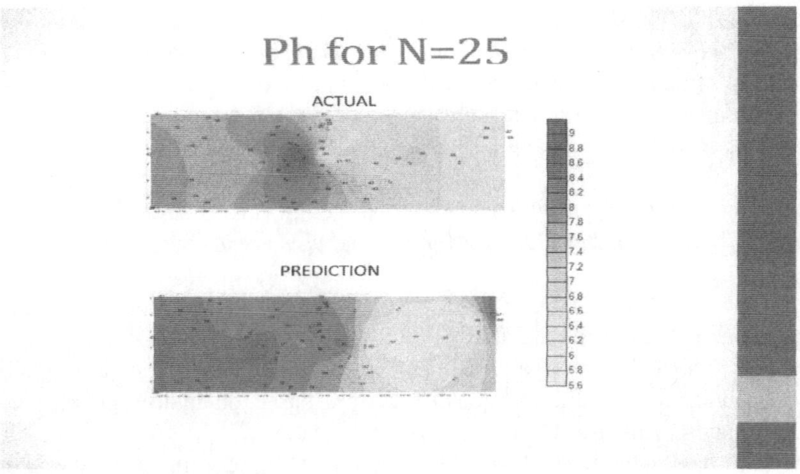

Figure 7. A mapping actual and prediction in study area for PH

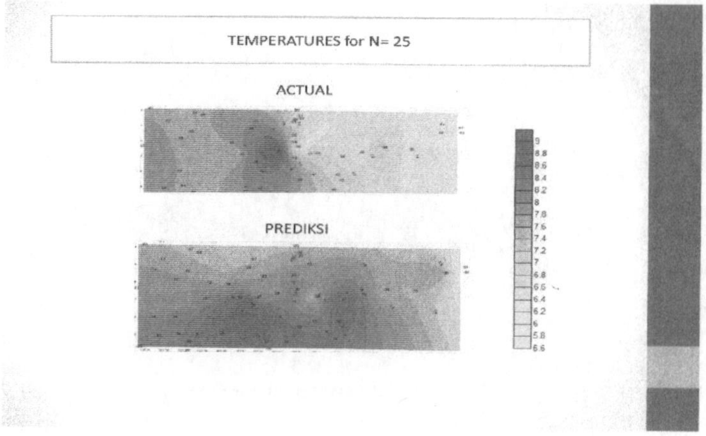

Figure 8. A mapping actual and prediction in study area for temperature

Conclusion

In this paper has been discussed about a spatial model for a stream networks of Citarik River with the environmental variables: potential of hydrogen (PH) and temperature. The following conclusions is based on the empirical results: The best fit of autocovariance functions for temperature and potential hydrogen (PH) of Citarik river is linear which also yields the smallest root mean squared prediction error (RMSPE). The spatial correlation values between the locations on upstream and on downstream of Citarik river exhibit decreasingly.

References

[1] Barry, R.P. and Ver Hoef, J.M. 1996. Blackbox kriging: Spatial prediction without specifying the variogram. *J. Agricult. Biol. Environ Stat.* 1, 297-322.

[2] Cressie, N.A.C 1993. *Statistics for Spatial data,* revised edition, John Wiley & Sons, Inc., New York.

[3] Cressie, Frey, B., Frey, B.H. and Smith, M. 2006. Spatial prediction on a river network. *Journal of Agricultural Biological and Environmental Statistics*, 11, 127-150.

[4] Curriero, F.C. 2006. On the use of non-Euclidean distance measures in geostatistics. *Math. Geology*, 38, 907-925.

[5] Curriero, F. C. 1997. *The use of non-Euclidean distances in geostatistics*, PhD Dissertation, Department of Statistics, Kansas State University.

[6] Garreta, V., Moniestier, P, and Ver Hoef, J.M. 2010. Spatial modelling and prediction on river network: up model, down model or hybrid?. *Environmetrics*, 21, 439-456.

[7] Higdon, D. 1998. A process-convolution approach to modelling temperatures in the North Atlantic Ocean, Journal of Environmental and Ecological Statistics 5, 173-190.

[8] Li, Jie 2009. *Spatial Multivariate design in the plane and stream networks, Dissertation,* University of Iowa.

[9] Little LS, Edwards D, Porter D.E. 1997. Kriging in estuaries: as the crow flies or as the fish swims? Journal of experimental marine biology and ecology, 213, pp. 1-11.

[10] Peterson, E.E., Theobald, D.M., and Ver Hoef, J.M. 2007. Geostatistical modelling on stream networks: developing valid covariance matrices based on hydrologic distance and stream flow. *Freshwater Biology*. 32.267-279.

[11] Peterson, E.E. and Ver Hoef J.M. 2010. A mixed model moving average approach to geostatistical modelling in stream network. *Ecology*, 91, 644-655.

[12] Rathbun, S. 1998. Spatial modelling in irregularly shaped regions: kriging estuaries. Environmetrics, 9, 109-129.
[13] Ver Hoef, J.M. and Peterson, E.E. 2010. A moving average approach for spatial statistical models of stream networks. *Journal of the American Statistical Association*, 105, 6-18.
[14] Ver Hoef, Peterson, E.E, and Theobald, D. 2006. Spatial statistical models that use flow and stream distance. *Environ Ecol Stat*. 13, 449-464.
[15] Ver Hoef, Peterson, E.E., Clifford, D., and Shah, R. 2014. SSN: An R package for spatial statistical modelling on stream networks. *Journal of Statistical Software*, 56, 1-45.

IOP Conf. Series: Materials Science and Engineering **332** (2018) 012002 doi:10.1088/1757-899X/332/1/012002

Appendix

This proof can be found in [8]

Where $W(x)$ is a white noise random process, with $E(W(x))=0$ and $\operatorname{var}(W(x))=1$, and $g(x|\boldsymbol{\theta})$ is the moving average function. The integral in Equation (1) is ordinary Riemann integral, but it should be really written as $Z(s,\omega)=\int_{-\infty}^{\infty} g(|x-s|\boldsymbol{\theta})W(x,\omega)\,dx=\int_{-\infty}^{\infty} g(|x-s|\boldsymbol{\theta})\,dW(x,\omega)$ which is a Lebesque integral.

$$
\begin{aligned}
c(h|\theta) &= \operatorname{cov}(Z(s),Z(s+h)) \\
&= \mathrm{E}\left(\int_{-\infty}^{\infty} g(x-s|\theta)W(x)\,dx\int_{-\infty}^{\infty} g(y-s-h|\theta)W(y)\,dy\right) \\
&= \int_{-\infty}^{\infty}\int_{-\infty}^{\infty} \mathrm{E}\left[g(x-s|\theta)g(y-s-h|\theta)W(x)W(y)\right]dxdy \\
&\quad \text{(Fubini's Theorem)} \\
&= \int_{-\infty}^{\infty} g(x-s|\theta)g(x-s-h|\theta)\,dx \\
&= \int_{-\infty}^{\infty} g(x|\theta)g(x-h|\theta)\,dx,
\end{aligned}
\tag{1}
$$

In order to Fubini's Theorem can be used that must be met:

$$
\begin{aligned}
&\mathrm{E}\left(\int_{-\infty}^{\infty}|g(x-s|\theta)W(x)\,dx\int_{-\infty}^{\infty}|g(y-s-h|\theta)W(y)|\,dy\right) \\
&= \int_{-\infty}^{\infty}\int_{-\infty}^{\infty} E|g(x-s|\theta)g(y-s-h|\theta)W(x)W(y)\,dxdy \\
&\quad \text{(Tonelli's Theorem)} \\
&= \int_{-\infty}^{\infty}\int_{-\infty}^{\infty}|g(x-s|\theta)g(y-s-h|\theta)|.\mathrm{E}|W(x)W(y)\,dxdy \\
&= \int_{-\infty}^{\infty}\int_{-\infty}^{\infty}|g(x-s|\theta)g(y-s-h|\theta)|\sqrt{EW^2(x)EW^2(y)}\,dxdy, \\
&= \int_{-\infty}^{\infty}\int_{-\infty}^{\infty}|g(x-s|\theta)g(y-s-h|\theta)|\,dxdy \\
&< \infty
\end{aligned}
$$

The Equation (1) can be as a function of distance h which is notated by $\rho(h)$. The moving average construction allows a valid autocovariance to be expressed by:

$$C(h|\boldsymbol{\theta}) = \begin{cases} \int_{-\infty}^{\infty} \left(g\left(x|\boldsymbol{\theta}\right)\right)^2 dx + v_j^2, h = 0 \\ \int_{-\infty}^{\infty} g\left(x|\boldsymbol{\theta}\right) g\left(x-h|\boldsymbol{\theta}\right) dx, h > 0 \end{cases} \quad (2)$$

Relationship between (1) and (2) can be formulated by [14]:

$$C\left(x|\boldsymbol{\theta}\right) = \begin{cases} \theta_0 + \theta_1, h = 0 \\ \theta_1 \rho\left(h/\theta_2\right), h > 0 \end{cases} \quad (3)$$

For example, $g(x) = \dfrac{1}{x+1}, x \geq 0$, using (2) and (3) is obtained:

$$C\left(x|\boldsymbol{\theta}\right) = \begin{cases} \theta_0 + \theta_1, h = 0 \\ \theta_1 \theta_2 \dfrac{\ln(\frac{h}{\theta_2}+1)}{h}, h > 0 \end{cases} \quad (4)$$

The Equation (4) is a valid autocovariance function for the moving average function, $g(x) = \dfrac{1}{x+1}, x \geq 0$. It can be derived for other functions to construct a valid autocovariance function.

Solving fuzzy shortest path problem by genetic algorithm

A Syarif[1], **K Muludi**[1], **R Adrian**[1], **M Gen**[2]

[1]Department of Computer Science, Faculty of Mathematics and Natural Sciences Lampung University, Indonesia, Jl. Brodjonegoro No. 1, Bandar Lampung,35145
[2]Fuzzy Logic Institute, Tokyo, Japan.

Corresponding author: admi_syarif@yahoo.com

Abstract: Shortest Path Problem (SPP) is known as one of well-studied fields in the area Operations Research and Mathematical Optimization. It has been applied for many engineering and management designs. The objective is usually to determine path(s) in the network with minimum total cost or traveling time. In the past, the cost value for each arc was usually assigned or estimated as a deteministic value. For some specific real world applications, however, it is often difficult to determine the cost value properly. One way of handling such uncertainty in decision making is by introducing fuzzy approach. With this situation, it will become difficult to solve the problem optimally. This paper presents the investigations on the application of Genetic Algorithm (GA) to a new SPP model in which the cost values are represented as Triangular Fuzzy Number (TFN). We adopts the concept of ranking fuzzy numbers to determine how good the solutions. Here, by giving his/her degree value, the decision maker can determine the range of objective value. This would be very valuable for decision support system in the real world applications.Simulation experiments were carried out by modifying several test problems with 10-25 nodes. It is noted that the proposed approach is capable attaining a good solution with different degree of optimism for the tested problems.

Keywords: Shortest Path Problem, Network Design Problem, Genetic Algorithm, Ranking Fuzzy Number, Decision Support System

1. Introduction

In business management, many important problems are expressed as network design problems. Among all classes of problems in network optimization, Shortest Path Problem (SPP) has been one of the most extensively disscused in computer science literature. It is commonly encountered in wide array of practical applications including transportation planning, communication, and production applications, salesperson routing and often imbedded within other types of network optimization problems [1], [2].

Given a weighed network graph, the SPP is to obtain the minimum total weight path(s) between nodes in the network. In the real world applications, the weight value represents the average delivery time of the commodities, reliability of transportation, and accessibility to the users, product deterioration, and so on. Thus, finding a good solution of SPP would be very important and could save millions of transportation and/or industrial cost, yet computationally difficult task. There have been several methods for solving traditional SPP including Dijkstra method [3]. However, due to the intrinsic difficulty, many authors introduced search methods based upon heuristics to give solutions within reasonable computational time [4], [5].

Traditionally, SPP was investigated in the area of Operations Research and Artificial Intelligence that is modelled by certain/deterministic cost values. However, for many real world applications, the deterministic model can be far from sufficient. For instance, in a transportation network, the path that has the least travel time, during the rush hours would have traffic jams that may lead to a drastic increase in travel time. One way of handling such uncertainty in decision making is by introducing fuzzy programming approach [6].

Artificial Intelligence (AI) has been discussed in many recent research fields. It has been successfully used to solve many difficult-real-world optimization problems. Among them, Genetic algorithm (GA) has been known as one of effective and popular methods. Since introduced by Holland [7], it has been implemented to give optimal or near optimal solution for many NP-hard optimization problems with less computational cost [8]. There are several advantages of using GA as follows: It deals with coding instead of decision variables. It requires no domain knowledge - only the payoff or objectives. In addition, traditional methods use deterministic transition rules to guide the search, such as hill-climbing, and neighbourhood search. Another difference between traditional methods and genetic algorithms is: the latter searches from a set of points, while the former from a single point. This makes GA more robust than traditional methods regarding their potential as optimization techniques [9]. In our previous works, we also have successfully implemented GA to solve various combinatorial optimization problems and found that GA can give very good solutions to the problems within reasonable computational time [10], [11], [12], [13].

In this paper, we develop new model of SPP called fuzzy Shortest Path Problem (f-SPP). Here, the cost values of the f-SPP are represented by using Triangular Fuzzy numbers (TFN). Our major efforts in this work include the adaption of the the rangking method for handing the fuzziness. We adopt the concept of the ranking Fuzzy numbers with integral value for the evaluation and selection. In this technique, the decision maker can determine his decision by giving flexible value for the degree of optimism [14].

The remainder of this paper is organized as follows: The description of the mathematical formulation of the f-SPP is given in Section 2. Section 3 describes a brief discussion design of algorithm. It includes ranking Fuzzy numbers, representation of chromosome, genetic operations and selection method. Section 4 presents the numerical experiments and results obtained by the proposed method. Finally, some concluding remarks are given in Section 5.

2. Mathematical Model

In this section, we shall present a comprehensive mathematical model of the problem. Suppose that we are given a graph *G* in which thr lenght of each arc has associated with a TFN. In some applications, that TFN may actually represent traveling time or some other values. For any two vertices, *s* and *t* in *G*, it is possible that there exist several paths from *s* to *t*. The problem involves finding path(s) from *s* to *t* that has the shortest traveling time.

The model of f-SPP can mathematically be stated as follows:

$$\textbf{min} \quad \sum_{i=1}^{m}\sum_{j=1}^{n}\tilde{c}_{ij}x_{ij} \quad \text{.......} \quad (1)$$

$$\textbf{s.t.} \quad \sum_{j=1}^{n}x_{ij}-\sum_{k=1}^{n}x_{ki}$$

$$=\begin{cases} 1 \quad (i=1) & \text{.......} \quad (2) \\ 0 \quad (i=2,3,\cdots,n-1) & \text{.......} \quad (3) \\ -1 \quad (i=n) & \text{.......} \quad (4) \end{cases}$$

$$x_{ij} = \begin{cases} 1, & \text{if link } (i, j) \text{ is included in the path} \\ 0, & \text{otherwise} \end{cases}$$

In the above model, directed graph **G=(V, A)** where **V** is a set of nodes, **A** is a set of links. $\tilde{c}_{ij}$ is a TFN representing fuzzy cost associated with each arc(i, j). Source node and destination node are node 1 and node n respectively. The length of a path is defined as the sum of the lengths of all individual arcs comprising the path.

3. Design of the Proposed Algorithm

3.1. Ranking Fuzzy Number (TFN)

When considering an optimization problem that its coefficients are represented with TFNs, the objective values of the problem will also be TFNs. In a fuzzy environment, an effective, efficient, and accurate ranking method becomes a very important decision making procedure. Many authors have investigated various ranking methods for fuzzy numbers [15]. The ranking fuzzy number with integral value technique was proposed by Liou & Wang [14]. The basic concept of this technique is to rank the TFN based on its total integral value that represents a "mean value". The left and right integral values are used to reflect the pessimistic and optimistic viewpoint of decision maker respectively. The total integral value is then computed based on this degree of optimism and each objective function values. A parameter α is given to adjust the degree of optimism.

Definition 1 [14]

Let $\tilde{A}$ be a TFN with its membership function $\mu_{\tilde{A}}$ as shown in Figure 1. A TFN is denoted by a triplet (a_1, a_2, a_3) where a_1, a_2, a_3 are real numbers. i.e. a minimum value, a modal value and a maximum value. The membership function of a TFN $\tilde{A}$, $\mu_{\tilde{A}}$ is defined as:

$$\mu_{\tilde{A}}(x) = \begin{cases} 0 & , x < a_1 \\ \dfrac{(x-a_1)}{(a_2-a_1)} & , a_1 \le x \le a_2 \\ \dfrac{(a_3-x)}{(a_3-a_2)} & , a_2 \le x \le a_3 \\ 0 & , x > a_3 \end{cases} \quad \text{...} \quad (5)$$

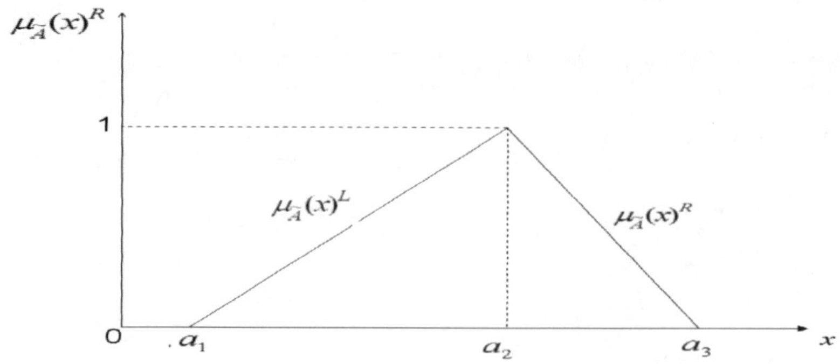

Figure 1. The membership function of TFN

From the definition of the above membership function, we can see that the left membership function $\mu_{\tilde{A}}(x)^L$ is continuous and strictly increasing. Its inverse function $g_{\tilde{A}}(x)^L$ would exist and continuous on interval [0, 1]. So we can compute its integral value on that interval. Similarly for the right membership function $\mu_{\tilde{A}}(x)^R$. We can compute both left and right integral values as follows:

$$I(\tilde{A})^L = \int_0^1 g_{\tilde{A}}(x)^L dy = \frac{1}{2}(a_1 + a_2) \quad (6)$$

$$I(\tilde{A})^R = \int_0^1 g_{\tilde{A}}(x)^R dy = \frac{1}{2}(a_2 + a_3) \quad (7)$$

Thus, the total integral value for triangular fuzzy number $\tilde{A}$ is

$$\begin{aligned} I_T^{\alpha}(\tilde{A}) &= \alpha I(\tilde{A})^R + (1-\alpha) I(\tilde{A})^L \\ &= \frac{1}{2}[\alpha\, a_3 + a_2 + (1-\alpha)\, a_1] \end{aligned} \quad (8)$$

The following definition is used in relation to the ranking of the TFN:

Definition 2: Suppose that $S = \{\tilde{A}_1, \tilde{A}_2, \cdots, \tilde{A}_n\}$ is a set of convex fuzzy numbers and that ranking fuzzy number R is a mapping from S to the real numbers. For any distinct $\tilde{A}_i, \tilde{A}_j \in S$ the ranking has the following properties.

1. if $R(\tilde{A}_i) < R(\tilde{A}_j)$ then $\tilde{A}_i < \tilde{A}_j$
2. if $R(\tilde{A}_i) = R(\tilde{A}_j)$ then $\tilde{A}_i = \tilde{A}_j$
3. if $R(\tilde{A}_i) > R(\tilde{A}_j)$ then $\tilde{A}_i > \tilde{A}_j$

Based on the above definition, thus for any fuzzy numbers $\tilde{A}_i$ and $\tilde{A}_j$ we shall have:

if $I_T^{\alpha}(\tilde{A}_i) < I_T^{\alpha}(\tilde{A}_j)$ then $\tilde{A}_i < \tilde{A}_j$
if $I_T^{\alpha}(\tilde{A}_i) = I_T^{\alpha}(\tilde{A}_j)$ then $\tilde{A}_i = \tilde{A}_j$
if $I_T^{\alpha}(\tilde{A}_i) > I_T^{\alpha}(\tilde{A}_j)$ then $\tilde{A}_i > \tilde{A}_j$.

3.2. Design of Genetic Algorithm (GA)

Generally, the behaviour of GAs depends on many uncertain factors. One of the important factors is the balance between exploitation and exploration in the search space. The balance is strongly affected by design strategy for GA and GA parameter values. Here, we would like to describe the consideration underlying the design of GA approach including representation method, genetic operators and selection method.

3.2.1. Chromosome Representation and Evaluation

It is well known that the first step in applying GA is to find a representation for a possible solution. An efficient design of the chromosome would lead to a good performance of GA.

For this problem, we adopt permutation-based representation. In this representation, the value of gene represents the priority of the node number for being included in the arc. There are several advantages of using this representation for network problems [9]. Those include, first, any permutation of the encoding corresponds to a path (legality). Most existing genetic operators can be easily applied to the encoding. Any path has a corresponding encoding (completeness). And, any point in solution space is accessible for genetic search.

The following Figure 2 shows the illustration of priority based representation.

7	2	5	1	4	3	9	8	6

Figure 2. Priority-based chromosome

In order to compute the fitness value, we decode the above chromosome using the decoding procedure as follows:

Procedure: Decoding
Step 1:
- number of nodes n,
- the alpha value
- the set of nodes S_i with all nodes adjacent to node i.
- the set of the TFN weight values denoted by (a_1, a_2, a_3)

Step 2
Determine path corresponding to the chromosome as follows:
initial source node $i \leftarrow 1$, $P_k \leftarrow \phi$;
while $S_i \neq \phi$ **do**
 select l from S_i with the highest priority;
 if $v_k(l) \neq 0$ **then** $v_k(l)=0$;
  $P_k \leftarrow P_k \cup \{x_{il}\}$;
  $i \leftarrow l$;
 else $S_i \leftarrow S_i \backslash \{l\}$
end
output the complete path P_k ;
end
Step 3
Compute the fuzzy objective values
Step 4
Determine the integral values of TFNs

3.2.2. Crossover and Mutation

The next step of GA is to select chromosome for replication by using genetic operator, in order to have new chromosome to be evaluated further. Basically, GA use two genetic operators, crossover and mutation.

Crossover is known as the most important recombination operator in GA. There have been many variants of crossover operations given in the literature for permutation based representation such as: PMX method (Partially Matched Crossover) [16], PX (Position-based crossover) [17] and WMX (Weight mapping crossover) [18].

In this paper, we adopt PMX crossover operations as follows:

Procedure : PMX
Step 1: Select a section of chromosome randomly

Selected sub string
↓

IOP Conf. Series: Materials Science and Engineering **332** (2018) 012003 doi:10.1088/1757-899X/332/1/012003

Parent 1	1	7	2	3	4	6	5	8
Parent 2	4	6	3	5	7	1	8	2

Step 2: Exchanged each sub string

Parent 1	1	7	3	5	7	6	5	8
Parent 2	4	6	2	3	4	1	8	2

Step 3: Determine the mapping of genes in each substring

3	5	7
2	3	4

$2 \leftrightarrow 3 \leftrightarrow 5$

$4 \leftrightarrow 7$

Step 4: Update chromosome with information given by Step 3

Offspring 1	1	4	3	5	7	6	2	8
Offspring 2	7	6	2	3	4	1	8	5

Mutation is usually done to prevent premature loss of information by exchanging the information within a chromosome. In this paper, we adopt the inversion mutation. It is done by selecting two positions within a chromosome at random and then inverts the sub-string between these two positions. We show the illustration of this mutation operation in Figure 3

Parent	4	6	3	5	7	1	8	2
offspring	4	6	7	5	3	1	8	2

Figure 3. Illustration of inversion mutation

3.2.3. Selection

When implementing GA, the way to select chromosome for the next generation is also very important issue. It should provide a balance between exploitation and exploration to search solution in search space. Here, we adopt one of the fitness-proportional methods called *roulette wheel* selection [2]. If the best chromosome is not selected for the next generation, we select one of the chromosome randomly and replace it with the best chromosome. We express our selection method as follows:

Procedure: Selection

Step 1: Calculate a cumulative probability a_p for each chromosome $X_p, (p = 1, 2, \cdots, pop\_size)$.

Step 2: Generate a random real number r in [0,1].

Step 3: If $r \leq a_1$, then select the first chromosome X_1; otherwise select the pth chromosome X_p, $(2 \leq p \leq pop\_size)$ such that $a_{p-1} < r < a_p$.

Step 4: Repeat Steps 2 and 3 *pop_size* times and obtain *pop_size* copies of chromosome.

Step 5: If the best chromosome is not selected in the next generation, replace one randomly from the new population by the best one.

3.2.4. Overall Procedure
The overall GA procedure used in this paper is illustrated as in Figure 4.

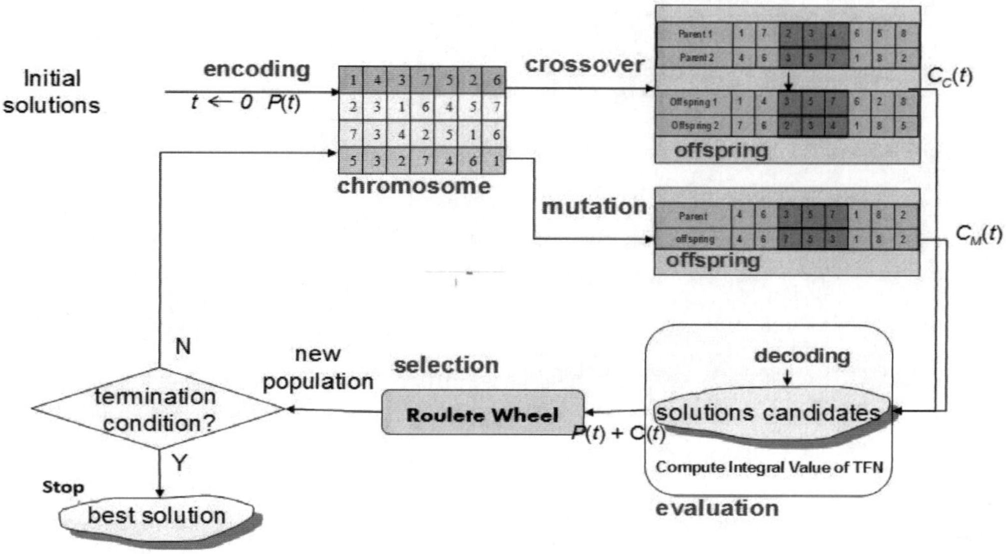

Figure 4. Overall procedure of proposed method

4. Experimental Design and Results

To explore how well the proposed GA can solve f-SPP. we have designed several numerical experiments. The proposed method was implemented by Using C language and run on PC Core i3. For the test problems, we modified the SPP test problems given [19], [2] and [20]. The size of test problems and GA parameters used for these experiments are set as presented in Table 1:

Table 1. Problem size and GA parameter values

Test Problem	Number of Node	Crossover Probability	Mutation Probability	Population Size	Maximimum generation
1	20	0.4	0.2	20	100
2	10	0.4	0.2	10	100
3	25	0.4	0.2	15	100

The first experiments were cunducted to see the effectiveness of GA with respect to the heterogeneity of the degree of optimism. To investigate this issue, for all test problems, we take a look at the variation of the degree of optimisms (0, 0.1, 0.2, …, 1). For each different degree of optimism, the experiment was done 10 (ten) times. We note the best, the average and the worst results. The experimental results are summarized in Table 2.

Table 2. Computational results for different alpha values

	Alpha										
Problem 1	**0.0**	**0.1**	**0.2**	**0.3**	**0.4**	**0.5**	**0.6**	**0.7**	**0.8**	**0.9**	**1.0**
Maximum	152,5	164,6	176,7	188,8	200,9	213	232,6	237,2	249,3	261,4	273,5
Minimum	112,5	122,1	131,7	141,3	150,9	160,5	170,1	179,7	189,3	198,9	208,5
Average	120,5	130,6	140,7	155,55	160,9	171	187,35	191,2	201,3	205,2	221,5
STDEV	16,86	17,91	18,97	22,94	21,08	22,13	27,85	24,24	25,29	19,76	27,41
Problem 2	**0.0**	**0.1**	**0.2**	**0.3**	**0.4**	**0.5**	**0.6**	**0.7**	**0.8**	**0.9**	**1.0**
Maximum	94	95,4	96,8	98,2	99,6	101	102,4	103,9	107,3	107,3	109
Minimum	92	93,7	95,4	97,1	98,8	100,5	102,2	103,8	105,2	106,6	108
Average	92,6	94,21	95,96	97,21	99,04	100,65	102,24	103,8	105,44	106,7	108,3
STDEV	0,96	0,82	0,72	0,34	0,38	0,24	0,08	0,042	0,66	0,29	0,483
Problem 3	**0.0**	**0.1**	**0.2**	**0.3**	**0.4**	**0.5**	**0.6**	**0.7**	**0.8**	**0.9**	**1.0**
Maximum	89	88,35	90,7	92,8	95,4	98	100,6	103,2	104,8	107,2	110
Minimum	85	87,6	90	92,5	94,5	96,75	99	101,3	103,5	105,8	108
Average	85,85	87,825	90,07	92,56	94,59	97	99,32	101,4	103,63	106,2	108,2
STDEV	1,31	0,36	0,22	0,12	0,28	0,52	0,67	0,617	0,41	0,67	0,632

To observe the speed of convergence of the proposed GA, we noted the objective value in each generation for the test problem 3. We compared the results for three different alpha values (0, 0.5 and 1). The results are shown in Figure 5.

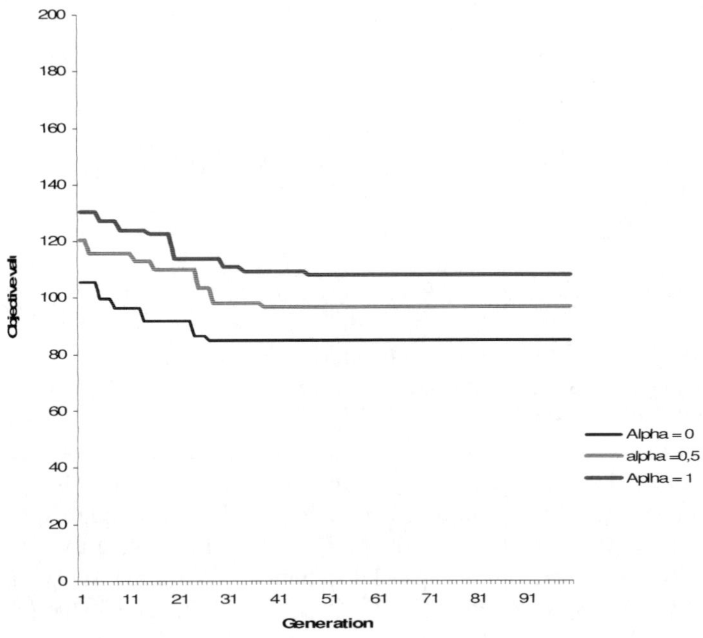

Figure 5. The comparative convergence behavior for different alpha values

We noted from the above results that, GA is very sensitive for overall optimization degrees. For different degree of optimism (alpha), GA can give different objective value all of the times. Thus by giving different alpha value, GA would be able to give decision maker solution for the problem. The above results also show that GA can convergence to the best solution very fast.

5. Conclusion

The contribution of this paper consists of introducing the model of shortest path problem with fuzzy coefficient (fSPP) and providing GA-based algorithm to solve it. Here, the proposed method use the fuzzy ranking strategy based on the alpha value/degree of optimism. The performance of the proposed method was evaluated by using modified several test problems given in the literature. The findings reported in this study show that, GA is very sensitive for overall degrees of optimism. Thus, the decision makers can have their solution based on their degree of optimism. Moreover, the results also show that GA can convergence to the best solution very fast (less than 50 generations).

Acknowledgments

The author acknowledge The Directorate of Higher Education, The Ministry of Research, Technology, and Higher Education, Republic of Indonesia, for financial support through research grant "Hibah Bersaing" 2015.

References

[1] Bazaraa M S, Jarvis J J and Sherali H D 1990 *Linear Programming and Network Flows*, 2nd ed. (New York: John Wiley and Sons)
[2] Gen M and Cheng R 1997 *Genetic Algorithm and Engineering Design*, 2nd ed. (New York: John Willey and Sons).
[3] Dijkstra E W 1959 A Note on Two Problems in Connection with Graph *Numer. Math* Vol 1 p. 269–271
[4] Evans J R and Minieka E 1992 *Optimization Algorithms for Networks and Graphs*. (New York: Marcel-Dekker)
[5] Ericsson M, Resende M G C and Pardalos P. M. 2002 A Genetic Algorithm for the Weight Setting Problem in OSPF Routing *J. Comb. Optim.* vol. 6 p. 299 - 333
[6] Li Y Z, Ida K and Gen M 1997 Improved Genetic Algorithm for Solving Multi-objective Solid Transportation Problem with Fuzzy Number *Japanese Sosiety for Fuzzy Theory and System*, vol. 9, no. 2, p. 239–250
[7] Holland J 1975 *Adaptation in Natural and Artificial Systems* (Michigan: University of Michigan Press)
[8] Michalewicz Z 1994 *Genetic Algorithms+Data Structures = Evolution Programs* Spring-Verlag.
[9] Gen M and Cheng R 2000 *Genetic ALgorithm and Engineering Optimization* (New York: John Willey and Sons)
[10] Syarif A, Yun Y S and Gen M 2002 Study on Multi-stage Logistic Chain Network: A Spanning Tree-based Genetic Algorithm Approach *Comput. Ind. Eng.* vol. 43 no. 1 p. 299-314
[11] Syarif A and Gen M 2003 Solving Exclusionary Side Constraint Transportation Problem: A Genetic Algorithm with Fuzzy Logic Approach *J. Intell. Manuf.* vol. 14 p. 389-399
[12] Syarif A and Gen M 2003 Double Spanning Tree-Based Genetic Algorithm for Two Stage Transportation Problem *Int. J. Knowl. Based Eng. Syst. Des.*, vol. 7, no. 4, p. 214–221
[13] Syarif A, Aristoteles, Aryanti D and Riska M 2016 Performance Evaluation of Various Genetic Algorithm (GA) Approaches For Knapsack Problem *ARPN J. Eng. Appl. Sci.*, vol. 11, no. 7, p.
[14] Liou T and Wang M J 1992 Ranking Fuzzy Numbers with Integral Value *Fuzzy Sets Syst.*, vol. 50, no. 3, p. 247–255
[15] Ezzati R, Enayati R, Mottaghi A and Saneifard T R 2011 A New Method for Ranking Fuzzy

IOP Conf. Series: Materials Science and Engineering **332** (2018) 012003 doi:10.1088/1757-899X/332/1/012003

Numbers without Concernig of Real Numbers *J. Pure Appl. Math.* vol. 2, no. 2, p. 256–270
[16] Goldberg D and Lingle R 1985 Alleles, loci and the Traveling Salesman Problem, in *Proc. of the First International Conference on Genetic Algorithm*, p. 154–159
[17] Syswerda G 1989 Uniform Crossover in Genetic Algorithms, in *Proc of the 3rd International. Conference on Genetic Algorithm,* Morgan Koufman p. 2-9
[18] Gen M, Lin L and Cheng R 2004 Bicriteria Network Optimization Problem using Priority-based Genetic Algorithm *IEEJ Trans. Electron. Information. Syst.* vol. 124, no. 10, p. 1972–1978
[19] Ahn C W and Ramakrishna R 2002 A Genetic Algorithm for Shortest Path Routing Problem and the Sizing of Populations *IEEE Trans. Evol. Comput* vol. 6, no. 6, p. 566–579
[20] Hashier T and Munakata J D 1993 A Genetic Algorithm Applied to the Maximum Flow problem in *Proc. of the 5th Int. Conf. on Genetic Algorithms*, p. 488–493.

Appendix: Test Problems

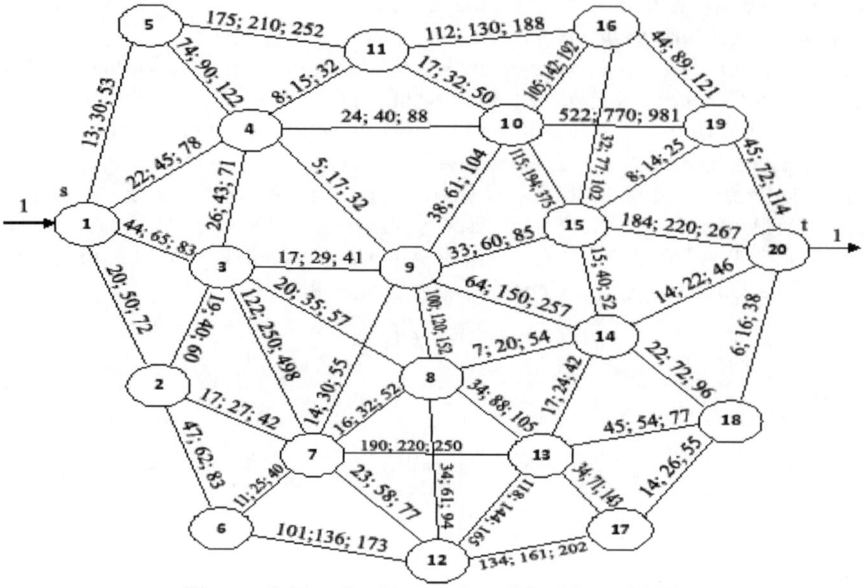

Figure 6. Test Problem 1 (modified from [19])

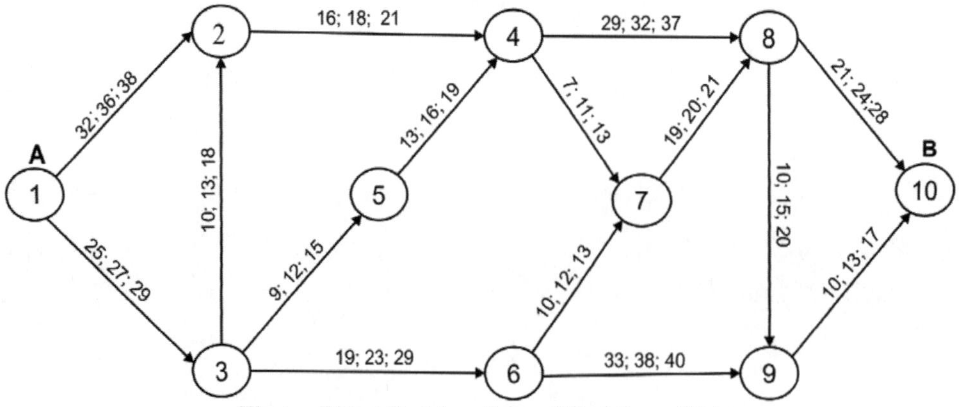

Figure 7 Test Problem 2 (modified from [2])

IOP Conf. Series: Materials Science and Engineering **332** (2018) 012003 doi:10.1088/1757-899X/332/1/012003

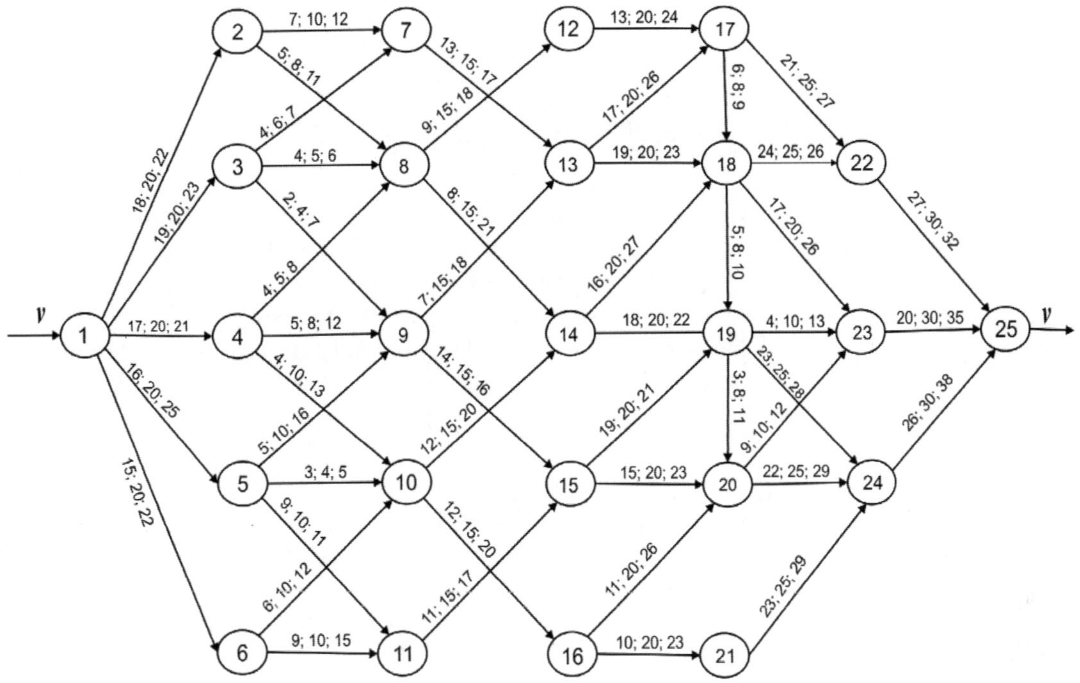

Figure 8 Test Problem 3 (modified from [20])

Nonlinear Diophantine equation $11^x + 13^y = z^2$

A Sugandha[1], A Tripena[2], A Prabowo[3] and F Sukono[4]

[1,2,3]Mathematics Department,Faculty of Mathematics and Science, Universitas Jenderal Soedirman, Purwokerto, Indonesia.
[4]Mathematics Department, Faculty of Mathematics and Science, Universitas Padjajaran, Indonesia.

Corresponding author: agus.sugandha@unsoed.ac.id

Abstract. This research aims to obtaining the solutions (if any) from the Non Linear Diophantine equation of $11^x + 13^y = z^2$. There are 3 possibilities to obtain the solutions (if any) from the Non Linear Diophantine equation, namely single, multiple, and no solution. This research is conducted in two stages: (1) by utilizing simulation to obtain the solutions (if any) from the Non Linear Diophantine equation of $11^x + 13^y = z^2$ and (2) by utilizing congruency theory with its characteristics proven that the Non Linear Diophantine equation has no solution for non negative whole numbers (integers) of x, y, z .

Keywords: non linear Diophantine equation, simulation, congruency theory

1. Introduction

The congruency theory was first found by Carl Friedrich Gauss (1977-1855) which is considered as one monumental findingand becomes the foundation of the modern number theory. Congruency is actually developed from the divisibility relation then further developed by Diophantine called the linear congruency theory. The linear congruency theory focuses on finding the solution (if any) from the equation in the form of $ax \equiv b \bmod(n)$. It means that if x_0 is the solution of the linear congruency $ax \equiv b \bmod(n)$ then $ax_0 \equiv b \bmod(n)$ is fulfilled. Fermat (1980) introduces the nonlinear Diophantine equation in the form $x^n + y^n = z^n$. Especially forn = 2, then all solutions are called the triple Pythagoras.

The forms of nonlinear Diophantine equation continuously develop in line with the involvement of researches on the nonlinear Diophantine equation. Acu [1] proves that (3,0,3) and (2,1,3) are the solutions of the nonlinear Diophantine equation $2^x + 5^y = z^2$.

Sroysang [3] finds that a single solution of the nonlinear Diophantine equation $3^x + 5^y = z^2$ is (1,0,2). Sroysang [7] then finds three solutions of the nonlinear Diophantine equation $2^x + 3^y = z^2$, that is, (0,1,2), (3,0,3), and (4,2,5). Tanakan [6] finds a single solution of the nonlinear Diophantine equation $19^x + 2^y = z^2$, that is, (0,3,3). The above research results lead to a question in order to find the other solution forms of the nonlinear Diophantine equation $11^x + 13^y = z^2$ where x, y, z are the non negative whole numbers.

2. Result and discussion

Theorem 1. Assuming that p is the prime number, then the Diophantine equation $p^y+1 = z^2$ has exactly 2 solutions of non negative whole numbers (p,y,z). Both solutions are (2,3,3) and (3,1,2).

Proof: for the case $y = 0$ and $z = 0$, there is a contradiction. For the case y, $z> 0$ then it may be written with $z^2 - 1 = (z - 1)(z + 1) = p^y$. Furthermore, it is obtained $2 = (z +1) - (z -1) = p^\beta - p^\alpha$ where $\alpha + \beta = y$ and $\alpha < \beta$. There are two possibilities may occur. If $p^\alpha = 1$ and $p^{\beta-\alpha} - 1 = 2$ then consequently$\alpha = 0$ and $0 + \beta = y$, that $\beta = y$. Furthermore, $p^{\beta-0} - 1 = p^\beta - 1 = p^y - 1 = 2$ that $p^y=3 =3^1$. Thus, it is obtained the solution $(p, y, z) = (3, 2, 1)$. On the other side, if $p^\alpha = 2$ and $p^{\beta-\alpha} - 1 = 1$, then $p^\alpha = 2^1$ that it is obtained $p = 2$ and $\alpha = 1$. Thus, itis consequently$p^{\beta-\alpha} - 1 = 1$ that it is obtained$\beta = 2$. Because$\alpha + \beta = y$ that it is obtained $y = 3$. Thus, it is obtained the solution $(p, y, z) = (2,3,3)$ and theorem 1 is proven.

Theorem 2. The Diophantine equation $p^y + 1 = z^2$ has no solution of positive whole numbers for the prime numbers $p> 3$.

Proof: based on theorem 1, it is obtained that at the Diophantine equation $p^y + 1 = z^2$, there are two solutions of non negative whole numbers: (2,3,3) and (3,1,2). Thus, it can be concluded that there is no other solution for each prime number p > 3.

Corollary 3. The Diophantine equation $1 + 13^y = z^2$ has no solution of non negative whole numbers.

Proof: Because 13 is the prime number $p> 3$, that according to theorem 2 then there is no solution of non negative whole number which fulfills the Diophantine equation $p^y + 1 = z^2$.

Corollary 4. The Diophantine equation $11^x + 1 = z^2$ has no solution of the non negative whole numbers.

Proof: From the Diophantine equation $p^y +1 = z^2$ if $y = x$ and $p = 11$ then it is obtained the Diophantine equation $11^x + 1 = z^2$ which has no solution of the non negative whole numbers, based on theorem 2, there is no fulfilling solution of the non negative whole numbers.

Case 1. For the case y = 1 then it is obtained that the Diophantine equation is $11^x + 13 = z^2$. By using congruency $13 \equiv 1 \pmod 4$. Furthermore, if x has the odd value, it is obtained $11^x \equiv 3 \pmod 4$ and if x has the even value, it is obtained $11^x \equiv 1 \pmod 4$. In addition, when z has the odd value, it is obtained $z^2 \equiv 1 \pmod 4$, and if z has the even value, it is obtained $z^2 \equiv 0 \pmod 4$. Because $13 \equiv 1 \pmod 4$, it is selected x with the odd value and z with the even value in such a way that $11^x +$

IOP Conf. Series: Materials Science and Engineering **332** (2018) 012004 doi:10.1088/1757-899X/332/1/012004

$13 = z^2 \Leftrightarrow 3(\text{mod}4) + 1(\text{mod}4) \equiv 0(\text{mod}4)$. Assuming that $x = 2k + 1$, k as the non negative whole number and assuming that $z = 2m$, with m as the non negative whole number, it is further obtained that

$11^x + 13 = z^2$

$11^{2k+1} + 13 = (2m)^2$

$11^{2k+1} + 13 = 4m^2$

$$m^2 = \frac{11^{2k+1} + 13}{4}$$

$$m = \sqrt{\frac{11^{2k+1} + 13}{4}}$$

Because

$m = \sqrt{\frac{11^{2k+1} + 13}{4}}$, with k as the non negative whole number. For k as the non negative whole number, it is obtained m as the non negative whole number that contradiction occurs. Consequently for case 1, the Diophantine equation $11^x + 13 = z^2$ has no solution.

To further generalization, for each y of the positive whole number $13^y \equiv 1$ (mod4).

Case 2. For x with the even value and z with the odd value, If x has the even value, it is obtained $11^x \equiv 1(\text{mod}4)$ andthen if z has the odd value, it is obtained $z^2 \equiv 1(\text{mod}4)$. See that $11^x + 13^y = z^2 \Leftrightarrow 1(\text{mod}4) + 1(\text{mod}4) \equiv 2(\text{mod}4)$. On the other side, it is known that if z has the odd value, then $z^2 \equiv 1(\text{mod}4)$. Thus, there is a contradiction. Consequently for case 2, the Diophantine equation $11^x + 13^y = z^2$ has no solution.

Case 3. For x with the even value and z with the even value, if x has the even value, then it is obtained $11^x \equiv 1(\text{mod}4)$ and if z has the even value, then it is obtained$z^2 \equiv 0(\text{mod}4)$. See that $11^x + 13^y = z^2 \Leftrightarrow 1(\text{mod}4) + 1(\text{mod}4) \equiv 2(\text{mod}4)$. On the other side, it is known that if z has the even value, then $z^2 \equiv 0(\text{mod}4)$. Thus, there is a contradiction. Consequently for case 3, the Diophantine equation $11^x + 13^y = z^2$ has no solution.

Case 4.For x with the odd value and z with the odd value, if x has the odd value, then it is obtained $11^x \equiv 3(\text{mod}4)$ and if z has the odd value, then it is obtained $z^2 \equiv 1(\text{mod}4)$. See that $11^x + 13^y = z^2 \Leftrightarrow 3(\text{mod}4) + 1(\text{mod}4) \equiv 0(\text{mod}4)$. On the other side, it is known that if z has the odd value, then $z^2 \equiv 1(\text{mod}4)$. Thus, there is a contradiction. Consequently for case 4, the Diophantine equation $11^x + 13^y = z^2$ has no solution.

Based on those four explanations, it can be concluded that at the Diophantine equation $11^x + 13^y = z^2$ there is no fulfilling non negative whole numbers (x, y, z). On the other words, the Diophantine equation $11^x + 13^y = z^2$ has no solution.

IOP Conf. Series: Materials Science and Engineering **332** (2018) 012004 doi:10.1088/1757-899X/332/1/012004

3. Conclusion

The nonlinear Diophantine equation $11^x + 13^y = z^2$ has no solution for non negative whole numbers of *x, y, z*.

References

[1] Acu D 2006 *General Mathematics* 15 **4**145
[2] Burton 1997*Elementary Number Theory*(New York: John Wiley & Sons)
[3] Fergy J 2013*Mathematical Journal of Interdisciplinary* Sciences 2 **1** 19
[4] Sroysang B 2012*International Journal of Pure and Applied Mathematics* 81 **4** 605
[5] Suvarnamani A 2010*International Journal of Mathematical Sciences and Application* 1 **3**1415
[6] Tanakan S 2014 *International Journal of Mathematical Contemp Mathematics Sciences* 9 **4** 159
[7] Sroysang B 2013, *International Journal of Pure and Applied Mathematics* 88**1**157
[8] Sroysang B 2014, *International Journal of Pure and Applied Mathematics* 91**1**131
[9] Sukirman 2005*PengantarTeoriBilangan*(Yogyakarta: HanggarKreator)

IOP Conf. Series: Materials Science and Engineering **332** (2018) 012005 doi:10.1088/1757-899X/332/1/012005

Strengthening economy through tourism sector by tourist arrival prediction

A Supriatna[1*], B Subartini[1], E Hertini[1], Sukono[1], Rumaisha[1], N Istiqamah[2]

Department of Mathematics, Faculty of Mathematics and Natural Sciences, Universitas Padjadjaran, INDONESIA

*Corresponding author: agus.supriatna@unpad.ac.id

Abstract. Tourism sector has a tendency to be proposed as a support for national economy to many countries with various of natural resources, such as Indonesia. The number of tourist is very related with the success rate of a tourist attraction, since it is also related with planning and strategy. Hence, it is important to predict the climate of tourism in Indonesia, especially the number of domestic or international tourist in the future. This study uses Seasonal Autoregressive Integrated Moving Average (SARIMA) time series method to predict the number of tourist arrival to tourism strategic areas in Nusa Tenggara Barat. The prediction was done using the international and domestic tourist arrival to Nusa Tenggara Barat data from January 2008 to June 2016. The established SARIMA method was $(0,1,1)(0,0,2)^{12}$ with MAPE error of 15.76. The prediction for the next six time periods showed that the highest number of tourist arrival is during September 2016 with 330,516 tourist arrivals. Prediction of tourist arrival hopefully might be used as reference for local and national government to make policies to strengthen national economy for a long period of time

1. Introduction

Indonesia, as the largest archipelago in the world is known as the second biggest mega biodiversity in the world for its natural resources, flora and fauna which are very highly diverse. Indonesia is located in the equator, which gives Indonesia the richness of tropical forests and diversity of local cultures. These potencies shows that Indonesia possesses a major resource and capital for the development of businesses and enhancing the tourism in Indonesia ([4]).

Lately, the tourism sectors become the pole of economic growth for countries with abundance of natural resources, such as Indonesia, Brazil, etc. This sector is affected by the number of domestic and international tourists, providing major income for the country. The number of tourist arrival to Indonesia is increasing annually, especially to potential tourism areas in Indonesia, i.e. Nusa Tenggara Barat (NTB). The increase of the number of tourist arriving to Nusa Tenggara Barat peaked significantly in 2012 (23.75%), hence noting that the year marked the start of a new era for tourism in Nusa Tenggara Barat. Thus, the prediction of tourist arrival to potential tourism areas in Nusa Tenggara Barat is needed.

Prediction of tourist arrival became an important element in decision making for tourism policies, especially in fulfilling product demands in tourism ([1]). The accuracy of prediction is important for the sake of planning efficiency ([6]), even though prediction shows results very close to facts in real life ([2]). Prediction could be divided into two types based on the data used in predicting; i.e.

qualitative and quantitative prediction. Quantitative prediction uses quantified data from the past, whereas qualitative prediction uses expert opinion/judgment through their previous experience ([7]).

Quantitative method consists of time series and causal. Time series is a data collected based on time period (t). Time series method includes naïve, moving average, exponential smoothing and ARIMA (Autoregressive Integrated Moving Average). Among those methods, ARIMA is the best mode because ARIMA does not require assumption about the current structural association or economic model ([1]).

This study applied the seasonal ARIMA or the commonly known as SARIMA (Seasonal Autoregressive Integrated Moving Average) to international tourist arrival data to Nusa Tenggara Barat. The arrival data of year 2008 to June 2016 was obtained from Department of Culture and Tourism Nusa Tenggara Barat. The SARIMA model has been used as an alternative solution for forecasting problems, e.g. by [8] on Forecasting Tourism Arrivals in South Africa, in [9] the comparative analysis of Holt Winter and SARIMA for statistical prediction of foreign tourist arrival to Kraton Yogyakarta is discussed.

Based on the background explained above, the problems could be formulated as follows:

1. How many tourists' arrivals to Nusa Tenggara Barat for the next six time periods?
2. What are the benefits for applying prediction methods in Nusa Tenggara Barat area?

This study aims to predict the number of tourist arrival to Nusa Tenggara Barat for the next six time periods. The result of study is hoped to be applied in policy making in tourism sector for the government in Indonesia, especially Nusa Tenggara Barat.

2. Research Methodology

This study used a four-step SARIMA method, consisting of model identification, parameter estimation, diagnostic tests and forecasting ([7]). These steps are presented in flowchart of SARIMA Modeling in Figure 1. The first step to be done was checking the data stationary. Transformation/differencing process would be done if the data were not stationary in terms of variance and mean. After the data was stationary, the next step would be model identification by looking at the Autocorrelation Function (ACF) plot and Partial Autocorrelation Function (PACF) do determine the possible temporary SARIMA model. The obtained model would be estimated for parameters and significance test would be done for each parameter. The insignificant model would be erased from the model. After all parameters had significant models, diagnostic test would be done to check whether the error was normally distributed and was not auto correlated. Kolmogorov-Smirnov test would be used to check the normality of the error; meanwhile, the test of autocorrelation for the error was L-Jung Box test. Should the model has not been yet obtained even after diagnostic tests, return to model identification step to look for another model. If all model assumptions had been fulfilled and if there were more than one model that could possibly be selected, the model with the least error was selected for forecasting. MAPE error was calculated then a conclusion was drawn.

Seasonal Autoregressive Integrated Moving Average (SARIMA) has a common form of $(p,d,q)(P,D,Q)^S$ and was written as follows:

$$\phi_p(\mathrm{B})\Phi_p(\mathrm{B}^S)(1-\mathrm{B})^d(1-\mathrm{B}^s)^D X_t = \theta_q(\mathrm{B})\Theta_Q(\mathrm{B}^S)e_t$$

e_t : error

X_t : Observed value at time of- t ($t = 1,2,...,n$)

$(1-\mathrm{B})^d$: Mathematical operation of non-seasonal differencing

$(1-\mathrm{B}^s)^D$: Mathematical operation of seasonal differencing

$\phi_p(\mathrm{B})$: Autoregressive Operator = $(1-\phi_1\mathrm{B}-\phi_2\mathrm{B}^2-...-\phi_p\mathrm{B}^p)$

$\Phi_p(\mathrm{B}^S)$: Seasonal Autoregressive Operator = $(1-\Phi_1\mathrm{B}^S-\Phi_2\mathrm{B}^{2S}-...-\Phi_P\mathrm{B}^{PS})$

$\theta_q(\mathrm{B})$: Seasonal *Moving Average* Operator = $(1+\theta_1\mathrm{B}+\theta_2\mathrm{B}^2+...+\theta_q\mathrm{B}^q)$

$\Theta_Q(B^S)$: Seasonal *Moving Average* Operator = $(1+\Theta_1 B^S + \Theta_2 B^{2S} + ... + \Theta_Q B^{QS})$ ([5])

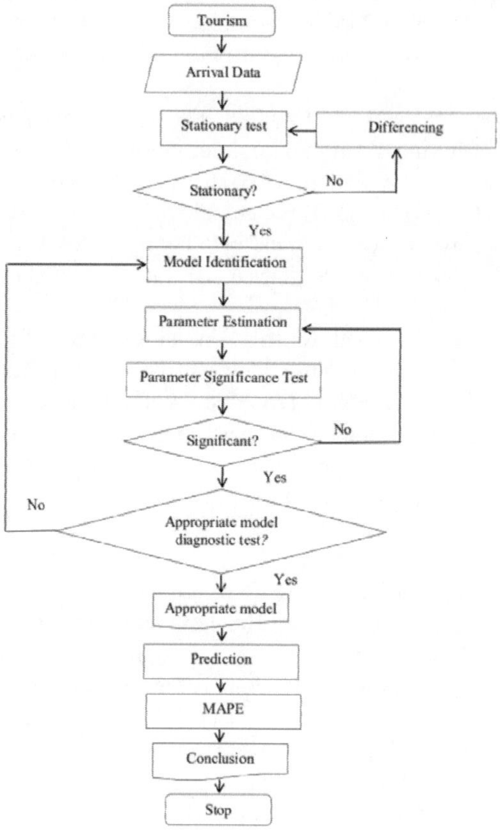

Figure 1. Flowchart of SARIMA Modeling

3. Results And Discussions

SARIMA process consisted of four steps, i.e. model identification, parameter estimation, diagnostic test and forecasting.

During the first step, data plotting to determine the data characteristics and test of stationary assumption in time series analysis was done. Time series data plot was presented in Figure 2.

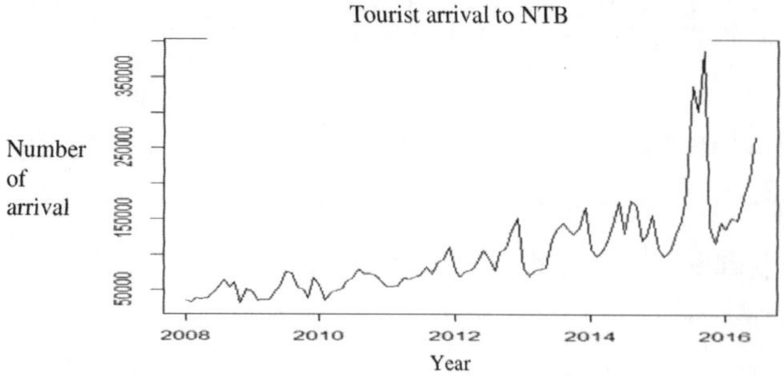

Figure 2. Tourist Arrival to NTB Year 2008-2016 Data Plot
(Reference: http://www.disbudpar.ntbprov.go.id/)

Data plot in Figure 2 showed that there was an increasing trend and seasonality pattern, determining that it was a seasonal ARIMA (SARIMA) model. Stationary assumption in terms of variance and mean was needed for SARIMA model. Test of stationary in variance could be conducted using Box-Cox transformation ([1]), with results shown in Figure 3.

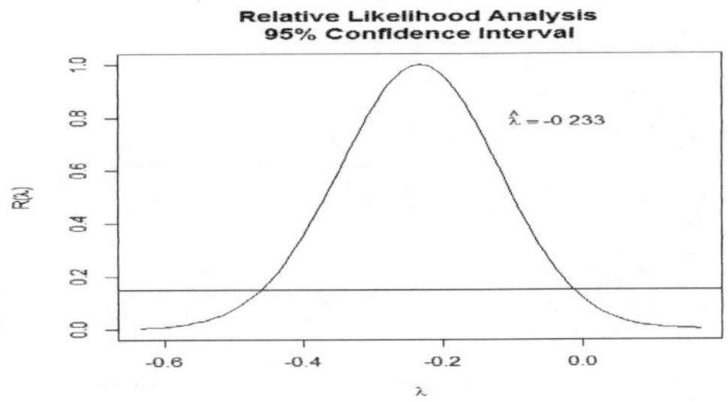

Figure 3. Box-Cox Transformation

Figure 3 showed that the acquired lambda was -0.2, thus a logarithmic transformation was done. After the transformation, an Augmented Dickey Fuller (ADF) test was done to check data stationary in terms of mean. ADF test showed that p-value = 0.2676 > 0.05 (Appendix 1), thus, it could be concluded that the data was not stationary in terms of mean and a first differencing process on non-seasonal order should be done.

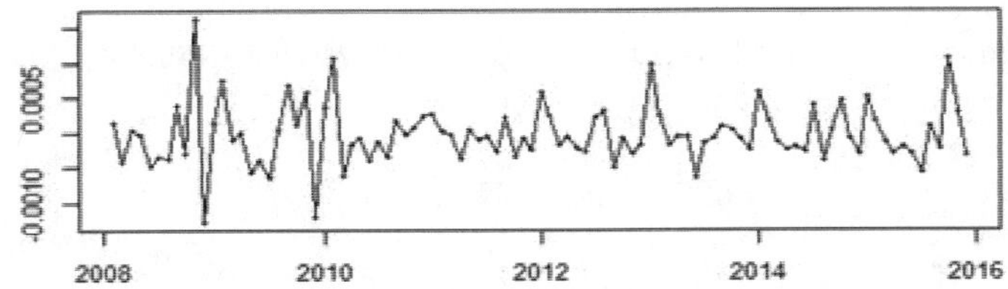

Figure 4. Non-seasonal order first differencing result

Figure 4 was a non-seasonal order first differencing result which showed that the data was stationary in terms of mean and to conform that it was stationary, the data could be re-tested with ADF test. The ADF test showed that the p-value = 0.01 < 0,05 (Appendix 2), meaning that the data was stationary.

Model Identification

After stationary was achieved, the ACF and PACF plot was checked to determine the temporary SARIMA model.

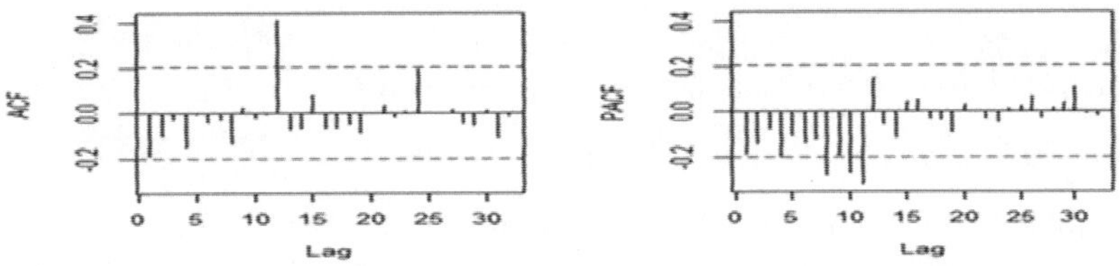

Figure 5. ACF and PACF Plot

Figure 5 showed that the ACF plot was cut on lag 1, lag 12 and lag 24, meanwhile the PACF plot showed a sinus wave; hence, the possible SARIMA model was SARIMA $(0,1,1)(0,0,1)^{12}$ and SARIMA $(0,1,1)(0,0,2)^{12}$.

Parameter Estimation

The parameter of the obtained models would be estimated using MLE method with software R 3.2.3 (Appendix 3), then would be tested for significance using t test. The result of parameter estimation was presented on Table 1.

Table 1. Parameter Estimation Result

Temporary Model	Parameter	S.E	Count t result	Significance test
SARIMA $(0,1,1)(0,0,1)^{12}$	θ_1 = -0.2083	0.1086	1.918	Significant
	Θ_1 = 0.3613	0.1371	2.635	Significant
SARIMA $(0,1,1)(0,0,2)^{12}$	θ_1 = -0.2720	0.1168	2.328	Significant
	Θ_1 = 0.3872	0.1558	2.485	Significant
	Θ_2 = 0.3656	0.1845	1.981	Significant

Table 1 showed that both models were significant since $t_{count} > t_{table} = 1.65$. The next step would be conducting diagnostic test on the models.

Diagnostic Test

Another assumption that must be fulfilled from SARIMA method was that the errors on the models were normally distributed and were not auto correlated. Test of normality on the error was conducted using Kolmogorov-Smirnov test and test for autocorrelation using L-jung Box test. The result of the calculation using software R 3.2.3 (Appendix 4) was shown on Table 2.

Table 2. Diagnostic Test Result

Model	P-value of Kolmogorov-Smirnov Test	P-value of L-Jung Box Test
SARIMA $(0,1,1)(0,0,1)^{12}$	0.9459	0.06828
SARIMA $(0,1,1)(0,0,2)^{12}$	0.9332	0.09853

Kolmogorov-Smirnov test showed the p-value for SARIMA $(0,1,1)(0,0,1)^{12}$ was 0.9459 and the p-value for SARIMA $(0,1,1)(0,0,2)^{12}$ was 0.9332; thus, it could be concluded that the H_0 (normally distributed error) was failed to be rejected since the p-value > 0.05 and that both models have normally distributed errors. Meanwhile, using L-Jung Box test, it was known that the p-value for SARIMA $(0,1,1)(0,0,1)^{12}$ was 0.06828 and the p-value for SARIMA $(0,1,1)(0,0,2)^{12}$ was 0.09853, hence, it could be concluded that the H_0 (error was not auto correlated) was failed to be rejected since the p-value > 0.05 and that both models have non-auto correlated errors.

Forecasting

Appropriate models were then evaluated by reviewing the error.

Table 3. AIC, RMSE and MAPE Error Result

Model	AIC	RMSE	MAPE
SARIMA $(0,1,1)(0,0,1)^{12}$	2264.78	57496.97	21.46042
SARIMA $(0,1,1)(0,0,2)^{12}$	2262.67	49293.35	18.95509

Based on error calculation on both models, the SARIMA model $(0,1,1)(0,0,2)^{12}$ was selected as the model with the least error and was mathematically written as:

$$(1-0.2720B)(1-B)X_t = (1-0.3872B^{12} - 0.3656B^{24})e_t$$

$$(1-B-0.2720B+0.2720B^2)X_t = e_t - 0.3872e_{t-12} - 0.3656e_{t-24}$$

$$X_t = X_{t-1} + 0.2720X_{t-1} - 0.2720X_{t-2} + e_t - 0.3872e_{t-12} - 0.3656e_{t-24}$$

Thus, the prediction result of tourist arrival was as shown below.

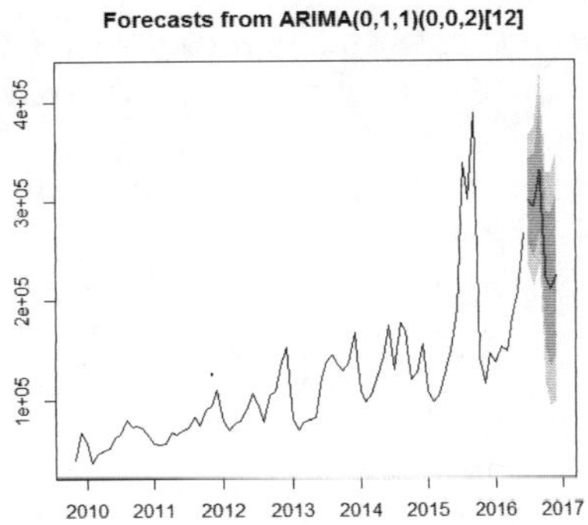

Figure 6. Arrival Prediction Result

Table 4. Prediction result from six time periods

Time	Prediction of Arrival
Jul-16	299.917
Aug-16	293.695
Sep-16	330.516
Oct-16	221.119
Nov-16	210.195
Dec-16	223.870

Table 4 showed that the predicted tourist arrival to NTB (Appendix 5) peaked on September 2016 and hit the lowest on November 2016.

4. Conclusion

The selected SARIMA model in this study was SARIMA $(0,1,1)(0,0,2)^{12}$. This model showed that the predicted tourist arrival to NTB during July 2016 was 299,917 tourist arrivals, August 2016 was 293,695 tourist arrivals, September 2016 was 330,516 tourist arrivals, October 2016 was 221,119 tourist arrivals, November 2016 was 210,195 tourist arrivals, December 2016 was 223,870 tourist arrivals with MAPE error of 15.76. This prediction of tourist arrival was hoped to be applied as a reference for local and national government in making more comprehensive and efficient policies in tourism sector, hence, strengthening the national economy as a long term effect.

References

[1] Bigovic M 2012 Demand forecasting within montenegrin tourism using box-jenkins methodology for seasonal ARIMA models *J. Tourism and Hospitaly Management* **18** 1 1-18.
[2] Clements M P and Hendry D F 2004 *A companion to economic forecasting* Blackwell Publishing Oxford.
[3] Dinas Budaya dan Pariwisata Provinsi NTB 2016 *Data kunjungan wisatawan ke NTB tahun*

2008 - Juni 2016 (http://www.disbudpar.ntbprov.go.id/).

[4] Fandeli C and Muhammad 2009 *Pengembangan ekowisata berbasis konservasi di taman nasional* Yogyakarta: UGM Press.

[5] Ibrahim N *et al.* 2016 *The fitting of SARIMA model on peads patients coming at outpatients medical laboratory (OPML)* Mayo hospital, Lahore, (https://www.researchgate.net/publication/299458058, accesed 5 September 2016)

[6] Li G et al. 2006 *Forecasting tourism demand using econometric models",* in Buhalis, D. and Costa, C.(Ed.), *Tourism management dynamics: trends, management and tools*, Butterworth-Heinemann, Oxford 219-220.

[7] Makridakis *et al* 1999 *Metode dan aplikasi peramalan, jilid 1* Jakarta: Binapura Aksara

[8] Saayman A and Saayman M 2010 Forcasting Tourist Arrivals In South Africa *Acta commerce* 2010 280-293.

[9] Widiarsih I N and Subekti R 2015 *Analisis Komparasi Holt Winter Dan Sarima Pada Peramalan Statistik Wisatawan Asing Kraton Yogyakarta* Paper Presented at Seminar Nasional Matematika dan Pendidikan Matematika, Universitas Negeri Yogyakarta, Yogyakarta, 2015.

IOP Conf. Series: Materials Science and Engineering **332** (2018) 012006 doi:10.1088/1757-899X/332/1/012006

Near field communication (NFC) model for arduino uno based security systems office system

A Chairunnas[1], I Abdurrasyid[2]

[1]Computer Science Study Program, Faculty of Mathematics and Natural Sciences, University of Pakuan, Indonesia
[2]Computer Science Study Program, Faculty of Mathematics and Natural Sciences, University of Pakuan, Indonesia

Corresponding author: b.x.lewandowski@magsci.edu

Abstract. Currently, many offices or companies that start growing rapidly in a company or office should have a very limited room to enter only people entitled to enter the room and use the facilities contained in it, for example, Files in it must have many files and documents very important because to reduce the abuse of files and irresponsible person. Because it will be made room door security system by using Near Field Communication on android smartphone. Software used is Arduino IDE. The tools used in this system are Arduino Uno R3, NFC shield, pear sensor, bell, led, servo, 16x2 LCD, and Near Field Communication (NFC) in android smartphone. This system runs based on 2 inputs of a new technology that is Near Field Communication (NFC) in android smartphone. And also use pear sensor to detect unauthorized person entering the room. If the correct password is entered then the door will open and the pear sensor will light off if wrong then the bell will light up.

1. Introduction

Near Field Communication (NFC) is a short-range wireless technology that can be used for data exchange between devices. NFC on Smartphone is a development of Bluetooth and RFID. NFC is a standard specification for smartphones and similar devices to build radio communications between devices by attaching or closing the two devices. NFC is designed to transmit various types of information, such as phone numbers, pictures, MP3 files, or digital authorization between two NFCs, between mobile phones or between mobile phones with RFID chip / reader cards that have close proximity to each other (Curran 2012).

Nowadays a lot of office offices or a company that began to grow rapidly in a company or office must have a room that is very limited to enter only the right people to enter it into the room and use existing facilities therein,for example the file space in it must be a lot of files and documents are very important because to reduce the misuse of a file, files and facilities in it.

This research will discuss about Near Field Communication (NFC) model for arduino uno office file security system with NFC input and process using arduino uno to generate servo output to move the door and buzzer for error notification if there is no access rights to enter indoors and a PIR Sensor that will detect that a person will enter into the room so that the light or LED will be automatically turns on.

2. Theoritical Basis

2.1 Definition of Near Field Communication Shield (NFC)

NFC Shield is a Near Field Communication interface for Arduino building around the popular NXP PN532 integrated circuit. NFC is a short-range radio technology that allows communication between devices held close together. The NFC traces its roots in RFID technology and open platform technologies are standardized in ECMA-340 and ISO / IEC 18092. The NFC Forum develops and declares compliance devices with NFC standards. NFC operates at 13.56 MHz ISM Frequency. NFC has 12Mhz crystal and 12 Volt power supply. Pin I / O from PN532 can be accessed easily using communication via SPI (seeedstudio 2015).

Figure 1. *NFC Shield*

2.2 Definition of Motor Servo

Servo motor is a motor with a closed feedback system where the position of the motor will be informed back to the control circuit in the servo motor. This motor consists of a DC motor, a series of gears, a potentiometer and a control circuit. Potentiometer serves to determine the angle limit of servo rotation. While the angle of the servo motor axis is set based on the width of the pulse sent through the signal foot of the motor cable. In general there are 2 types of servo motors. That is a standard servo motor and continuous motor servo. Standard type servo motors can only rotate 180 degrees (Suranata 2015).

Figure 2. Motor Servo

2.3 Definition of Light Emitting Dioda (LED)

Light Emitting Diode or often abbreviated as LED is an electronic component that can emit monochromatic light when given the forward voltage. LED is a family of Diodes made of semiconductor materials. The colors of the light emitted by the LEDs depend on the type of semiconductor material they use.

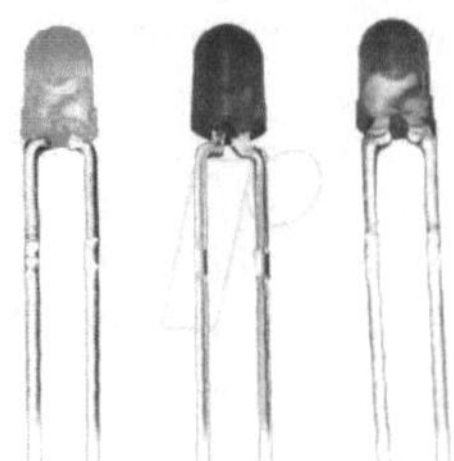

Figure 3. *LED*

2.4 Definition of Switch

Switches are fundamental components in electrical circuits and control system circuits. Its main function is to connect, disconnect and change the direction of the connection on electricity.

IOP Conf. Series: Materials Science and Engineering **332** (2018) 012006 doi:10.1088/1757-899X/332/1/012006

Figure 4. Switch

2.5 Definition of Relay

Relays are electro-mechanical components used to operate a set of switch contacts, utilizing electric power as a source of energy. The current required by the relay module is 5V 15-20mA, equipped with high current, AC250V 10A, DC30V 10A.

Figure 5. Relay

2.6 Definition of Arduino Uno

Arduino Uno is a microcontroller board in which there is a microcontroller, the use of different types of microcontroller - depending on the specifications. In Arduino Uno used ATmega 328 based microcontroller Has 14 input pin from digital output where 6 pin input can be used as PWM output and 6 input analog pin, 16 MHz crystal oscillator, USB connection, power jack, ICSP header, and reset button. To support the microcontroller in order to use, simply connect only the Arduino Uno Board to the computer using a USB or power cord with AC-to-DC adapter or battery to run it (Sainsmart 2015).

Figure 6. Arduino Uno

2.7 Definition of Passive Infra-Red Sensor (PIR)

PIR motion sensor (Passive Infra-Red) is a sensor that works for motion detection that works by detecting a difference / change in temperature now and before. Motion sensor using pear module is very simple and easy to apply because PIR module only requires 5V DC input voltage effective enough to detect movement up to distance 5 meter. When it does not detect motion, the output module is LOW. And when it detects movement, the output will change to HIGH.

Figure 7. *Passive Infra Red*

2.8 Definition of LCD 16 x 2

LCD (Liquid Crystal Display) is a type of display media that uses liquid crystals as the main viewer. LCDs are used in various fields such as electronic devices such as televisions, calculators, or computer screens. LCD is very functional as a viewer that will be used to display the working status of the tool. (Sainsmart 2015).

Figure 8. *LCD*

2.9 Buzzer

Buzzer is an electronic component that serves to convert electrical vibrations into vibrations of sound. Basically the buzzer working principle is almost the same as the loud speaker, so the buzzer also consists of a coil attached to the diaphragm and then the coil is flowed so that it becomes an electromagnet, the coil will be pulled in or out, depending on the direction of current and the polarity of the magnet, mounted on the diaphragm then each movement of the coil will move the diaphragm back and forth so as to make the vibrating air that will produce sound

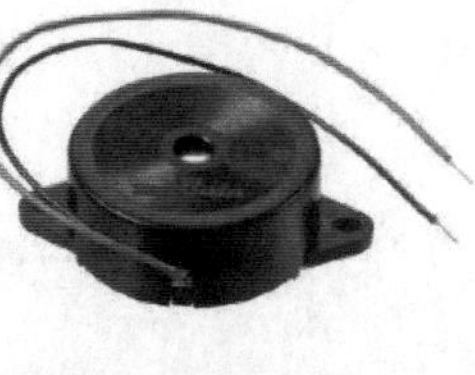

Figure 9. *Buzzer*

3 Research Methodology

The research method used is to use research methods in the field of hardware programing shown in Figure 10.

3.1 Planning of Research Design (Project Planning)

In the planning of the research project, there are several important things that must be determined and considered, among others:

1. Determination of Research topics.
2. Estimation of equipment and materials needs.
3. Budget Estimates.
4. Possible application of the application to be designed.

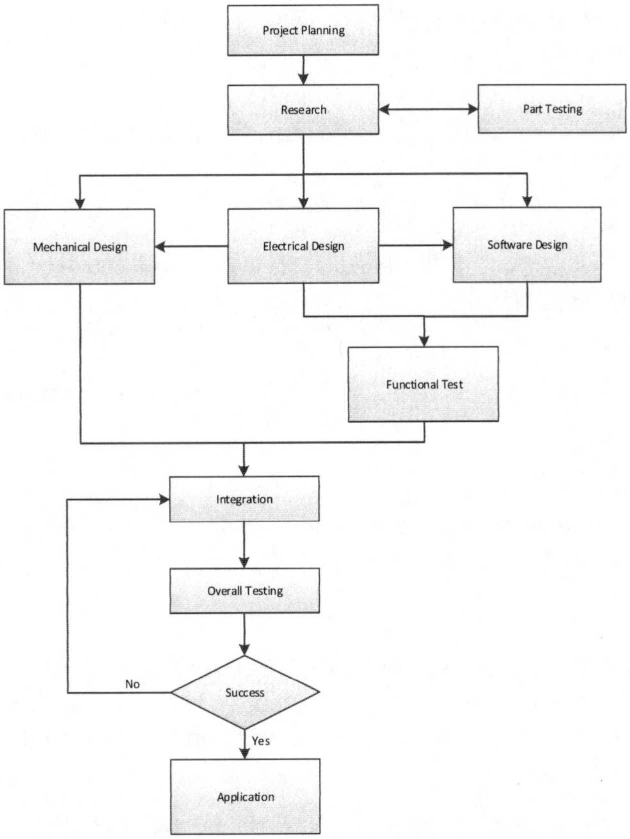

Figure 10. Hardware Programming Research Methods

3.2 Research

Initial research of the application to be made, from selection and testing of components (tools and materials) to be used, the possibility of the initial and final design of "Near Field Communication (NFC) Utilization Model on Android Smartphone for Space File Security System in Arduino Based Office Uno."

3.3 Testing Components (Part Testing)

In testing component done testing tool to work function of component based on requirement of application to be designed.

3.4 Mechanical System Design (Mechanical Design)

In hardware design, mechanical design is an important thing to consider. In general, the application needs to mechanical design, among others:

1. The shape and size of PCB (Printed Circuit Board)
2. Flexibility to the environment
3. Placement of electronic modules
4. Testing of mechanical systems that have been designed
5. Design the size of interface system

3.5 Mechanical System Design (Mechanical Design)

In the design of electrical and mechanical systems there are several things that must be considered, among others:

1. Power supply and power division for each component
2. Voltage and current requirements for microcontroller, sensor and actuator
3. Design of circuit sequence

3.6 Software Design

The design of the software used in this research is using Arduino IDE, MS office, Visio and Fritzing software (Sketchup 2016).

3.7 Functional Test (Function Test)

Functional tests include system functional testing that has been integrated between electrical design and software design.

3.8 Integration or Assembly (Integration)

The electrical modules that are integrated with the software inside the controller, are integrated in the mechanical structure that has been designed. Then performed a functional test of the whole system.

3.9 Overall System Functional Tests (Overall Testing)

At this stage testing the function of the whole system. Whether it can work in accordance with the concept or not. If there is a system that cannot work properly then the process should be re-assembled on each system design.

3.10 Application

Application to improve the performance of the applications that have been designed. Optimization is emphasized on mechanical design for maximum usage as well as optimal.

4 Planning

4.1 Planning of Research Design (Project Planning)

The research project planning stage is the activity step of the system creation process. Components needed in the design system are Two Arduino Uno, NFC shield, LCD, relays, servo, PIR sensor, buzzer, smartphone and LED.

4.2 Research

After system planning, then proceed with the initial research of the system to be created. In the research phase, the initial design of the mechanical circuit and the components of this security system model to ensure that all components can run optimally. This system uses one Arduino Uno. Input system using NFC on android smartphone. Control system using NFC shield with radio frequency ISM 13.56 MHz. The system output is the door lock lever using servo and led.

4.3 Testing Components (Part Testing)

At this stage testing the components that will be used using multimeter. Testing using Arduino serial monitoring is done by looking at the output of each component connected to the Arduino via USB connection. Tests using multimeters include testing the input and output voltages of each component.

4.4 Mechanical System Design

Here the mechanical design of the system as in Figure 11 below:

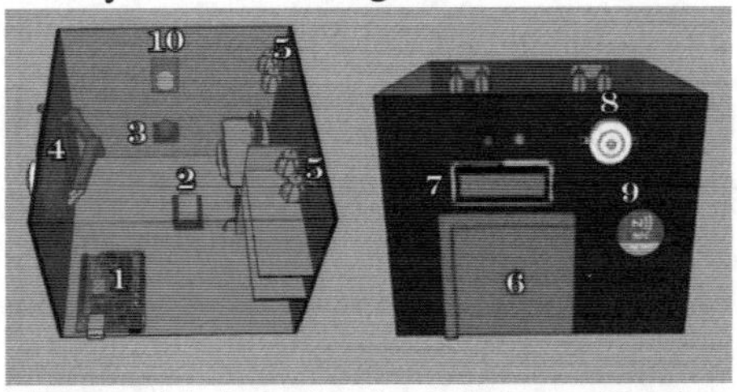

Figure 11. Mechanical System Design

The prototype or model is made of acrylic, with a height of ± 30cm and a length of ± 30cm, the door lock lever using Solenoid.

4.5 Electronic Design (Electronics Design)

The schematic design of the circuit using the Fritzing software is based on the block diagram in Figure 12 below.

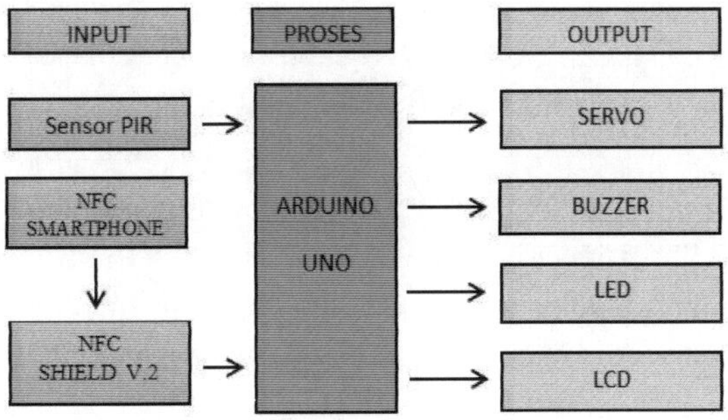

Figure 12. Block Diagram

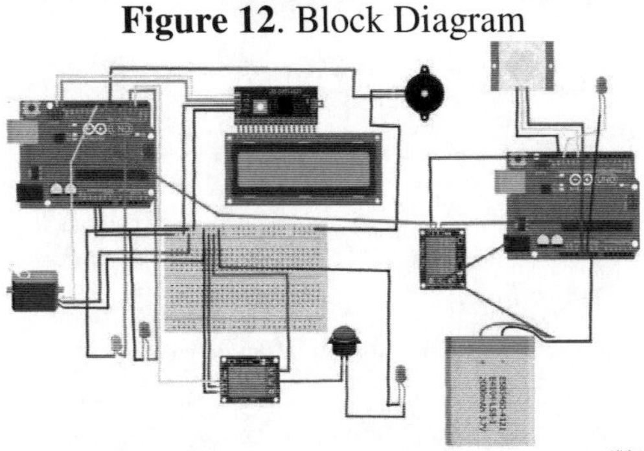

Figure 13. Network Schematic

The voltage source uses a 12V battery that will supply the current to each component. The voltage that goes into the relay and solenoid is 11,45V. While the voltage flowing to Arduino and NFC shield as much as 5V.

4.6 Software Design

The design of the system software is made with the Processing Language Processing in Arduino Uno based on the flowchart in Figure 14 below:

IOP Conf. Series: Materials Science and Engineering **332** (2018) 012006 doi:10.1088/1757-899X/332/1/012006

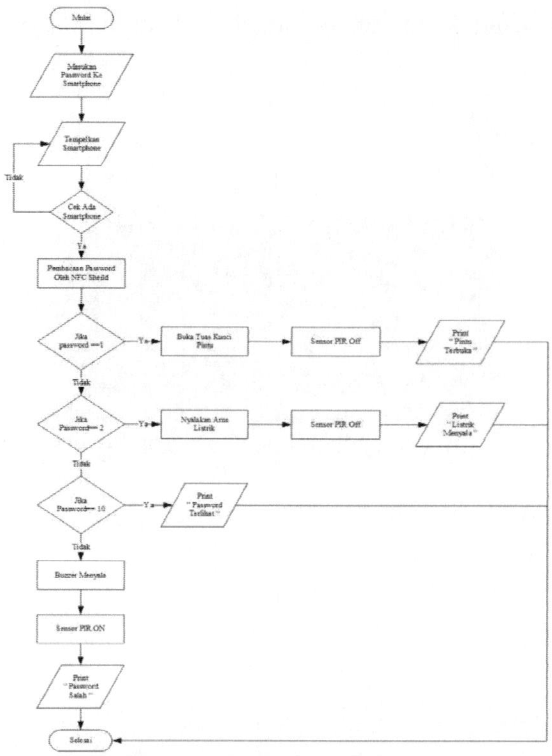

Figure 14. System Flowchart

4.7 Functional Tests

Functional tests are performed on software that has been designed. This test process is done to improve the performance of the software in controlling the electrical design and eliminate and anticipate errors of software made. When the software system has been completed tested then go into the assembly process.

4.8 Assembly

In this process the assembly process is based on the design process, both mechanical design, electronic and software design.

5 Discussion

5.1 Research Results

From the results of this study the authors complete some things that become reference and reference in order to get maximum results with model design is made of acrylic with a height of ± 20cm and length ± 25cm and use two arduino who communicate with each other. Using two arduino paired NFC shield as input from NFC smartphone and another arduino paired with PIR sensor as input from motion or sensor to detect a movement. Use two arduino because NFC shield cannot be added any input other than NFC shield itself, therefore using this two interlocking arduino and arduino one not installed NFC shield then additional input is mounted on this arduino.

Furthermore, in doing communication using Arduino Uno tool. Arduino uno with 16x2 LCD connected via i2c (Inter Integrated Circuit). In the use of NFC shield the tool will be connected to the Arduino Uno via SPI communication (Serial Peripheral Interface) and relay1 is connected to pin 6, riley2 is connected to pin 2, the buzzer is connected to pin 7, the red led is connected to pin 3, the led LED is connected to pin 4 , the servo is connected to pin 9 of the Arduino Uno while in the arduino another PIR sensor is connected to pin 2, the red led is connected to the pin 13 By maximizing the functions of the Tool to provide security to the file room there are 3 (three) security systems:

1. Security of doors by using servo serves as open door closes automatically and locks the door.
2. PIR sensor function to detect a movement in the room.
3. Security of the use of facilities in the room by turning off all electric current contained in the room by using relays to arrange it.

The tool is accessed by using a NFC smartphone with NFC Shield V.2. Function of NFC smartphone to enter or input the password into the tool, the password already entered on the smartphone will be read by NFC shield by attaching the NFC smartphone to the NFC shield after pasted then there is a command to tab the smartphone screen with one tab only if the smartphone screen not tab then there will be no communication between NFC smartphone with NFC shield or password that will not be read by the NFC shield it will fail and can be tried again by reattaching the NFC smartphone to NFC shield.

If the password entered is 1 (one) then if the correct door will open and the PIR sensor will not turn on and if the password entered on the smartphone is 2 (two) then the power will turn on and led is indicator that the electricity in the room is on and can use the facilities contained in the file room and if we input password 2 (two) back then the power will die so password 2 (two) it's function to turn on and off the electricity.

If the password entered is 3 (three) then LCD will display the existing password just in case if user forgot the password to open, turn on the power and turn off the power. If the password is entered in addition to 1 (one), 2 (two) and 3 (three) then the password is false then the door and the electricity will not turn on and open then buzzer will light indicates that the password entered is wrong and if someone is successful enter the room without going through the main door or without entering the password then the PIR sensor will always detect the presence of anyone who is in the room, if there is someone who enter it without permission it will be captured by the PIR sensor and led will also light up as an indicator that an unauthorized person managed to get into the file room of the office.

5.2 Overall System Functional Testing (Overall Testing)

This stage is done testing the function of the whole system. Whether it can work in accordance with the concept or not. If there is a system that cannot work properly, it must be done re-assembly process every part of the system. These tests include structural, functional and validation testing.

Structural Testing. At this stage a test is conducted to determine whether the circuit paths are connected properly so that the system can work properly. This test is done by testing the circuit paths using multimeter.

Functional Testing. At this stage a test is conducted to determine whether the voltage flowing in the circuit is in accordance with the required. This test is done by testing the output voltage of each component by using multimeter and program.

Validation Test. This stage is done to test the distance from the NFC range as system input. By attaching a NFC shield antenna with a NFC smartphone, testing will be done in the absence of obstructions and barriers. This test is done 10 times, it is assumed n = 10 times test, can describe the optimal test result.

6 Conclusions and Suggestions

6.1 Conclusion

The conclusion of this research, the model of utilization of near field communication (NFC) on smartphone for security system in arduino uno based file room. The system uses Arduino Uno, NFC shield, servo, relay, buzzer, LCD, led and smartphone. Input system using the latest technology that is Near Field Communication (NFC) on android smartphone. The system can receive password input and in continue with NFC input on the smartphone cannot vice versa. NFC Shield connection distance and android smartphone that can be achieved at no barrier that is a maximum of 4cm with an average response time of 0.77 seconds. when blocked by acrylic maximum distance is 3cm with an average response time of 0.63 seconds. The system output is the door lock lever using servo and switching the electric current inside using relay.

6.2 Suggestions

The file system security model using Near Field Communication (NFC) on Android smartphones needs further development to improve security by adding functionality and system features such as cameras or indoor sensors anticipating people entering the premises not through the main door so that the security is layered from outside and from within the room as well as adding layered passwords and for obstacle forgotten passwords that for user error can change the password through the internet and no need to re-upload the program into the microcontroller.

References

[1] Aisyah, Q, S. Nasution, M, S & Jati, N, A. 2015. Perancangan dan implementasi sistem akses kontrol pada pintu berbasis teknologi Near Field Communication dengan Mikrokontroler Arduino Uno. Skripsi. Program Studi S1 Teknik Komputer, Fakultas Teknik Elektro, Universitas Telkom, Bandung

[2] Arduino.cc. 2015. *Datasheet* Arduino Uno, Italy

[3] Curran, K. Millar, A & Garvey, Mc. C. 2012. *Near field communication*. (*IJECE Vol. 2 :371-382*)

[4] Marnis, Yeni. 2005. Implementasi Sensor PIR (*Passive Infrared Receiver)* KC7783R Pada Sistem Pengamanan Ruangan Berbasis Mikrokontroler Dengan Keluaran Suara. Program Studi Pascasarjana, Universitas Andalas, Padang. Horowitz G T and Maldacena J 2004 The black hole final state *J. High Energy Phys.* JHEP02(2004)008

[5] Mahfudin, Dicky. 2013. Model Sistem Keamanan Pintu Menggunakan Password dan *Near Field Communication* (NFC) Pada *Smartphone* Android. Program Studi S1 Ilmu Komputer, Universitas Pakuan, Bogor

[6] Pranoto, H. Ulvan, A & Melvi. 2015. *Near Field Communication (NFC)* pada Sistem Tertanan *Raspberry Pi* sebagai Pendeteksi Identitas Otomatis dan *e-ticket* Pada *Smart Transport.* Penelitian ini didukung oleh MP3EI 2015. Penghargaan juga atas kontribusi dari Direktorat Jenderal Pendidikan Tinggi, dalam mendukung pada publikasi

[7] Sainsmart. 2015. *Datasheet Keypad,* LCD, *relay,* Lenexa, Kansas. Amerika serikat

[8] Seeedstudio. 2015. *Datasheet* NFC Shield. Xili Town, Nanshan District, Shenzhen. China

[9] Setiawan, A. 2014. Prototipe sistem keamanan pintu geser menggunakan *Radio Frequency Identification* (RFID) dan *Bluetooth.* Skripsi. Program Studi Ilmu Komputer FMIPA Universitas Pakuan, Bogor.

Risk adjustment model of credit life insurance using a genetic algorithm

A Saputra[*], Sukono, E Rusyaman

Department of Mathematics, Faculty of Mathematics and Natural Sciences, Universitas Padjadjaran, INDONESIA

*Corresponding authors: andrispt@yahoo.com

Abstract. In managing the risk of credit life insurance, insurance company should acknowledge the character of the risks to predict future losses. Risk characteristics can be learned in a claim distribution model. There are two standard approaches in designing the distribution model of claims over the insurance period i.e, collective risk model and individual risk model. In the collective risk model, the claim arises when risk occurs is called individual claim, accumulation of individual claim during a period of insurance is called an aggregate claim. The aggregate claim model may be formed by large model and a number of individual claims. How the measurement of insurance risk with the premium model approach and whether this approach is appropriate for estimating the potential losses occur in the future. In order to solve the problem Genetic Algorithm with Roulette Wheel Selection is used.

Keyword: risk, claims, aggregation claims, collective risk model, genetic algorithm

1. Introduction

A type of life insurance quite developed in Indonesia is credit life insurance. Every customer of this insurance policy holder receives compensation afforded by the insurance company in case of death in accordance with the contract of agreement. In providing guarantee to the policy holders, the insurance company must prepare funds to anticipate the occurrence of claims from policyholders [2].

In managing risk insurance companies should acknowledge characteristic of the risk to predict future losses. These risk characteristics can be learned in a claim distribution model. There are two standard approaches to establish a model of claim distribution over the insurance period e.g. the collective risk model and individual risk model.

In collective risk model, claim that arise when risk occurs is called individual claim, the accumulation of individual claims during period of insurance is called an aggregate claim. The aggregate claim model can be developed using the large model and the number of individual claims.

Model-based approach should be considered as context of the objectives of each given problem. Many problems in actuarial science involve development of mathematical models that can be used to predict or forecast future insurance costs. The model is a simplified mathematical description built on the knowledge and experience of an actuary combined with data from the past. Data guides the actuary in selecting the model form as well as calibrating an unknown number, usually called a parameter. This model provides a balance between simplicity and conformity with existing data. Simplicity is measured in terms of unknown parameter factors (the fewer the simpler); Compliance with data is measured in terms of discrepancies between data and models. The selection model is based on a balance between two criteria, namely conformity and simplicity.

Determination of using the model should consider several things such as decision-making based on empirical evidence. The empirical approach assumes that the future can be expected just as it has happened in the past, perhaps adapted to trend such as inflation[11]. How to measure insurance risk with the premium model approach and whether this approach is appropriate for estimating the potential losses incurred in the future.

2. Research methodology

The models used in the development of claims distribution model and the calculation of risk of term life insurance using the claim distribution model, Collective Modified Value-at-Risk Model (*CMVaR*).

2.1. Claim distribution model

The claim distribution is assumed to follow the geometric distribution model. If N is a random variable of the claim frequency with a value of n, and p is the probability parameter of the occurring claim, then the probability function of the claim event is [1],[13].

$$\eta(n,p) = \begin{cases} pq^{n-1}; n = 1,2,3,... \\ 0; \quad n \text{ otherwise} \end{cases}, \tag{1}$$

where $q = 1 - p$ is the parameter of no claim. The generating function of equation (1) is [9], [4]:

$$M_N(t) = \frac{pe^t}{(1-qe^t)}$$

For large claim it is assumed to follow the Gamma distribution. If X is given a large random variable of claim with value x, and α and β is the parameter of gamma distribution, then the probability density function of the distribution of the claim is [1], [12]:

$$f(x) = \begin{cases} \frac{1}{\Gamma(\alpha)\beta^{\alpha}} x^{n-1} e^{-x/\beta}; x > 0 \\ 0; x \text{ otherwise} \end{cases}, \tag{2}$$

whereas the moment generating function to k of the equation (2) is

$$m_k = M_X(k) = (1-\beta k)^{-\alpha}; k = 1,2,3,4,....$$

2.2. Collective risk model

For collective risk models, a random process assumption that claims to be sustainable in the insurance policy portfolio can be used. Generally can be formulated [10] by defining the random variable S is the number of claims occurring from the risk within a year. Random variable N is the number of claims of risk in this year, *Xi'* random variable is the number of claims incurred. The number of aggregate claims is the number of individual claims, as the formula below:

$$S = \sum_{i=1}^{N} X_i \tag{3}$$

while the conditional Expectation value is: $E[S \mid N = n] = E\left[\sum_{i=1}^{N} X_i\right] = \sum_{i=1}^{N} E[X_i] = nm1$, and apply $n = 0,1,2,...,$ $E[S \mid N] = Nm1$, then the unconditional expectation is obtained:

$$E[S] = E[Nm_1] = E[N]m_1$$

thus, the variance is:

$Var[S \mid N = n] = Var\left[\sum_{i=1}^{N} X_i\right] = \sum_{i=1}^{N} Var[X_i] = n(m_2 - m_1^2)$, and apply n = 0,1,2,...

$Var[S \mid N] = N(m_2 - m_1^2)$, therefore this conditional variance of (3) is :

$$Var[S] = E[Var(S \mid N)] + Var[E(S \mid N)] = E[N](m_2 - m_1^2) + Var[N]m_1^2$$

If it is assumed that $\{X_i\}_{i=1}^{\infty}$ distributed identically and independently, the moment function of equation (3) can be expressed as [1]

$$M_s(t) = E[\{M_x(t)\}^N] = E[Exp\{\log\{M_x(t)^N\} = M_N(t)\log M_x(t)]$$

where M_S is expressed in terms of $M_N(t)$ and $M_X(t)$, and S is the random variable to the magnitude of the aggregate claim.

2.3. Collective value-at-risk

According to Dowd [5] and Yildrim [14] and Sukono [12], Value-at-Risk (*VaR*) is a measure of market risk, which means when invested initial capital of V_o, on market assets with normal distributed returns with μ and standard deviation σ, will have maximum potential loss is $VaR = -V_0(\mu + z_c\sigma)$, where z_c percentile and standard normal distribution at significance level (1 – c) %. Collective Value-at-Risk (*CVaR*) is the development of the application of Value-at-Risk principle which is an alternative measurement of aggregate claim risk as equation (3). Then *CVaR* can be formulated as follows:

$$CVaR = -\tilde{N}_0(E[S] + z_c\{VaR[S]\}^{1/2}) = -\tilde{N}_0(E[N]m_1 + z_c\{E[N\}(m_2 - m_1^2) + Var[N]m_1^2)^{1/2}$$

where is : $\tilde{N}_0$ is a lot of claims that occur, the sign minus $(-)$ states losses (losses) [3].

2.4. Collective modified value-at-risk

According to Dowd [5], Guarda, et.al [7] and Iqbal, et.al [8], market assets are not normally distributed, their expansion by providing a factor adjustment to the estimated percentile, with the adjustment given from normality is small called Cornish Fisher. The expansion of Cornish Fisher is expressed as follows:

$$CMVaR = \tilde{N}_0\{E[S]\} + (z_c + \frac{1}{6}(z_c^2 - 1)\varsigma[S] + \frac{1}{12}(z_c^3 - 3z_c)K[S] - \frac{1}{36}(2z_c^3 - 5z_c)(\varsigma[S])^2)(Var[S])^{1/2}\}$$

Where :

E[S]	= mean	Var[S]	= variance
ζ[S]	= skewness	κ[S]	= kurtosis
Z_c	= normal distribution percentile		

2.5. Genetic algorithms

The purpose of the algorithm is to dynamically determine the distribution of life insurance claims and certain restrictions to minimize risk. The claim distribution structure using genetic algorithms (GA) can be seen in the following figure:

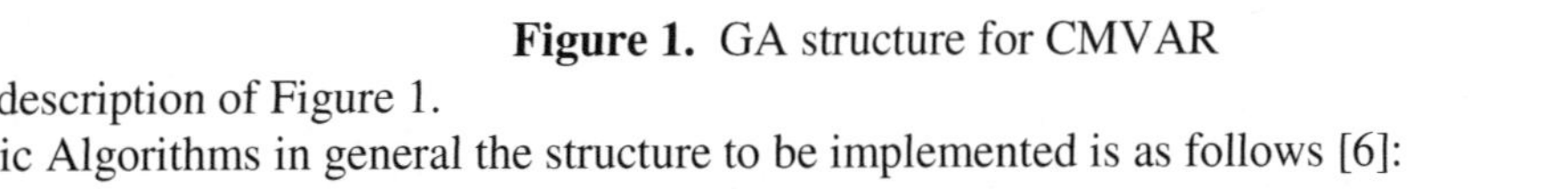

Figure 1. GA structure for CMVAR

Brief description of Figure 1.

Genetic Algorithms in general the structure to be implemented is as follows [6]:

a. Generate the initial population

This process is a process used to generate the initial population at random to obtain an initial solution. This initial population is generated randomly so that the initial solution is obtained. The population itself consists of a number of chromosomes that present the desired solution.

b. Evaluation of fitness
This process is a process for evaluating each population by calculating the fitness value of each chromosome and evaluating it until it is met the stop criteria. An individual is evaluated based on a certain function as a measure of its performance. In natural evolution, individuals with low fitness value will die.

c. Selection:
Selection process is a process to determine which individuals will be selected for crossover. There are several types of selection methods commonly used, such as: The method that mimics a roulette-wheel game where each chromosome occupies a circle cut on the roulette wheel proportionately according to its fitness value. Selection Ranks the process begins by ranking or chromosome sequencing in the population based on their fitness then assigning a new fitness value based on the sequence. In this problem the selection method used is the roulette wheel selection.

d. Crossover
This crossover process is a process to increase the diversity of strings in one population. Crossover operators have the most important role in the genetic algorithm because there is a process of marriage (crosses) of genes between two individuals (parent) that produce two new individuals (offspring) in the next generation.

e. Mutation
Mutation is the process of changing the value of one or more genes in a chromosome. Mutations create new individuals by modifying one or more genes in the same individual. Mutations act to replace lost genes of the population during the selection process and provide genes that are absent in the initial population.

f. Criterion stopped
Stop criteria are the criteria used to stop the genetic algorithm process which is the goal to be achieved from the process.

g. Results
The result is an optimum solution obtained by using Genetic Algorithm.

The application of roulette wheel selection method in credit risk adjustment modeling is as to :

a. Generate initial population
All possible risks of claims are employed by permutation coding techniques which are subsequently taken by some random claims. Claims taken randomly form the first population in the first generation and so on down to the last generation. The population is a set of chromosomes consisting of k genes representing x_i = 1,2, ... 64, numbers, where x_i is the number of claims occurring on i. The population was initially randomly using native code.

x_1	x_2	....	x_{64}

Figure 2. Chromosome structure

b. Determining fitness value
After generating the initial population the next step is to determine the fitness value of each individual formed. Each individual is calculated the risk value of his claim. To calculate the fitness value by the formula:

$$eval(v_k) = f(x), k = 1, 2, ..., 64$$

To find the minimum value then:

$$eval(v_k) = \frac{1}{x}$$ where x = the claim value of each individual

With the help of matlab software, determined the fitness value of individuals big claims and many claims. Fitness value as follows

Table 1. Fitness Value

Individual	Fitness Value Claim Large	Fitness Value Many Claims	Individual	Fitness Value Claim Large	Fitness Value Many Claims
1	0.0000000401	1.00	33	0.0000000019	0.08
2	0.0000002042	1.00	34	0.0000000073	0.50
3	0.0000000120	0.33	35	0.0000000257	0.25
4	0.0000000191	0.33	36	0.0000000064	0.17
5	0.0000000067	0.20	37	0.0000000076	0.25
6	0.0000000150	0.50	38	0.0000000104	0.50
7	0.0000000026	0.25	39	0.0000000091	0.25
8	0.0000000123	0.50	40	0.0000000043	0.14
9	0.0000000031	0.09	41	0.0000000045	0.20
10	0.0000000038	0.17	42	0.0000000052	0.25
11	0.0000000067	0.20	43	0.0000000060	0.25
12	0.0000000022	0.11	44	0.0000000135	0.33
13	0.0000000054	0.14	45	0.0000000022	0.11
14	0.0000000022	0.13	46	0.0000000204	0.50
15	0.0000000063	0.20	47	0.0000000091	0.20
16	0.0000000117	0.20	48	0.0000000468	1.00
17	0.0000000118	0.33	49	0.0000000189	0.33
18	0.0000000060	0.20	50	0.0000000563	1.00
19	0.0000000027	0.14	51	0.0000000222	0.50
20	0.0000001828	1.00	52	0.0000000086	0.25
21	0.0000000024	0.14	53	0.0000001576	1.00
22	0.0000000029	0.25	54	0.0000000327	0.50
23	0.0000000096	0.33	55	0.0000000466	1.00
24	0.0000000040	0.14	56	0.0000000124	0.33
25	0.0000000044	0.17	57	0.0000006475	1.00
26	0.0000000030	0.20	58	0.0000000314	1.00
27	0.0000000025	0.11	59	0.0000000289	1.00
28	0.0000000066	0.25	60	0.0000006802	1.00
29	0.0000000043	0.25	61	0.0000000995	1.00
30	0.0000000059	0.25	62	0.0000000099	0.50
31	0.0000000021	0.08	63	0.0000000255	1.00
32	0.0000000036	0.11	64	0.0000000736	1.00

From the fitness value of each individual, the best fitness values are taken to be maintained and brought to the next generation. In this case for the largest claim the largest fitness value is the 60th individual is 0.0000006802 and for many claims of fitness value 1. Individuals with the best fitness value of the first generation population will be retained and taken to the next generation

c. Selection

After a population assessment operation, the best chromosomes are selected by using wheel selection that is associated with each chromosome of selection probability, noted, Pi with the following formula:

$$P_i = \frac{1}{N-1}\left[1 - \frac{h_i}{\sum_{i \in Pop} h_i}\right]$$

Each chromosome is reproduced with probability. Some chromosomes are "more" reproduced and "bad" ones are removed. Next randomly select individuals from the population to serve as parent. The parent will be in the process of crossing with another selected individual.
With the help of matlab software got parent selected randomly. Here are the individual results selected as the parent.

Table 2. Claim Large Selection

Parent	Individual	Parent	Individual	Parent	Individual	Parent	Individual
Parent 1	5	Parent 1	11	Parent 1	55	Parent 1	7
Parent 2	24	Parent 2	27	Parent 2	63	Parent 2	12
Parent 1	47	Parent 1	19	Parent 1	21	Parent 1	59
Parent 2	56	Parent 2	41	Parent 2	34	Parent 2	28
Parent 1	20	Parent 1	4	Parent 1	39	Parent 1	48
Parent 2	33	Parent 2	42	Parent 2	25	Parent 2	64
Parent 1	38	Parent 1	14	Parent 1	35	Parent 1	1
Parent 2	52	Parent 2	6	Parent 2	32	Parent 2	8
Parent 1	30	Parent 1	46	Parent 1	37	Parent 1	36
Parent 2	44	Parent 2	31	Parent 2	53	Parent 2	43
Parent 1	3	Parent 1	10	Parent 1	16	Parent 1	54
Parent 2	29	Parent 2	17	Parent 2	49	Parent 2	57
Parent 1	13	Parent 1	23	Parent 1	60	Parent 1	62
Parent 2	18	Parent 2	58	Parent 2	15	Parent 2	45
Parent 1	26	Parent 1	9	Parent 1	51	Parent 1	61
Parent 2	50	Parent 2	40	Parent 2	2	Parent 2	22

Table 3. Many Large Selection

Parent	Individual	Parent	Individual	Parent	Individual	Parent	Individual
Parent 1	22	Parent 1	62	Parent 1	32	Parent 1	33
Parent 2	37	Parent 2	52	Parent 2	40	Parent 2	48
Parent 1	47	Parent 1	3	Parent 1	59	Parent 1	49
Parent 2	60	Parent 2	14	Parent 2	9	Parent 2	18
Parent 1	4	Parent 1	41	Parent 1	11	Parent 1	38
Parent 2	20	Parent 2	19	Parent 2	30	Parent 2	36
Parent 1	54	Parent 1	50	Parent 1	23	Parent 1	55
Parent 2	31	Parent 2	51	Parent 2	10	Parent 2	43
Parent 1	57	Parent 1	12	Parent 1	39	Parent 1	27
Parent 2	16	Parent 2	25	Parent 2	45	Parent 2	29
Parent 1	13	Parent 1	46	Parent 1	26	Parent 1	15
Parent 2	34	Parent 2	61	Parent 2	44	Parent 2	35
Parent 1	1	Parent 1	21	Parent 1	7	Parent 1	42
Parent 2	17	Parent 2	56	Parent 2	28	Parent 2	6
Parent 1	63	Parent 1	64	Parent 1	2	Parent 1	53
Parent 2	8	Parent 2	5	Parent 2	24	Parent 2	58

d. Crossover

After using the selection method for the selection of two individuals, we apply the crossover operator to a point on this pair. This operator divides each parent into two parts in the same position, chosen at random. Child 1 was made part of the first parent and the second part of the second parent when child 2 consisted of the second part of the first parent and the first part of the second parent

e. Mutation

This operation provides the genetic algorithm of ergodicity property which indicates that it will most likely reach all parts of the space belonging to the state, without traveling all in the resolution process. It's usually to draw a random gene in a chromosome and replace it with a random value.

f. Conditions for convergence
At this level, the final generation is considered. If the result is favorable then the optimal chromosome is obtained. Otherwise, the evaluation and reproduction steps are repeated until a certain number of generations, until the convergence criteria are established or until the convergence criteria are met.

3. Main result and conclusions

With the help of matlab software, Genetic Algorithm output process obtained results as follows:

$\tilde{N}_0$	= 64
p	= 0.26667
m_1	= 2.3702E+008
m_2	= 5.6321E+017
m_3	= 2.5115E+028
m_4	= 5.2923E+040
ζ	= 0.9531
κs	= 2.8623

D	= 0.33984 (many claims)
D	= 0.14049 (claim value)
α	= 180.79
B	= 0.1012
σ_s	= 1. 9182E+009
σ_s^2	= 7.3285E+018
c	= 0.01
$Z_{0.01}$	= −2.3263

3.1. Results

In this section the results of the data processing analysis will be explained comprising the estimation of the frequency distribution model of the claim, the estimated claim distribution model, and the collective risk estimation of the assurance claim.

3.1.1 Estimation of claims frequency distribution model

Referring to (1) and after following the identification stages of the distribution model, distribution model parameter estimation, and test of distribution model matching. The frequency of claims obtained, a random sequence of events that can be predicted through a discrete distribution model.

The identification of frequency distribution model of the claim is made by curve matching, result of curve matching, the late indicates the model suitable for the frequency of the claim is following the geometric distribution. While estimation of frequency distribution model of claim model is performed by using Maximum Likelihood Estimator method. From result of processing obtained by parameter estimator p = 0.26667. The probability density function for geometric distribution is as follows:

$$\eta(n, 0.26667 = (0.26667)(1-0.26667)^{n-1}; n = 1, 2, 3, ...$$

Estimator of the frequency distribution model of this claim, subsequently used to estimate the collective risk of insurance claims.

3.1.2. Estimate the large distribution model of the claim

From the estimator values, namely: α = 180.79 and β = 0.1012. Furthermore, the matching test will be done by comparing the result of output of D value with table *D* of 95% confidence level (*D1*-95%) its critical value is 0.37543. Since the value of *D* <*D1*-95% (0.14049 <0.37543), then *Ho* is accepted. This means that the claim data follows the gamma distribution. The gamma distribution is as follows:

$$f(x) = \frac{1}{\Gamma(180.79)(0.1012)^{180.79}} x^{180.79-1} e^{-(x/0.1012)}; x > 0$$

3.1.3. Estimated collective risk of insurance claims

IOP Conf. Series: Materials Science and Engineering **332** (2018) 012007 doi:10.1088/1757-899X/332/1/012007

The collective risk estimates of insurance claims include aggregate claims and collective risk measurement. Collective risk measurement is based on Collective Risk, Collective Value-at-Risk, and Collective Modified Value-at-Risk. Estimates of aggregate claims expectations and collective risk measurements are performed by estimating the frequency distribution model of claims as equations (10) and (11). Estimator values of the parameters $E[N]$ and Var$[N]$, as well as the values of m_1, m_2, m_3 and m_4, are used in calculating aggregate claims estimation estimates and collective risk measurements. As the above table values the moments obtained from the Genetic Algorithm output, ie m_1 = 2.3702e + 008, m_2 = 5.6321e + 017, m_3 = 2.5115e + 028, m_4 = 5.2923e + 040, σ_s = 1.9182e + 009, = 7.3285 + 018, aggregate claim risk value as measured by Collective Value-at-Risk (*CVaR*) at significance level α = 0.01 (standard normal distribution $Z_{0.01}$= 2.3263), is as follows:

$$CVaR = -64\{-2.3263x7.3285+018)+(1.\ 9182E+009)\} = 280.280.000.000$$

From filing claims data of 64 events, with a claim value of 10,260,884,880 IDR. Based on the result of value analysis σ_s = 1,918,200,000 IDR per month or in 1 (one) year of 23,018,400,000 IDR. Based on the estimated *CVaR* estimator of 280,280,000,000 IDR, this value is an assumption of the value of the claim that will occur. If the risk of aggregate claims is measured by Collective Modified Value-at-Risk *(CMVaR)* using Genetic Algorithms, the best result is 226,484,957,615 IDR,

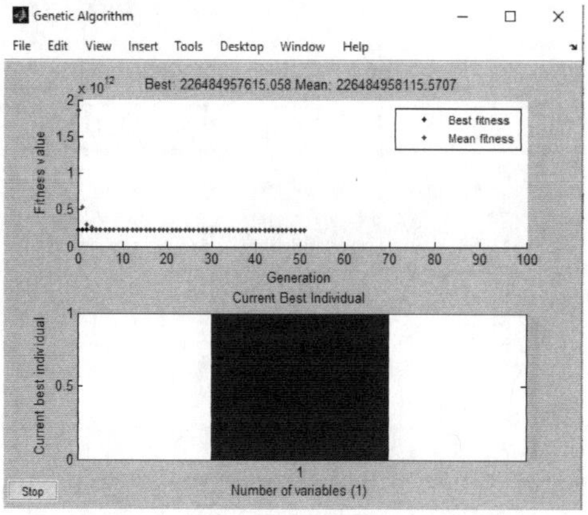

Figure 3. Result Genetic Algorithm

While if calculated manually in get the results as follows:

$$CMVaR = -64\{(1.\ 9182e+009)+((-2.3263+\frac{1}{6}(-2.3263^2-1)(0.9531))+\frac{1}{12}(-2.3263^3-3(-2.3263)(2.8623))-\frac{1}{36}(2(-2.3263^3)-5(-2.3263)(0.9531)^2)(7.3285+018)^{1/2}\}$$

$$CMVaR = 331{,}500{,}000{,}000 \text{ IDR}$$

Which risk model will be chosen by the insurance company will be determined by considering other factors.

3.2. Conclusion

After analyzing the model study of credit life insurance risk adjustment, from the secondary data of one insurance company with the assistance of Matlab software, it can be concluded that the frequency of claims follows geometric distribution with p = 0.26667. For the large claims follow the gamma distribution with estimators α = 180.79 and β = 0.1012. From that result the aggregate claim model formed from the frequency distribution of claims and the magnitude of individual claims, is a geometric compound distribution. The geometric compound distribution is determined by moments m_1 = 2.3702E + 008, m_2 = 5.6321E + 017, m_3 = 2.5115E + 028, m_4 = 5.2923E + 040. Value $\hat{p}, m_1, m_2, m_3,$ and, m_4 with a significance level of 5% obtained a value of *CMVaR* of

Rp.331.500.000.000. The *CMVaR* value is an alternative risk measure, especially the large insurance claims not following the normal distribution

References

[1] Dickson, D.C.M., 2005, Insurance Risk and Ruin. Cambridge: Cambridge University Press.
[2] Dionne, G., 2013, Risk Management History, Definiton and Crtique, Cirrelt-2013-17, March 2013, pp1-22.www.cirrelt.ca.
[3] Dokov, S., Stoyanov, S.S.V. & Rachev, S.T., 2007, Computing VaR and AVAR of Skewed-T Distribution, Working Paper, University of Karlshure, Germany, https://statistik.ets.kit.edu/download/doc.../skewedTAVAR_dEC11.PDF, DIAKSES 5 Desember 2011.
[4] Djuric, Z., 2013, Collective Risk Model in Non-Life Insurance, Economic Horizons, May – August 2013, Volume 15, Number 2, pp. 167 – 175, UDC: 33 eISSN 2217-9232, UDC: 005.334:368.025.6 ; 347.426.6, doi: 10.5937/ekonhor1302163D
[5] Dowd, K., 2002, An Introduction to Market Risk Measurement, John Wiley & Sons, Inc., New Delhi, India
[6] Elhachloufi, M, Z.Guennoun, F. Hamza, 2016, Minimization of Value at Risk of Financial Assets Portfolio using Genetic Algorithms and Neural Networks, Journal of Applied Finance & Banking, vol. 6, no. 2.
[7] Guarda, P., A. Rouabah, and. J. Theal, 2012, An MVaR Framework to Capture Extreme Events in Macro Prudential Stress Tests, Working Paper Series, NO 1464/AUGUST 2012, PP.1-46
[8] Iqbal, et al., 2013, Collective Modified Value-at-Risk in Life Insurance When the Number adn Amount Claims Has Poisson and Lognormal Distributions, Malaysia: SORIC2013
[9] Klugman, S.A., H.H. Panjer, and G.E. Willmot, 1998, Loss Models: From Data to Decision, Wiley Series in Probability and Statistics, John Wiley & Sons, Inc., New York.
[10] Nino, S and C.G. Paolo, 2010 A. Collective Risk Model for Claims Reserve Distributions, 29th Internasional Congress of Actuaries – ICA 2010, Cape Town – March 7-12th 2010, pp. 1-22.
[11] Sukono, Nahar, J., Putri, F. T., Subiyanto, Mamat, M., & Supian, S., 2017. Estimation of outstanding claims reserving based on inflation risk on car insurance companies by using the bootstrap method. *Far East Journal of Mathematical Sciences, 102*(4), 687-706. doi:10.17654/MS102040687
[12] Sukono, Suyudi, M., Islamiyati, F., and Supian, S., 2017. Estimation Model of Life Insurance Claims Risk for Cancer Patients by Using Bayesian Method. IOP Conf. *Series: Material Science and Engineering* 166 (2017) 012022 doi:10.1088/1757-899X/166/1/012022. Pp.1-9.
[13] Valecky, J., 2016, Modelling Claim Frequency in Vehicle Insurance, ACTA Universitatis Agriculture et Silviculture Mendeliane Brunensis, Volume 64, Numer 2, 2016, pp. 683-689, http://dx.doi.org/10.11118/actaun201664020683
[14] Yildrim, I., 2015, Financial Risk Measurements for Turkish Insurance Companies Using VaR Models, Journal of Financial Risk Management, 2015, 4, pp 158-167, Published Online September 2015 in SciRes. http://www.scirp.org/journal/jfrm.http://dx.doi.org/ 10.4236/jfrm.2015.43013.

IOP Conf. Series: Materials Science and Engineering **332** (2018) 012008 doi:10.1088/1757-899X/332/1/012008

Identification of jasmine flower (*Jasminum sp.*) based on the shape of the flower using sobel edge and *k*-nearest neighbour

A Qur'ania* and I Sarinah

Computer Science, Faculty of Mathematics and Natural Sciences, Pakuan University Jalan Pakuan PO BOX 452 Bogor 16143, Indonesia

*Corresponding author: qurania@unpak.ac.id

Abstract. People often wrong in knowing the type of jasmine by just looking at the white color of the jasmine, while not all white flowers including jasmine and not all jasmine flowers have white. There is a jasmine that is yellow and there is a jasmine that is white and purple.The aim of this research is to identify Jasmine flower (*Jasminum sp.*) based on the shape of the flower image-based using Sobel edge detection and k-Nearest Neighbor. Edge detection is used to detect the type of flower from the flower shape. Edge detection aims to improve the appearance of the border of a digital image. While k-Nearest Neighbor method is used to classify the classification of test objects into classes that have neighbouring properties closest to the object of training. The data used in this study are three types of jasmine namely jasmine white (*Jasminum sambac*), jasmine gambir (*Jasminum pubescens*), and jasmine japan (*Pseuderanthemum reticulatum*).Testing of jasmine flower image resized 50 x 50 pixels, 100 x 100 pixels, 150 x 150 pixels yields an accuracy of 84%. Tests on distance values of the k-NN method with spacing 5, 10 and 15 resulted in different accuracy rates for 5 and 10 closest distances yielding the same accuracy rate of 84%, for the 15 shortest distance resulted in a small accuracy of 65.2%.

1. Introduction

Jasmine (*Jasminum sp.*) is a famous ornamental flower with a fragrant aroma and one raw material for perfumes and commodities of high economic value. Around the world is known to 200 types of jasmine but often cultivated and known in Indonesia only some types, including jasmine white (*Jasminum sambac*), jasmine gambir (*Jasminum officinale*), Jasmine Star (*Jasminum multiflorum*), jasmine king (*Jasminum rex*) jasmine australia (*Jasminum simplicifolium*), japanese jasmine (*Pseuderanthemum reticulatum*), jasmine primrose (*Jasminum menshi*), and japanese jasmine (*Pseuderanthemum reticulatum*). The data used in this study are three types of jasmine namely jasmine white (*Jasminum sambac*), jasmine gambir (*Jasminum pubescens*), and jasmine japan (*Pseuderanthemum reticulatum*). Edge detection (edge detection) is used to detect the type of interest from the flower shape. This process aims to improve the appearance of the border on a digital image. Another purpose of edge detection is to identify an area in a digital image where there is a major change in intensity. With edge detection on an image processing system will find surface or line marks that arise from the object [3].

The *k*-Nearest Neighbor (k-NN) method is used for classifying the classification of a test object into a class that has the closest neighbor property to the training object. In this study, the k-nearest neighbor classification uses the closest distance calculation with the euclidean distance method. The classification with k-Nearest Neighbor aims to select the training image with the smallest distance to the tested image [9]. Previous studies of the classification of Japanese cinnamon flower (*Adenium sp.*) Based on crown image using color feature extraction and edge detection 96% percentage based on the combined features of sobel edge detection and histogram hue [2]. Detection of interest type using the value of HSV from the image of the crown of flowers managed to detect objects with an average of 86.67% [8]. Research on the introduction of flower type based on color features using artificial neural network and fuzzy with the average result accuracy of 94.44% [1]. Based on previous research, this research was conducted with the aim of identification of jasmine flowers based on flower shape using extraction of sobel edge detection feature and classified using k-NN.

2. Methods

Methods in this study consisted of data, pre-processing, form extraction by using the Sobel operator that was introduced by Irwin Sobel in 1970. The advantage of this sobel operator is its ability to reduce noise before performing edge detection calculations [10], color extraction, identification and trial as shown in Figure 1.

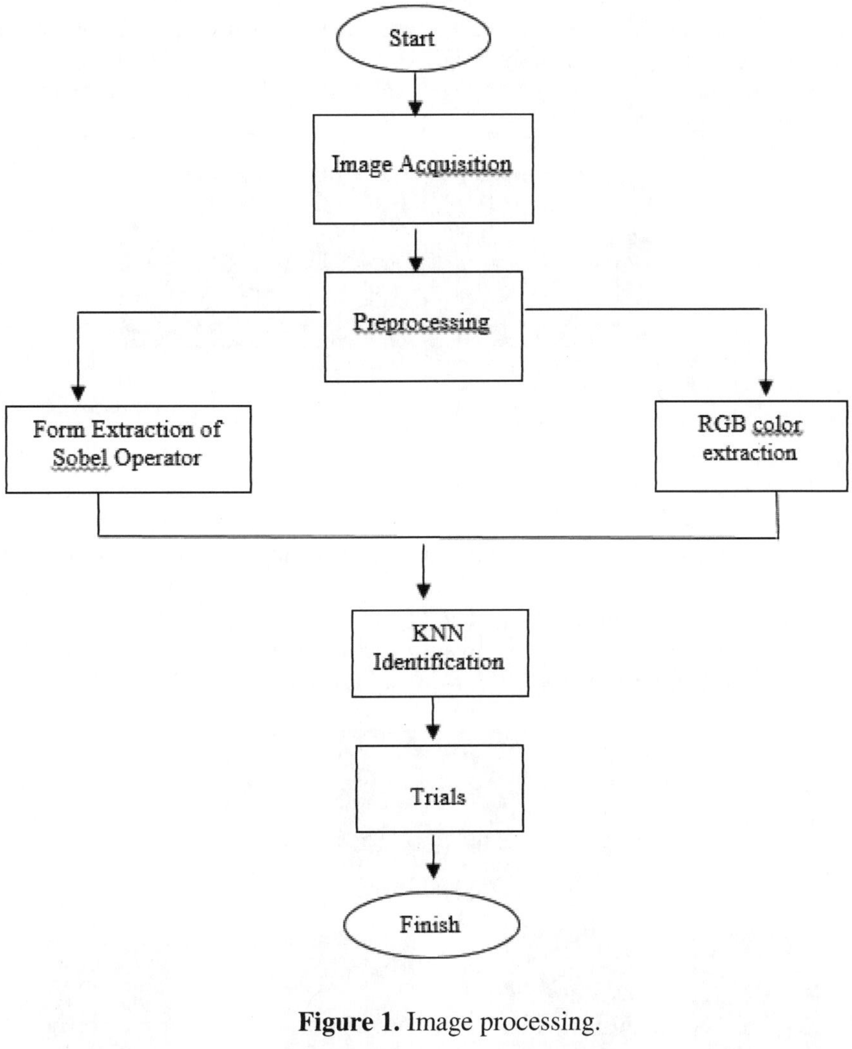

Figure 1. Image processing.

The algorithm of the sobel method [7] in detecting the edge of a digital image in the form of grayscale image is as follows:

IOP Conf. Series: Materials Science and Engineering **332** (2018) 012008 doi:10.1088/1757-899X/332/1/012008

1. Convolute grayscale image with horizontal Sobel (**Sx**) kernel is matriks transform of sobel kernel vertikal (**Sy**) and Sobel kernel vertikal (**Sy**)

$$S_y = \begin{bmatrix} 1 & 2 & 1 \\ 0 & 0 & 0 \\ -1 & -2 & -1 \end{bmatrix}$$

2. Calculate the Gradient by the formula $G=\sqrt{Sx^2 + Sy^2}$
3. The output image is the result of a large gradient (G).

2.1.Image Acquisition

Flower image acquisition phase is data collection of flowers taken in BALITRO. Flowers taken there are 3 types of jasmine white (Jasminum Sambac), jasmine japan (Pseuderathemum Reticulatum), and jasmine gambir (Jasminum Pubescens) with each flower taken 3 petals of the same type. Image acquisition performed a total of 33 times on each type of flower using a camera phone with a pixel quality 5MP do during the day centred on the petals and black paper background. Total data of jasmine flowers as much as 99 which divided into 90 data train and 9 test data. Examples of the types of flowers identified are shown in Figure 2.

Figure 2. Type of jasmine flowers.

2.2. Pre-processing and Identification

Image processing aims to process information contained in an image as the need for automatic object recognition and analysis [4]. The initial stage to get the information that the preprocessing stage color image will be resized to a uniform size resize the image into four types, namely the pixel size 30 x 30, 50 x 50, 100 x 100, and 150 x 150 and edge detection is performed to detect the type of flower from the characteristic of the flower shape. This process aims to improve the appearance of the border on a digital image. Another purpose of edge detection is to identify an area in a digital image where there is a major change in intensity. With edge detection in an image processing system (computer-based) will find any signs of surface or contour arising from the object [3] as shown in Figure 3.

Figure 3. Difference of resizing size.

Resized images 50 x 50 are calculated values of Red, Gren, and Blue (RGB), RGB skewness value, RGB kurtosis and RGB variance that will be stored as training data. For identification using image of test data to be calculated using *k*-Nearest Neighbor (k-NN) method with value *k* is 5. The *k*-Nearest Neighbour method is used to categorize test object into a class that has the closest neighbouring property (Neighbourhood) to the training object. In this study, the *k*-nearest neighbour classification uses the closest distance calculation with the euclidean distance method. Classification with *k*-Nearest Neighbour aims to select the training image with the smallest distance to the test image [9]. The *k*-NN algorithm is as follows [5].

1. Determine the value of *k*.
2. Calculate the distance between the new data into each labelled data.
3. Determine the labelled data k that has the most minimal distance.
4. Classify the new data into the labelled data of the majority.

When compared with other classification methods, this method has a high enough degree of accuracy because the incoming data will be classified based on similarity of traits that exist from previous data that has been classified. However, the *k*-NN algorithm needs to determine the value of the paremeter *k* (the number of nearest neighbours) and the learning based on the distance is unclear as to what kind of distance to use and which attribute should be used to obtain the best result [6].

3. Result

The results of feature extraction of color from the image with a pixel size of 50 x 50 generates RGB values, RGB skewness, RGB kurtosis, and RGB variance and will be saved into a data sheet that will become a reference as training data. The results of the color extraction process are shown in Figures 4 and 5.

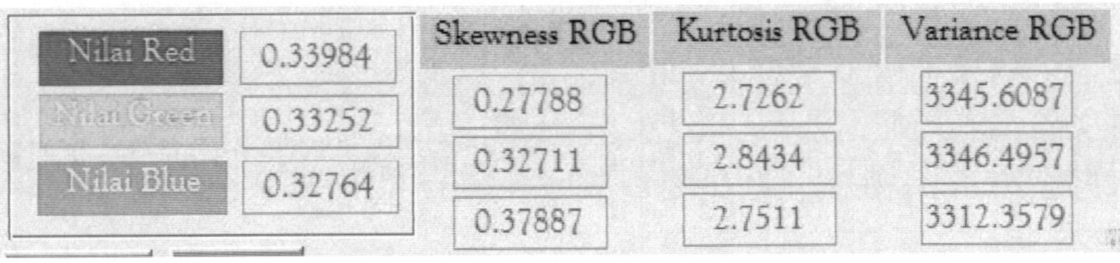

Figure 4. RGB color extraction results.

1	CITRA	R	G	B	C1	C2	C3	K1	K2	K3	V1	V2	V3	KATEGORI
2	1	0.33797	0.33395	0.32808	0.78915	0.85238	0.80548	3.6636	3.951	3.7423	2611.36	2696.19	2687.55	Melati Putih
3	2	0.33778	0.3341	0.32813	0.81178	0.87862	0.83524	3.829	4.0434	3.8829	2566.46	2649.4	2645.22	Melati Putih
4	3	0.33735	0.33463	0.32802	0.5332	0.6448	0.5773	2.7067	3.2331	2.9309	2517.88	2604.04	2590.27	Melati Putih
5	4	0.33998	0.33291	0.32711	0.61928	0.68942	0.64546	2.8316	2.9792	2.9302	2614.52	2651.08	2703.73	Melati Putih
6	5	0.34027	0.33284	0.32689	0.59401	0.67932	0.646	2.6828	2.8583	2.8204	2598.11	2630.22	2679.16	Melati Putih
7	6	0.3403	0.33038	0.32932	0.46039	0.53779	0.50959	2.38	2.3689	2.2691	2861.33	2905.47	2929.46	Melati Putih
8	7	0.34102	0.3298	0.32918	0.43687	0.5002	0.48234	2.4021	2.3203	2.2825	2566.34	2601.79	2618.14	Melati Putih
9	8	0.33957	0.33038	0.33004	0.44732	0.52042	0.50386	2.3797	2.4032	2.3821	2490.5	2525.83	2539.18	Melati Putih
10	9	0.33961	0.33053	0.32986	0.49536	0.5646	0.52737	2.4661	2.4376	2.3608	2512.9	2548.71	2561.66	Melati Putih
11	10	0.33961	0.33053	0.32986	0.49536	0.5646	0.52737	2.4661	2.4376	2.3608	2512.9	2548.71	2561.66	Melati Putih
12	11	0.33903	0.33063	0.33034	0.36812	0.39398	0.37409	2.7212	2.7065	2.6009	2891.27	2932.95	2930.34	Melati Putih
13	12	0.33964	0.33028	0.33008	0.8577	0.97785	0.92354	3.7495	4.1906	3.9883	2308.05	2317.62	2332.93	Melati Putih
14	13	0.3382	0.33341	0.32839	0.15211	0.1734	0.16717	2.1566	2.3137	2.1362	3435.54	3495.39	3481.13	Melati Putih
15	14	0.33747	0.3349	0.32763	0.48655	0.53311	0.5704	3.2766	3.2648	3.193	2998.01	3023.6	2988.04	Melati Putih
16	15	0.33777	0.33484	0.32739	0.45847	0.50321	0.51028	3.2899	3.3072	3.1439	3016.76	3051.15	3012.31	Melati Putih
17	16	0.33836	0.33466	0.32698	0.48225	0.47828	0.49873	3.2093	3.0688	3.0172	2989.52	3017.05	2976.71	Melati Putih
18	17	0.33363	0.33893	0.32744	0.55535	0.62952	0.66954	4.0262	4.4046	4.3126	2321.63	2357.62	2410.47	Melati Putih

Figure 5. The value of color extraction in the form of data sheets.

IOP Conf. Series: Materials Science and Engineering **332** (2018) 012008 doi:10.1088/1757-899X/332/1/012008

The result of feature extraction using edge detection with a sobel operator will result in a binary value. Number 1 is the value of the edge of the image of jasmine and the number 0 is the value of the background image jasmine. The results of sobel edge detection process is shown in Figure 6.

0	0	0	0	0	0	0	0	0	0	0	0	0	0	0	0	0	0	0	0	0	0	0	0
0	0	0	0	0	0	0	0	0	0	0	0	0	0	0	1	1	0	0	0	0	0	0	0
0	0	0	0	0	0	0	0	0	0	0	0	0	0	0	0	0	0	0	0	0	0	0	1
0	0	0	0	0	0	0	0	0	0	0	0	0	0	0	0	0	0	0	0	0	0	0	1
0	0	0	0	0	0	0	0	0	0	0	0	0	0	0	0	0	0	0	0	0	0	0	1
0	0	0	0	0	0	0	0	0	0	0	0	0	0	0	0	0	0	0	0	0	0	0	0
0	1	0	0	0	0	0	0	0	0	0	0	0	0	0	0	0	0	0	0	0	0	0	0
0	1	0	0	0	0	0	0	0	0	0	0	0	0	0	0	0	0	0	0	0	0	0	0
0	0	1	0	0	0	0	0	0	0	0	0	0	0	0	0	0	0	0	0	0	0	0	0
0	0	0	0	0	0	0	0	0	0	0	0	0	0	0	0	0	0	0	0	0	0	0	0
0	1	0	0	0	0	0	0	0	0	0	1	1	1	0	0	0	0	0	0	0	0	0	1
1	0	0	0	0	0	0	0	0	0	1	0	0	1	0	0	0	0	0	0	0	0	0	1
0	0	0	0	0	0	0	0	0	0	1	0	1	0	0	0	0	0	0	0	0	0	0	1
0	0	0	0	0	0	0	0	0	0	0	1	0	0	0	0	0	0	0	0	1	1	1	0
0	0	0	0	0	0	0	0	0	0	0	0	0	0	0	0	0	0	0	0	0	0	0	0
0	0	0	0	0	0	0	0	0	0	0	0	0	0	0	0	0	0	0	0	1	0	0	0
0	0	0	0	0	0	0	0	0	0	0	0	0	0	0	0	0	0	0	0	0	1	0	0
1	0	0	0	0	0	0	0	0	0	0	0	0	0	0	0	0	0	0	0	0	0	1	0
0	1	1	0	0	0	0	0	0	0	0	0	0	0	0	0	0	0	0	0	0	0	1	0
0	0	0	1	0	0	0	0	0	0	0	0	0	0	0	0	0	0	0	0	0	0	0	0
0	0	0	1	0	0	0	0	0	0	0	0	0	0	0	0	0	0	0	0	0	0	0	0
0	0	0	1	0	0	0	0	0	0	0	0	0	0	0	0	0	0	0	0	0	0	0	0
0	0	0	1	0	0	0	0	0	0	0	0	0	0	0	0	0	0	0	0	0	0	0	0

Figure 6. The results of edge detection sobel.

An example of the identification results is shown in Figure 7.

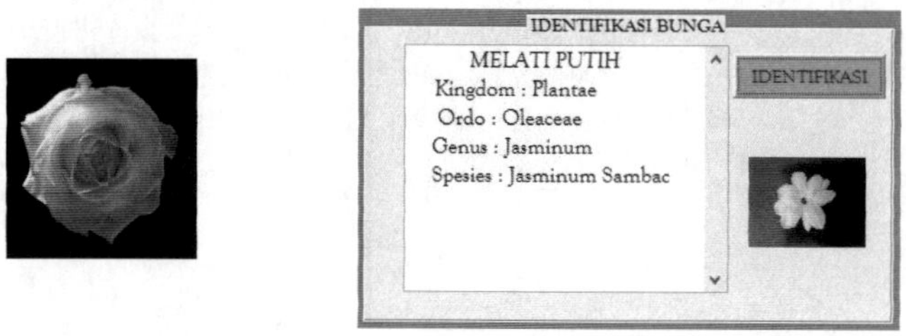

White Rose Result of identification

Figure 7. Results of identification of white roses.

The test conducted on 9 test data and produce valid data as much as 7, and the data is invalid 2. The factor of possible testing failure is the color of the jasmine white flowers that are similar to jasmine gambier flowers and the shape of the jasmine white flowers are almost similar to jasmine japan flowers.

From the validation test on the identification of jasmine flower using RGB color feature extraction method and extraction of the characteristic of the sobel operator and the identification method k-Nearest Neighbor obtained the accuracy level with 84% percentage with k = 5 and 10 value, while for the value k = 15 obtained accuracy of 65.2%. Test results with k values of 5 and 10 are shown as in Table 1.

Table 1. Test results of identification.

No	Picture of Flowers	Result of Identification	Description
1	Jasmine White	Jasmine White	Matching
2	Jasmine White	Jasmine White	Matching
3	Jasmine White	Jasmine White	Matching
4	Jasmine Japan	Jasmine Japan	Matching
5	Jasmine Japan	Jasmine White	No Matching
6	Jasmine Japan	Jasmine Japan	Matching
7	Jasmine Gambir	Jasmine White	No Matching
8	Jasmine Gambir	Jasmine Gambir	Matching
9	Jasmine Gambir	Jasmine Gambir	Matching

Testing validation there are nine test data that is three test data of jasmine white, three test data of japan jasmine and three test data of jasmine gambir using confusion matrix. The image of jasmine white is well detected, while jasmine japan image produce two detected images into its type and one image is detected into the type of jasmine white, this is because jasmine japan has oval shaped petals that are almost similar to the petals shape of jasmine white, jasmine gambir image produce two images into its type and one image is detected into jasmine white, this is because jasmine gambir have the same colour with jasmine white. Validation results are shown in the Table 2.

Table 2. Results of validation.

Test data	Train Data			
	Jasmine White	Jasmine Japan	Jasmine Gambir	Total
Jasmine White	3	0	0	3
Jasmine Japan	1	2	0	3
Jasmine Gambir	1	0	2	3
Total	5	2	2	9

4. Conclusion

Identification of jasmine flowers (Jasminum sp.) using RGB color feature extraction takes on RGB value, RGB skewness, RGB kurtosis, and RGB variance, while for edge detection using a sobel operator. The image used consisted of 90 training data and 9 test data. Image taken using a camera phone with a 5MP quality. Image preprocessing is done to equalize the size to the pixel size of 30 x 30, 50 x 50, 100 x 100 and 150 x 150 pixel generates an accuracy rate of 84%. The k-NN method for identification with *k* values of 5, 10 and 15 yielded different accuracy rates for 5 and 10 of the closest distance resulted in the same accuracy of 84%, for the nearest 15 resulted in a small accuracy of 65.2%.

References

[1] Fitriyani Ismi 2014 Studi komparasi kinerja jaringan saraf tiruan dan fuzzy untuk pengenalan jenis bunga berdasarkan fitur warna
[2] Ichsani Urai Nur et al 2015 Klasifikasi jenis bunga kamboja jepang (*Adenium sp.*) berdasarkan citra mahkota menggunakan ekstraksi fitur warna dan deteksi tepi.
[3] Kalo A Y 2014 Information Retrieval Berbasis Citra Berdasarkan Logo Perusahaan Dengan Menggunakan Metode Edge Matching Dan Euclidean Distance Tugas Akhir Teknik dan Ilmu Komputer UNIKOM Bandung
[4] Munarto, Permata, Salsabilla. 2014. Klasifikasi kualitas biji jagung manis berdasarkan fitur warna menggunakan fuzzy logic. Jurusan Teknik Elektro, Fakultas Teknik,Universitas Sultan Ageng Tirtayasa, Banten
[5] Ndaumanu, Kusrini, Arief 2014. Analisis Prediksi Tingkat Pengunduran Diri Mahasiswa dengan Metode K-Nearest Neighbor. *Jatisi, Vol. 1, No. 1.*
[6] Permatasari, Debby 2012 Sistem Klasifikasi Kualitas Biji Jagung Berdasarkan Tekstur Berbasis Pengolahan Citra Digital. Skripsi. Jurusan Teknik Telekomunikasi Fakultas Elektro dan Komunikasi, Institut Teknologi Telkom Bandung.
[7] Putra D 2010 *Pengolahan Citra Digital* Yogyakarta: Andi Offset.
[8] Rakhmawati Rizqa Puji 2013 Sistem pendeteksi jenis bunga menggunakan nilai hsv dari citra mahkota bunga
[9] Royce1 Eduard et al, 2012 Sistem pendeteksi senyum berdasarkan metode edge detection, histogram equalization dan nearest neighbor.
[10] Sigit, R., Basuki et al, D 2007 *Step By Step Pengolahan Citra Digital* Yogyakarta: Andi Offset

IOP Conf. Series: Materials Science and Engineering **332** (2018) 012009 doi:10.1088/1757-899X/332/1/012009

Decision support system for determining Bantuan Siswa Miskin (BSM) receivers with profile matching method

B H Situmorang*, **E Pibriana**, **E T Tosida**

Department of Computer Science, Faculty of Mathematic and Natural Sciences, Pakuan University, Indonesia.

*Corresponding author: boldson_skom@yahoo.com

Abstract. Bantuan Siswa Miskin (BSM) is a National Programs aimed at eliminating the barriers of poor students participating to school by helping poor students gain access to appropriate education services, prevent dropping out of school, attract poor students back to school, assis students in providing for learning activities, support the Nine Years Basic Education (and even up to senior high school) program, as well as helping to smooth the school programs [1]. Decision Support System is made by applying Profile Matching method to assist teachers or school operators in SMP PGRI Ciasmara in selecting prospective recipients of BSM program and providing recommendations in decision making. Profile Matching is used to compare the actual data value of a profile to be assessed by the expected profile value, so that it can be known the difference of competence (also called GAP). If the resulting value of GAP is smaller then the weight of value will be greater, which means it has a greater chance to be recommended as a potential recipient of the BSM program. Decision Support System for determining BSM receivers is only choosing the right alternatives to receive BSM according to the BSM quota given to SMP PGRI Ciasmara. The right alternatives to receive this BSM is the highest ranking alternatives.

Keyword: Bantuan Siswa Miskin (BSM) , decission support system, profile matching

1. Introduction

The BSM program is a government program funded from the state budget as the responsibility of the government in order to improve access to better education, so that education can be reached by people with weak economy. SMP PGRI Ciasmara is one of the junior high schools to receive BSM program funding. In determining the students who are entitled to get BSM funds, SMP PGRI Ciasmara still use the conventional system that is by doing selection based on eligibility criteria. Implementation of this conventional system has resulted in the mistake of the wrong target on the determination of BSM fund candidates because the total value of each aspect is obtained by calculating the average value in each aspect and then followed by the calculation process to determine the ranking.

Decision Support System (DSS) to determine BSM fund recipients in SMP PGRI Ciasmara is needed to support efficient and accurate decision-making. The process of determining BSM fund recipients in the DSS uses Profile Matching method which aims to compare the actual data value of a profile to be assessed by the expected profile value, so as to know the difference of competence (also called GAP). If the resulting value of GAP is smaller then the weight of value will be greater, which means it has a greater chance to be recommended as a potential recipient of the BSM program.

The method of profile matching has been widely applied to previous studies. H. J. Pahk, J. S. Park, I. Yeo used the profile matching technique for the long guideways, which can be applied to most

IOP Conf. Series: Materials Science and Engineering **332** (2018) 012009 doi:10.1088/1757-899X/332/1/012009

straightness measurements for machine tools and measuring machines [1]. The implementing of profile-matching method in wireless fingerprinting [2] provides a more reliable solution especially in environments with sparse distributed access points and a more accurate initial position.

2. Material and Methods

Assessment criteria used to determine BSM candidates based on the principal decision letter of SMP PGRI Cismara number 380/III/5/O.2016 on the determination of BSM recipient students, including:

1. Economic Aspects (the weight value is 40%), reviewed by parent's income (C1) and the number of parental dependents (C2).
2. Academic Aspects (the weight value is 35%), reviewed by average rate of raport (C3) and rank of students (C4).
3. Non-Academic Aspects (the weight value is 25%), reviewed based on Student Morals (C5), Student Personality (C6), Extra Curricular which includes Scouting (C7), Troop flag raiser (C8), and Sports (C9).

   Sub-Criteria used in this study have been determined based on interviews with Principal SMP PGRI Ciasmara as shown in the table.

Table 1. The weight value of parent's income

Parent's Income (IDR)	Value	Weight
<= 500.000	5	Very Important
500.000–1.000.000	4	Important
1.000.000–2.000.000	3	QuiteImportant
2.000.000–3.000.000	2	Less Important
> =3.000.000	1	Not Important

Table 2. The weight value of parental dependents

The number of parental dependents	Value	Weight
>5	5	Very Important
4	4	Important
3	3	QuiteImportant
2	2	Less Important
1	1	Not Important

Table 3. The weight value of average rate of raport

Average rate of raport	Value	Weight
85 – 100	5	Very Important
75 – 84	4	Important
71 – 74	3	QuiteImportant
60 – 70	2	Less Important
<59	1	Not Important

Table 4. The weight value of rank of students

Rank of students	Value	Weight
1-3	5	Very Important
4-5	4	Important
6-8	3	Quite Important
9-10	2	Less Important
>10	1	Not Important

Table 5. The weight value of student morals

Student Morals	Value	Weight
A (Excellent)	5	Very Important
B (Good)	4	Important
C (Pass)	3	Quite Important
D (Poor)	2	Less Important
E (Bad)	1	Not Important

Table 6. The weight value of student personality

Student Personality	Value	Weight
A (Excellent)	5	Very Important
B (Good)	4	Important
C (Pass)	3	Quite Important
D (Poor)	2	Less Important
E (Bad)	1	Not Important

Table 7. The weight value of scouting

Scouting	Value	Weight
A (Excellent)	5	Very Important
B (Good)	4	Important
C (Pass)	3	Quite Important
D (Poor)	2	Less Important
E (Bad)	1	Not Important

Table 8. The weight value of troop flag raiser

Troop Flag Raiser	Value	Weight
A (Excellent)	5	Very Important
B (Good)	4	Important
C (Pass)	3	Quite Important
D (Poor)	2	Less Important
E (Bad)	1	Not Important

Table 9. The weight value of sports

Sports	Value	Weight
A (Excellent)	5	Very Important
B (Good)	4	Important
C (Pass)	3	Quite Important
D (Poor)	2	Less Important
E (Bad)	1	Not Important

Flowchart of the implementation process of the Profile Matching method shown in Figure 1.

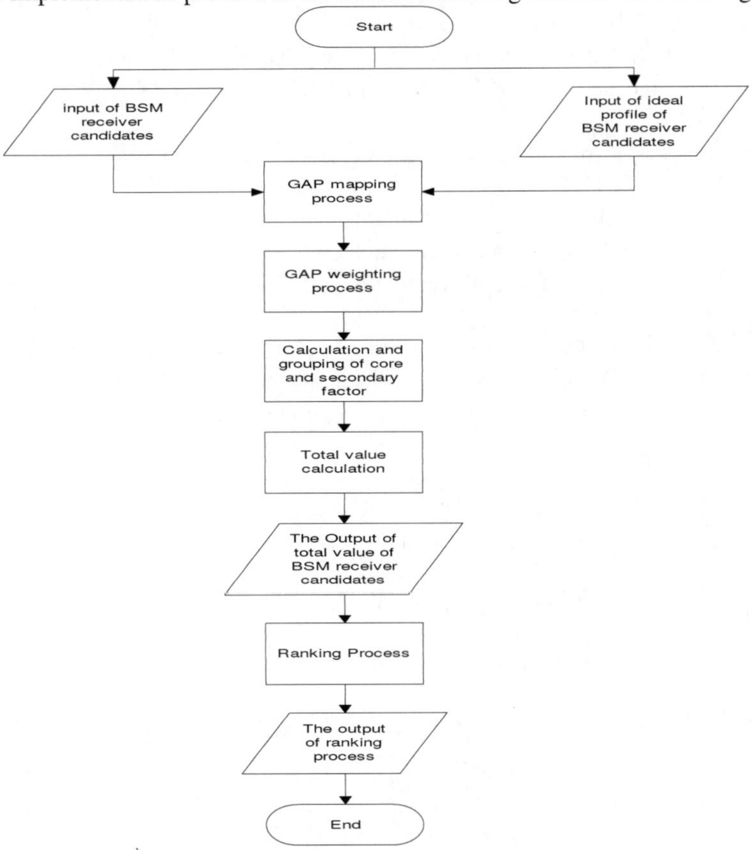

Figure 1. Flowchart of the implementation process of the Profile Matching method

The profile matching method is to compare the actual data value of a profile to be assessed by the expected profile value, so it can be known the difference of competence (called GAP).

$$\text{GAP} = \textit{Value}\ \text{Atribut} - \textit{Value}\ \text{Target} \ldots\ldots\ldots (1)$$

If the GAP results are smaller then the value of the weight is greater, it means that there is a greater chance to be recommended as a potential recipient of BSM.

After the weight of the gap value of all aspects is determined in the same way, each aspect is subdivided into two groups: the Core Factor group (the main factor) and the Secondary Factor (supporting factors). The core factor calculation is shown in equation (2) and the secondary factor calculation is shown in equation (3).

$$ACF = \frac{\sum VC}{\sum CI} \quad \ldots\ldots\ldots (2)$$

ACF: The average value of core factor
VC: total value of core factor
CI: Number of core factor items

$$ASF = \frac{\sum VS}{\sum SI} \ldots\ldots\ldots\ldots (3)$$

ASF: The average value of secondary factor
VS: Total value of secondary factor values

SI: Number of secondary factor items

After the average value of core factor and secondary factor are obtained, then the total persentage value of core factor and secondary factor will be calculated based on equation (4).

$$T = (x)\% \, ACF + (y)\% \, ASF \ldots\ldots\ldots\ldots (4)$$

T: The total value of each aspect
ACF: The average value of core factor
ASF: The average value of secondary factor
(x)%: Inputed percent value
(y)%: Inputed percent value

The final process of the implementation of Profile Matching method is the ranking of recipient of BSM recipient candidates. Ranking process is calculated based on equation (5).

$$FR = (x)\% \, EA + (y)\% \, AA + (z)\% \, NAA \ldots\ldots\ldots\ldots\ldots (5)$$

FR: Final Result
EA: Value of Economic Aspects
AA: Value of Academic Aspect
NAA: Value of Non-Academic Aspect
(x)%: Inputed percent value
(y)%: Inputed percent value
(z)%: Inputed percent value

DSS is made through the System Development Life Cycle (SDLC) method as shown in Figure 2.

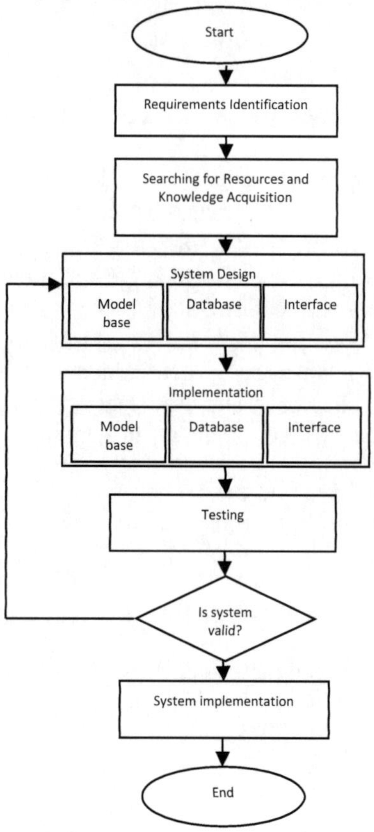

Figure 2. System Development Life Cycle [3]

IOP Conf. Series: Materials Science and Engineering **332** (2018) 012009 doi:10.1088/1757-899X/332/1/012009

3. Result & Discussion

Decision Support System for determining BSM recipients is made using PHP and MySQL programming languages. The user interface on Decision Support System to determine the BSM recipient candidates includes login page and main page in which there is an optional menu of student data, assessment criteria, sub criteria, change password, and exit.

Testing are performed after a Decision Support System is completed to ensure that the system meets the needs of users. Test case is done by entering data of BSM candidates from SMP PGRI Ciasmara into the Decision Support System application and the resulting output is the list of candidates receiving BSM rankings as shown in Table 10.

Table 10. The list of candidates receiving BSM rankings

#	ID	Name	Class	Result	Ranking
1	151607150	Triatna	7.D	4.61833	1
2	151607057	Rizky Jaenudin	7.B	4.54	2
3	151607034	Wika Destia. D	7.A	4.52	3
4	151607092	Muhammad Jandan	7.C	4.45833	4
5	151607043	Liska Juniar	7.B	4.45333	5
6	151607078	Ernawati	7.C	4.40667	6
7	151607151	Unih	7.D	4.40333	7
8	131407026	Rahmat Gunawa	9.D	4.32333	8
9	131407082	Ririn Rahmayanti	9.D	4.31	9
10	151607028	Rosdiyana	7.A	4.29333	10

The comparison between the ranking results of BSM recipient candidates in Decision Support System application using the profile matching method with conventional system implemented by SMP PGRI Ciasmara is shown by Table 11.

Table 11. The comparison between the ranking results in Decision Support System application with conventional system implemented by SMP PGRI Ciasmara

#	ID	Name	Sex	Class	The Result of DSS Application	The Ranking Result of DSS Application	The Result of The System Implemented by SMP PGRI Ciasmara	The Ranking Result of The System Implemented by SMP PGRI Ciasmara
1	151607150	TRIATNA	F	7.D	4.61833	1	4.65	1
2	151607057	RIZKY JAENUDIN	M	7.B	4.54	2	4.425	3
3	151607034	WIKA DESTIA DANTIYANTI	F	7.A	4.52	3	4.4	4
4	151607092	MUHAMAD JANDAN	M	7.C	4.45833	4	4.45	2
5	151607043	LISKA JUNIAR	F	7.B	4.45333	5	4.3	7
6	151607078	ERNAWATI	F	7.C	4.40667	6	4.35	5
7	151607151	UNIH	F	7.D	4.40333	7	4.425	3
8	131407026	RAHMAT GUNAWAN	M	9.D	4.32333	8	4.325	6
9	131407082	RIRIN RAHMAYANTI	F	9.D	4.31	9	4.3	7
10	151607018	MUHAMMAD IKBAL L	M	7.A	4.29333	10	4.1	9

Based on comparison of alternative ranking result as many as 119 students can be concluded that the result of ranking between conventional system implemented by SMP PGRI Ciasmara with system that use method matching profile reaching valid level 1% only because SMP PGRI Ciasmara only use calculation in equation 6.

$$\text{Total} = (\text{Economic Aspects}/2) * 40\% + (\text{Academic Aspects}/2) * 35\% + (\text{Non} - \text{Academic Aspects})/5 * 25\% \ldots\ldots (6)$$

The correlation test was conducted to measure the closeness of the relationship between the calculation result using the profile matching method and the calculation result implemented by SMP PGRI Ciasmara. The level of closeness of the relationship between these two methods is in perfect correlation and the relationship belongs to a positive linear correlation because its correlation value is 0.986267. The correlation test is shown in the graph in Figure 3.

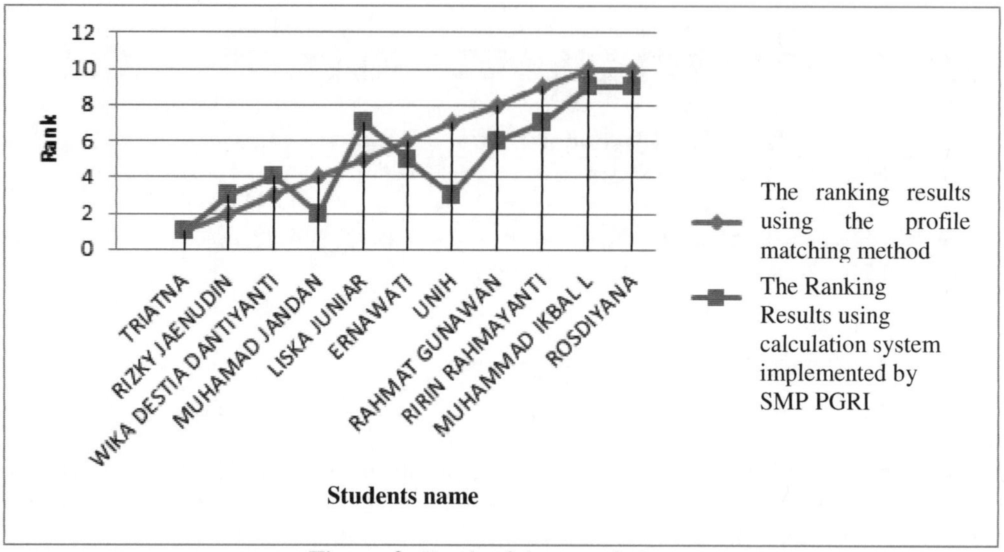

Figure 3. Graph of the correlation test

4. Conclusion

Decision support system for determining BSM recipients is made using PHP and MySQL programming language by applying Profile Matching method. Based on comparison of alternative ranking result as many as 119 students can be concluded that the result of ranking between conventional system implemented by SMP PGRI Ciasmara with system that use method matching profile reaching valid level 1% only because SMP PGRI Ciasmara does not do GAP mapping process, weighting GAP and grouping core factor and secondary factor but only calculate the mean value in every aspect. The level of closeness of the relationship between these two methods is in perfect correlation and the relationship belongs to a positive linear correlation because its correlation value is 0.986267.

References

[1] H J Pahk, J S Park, I Yeo . 2015. Development of straightness measurement technique using the profile matching method. International Journal of Machine Tools and Manufacture (Elsevier) vol. 37 p. 135.
[2] Li Y, Zhuang Y, Lan H, Niu X, El-Sheimy N. 2016. A profile-matching method for wireless positioning. IEEE communications letters vol. 20, p. 2514-2517.
[3] Grzybowska H, Barceló J. 2012. Decision support system for real-time urban freight management. Procedia-social and behavioral sciences (Elsevier) vol. 39 p. 712.

IOP Conf. Series: Materials Science and Engineering **332** (2018) 012010 doi:10.1088/1757-899X/332/1/012010

A new two-scroll chaotic attractor with three quadratic nonlinearities, its adaptive control and circuit design

C-H Lien[1*], S Vaidyanathan[2], A Sambas[3], Sukono[4], M Mamat[5], W S M Sanjaya[6], Subiyanto[7]

[1]Department of Marine Engineering, National Kaohsiung Marine University, Kaohsiung, Taiwan 811, R.O.C.
[2]Research and Development Centre, Vel Tech University, Avadi, Chennai, India.
[3]Department of Mechanical Engineering, Universitas Muhammadiyah Tasikmalaya, Indonesia.
[4]Department of Mathematics, Faculty of Mathematics and Natural Sciences, Universitas Padjadjaran, Indonesia.
[5]Faculty of Informatics and Computing, Universiti Sultan Zainal Abidin, Kuala Terengganu, Malaysia.
[6]Department of Physics, Universitas Islam Negeri Sunan Gunung Djati, Bandung, Indonesia
[7]Department of Marine Science, Faculty of Fishery and Marine Science, Universitas Padjadjaran, Indonesia.

*Email: chlien@mail.nkmu.edu.tw

Abstract. A 3-D new two-scroll chaotic attractor with three quadratic nonlinearities is investigated in this paper. First, the qualitative and dynamical properties of the new two-scroll chaotic system are described in terms of phase portraits, equilibrium points, Lyapunov exponents, Kaplan-Yorke dimension, dissipativity, etc. We show that the new two-scroll dissipative chaotic system has three unstable equilibrium points. As an engineering application, global chaos control of the new two-scroll chaotic system with unknown system parameters is designed via adaptive feedback control and Lyapunov stability theory. Furthermore, an electronic circuit realization of the new chaotic attractor is presented in detail to confirm the feasibility of the theoretical chaotic two-scroll attractor model.

1. Introduction

In the last few decades, chaotic and hyperchaotic systems have been applied in several areas of science and engineering [1-2]. Some important applications of chaotic systems can be listed out such as chemical reactors [3-5], oscillators [6-8], neural networks [9-10], memristors [11-12], ecology [13-14], robotics [15-16], Tokamak reactors [17-18], finance [19-20], etc.

In the chaos literature, there is good interest in shown in the modeling of chaotic systems with multi-scroll attractors such as two-scroll attractors [21-25], three-scroll attractors [26-28], four-scroll attractors

IOP Conf. Series: Materials Science and Engineering **332** (2018) 012010 doi:10.1088/1757-899X/332/1/012010

[29-30], etc. There are also many chaotic systems with quadratic nonlinearities in the chaos literature [31-36].

In this work, we derive a new 3-D new dissipative chaotic system with three quadratic nonlinearities in this paper. The new chaotic system displays a two-scroll chaotic attractor.

This paper is organized as follows. Section 2 describes the new two-scroll chaotic system with three quadratic nonlinearities. This section also details dynamical properties such as phase portraits, Lyapunov exponents and Kaplan-Yorke dimension. Section 3 describes the global chaos control of the new chaotic system with unknown parameters. In Section 4, we use MultiSIM to build an electronic circuit realization of the new two-scroll chaotic system. The circuit experimental results of the new chaotic attractor show agreement with the numerical simulations. Section 5 contains the conclusions.

2. A new two-scroll chaotic system with three quadratic nonlinearities

In this paper, we design a new two-scroll chaotic system with three quadratic nonlinearities given by

$$\begin{cases} \dot{x}_1 = a(x_2 - x_1) + x_2 x_3 \\ \dot{x}_2 = b x_1 - x_2 + c x_1 x_3 \\ \dot{x}_3 = -x_3 - x_1 x_2 \end{cases} \tag{1}$$

where x_1, x_2, x_3 are state variables and a, b, c are positive constants.

In this paper, we show that the system (1) is *chaotic* for the parameter values

$$a = 10, \quad b = 20, \quad c = 30 \tag{2}$$

For numerical simulations, we take the initial values of the system (1) as

$$x_1(0) = 0.1, \quad x_2(0) = 0.1, \quad x_3(0) = 0.1 \tag{3}$$

Figure 1 shows the phase portraits two-scroll strange attractor of the new chaotic system (1) for the parameter values (2) and initial conditions (3). Figure 1 (a) shows the 3-D phase portrait of the new chaotic system (1). Figures 1 (b)-(c) show the projections of the new chaotic system (1) in (x_1, x_2), (x_2, x_3) and (x_1, x_3) coordinate planes, respectively.

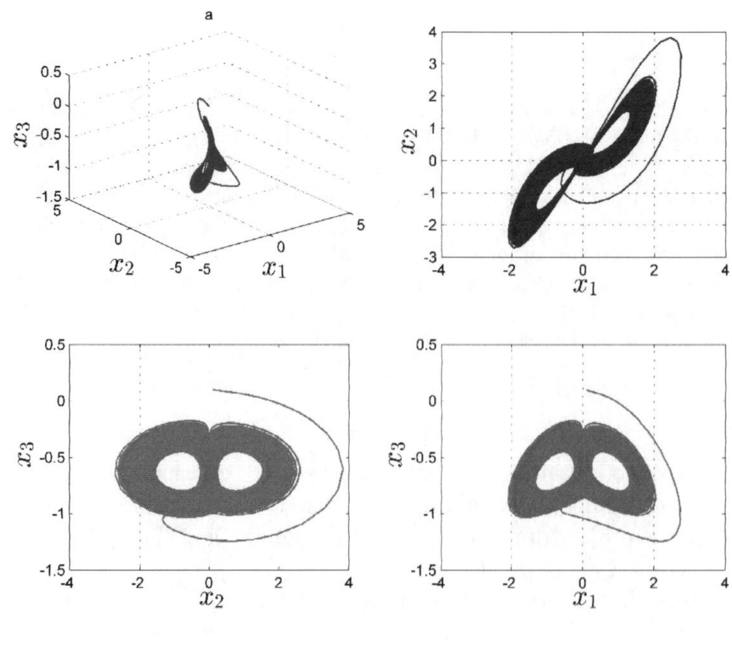

Figure 1. Phase portraits of the new chaotic system (1) for $a = 10, \ b = 20, \ c = 30$

For the rest of this section, we take the parameter values as in the chaotic case (2).

The equilibrium points of the new chaotic system (1) are obtained by solving the system of equations

$$\begin{cases} a(x_2 - x_1) + x_2 x_3 = 0 \\ b x_1 - x_2 + c x_1 x_3 = 0 \\ -x_3 - x_1 x_2 = 0 \end{cases} \tag{4}$$

Solving the equations in (4) we obtain the equilibrium points of the system (1) as

$$E_0 = \begin{bmatrix} 0 \\ 0 \\ 0 \end{bmatrix}, \quad E_1 = \begin{bmatrix} 0.7689 \\ 0.8207 \\ -0.6311 \end{bmatrix}, \quad E_2 = \begin{bmatrix} -0.7689 \\ -0.8207 \\ -0.6311 \end{bmatrix} \tag{5}$$

It is easy to verify that E_0 is a saddle point, while E_1 and E_2 are saddle-focus points.

For the parameter values as in the chaotic case (2) and the initial state as in (3), the Lyapunov exponents of the new 3-D system (2) are determined using Wolf's algorithm as

$$L_1 = 0.4260, \quad L_2 = 0, \quad L_3 = -12.4260 \tag{6}$$

Since $L_1 > 0$, the new 3-D system (1) is chaotic. Thus, the system (1) exhibits a two-scroll chaotic attractor. Also, we note that the sum of the Lyapunov exponents in (6) is negative. This shows that the new two-scroll chaotic system (1) is dissipative.

The Kaplan-Yorke dimension of the new 3-D system (1) is determined as

$$D_{KY} = 2 + \frac{L_1 + L_2}{|L_3|} = 2.0343, \tag{7}$$

which indicates the complexity of the new two-scroll chaotic system (1).

Figure 2 shows the Lyapunov exponents of the new chaotic system (1) with a strange attractor.

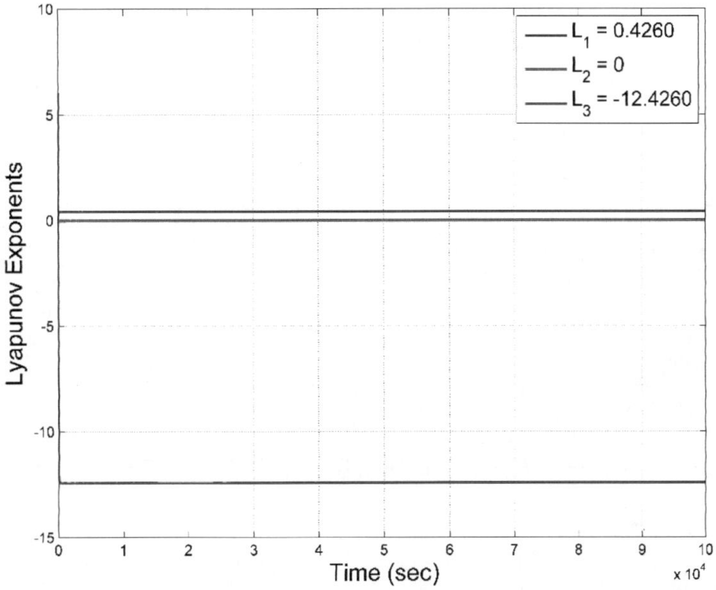

Figure 2. Lyapunov exponents of the new chaotic system (1) for $a = 10, \quad b = 20, \quad c = 30$

3. Global chaos control of the new two-scroll chaotic system via adaptive control method

In this section, we devise adaptive controller so as to globally stabilize all the trajectories of the new two-scroll chaotic system. The main result is proved via Lyapunov stability theory.

In this section, we consider the controlled chaotic system given by

$$\begin{cases} \dot{x}_1 = a(x_2 - x_1) + x_2 x_3 + u_1 \\ \dot{x}_2 = bx_1 - x_2 + cx_1 x_3 + u_2 \\ \dot{x}_3 = -x_3 - x_1 x_2 + u_3 \end{cases} \tag{8}$$

where x_1, x_2, x_3 are the states and a, b are unknown parameters.

We consider the adaptive control defined by

$$\begin{cases} u_1 = -\hat{a}(t)(x_2 - x_1) - x_2 x_3 - k_1 x_1 \\ u_2 = -\hat{b}(t)x_1 + x_2 - \hat{c}(t)x_1 x_3 - k_2 x_2 \\ u_3 = x_3 + x_1 x_2 - k_3 x_3 \end{cases} \tag{9}$$

where k_1, k_2, k_3 are positive gain constants.

Substituting (9) into (8), we obtain the closed-loop system

$$\begin{cases} \dot{x}_1 = [a - \hat{a}(t)](x_2 - x_1) - k_1 x_1 \\ \dot{x}_2 = [b - \hat{b}(t)]x_1 + [c - \hat{c}(t)]x_1 x_3 - k_2 x_2 \\ \dot{x}_3 = -k_3 x_3 \end{cases} \tag{10}$$

We define the parameter estimation errors as

$$\begin{cases} e_a(t) = a - \hat{a}(t) \\ e_b(t) = b - \hat{b}(t) \\ e_c(t) = c - \hat{c}(t) \end{cases} \tag{11}$$

Using (11), we can simplify (10) as

$$\begin{cases} \dot{x}_1 = e_a(x_2 - x_1) - k_1 x_1 \\ \dot{x}_2 = e_b x_1 + e_c x_1 x_3 - k_2 x_2 \\ \dot{x}_3 = -k_3 x_3 \end{cases} \tag{12}$$

Differentiating (11) with respect to t, we obtain

$$\begin{cases} \dot{e}_a(t) = -\dot{\hat{a}}(t) \\ \dot{e}_b(t) = -\dot{\hat{b}}(t) \\ \dot{e}_c(t) = -\dot{\hat{c}}(t) \end{cases} \tag{13}$$

Next, we consider the Lyapunov function defined by

$$V(x_1, x_2, x_3, e_a, e_b, e_c) = \frac{1}{2}\left(x_1^2 + x_2^2 + x_3^2 + e_a^2 + e_b^2 + e_c^2\right) \tag{14}$$

which is positive definite on $\boldsymbol{R}^6$.

Differentiating V along the trajectories of (12) and (13), we obtain

$$\dot{V} = -k_1 x_1^2 - k_2 x_2^2 - k_3 x_3^2 + e_a\left[x_1(x_2 - x_1) - \dot{\hat{a}}\right] + e_b\left[x_1 x_2 - \dot{\hat{b}}\right] + e_c\left[x_1 x_2 x_3 - \dot{\hat{c}}\right] \tag{15}$$

In view of the equation (15), we take the parameter update law as

$$\begin{cases} \dot{\hat{a}} = x_1(x_2 - x_1) \\ \dot{\hat{b}} = x_1 x_2 \\ \dot{\hat{c}} = x_1 x_2 x_3 \end{cases} \tag{16}$$

Theorem 1. The novel two-scroll chaotic system (8) is globally and exponentially stabilized by the adaptive control law (9) and the parameter update law (16), where k_1, k_2, k_3 are positive constants.

Proof. The Lyapunov function V defined by (14) is quadratic and positive definite on R^6.

By substituting the parameter update law (16) into (15), we obtain the time-derivative of V as

$$\dot{V} = -k_1 x_1^2 - k_2 x_2^2 - k_3 x_3^2 \tag{17}$$

which is negative semi-definite on R^6.

Thus, by Barbalat's lemma [37], it follows that the closed-loop system (15) is globally exponentially stable for all initial conditions $x(0) \in R^3$. This completes the proof. ■

For numerical simulations, we take the gain constants as $k_i = 10$ for $i = 1, 2, 3$.

We take the parameter values as in the chaotic case (2), i.e. $a = 10$, $b = 20$ and $c = 30$.

We take the initial conditions of the states of the novel chaotic system (8) as $x_1(0) = 12.3$, $x_2(0) = 7.4$ and $x_3(0) = 19.2$. We take the initial conditions of the parameter estimates as $\hat{a}(0) = 4.7$, $\hat{b}(0) = 10.4$ and $\hat{c}(0) = 5.8$.

Figure 3 shows the time-history of the controlled states x_1, x_2, x_3. Thus, Figure 3 illustrates the control law stated in Theorem 1 for the global chaos control of the novel chaotic system (8).

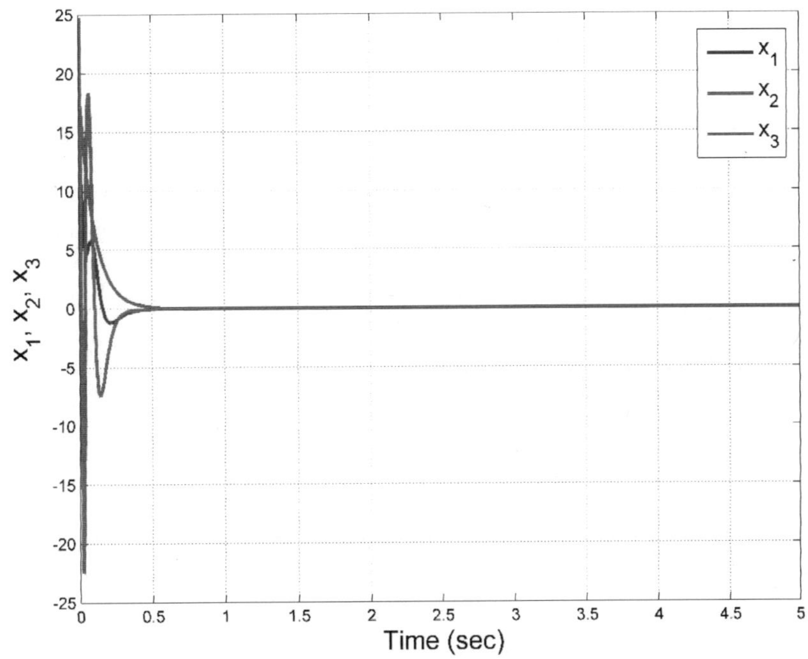

Figure 3. Time-history of the controlled chaotic system (8)

IOP Conf. Series: Materials Science and Engineering **332** (2018) 012010 doi:10.1088/1757-899X/332/1/012010

4. Circuit implementation of the new two-scroll chaotic system

In this section, the new two-scroll chaotic system (1) is designed as an electronic circuit as seen on **Figure 4** and set in MultiSIM. As seen on **Figure 4**, 3 integrators, 3 multipliers and 2 inverters were used in the circuit in order to implement 3 differential equations that make up the chaotic system. By applying Kirchhoff's circuit laws, the corresponding circuital equations of the designed circuit can be written as::

$$\left.\begin{aligned} \dot{x}_1 &= \frac{1}{C_1R_1}x_2 - \frac{1}{C_1R_2}x_1 + \frac{1}{10C_1R_3}x_2x_3 \\ \dot{x}_2 &= \frac{1}{C_2R_4}x_1 - \frac{1}{C_2R_5}x_2 + \frac{1}{10C_2R_6}x_1x_3 \\ \dot{x}_3 &= -\frac{1}{C_3R_7}x_3 - \frac{1}{10C_3R_8}x_1x_2 \end{aligned}\right\} \tag{18}$$

In system (18), the variables x_1, x_2, and x_3 are the outcomes of the integrators U1A, U2A, U3A. The circuit components have been selected as: $R_1 = R_2 = R_3 = R_8 = 40$ kΩ, $R_5 = R_7 = 400$ KΩ, $R_4 = 20$ KΩ, $R_6 = 1.33$ KΩ, $R_9 = R_{10} = R_{11} = R_{12} = 100$ KΩ, $C_1 = C_2 = C_3 = 1$ nF. The supplies of all active devices are ±15 Volt. The obtained results are presented in **Figures 5** (a) - (c), which show the phase portraits of the chaotic attractor in x_1-x_2, x_2-x_3 and x_1-x_3 planes, respectively. Numerical simulations (**see Figure 1**) are similar with the circuital ones (**see Figure 5**).

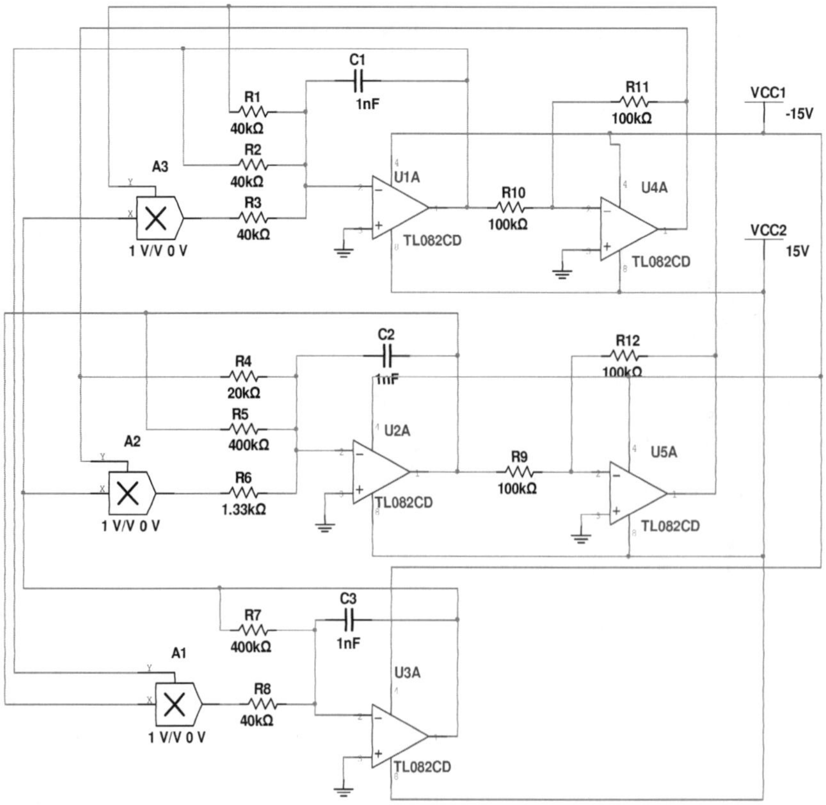

Figure 4 Circuit design for new two-scroll chaotic system (1) by MultiSIM

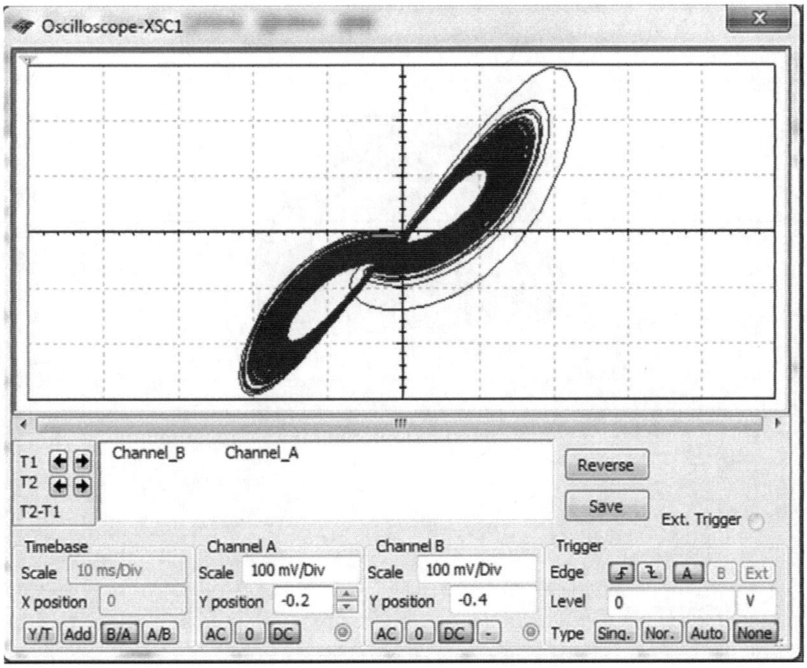
Oscilloscope-XSC1
T1
T2
T2-T1
Channel_B
Channel_A
Reverse
Save
Ext. Trigger
Timebase
Scale
10 ms/Div
X position
0
Y/T
Add
B/A
A/B
Channel A
Scale
100 mV/Div
Y position
-0.2
AC
0
DC
Channel B
Scale
100 mV/Div
Y position
-0.4
AC
0
DC
-
Trigger
Edge
A
B
Ext
Level
0
V
Type
Sing.
Nor.
Auto
None

(a)

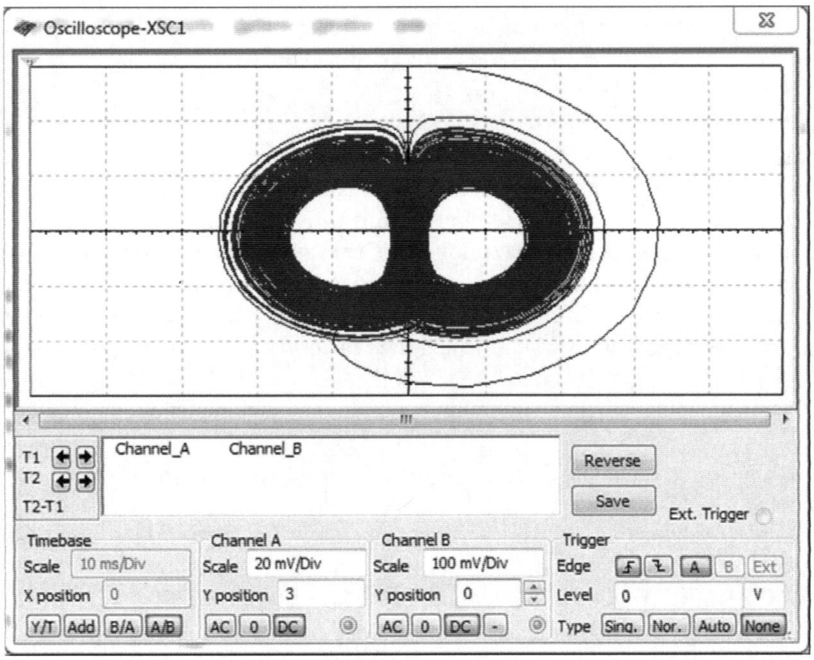
Oscilloscope-XSC1
T1
T2
T2-T1
Channel_A
Channel_B
Reverse
Save
Ext. Trigger
Timebase
Scale
10 ms/Div
X position
0
Y/T
Add
B/A
A/B
Channel A
Scale
20 mV/Div
Y position
3
AC
0
DC
Channel B
Scale
100 mV/Div
Y position
0
AC
0
DC
-
Trigger
Edge
A
B
Ext
Level
0
V
Type
Sing.
Nor.
Auto
None

(b)

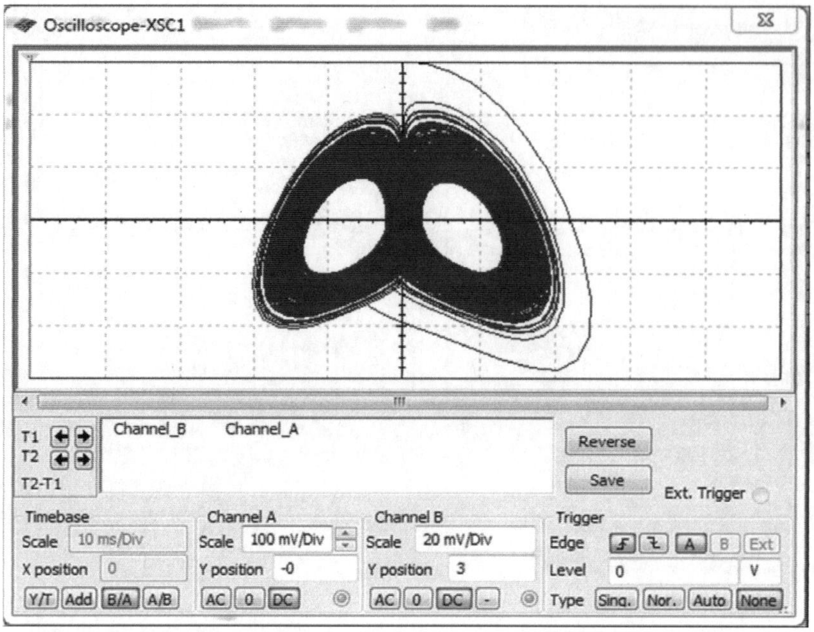

(c)

Figure 5 The phase portraits of new two-scroll chaotic system (1) observed on the oscilloscope in different planes (a) x_1-x_2, (b) x_2-$x_3$plane and (c) x_1-x_3 plane by MultiSIM

5. Conclusions

This work described a new two-scroll chaotic system with three quadratic nonlinearities. First, the qualitative properties of the new two-scroll chaotic system are detailed. Dynamical behaviors of the new two-scroll chaotic system with three quadratic nonlinearities are investigated through equilibrium points, projections of chaotic attractors, Lyapunov exponents and Kaplan–Yorke dimension. In addition, the adaptive control scheme of the new two-scroll chaotic system is shown via adaptive control approach. Furthermore, an electronic circuit realization of the new two-scroll chaotic system using the electronic simulation package MultiSIM confirmed the feasibility of the theoretical model.

References

[1] Azar A T and Vaidyanathan S 2015 *Chaos Modeling and Control Systems Design* (Berlin: Springer)
[2] Azar A T and Vaidyanathan S 2016 *Advances in Chaos Theory and Intelligent Control* (Berlin: Springer)
[3] Xu C and Wu Y 2015 *Applied Mathematical Modelling* **39** 2295-2310
[4] Bodale I and Oancea V A 2015 *Chaos Solitons and Fractals* **78** 1-9
[5] Vaidyanathan S 2015 *International Journal of ChemTech Research* **8** 73-85
[6] Sambas, A., WS, M. S., Mamat, M., and Prastio, R. P 2016 *Advances in Chaos Theory and Intelligent Control* (Berlin: Springer)
[7] Vaidyanathan S 2015 *International Journal of PharmTech Research* **8** 106-116
[8] Vaidyanathan S 2016 *International Journal of ChemTech Research* **9** 297-304
[9] Akhmet M and Fen M O 2014 *Neurocomputing* **145** 230-239
[10] Lian S and Chen X 2011 *Applied Soft Computing* **11** 4293-4301

[11] Yang J, Wang L, Wang Y and Guo T 2017 *Neurocomputing* **227** 142-148
[12] Bao B C, Bao H, Wang N, Chen M and Xu Q 2017 *Chaos Solitons and Fractals* **94** 102-111
[13] Voorsluijs V and Decker Y D 2016 *Physica D: Nonlinear Phenomena* **335** 1-9
[14] Chattopadhyay J, Pal N, Samanta S, Venturino E and Khan Q J A 2015 *Biosystems* **138** 18-24
[15] Iqbal S, Zang X, Zhu Y and Zhao J 2014 *Robotics and Autonomous Systems* **62** 889-909
[16] Sambas, A., Vaidyanathan, S., Mamat, M., Sanjaya, W. M., and Rahayu, D. S 2016 *Advances and Applications in Chaotic Systems* (Berlin: Springer).
[17] Punjabi A and Boozer A 2014 *Physics Letters A* **378** 2410-2416
[18] Vaidyanathan S 2015 *International Journal of ChemTech Research* **8** 818-827
[19] Jajarmi A, Hajipour M and Baleanu D 2017 *Chaos Solitons and Fractals* **99** 285-296
[20] Huang C and Cao J 2017 *Physica A* **473** 262-275
[21] Lorenz E N 1963 *J. Atmospheric Sciences* **20** 130-141
[22] Chen G and Ueta T 1999 *International Journal of Bifurcation and Chaos* **9** 1465-1466
[23] Lu J and Chen G 2002 *International Journal of Bifurcation and Chaos* **12** 659-661
[24] Tigan G and Opris D 2008 *Chaos, Solitons and Fractals* **36** 1315-1319
[25] Vaidyanathan S and Rajagopal K 2016 *International Journal of Control Theory and Applications* **9** 151-174
[26] Vaidyanathan S 2016 *International Journal of Control Theory and Applications* **9** 1-20
[27] Wang Z, Sun Y, Van Wyk B J, Qi G and Van Wyk M A 2009 *Brazilian Journal of Physics* **39** 547-553
[28] Pan L, Zhou W, Fang J and Li D 2010 *International Journal of Nonlinear Science* **10** 462-474
[29] Yu F and Wang C 2014 *Optik* **125** 5920-5925
[30] Sampath S, Vaidyanathan S, Volos C K and Pham V T 2015 *Journal of Engineering Science and Technology Review* **8** 1-6
[31] Genesio R and Tesi A 1992 *Automatica* **28** 531-548
[32] Vaidyanathan S and Madhavan K 2013 *International Journal of Control Theory and Applications* **6** 121-137
[33] Vaidyanathan S 2016 *International Journal of Control Theory and Applications* **9** 199-219
[34] Sambas, A., Mamat, M., Mada Sanjaya, W. S., Salleh, Z., and Mohamad, F. S 2015 *Advanced Studies in Theoretical Physics* **9** 379-394
[35] Pandey A, Baghel R K and Singh R P 2012 *IOSR Journal of Electronics and Communication Engineering* **1** 16-22
[36] Sambas, A., Mamat, M., and WS, M. S 2016 *International Journal of Control Theory and Applications* **9** 365-382
[37] Khalil H K 2002 *Nonlinear Systems* (New York: Prentice Hall)

IOP Conf. Series: Materials Science and Engineering **332** (2018) 012011 doi:10.1088/1757-899X/332/1/012011

Handling Uncertain Gross Margin and Water Demand in Agricultural Water Resources Management using Robust Optimization

D. Chaerani, E. Lesmana, N. Tressiana

Department of Mathematics
Faculty of Mathematics and Natural Sciences Universitas Padjadjaran
Jalan Raya Bandung Sumedang KM 21 Jatinangor Sumedang Indonesia

E-mail: d.chaerani@unpad.ac.id, eman.lesmana@unpad.ac.id, n.tressiana@gmail.com

Abstract. In this paper, an application of Robust Optimization in agricultural water resource management problem under gross margin and water demand uncertainty is presented. Water resource management is a series of activities that includes planning, developing, distributing and managing the use of water resource optimally. Water resource management for agriculture can be one of the efforts to optimize the benefits of agricultural output. The objective function of agricultural water resource management problem is to maximizing total benefits by water allocation to agricultural areas covered by the irrigation network in planning horizon. Due to gross margin and water demand uncertainty, we assume that the uncertain data lies within ellipsoidal uncertainty set. We employ robust counterpart methodology to get the robust optimal solution.

1. Introduction

Refers to Sabouni and Mardani in [12], an agricultural irrigation network is comprised of a water supply source (river), a water distribution network (primary and branch irrigation canals), and demand centers (the areas that are covered by the irrigation network). Optimization of an irrigation system for the allocation of available water and cultivable area is one of the best methods of increasing a farmer's income and can be facilitated by an irrigation water distribution network. Irrigation water that is allocated to various crops is usually based on predictive factors such as the water demand for each crop and the availability of water for an irrigation network. Utilization of water resources can be done almost on every aspect of human life for both daily necessities and for businesses that use basic water or as a support, including business in agriculture. The fulfillment of raw water for agriculture is carried out with the development of an irrigation system, as mentioned in [13]. Referring to [10], farmers have the discretion to conduct farming enterprises that generate the highest financial returns.

The optimization of irrigation systems for the allocation of available water and land is one of the best methods to increase farmers' income and can make farmland well facilitated by irrigation water distribution networks, see [12]. Water allocated to various farms is usually based on predictable factors such as water demand for every farmland and water availability for irrigation networks. Uncertainty is an inherent aspect of predicting these factors. The main concern to reduce costs or increase profits from agricultural irrigation systems, without considering uncertainty, can cause many problems related to their future development. Possible consequences of not paying attention to uncertainty include the formation of less profits from expectations and possible system failures where non-fulfillment of demand or other constraints, see [14]. Other study about water distribution management and its robust version can be seen in [9] and [4].

Uncertainty in predicting data is an important issue in water resources management. There are several classical methods of overcoming the uncertainty parameters, among them Sensitivity Analysis and Stochastic Optimization. However, Sensitivity Analysis is just a tool for analyzing the goodness of the solution. It is not very helpful to generate a robust solution to data changes. In Stochastic Optimization, opportunity constraints can damage the convex nature and significantly increase the complexity of the original problem as stated in [15]. In this thesis, the authors use Robust Optimization approach of [1], which decision maker must choose strategy without knowing exact value taken by uncertain parameter.

2. Problem Description

2.1. Agricultural Water Resources Management

Water resources management is the activity of planning, developing, distributing and managing the optimal use of water resources. Management of water resources can be done almost on every aspect of human life for both daily necessities and for businesses that use basic water or as a support, including business in agriculture. Surface water and groundwater are the main water resources used to meet agricultural needs. But now most of the needs still rely on surface water sources, therefore the surface water source needs to be managed properly so as to provide benefits for the development of the agricultural sector. Surface water is all water found on the ground (rivers, lakes, springs, waterlogging). Flooded and flooded surface waters (lakes, reservoirs, swamps) and some subsurface water will collect and flow to form rivers and ends at sea. The process of water travel in the mainland occurs in the components of the hydrological cycle that form the system of Watershed (DAS). In this paper, the distribution of water from water resources to farms is illustrated as in Figure 1.

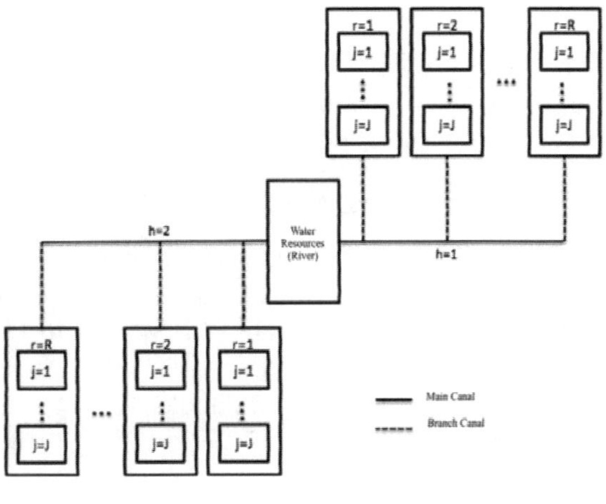

Figure 1. Illustration of Water Distribution from River to Agricultural Fields

In Figure 1, the river as water resource is illustrated to have two branches h which is the main channel for the left side of the river and the right side of the river. The main channels h distribute water to areas where r in each region r there are branch channels to drain water to some agricultural land j in r area. According to Rossiter and Wambeke [11], gross margin is the average amount of income minus the average amount of all costs incurred on a certain land area within a certain period. Basically, gross margin for the farmer is the income from total agricultural production minus operational cost for agriculture.

The Linear Programming Model for water resource management in this paper refers to Sabouni and Mardani, 2013 [12] with objective function is to maximize the total profit of farmers in a horizon planning. The allocation of net flow of water on each farm land j in the r area of the main channel h denoted by q_{hrjt}. The total area allocation of agricultural land j in area r in horizon planning t for main channel h denoted by y_{hrjt}. The gross margin of the water allocation to land j in the r region of the channel h in the horizon planning t denoted by n_{hrjt}.

The objective function is obtained by calculating the difference between the gross margin of the water allocation to land j in the r region of the channel h in the horizon planning t and p_{hrjt}, i.e., the price of water allocated to land j in the r region of the channel h in horizon planning t, goals can be formulated as follows.

$$\max \sum_{h=1}^{H}\sum_{r=1}^{R}\sum_{j=1}^{J}\sum_{t=1}^{T} n_{hrjt}q_{hrjt} - \sum_{h=1}^{H}\sum_{r=1}^{R}\sum_{j=1}^{J}\sum_{t=1}^{T} p_{hrjt}q_{hrjt} \qquad (1)$$

Furthermore, it is assumed that the water allocated to each land j is limited by the available irrigation water. Therefore, the first constraint is the amount of water allocated to each agricultural land j in the r area of the main channel h and the water lost from the main channel h for the agricultural land j in the r area with the total quantity parameter of water lost in the main channel h and the branch channel in region r is denoted by δ_{hr} and should not exceed k_t ie the quantity of water available for each horizon plan t, so that the first constraint function can be formulated as follows:

$$\sum_{h=1}^{H}\sum_{r=1}^{R}\sum_{j=1}^{J} (1+\delta_{hr})\, q_{hrjt} \le k_t, \qquad \forall t \qquad (2)$$

Then a second constraint function is created. the amount of water allocated to land j in the r region of the main channel h in the horizontal and wasted water planning in the main channel h and the branch channel in r area in horizon planning t should not exceed the value of c_{ht}, i.e., capacity the main channel h on horizon planning t, so obtained.

$$\sum_{r=1}^{R}\sum_{j=1}^{J} (1+\delta_{hr})\, q_{hrjt} \le c_{ht}, \qquad \forall h,t \qquad (3)$$

Subsequently, a third constraint function is formed which is the sum of the total water allocated to land j in the r region of the main channel h in the horizontal and wasted water planning in the branch channel in r area with the water level wasted in the branch channel at r region of the canal the primary h for the land j is denoted by, should not exceed the value of b_{hrjt} is the capacity of the branch channel in r region for the total area of agricultural land j in horizon planning t, so that it is obtained:

$$\sum_{j=1}^{J} (1+\xi_{hrj})\, q_{hrjt} \le b_{hrjt}, \qquad \forall h,r,t \qquad (4)$$

The fourth constraint function, i.e., water allocated to land j in the r area of the main channel h in horizon planning t shall be more than or equal to the minimum demand for irrigation water and less than or equal to the maximum demand for irrigation water, meaning that any land must obtain the amount of water sufficient, not less or more than the limit of its request, so obtained:

$$(d_{\min})_{hrjt} \leq q_{hrjt} \leq (d_{\max})_{hrjt}, \qquad \forall h, r, j, t \tag{5}$$

The total area of agricultural land j in area r in the horizon year plan t for the main channel h denoted by y_{hrjt} is influenced by the irrigation demand unit of the agricultural land j for region r on the main channel h denoted by w_{hrj},, and the flow water allocated to land j in region r of the main channel h in horizon planning t, is formulated into the fifth constraint function as follows

$$w_{hrj} y_{hrjt} = q_{hrjt}, \forall h, r, j, t \tag{6}$$

To obtain the value of the irrigation demand unit of the j plant in the r area for the main channel h, suppose ni_{hrj} is the irrigation network needed for the j plant in the r area on the main channel h and SR is an irrigation efficiency worth $0 < SR \leq 1$, it can be calculated as follows:

$$w_{hrj} = \frac{ni_{hrj}}{SR} \tag{7}$$

At the last constraint all decision variables must be non-negative, so they are obtained

$$q_{hrjt}, y_{hrjt} \geq 0, \quad \forall h, r, j, t \tag{8}$$

Thus the complete the linear programming model for water resource management in agriculture (Sabouni and Mardani [12]) is formulated in the following equations:

$$\max \sum_{h=1}^{H}\sum_{r=1}^{R}\sum_{j=1}^{J}\sum_{t=1}^{T} n_{hrjt} q_{hrjt} - \sum_{h=1}^{H}\sum_{r=1}^{R}\sum_{j=1}^{J}\sum_{t=1}^{T} p_{hrjt} q_{hrjt} \tag{9}$$

$$s.t \sum_{h=1}^{H}\sum_{r=1}^{R}\sum_{j=1}^{J} (1+\delta_{hr})\, q_{hrjt} \leq k_t, \qquad \forall t \tag{10}$$

$$\sum_{r=1}^{R}\sum_{j=1}^{J} (1+\delta_{hr})\, q_{hrjt} \leq c_{ht}, \qquad \forall h, t \tag{11}$$

$$\sum_{j=1}^{J} (1+\xi_{hrj})\, q_{hrjt} \leq b_{hrjt}, \qquad \forall h, r, t \tag{12}$$

$$(d_{\min})_{hrjt} \leq q_{hrjt} \leq (d_{\max})_{hrjt}, \quad \forall h, r, j, t \tag{13}$$

$$w_{hrj} y_{hrjt} = q_{hrjt}, \forall h, r, j, t \tag{14}$$

$$q_{hrjt}, y_{hrjt} \geq 0, \forall h, r, j, t. \tag{15}$$

3. Methodology: Robust Optimization

In this section we present a brief discussion about the theory of Robust Optimization (RO) as proposed by [1] and discussed in [6], [7], also discussed in [3] and [2]. Let $c \in R^n, b \in R^m$ and $A \in R^m \times R^n$ be the parameters of the canonical linear programming problem

$$\min_x \left\{ c^T x \;:\; Ax \leq b \right\} \tag{LP}$$

The main idea of Robust Optimization is we are interested in a version of (LP) problem in which the data (c, A, b) are uncertain, but are known to reside in an uncertain set U. In this case the

(LP) is not a single deterministic problem, but a family of problems, one for each realisation $(c, A, b) \in U$:

$$\min_{x \in R^n} \left\{ c^T x \,:\, Ax \leq b \right\}_{(c,A,b)\in U} \qquad \text{(ULP)}.$$

We assume that the decision environment is such that

(i) Decision is offline: the entire decision vector x is to be fixed prior to knowing which value the actual parameters take "here-and-now-decision".

(ii) In the dynamic (ULP) case only part of the variables $(x_1, \ldots, x_n)$ need to be determined off-line. The rest may be determined after some of the uncertain parameters become known ("wait and see" decision).

(iii) The prior information on the data (c, A, b) is crude and is captured by a compact uncertainty set U.

(iv) The inequality constraints $Ax \leq b$ are hard constraints, i.e., they must all be satisfied whenever the uncertain parameters reside in U.

Under the above conditions the robust optimization approach converts the uncertain family of the problems (ULP) into the following single deterministic problem, which we call *Robust Counterpart*:

$$\pi^* = \min_{x \in R^n} \left\{ c^T x \,:\, Ax \leq b, (c, A, b) \in U \right\} \qquad \text{(RC)}.$$

A vector x^* is the called a *robust optimal solution* if for all realizations $(c, A, b) \in U$, x^* is feasible and the value of objective function is guaranteed to be at most π^*. Problem (RC) can be written equivalently as a problem with a linear certain objective function and only uncertain constraints as follows:

$$\min\ t \tag{16}$$
$$s.t\ \ c^T x - t \leq 0, \tag{17}$$
$$a_i^T x - b_i \leq 0, i = 1, \ldots, m, \forall (c, A, b) \in U \tag{18}$$

3.1. The Treatment of Robust Linear Optimization (RLO)

We cite from [6] that for the treatment of Robust Linear Optimization (RLO) we will assume without loss of generality that,

(i) Firstly, the objective $c^T x$ is certain. If there is uncertainty in the objective, we can reformulate the problem such that this uncertainty appears in a constraint see (16).

(ii) Secondly, the right-hand-side b is certain. If b is uncertain, we can introduce an extra variable x_{n+1} and change the problem into

$$min\ c^T x \tag{19}$$
$$s.t\ \ a_i^T x - b_i x_{n+1} \leq 0, \tag{20}$$
$$x_{n+1} = 1, i = \ldots, m, \forall (A, b) \in U. \tag{21}$$

This assumption is made to make the notation for the uncertainty region and the resulting robust counterpart formulations easier.

(iii) Third, the robustness with respect to U can be formulated constraint-wise and the last, the uncertainty set U is closed and convex.

Notes that because this inequality on x is of the semiinfinite type (it should be satisfied for an infinite number of values of the parameters A and b), it cannot directly be efficiently solved by standard optimization methods. The challenge is to find for which types of uncertainty sets problem (RC) can be reformulated into a tractable optimization problem.

3.2. Types of uncertainty sets

The challenge is to find U for which types of uncertainty sets problem can be reformulated into a tractable optimization problem. Since the robustness with respect to U can be formulated constraint-wise, thus we can reformulate each constraint which involves the uncertain data. Refers to [6], consider the canonical robust semi-infinite constraint

$$a^T x - b \leq 0, \forall (a, b) \in U. \tag{22}$$

Here a is a vector in R^n and b which are the general representatives of a_i and b_i.

Similarly, U stands for U_i. Its convenient to describe the uncertainty parameters a and b and the uncertainty set U in terms of a primitive factor $\zeta \in R^L$. Namely,

$$a = \bar{a} + Q\zeta, b = \bar{b} + q^T \zeta \tag{23}$$

where $\bar{a} \in R^n, Q \in R^{n \times L}, \bar{b} \in R$ and $q \in R^L$ and

$$U = \left\{ \begin{pmatrix} a = \bar{a} + Q\zeta \\ b = \bar{b} + q^T \zeta \end{pmatrix} : \zeta \in Z \right\} \tag{24}$$

where $Z \subset R^L$ is the uncertainty set for the primitive factors. The fixed vector $\bar{a}$ and the scalar $\bar{b}$ will thereafter be called nominal. In the view of representation (24) the alternative formulation of (22). Replacing a and b by their expressions in ζ, we have

$$(\bar{a}^T x - \bar{b}) + (Q^T x - q)^T \zeta \leq 0, \forall \zeta \in Z. \tag{25}$$

3.2.1. Case 1: Using Interval or Box Uncertainty Set In general, a tractable robust counterpart formulation for RLO of (22) with interval (or box) uncertainty regions can be stated as the following formulation.

$$(\bar{a} + P\zeta)^T x \leq b, \forall \zeta \, : \, \|\zeta\|_\infty \leq \mu. \tag{26}$$

This problem is a semi-infinite programming problem.

3.2.2. Case 2: Using Ellipsoidal Uncertainty Set Refers to [6], in the case of ellipsoidal uncertainty, the robust counterpart of (22) becomes

$$\begin{aligned} &(\bar{a} + P\zeta)^T x \leq b, \forall \zeta \, : \, \|\zeta\|_2 \leq \mu. \\ &\equiv \bar{a}^T x + \max_{\zeta \, : \, \|\zeta\|_2 \leq \mu} (P^T x)^T \zeta \leq b. \end{aligned} \tag{27}$$

It can easily be verified that

$$\max_{\zeta \, : \, \|\zeta\|_2 \leq \mu} (P^T x)^T = \mu \left\| P^T x \right\|_2. \tag{28}$$

Hence x satisfies (27) if and only if

$$\bar{a}^T x + \mu \left\| P^T x \right\|_2 \leq b. \tag{29}$$

Now the constraint does not have the semi-infinite structure as the original one in (27). This is a conic quadratic programming problem. Again as in the case of box uncertainty, in the final robust counterpart an extra safety term $\mu \left\| P^T x \right\|_2$ is added to account for uncertainty and it depends on the value of x.

IOP Conf. Series: Materials Science and Engineering **332** (2018) 012011 doi:10.1088/1757-899X/332/1/012011

4. Result and Discussion

4.1. Data Uncertainty In Agricultural Water Resources Management

Assume that the model of water resource management problems in agriculture contains uncertainty on gross margin and water demand, i.e.,

$$n_{hrjt}, (d_{\min})_{hrjt}, (d_{\max})_{hrjt} \in U. \tag{30}$$

Under Robust Optimization assumption the right-hand side must be a certain parameter, thus define a new variable $x_0 = 1$. Based on these assumptions, the model of the water resource management problem can be reformulated into a robust counterpart model as follows:

$$\max \omega \tag{31}$$

$$s.t \left[\sum_{h=1}^{H}\sum_{r=1}^{R}\sum_{j=1}^{J}\sum_{t=1}^{T} n_{hrjt}q_{hrjt} - p_{hrjt}q_{hrjt}\right] - \omega \leq 0, \tag{32}$$

$$\sum_{h=1}^{H}\sum_{r=1}^{R}\sum_{j=1}^{J}(1+\delta_{hr})q_{hrjt} \leq k_t, \forall t \tag{33}$$

$$\sum_{r=1}^{R}\sum_{j=1}^{J}(1+\delta_{hr})q_{hrjt} \leq c_{ht}, \forall h, t \tag{34}$$

$$\sum_{j=1}^{J}(1+\xi_{hr})q_{hrjt} \leq b_{hrj_t}, \forall h, r, t \tag{35}$$

$$q_{hrjt} - (d_{\max})_{hrjt}x_0 \leq 0, \forall h, r, j, t \tag{36}$$

$$q_{hrjt} - (d_{\min})_{hrjt}x_0 \geq 0, \forall h, r, j, t \tag{37}$$

$$w_{hrjt}y_{hrjt} - q_{hrjt} = 0, \forall h, r, j, t \tag{38}$$

$$q_{hrjt}, y_{hrjt} \geq 0, \forall h, r, j, t \tag{39}$$

$$x_0 = 1 \tag{40}$$

$$n_{hrjt}, (d_{\min})_{hrjt}, (d_{\max})_{hrjt} \in U. \tag{41}$$

4.2. Robust Version for Water Management Problem with Box Uncertainty Set

For water resource management problems in agriculture with uncertainty as described by simple intervals as follows:

$$U = \left\{ (n, d_{\max}, d_{\min}) : \begin{array}{c} (n_{hrjt})_{\min} \leq n_{hrjt} \leq (n_{hrjt})_{\max} \\ \left((d_{\max})_{hrjt}\right)_{\min} \leq (d_{\max})_{hrjt} \leq \left((d_{\max})_{hrjt}\right)_{\max}, \forall h, r, j, t \\ \left((d_{\min})_{hrjt}\right)_{\min} \leq (d_{\min})_{hrjt} \leq \left((d_{\min})_{hrjt}\right)_{\max} \end{array} \right\} \tag{42}$$

Based on (Hertog, 2015), we may select the set of U uncertainties that satisfy representatives of (19) with the following nominal data conditions.

$$\begin{aligned} \bar{n}_{hrjt} &= \frac{(n_{hrjt})_{\max} + (n_{hrjt})_{\min}}{2} \\ (Q_{hrjt})_n &= diag\left(\frac{(n_{hrjt})_{\max} - (n_{hrjt})_{\min}}{2}\right) \end{aligned} \tag{43}$$

$$\begin{aligned} (\bar{d}_{\max})_{hrjt} &= \frac{((d_{\max})_{hrjt})_{\max} + ((d_{\max})_{hrjt})_{\min}}{2} \\ (Q_{hrjt})_{d_{\max}} &= diag\left(\frac{((d_{\max})_{hrjt})_{\max} - ((d_{\max})_{hrjt})_{\min}}{2}\right) \end{aligned} \tag{44}$$

$$\begin{aligned}(\bar{d}_{\min})_{hrjt} &= \frac{((d_{\min})_{hrjt})_{\max}+((d_{\min})_{hrjt})_{\min}}{2}\\ (Q_{hrjt})_{d_{\min}} &= diag\left(\frac{((d_{\min})_{hrjt})_{\max}-((d_{\min})_{hrjt})_{\min}}{2}\right)\end{aligned} \tag{45}$$

Notice the constraint equation containing uncertain parameters n_{hrjt}, e.g.

$$n_{hrjt} = \bar{n}_{hrjt} + (Q_{hrjt})_n(\zeta_{hrjt})_n. \tag{46}$$

Now we derive first the part of (25) that contains the uncertainty parameter as follows.

$$\begin{aligned}\sum_{h=1}^{H}\sum_{r=1}^{R}\sum_{j=1}^{J}\sum_{t=1}^{T}(n_{hrjt}q_{hrjt} - p_{hrjt}q_{hrjt}) &= \sum_{h=1}^{H}\sum_{r=1}^{R}\sum_{j=1}^{J}\sum_{t=1}^{T}\left((\bar{n}_{hrjt} + (Q_{hrjt})_n(\zeta_{hrjt})_n)\, q_{hrjt} - p_{hrjt}q_{hrjt}\right)\\ &= \left[\sum_{h=1}^{H}\sum_{r=1}^{R}\sum_{j=1}^{J}\sum_{t=1}^{T}(\bar{n}_{hrjt}q_{hrjt} - p_{hrjt}q_{hrjt} + (Q_{hrjt})_n(\zeta_{hrjt})_n q_{hrjt})\right]\\ &= \left(\sum_{h=1}^{H}\sum_{r=1}^{R}\sum_{j=1}^{J}\sum_{t=1}^{T}(\bar{n}_{hrjt}q_{hrjt} - p_{hrjt}q_{hrjt})\right) +\\ &\quad \max_{\zeta_n:|(\zeta_{hrjt})_n|\le\sigma}\sum_{h=1}^{H}\sum_{r=1}^{R}\sum_{j=1}^{J}\sum_{t=1}^{T}(Q_{hrjt})_n(\zeta_{hrjt})_n q_{hrjt}\end{aligned} \tag{47}$$

Thus the robust counterpart of (25) is obtained as follows:

$$\begin{aligned}&\left[\left(\sum_{h=1}^{H}\sum_{r=1}^{R}\sum_{j=1}^{J}\sum_{t=1}^{T}(\bar{n}_{hrjt}q_{hrjt} - p_{hrjt}q_{hrjt})\right) + \sigma\left\|\sum_{h=1}^{H}\sum_{r=1}^{R}\sum_{j=1}^{J}\sum_{t=1}^{T}(Q_{hrjt})_n q_{hrjt}\right\|_1\right] - \omega =\\ &\left[\left(\sum_{h=1}^{H}\sum_{r=1}^{R}\sum_{j=1}^{J}\sum_{t=1}^{T}(\bar{n}_{hrjt}q_{hrjt} - p_{hrjt}q_{hrjt})\right) + \sigma\left|\sum_{h=1}^{H}\sum_{r=1}^{R}\sum_{j=1}^{J}\sum_{t=1}^{T}(Q_{hrjt})_n q_{hrjt}\right|\right] - \omega \le 0\end{aligned} \tag{48}$$

Use the same way for constraints that contain uncertain parameters

$$(d_{\max})_{hrjt} = (\bar{d}_{\max})_{hrjt} + (Q_{hrjt})_{d_{\max}}(\zeta_{hrjt})_{d_{\max}} \tag{49}$$

$$[(d_{\min})_{hrjt} = (\bar{d}_{\min})_{hrjt} + (Q_{hrjt})_{d_{\min}}(\zeta_{hrjt})_{d_{\min}} \tag{50}$$

then we obtain the robust counterpart as follows:

$$\begin{aligned}q_{hrjt} - (d_{\max})_{hrjt}x_0 &= q_{hrjt} - \left((\bar{d}_{\max})_{hrjt} + (Q_{hrjt})_{d_{\max}}(\zeta_{hrjt})_{d_{\max}}\right)x_0\\ &= q_{hrjt} - (\bar{d}_{\max})_{hrjt}x_0 - (Q_{hrjt})_{d_{\max}}(\zeta_{hrjt})_{d_{\max}}x_0\\ &= q_{hrjt} - (\bar{d}_{\max})_{hrjt}x_0 - \max_{\zeta_n:|(\zeta_{hrjt})_n|\le\varepsilon}\left|(Q_{hrjt})_{d_{\max}}(\zeta_{hrjt})_{d_{\max}}x_0\right|\\ &= q_{hrjt} - (\bar{d}_{\max})_{hrjt}x_0 - \varepsilon\left\|(Q_{hrjt})_{d_{\max}}x_0\right\|_1\\ &= q_{hrjt} - (\bar{d}_{\max})_{hrjt}x_0 - \varepsilon\left|(Q_{hrjt})_{d_{\max}}x_0\right| \le 0\end{aligned} \tag{51}$$

and

$$\begin{aligned}
q_{hrjt} - (d_{\min})_{hrjt} &= q_{hrjt} - \left((\bar{d}_{\min})_{hrjt} + (Q_{hrjt})_{d_{\min}} (\zeta_{hrjt})_{d_{\min}} \right) x_0 \\
&= q_{hrjt} - (\bar{d}_{\min})_{hrjt} x_0 - (Q_{hrjt})_{d_{\min}} (\zeta_{hrjt})_{d_{\min}} x_0 \\
&= q_{hrjt} - (\bar{d}_{\min})_{hrjt} x_0 - \max_{\zeta_n : \left|(\zeta_{hrjt})_n\right| \le \tau} \left| (Q_{hrjt})_{d_{\min}} (\zeta_{hrjt})_{d_{\min}} x_0 \right| \\
&= q_{hrjt} - (\bar{d}_{\min})_{hrjt} x_0 - \tau \left\| (Q_{hrjt})_{d_{\min}} x_0 \right\|_1 \\
&= q_{hrjt} - (\bar{d}_{\min})_{hrjt} x_0 - \tau \left| (Q_{hrjt})_{d_{\min}} x_0 \right| \ge 0 \qquad (52)
\end{aligned}$$

So that obtained model robust counterpart for water resource management problem in agriculture with box uncertainty approach as follows.

$$\begin{aligned}
&\max \omega \\
&\quad s.t \\
&\left[\left(\sum_{h=1}^{H} \sum_{r=1}^{R} \sum_{j=1}^{J} \sum_{t=1}^{T} (\bar{n}_{hrjt} q_{hrjt} - p_{hrjt} q_{hrjt}) \right) + \sigma \left| \sum_{h=1}^{H} \sum_{r=1}^{R} \sum_{j=1}^{J} \sum_{t=1}^{T} (Q_{hrjt})_n q_{hrjt} \right| \right] - \omega \le 0 \\
&\sum_{h=1}^{H} \sum_{r=1}^{R} \sum_{j=1}^{J} (1 + \delta_{hr}) q_{hrjt} \le k_t, \forall t \\
&\sum_{r=1}^{R} \sum_{j=1}^{J} (1 + \delta_{hr}) q_{hrjt} \le c_{ht} \\
&q_{hrjt} - (\bar{d}_{\max})_{hrjt} x_0 - \varepsilon \left| (Q_{hrjt})_{d_{\max}} x_0 \right| \le 0, \forall h, r, j, t \qquad (53) \\
&q_{hrjt} - (\bar{d}_{\min})_{hrjt} x_0 - \tau \left| (Q_{hrjt})_{d_{\min}} x_0 \right| \ge 0, \forall h, r, j, t \\
&w_{hrjt} y_{hrjt} - q_{hrjt} = 0, \forall h, r, j, t \\
&q_{hrjt}, y_{hrjt} \ge 0, \forall h, r, j, t \\
&x_0 = 1
\end{aligned}$$

4.3. Robust Version for Water Management Problem with Ellipsoidal Uncertainty Set

For the case of ellipsoidal uncertainty set, assume that the uncertain data lies in n_{hrjt}, $(d_{\min})_{hrjt}$ and $(d_{\max})_{hrjt}$. The following is how we define the robust counterpart for n_{hrjt}.

$$\begin{aligned}
&\left[\sum_{h=1}^{H} \sum_{r=1}^{R} \sum_{j=1}^{J} \sum_{t=1}^{T} (n_{hrjt} q_{hrjt} - p_{hrjt} q_{hrjt}) \right] - \omega \le 0, \\
&\equiv \left[\left(\sum_{h=1}^{H} \sum_{r=1}^{R} \sum_{j=1}^{J} \sum_{t=1}^{T} (\bar{n}_{hrjt} q_{hrjt} - p_{hrjt} q_{hrjt}) \right) + \sigma \left\| \sum_{h=1}^{H} \sum_{r=1}^{R} \sum_{j=1}^{J} \sum_{t=1}^{T} (Q_{hrjt})_n q_{hrjt} \right\|_2 \right] - \omega \qquad (54) \\
&= \left[\left(\sum_{h=1}^{H} \sum_{r=1}^{R} \sum_{j=1}^{J} \sum_{t=1}^{T} (\bar{n}_{hrjt} q_{hrjt} - p_{hrjt} q_{hrjt}) \right) + \sigma \sqrt{ \sum_{h=1}^{H} \sum_{r=1}^{R} \sum_{j=1}^{J} \sum_{t=1}^{T} ((Q_{hrjt})_n q_{hrjt})^2 } \right] - \omega \le 0.
\end{aligned}$$

The following is how we define the robust counterpart for $(d_{\max})_{hrjt}$:

$$\begin{aligned}
q_{hrjt} - (d_{\max})_{hrjt} &\equiv q_{hrjt} - (\bar{d}_{\max})_{hrjt} x_0 - \varepsilon \left\| (Q_{hrjt})_{d_{\max}} x_0 \right\|_2 \\
&= q_{hrjt} - (\bar{d}_{\max})_{hrjt} x_0 - \varepsilon \sqrt{ (Q_{hrjt})_{d_{\max}} x_0^2 } \ge 0 \qquad (55)
\end{aligned}$$

The following is how we define the robust counterpart for $(d_{\min})_{hrjt}$.

$$
\begin{aligned}
q_{hrjt} - (d_{\min})_{hrjt} &\equiv q_{hrjt} - (\bar{d}_{\min})_{hrjt} x_0 - \tau \left\| (Q_{hrjt})_{d_{\min}} x_0 \right\|_2 \\
&= q_{hrjt} - (\bar{d}_{\min})_{hrjt} x_0 - \tau \sqrt{(Q_{hrjt})_{d_{\min}} {x_0}^2} \leq 0.
\end{aligned} \tag{56}
$$

Thus the complete formulation for Robust Version for Water Management Problem with Ellipsoidal Uncertainty Set can be written as follows.

$$
\begin{aligned}
&\max \omega \\
&s.t \left[\left(\sum_{h=1}^{H} \sum_{r=1}^{R} \sum_{j=1}^{J} \sum_{t=1}^{T} (\bar{n}_{hrjt} q_{hrjt} - p_{hrjt} q_{hrjt}) \right) + \sigma \sqrt{\sum_{h=1}^{H} \sum_{r=1}^{R} \sum_{j=1}^{J} \sum_{t=1}^{T} ((Q_{hrjt})_n q_{hrjt})^2} \right] - \omega \leq 0 \\
&\sum_{h=1}^{H} \sum_{r=1}^{R} \sum_{j=1}^{J} (1 + \delta_{hr}) q_{hrjt} \leq k_t, \forall t \\
&\sum_{r=1}^{R} \sum_{j=1}^{J} (1 + \delta_{hr}) q_{hrjt} \leq c_{ht}, \forall h, t \\
&\sum_{j=1}^{J} (1 + \xi_{hr}) q_{hrjt} \leq b_{hrj_t}, \forall h, r, t \\
&q_{hrjt} - (\bar{d}_{\max})_{hrjt} x_0 - \varepsilon \sqrt{\left((Q_{hrjt})_{d_{\max}} x_0 \right)^2} \leq 0, \forall h, r, j, t \\
&q_{hrjt} - (\bar{d}_{\min})_{hrjt} x_0 - \tau \sqrt{\left((Q_{hrjt})_{d_{\min}} x_0 \right)^2} \geq 0, \forall h, r, j, t \\
&w_{hrjt} y_{hrjt} - q_{hrjt} = 0, \forall h, r, j, t \\
&q_{hrjt}, y_{hrjt} \geq 0, \forall h, r, j, t \\
&x_0 = 1
\end{aligned} \tag{57}
$$

4.4. Numerical Experiment

The data used is secondary data. For water availability data on water resources, the quantity of water in the main canal, the quantity of water in branch channels, and the water demand unit is referenced from Sabouni and Mardani [12], for the price of water referenced from [5], and for gross margin of each plant is referenced from [8].

The data quantity of water available for irrigation networks in two years of horizon planning are $k_1 = 470 \times 10^6 m^3, k_2 = 343 \times 10^6 m^3$. The irrigation flow required for each crop in a different area is presented in the following Table 1.

From the irrigation flow data required for each plant in different areas, with an irrigation efficiency of 35 percentage, an irrigation demand unit of plant j in r area for main channel h (w_{hrj}) as shown in Table 2. Main channel capacity for left side (c_{1t}) is $1.577 \times 10^6 m^3$ per year and right side (c_{2t}) is $473 \times 10^6 m^3$ per year. The main channel capacity is assumed to be constant for each year. The capacity of branch canals for each region (b_{hrjt}) is presented in Table 3. Assume 1 percentage of water allocated per kilometer is lost on the main channel h. The distance between the starting point of each branch's branch and the dam that divides the main channel is presented in Table 4. The water lost in each crop is assumed to be 15 percentage. The price of water for agriculture (p_{hrjt}) is 0.8 USD/1000m^3 or 25,8065 $rial/m^3$. Revenues earned from each crop (n_{hrjt}) and water demand from each crop are presented on Table 5.

Table 1. Water Distribution for different area and plants

Area	Plants (m^3/ha)			
	Wheat $j = 1$	Barley $j = 2$	Rice $j = 3$	Alfalfa $j = 4$
Mobarake $r = 1$	4.970	4.460	7.090	9.280
Nadjafabad $r = 2$	4.270	3.590	7.940	9.310
Lenjan $r = 3$	4.270	3.590	6.890	4.460
Flavarjan $r = 4$	4.910	6.400	6.890	4.460

Table 2. Unit of irrigation demand from plant j in region r for main channel h

Main Canal	Area	The required irrigation flow / irrigation efficiency (m^3/ha)			
		Wheat $j = 1$	Barley $j = 2$	Rice $j = 3$	Alfalfa $j = 4$
Left side	Mobarake $r = 1$	14.200,00	12.742,86	20.257,14	26.514,29
$h = 1$	Nadjafabad $r = 2$	12.200,00	10.257,14	22.685,71	26.600,00
	Lenjan $r = 3$	12.200,00	10.257,14	19.685,71	12.742,86
	Falavarjan $r = 4$	14.028,57	18.285,71	19.685,71	12.742,86
Right side	Mobarake $r = 1$	14.200,00	12.742,86	20.257,14	26.514,29
$h = 2$	Nadjafabad $r = 2$	-	-	-	-
	Lenjan $r = 3$	-	-	-	-
	Falavarjan $r = 4$	14.028,57	18.285,71	19.685,71	12.742,86

Table 3. Water Distribution for different area and plants

Main Canal	Kapasitas ($m^3 \times 10^6$)			
	Mobarakeh	Nadjafabad	Lenjan	Flavarjan
Leftside	79	349	79	349
Rightside	213	-	-	213

Table 4. The distance between the starting point and the dam.

Main Canal	Distances (*kilometer*)			
	Mobarakeh	Nadjafabad	Lenjan	Flavarjan
Leftside	10,3	52,1	24,2	18,2
Rightside	30,6	-	-	35,1

IOP Conf. Series: Materials Science and Engineering **332** (2018) 012011 doi:10.1088/1757-899X/332/1/012011

Table 5. The distance between the starting point and the dam.

Crops	Revenues $(rial/m^3)$			
	Mobarakeh	Nadjafabad	Lenjan	Flavarjan
Leftside	10,3	52,1	24,2	18,2
Rightside	30,6	-	-	35,1

The nominal model for water resource management problems in agriculture in numerical simulations involves 162 constraints and 96 variables, while for the Robust Optimization model involving 164 constraints for box uncertainty approach and 163 constraints for ellipsoidal uncertainty approach and both involving 97 variables with additional variables ω as delimiter of purpose function. In the nominal model it takes 405 milliseconds and uses 11.54 MiB memory, the Robust Optimization Model with the uncertainty box takes 381 milliseconds using 12 MiB memory, and in Robust Optimization Model with ellipsoidal uncertainty it takes 144 milliseconds using 6.79 MiB. In this numerical experiment, the result for box and ellipsoidal uncertainty give the same solution, as seen in Figure 2 and 3. The objective function result can be seen in Figure 4.

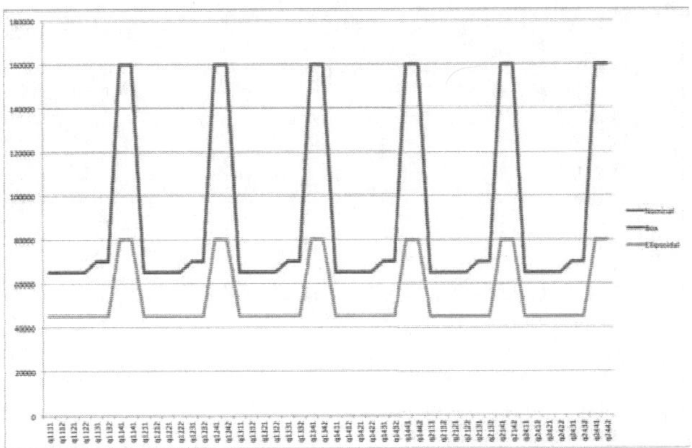

Figure 2. Water Allocation Calculation Result for Nominal, Box and Ellipsoidal Model

5. Conclusion and Future Research

Robust optimization is applied to water resource management issues in agriculture with uncertainty of gross margin of water demand. Thethe robust counterpart form becomes a linear optimization problem in case of box uncertainty set and becomes conic quadratic optimization model in case of ellipsoidal uncertainty set. This show us that the robust counterpart is guaranteed computationally tractable. Referring to Sabouni and Mardani [12] this study can be continued taking into consideration the uncertainties in the quantity of water available in water resource management issues in agriculture. In addition, referring to Hertog [6] on Robust Optimization, the problem of water resources management in agriculture can be continued with the formulation of Adjustable Robust Counterpart.

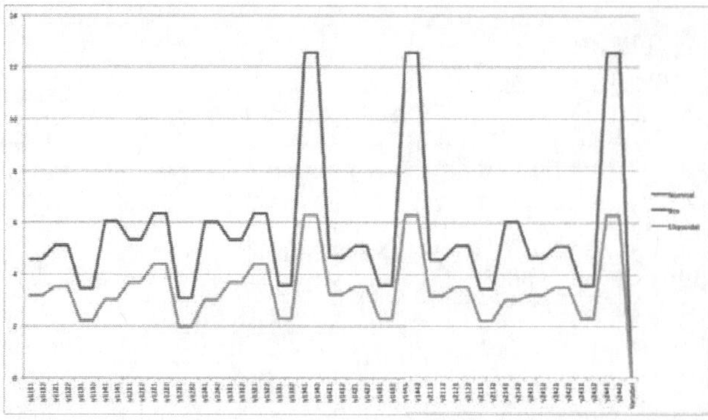

Figure 3. Area Allocation Calculation Result for Nominal, Box and Ellipsoidal Model

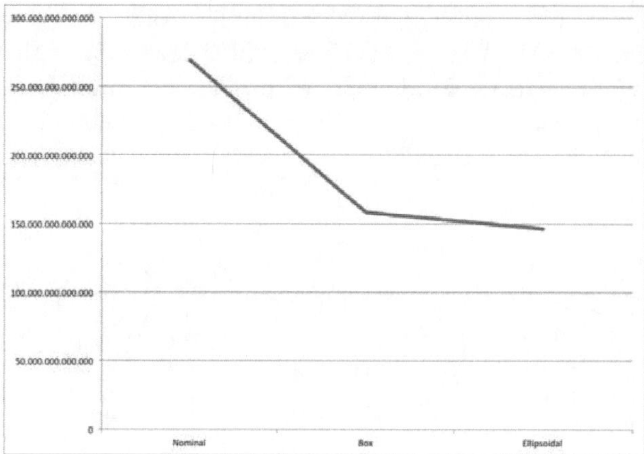

Figure 4. Maximum Profit for Nominal, Box and Ellipsoidal Model

Acknowledgement
This research is funded by Competency Based Research or *Penelitian Berbasis Kompetensi* from Indonesian Ministry of Research Technology and Higher Education for the year of 2017 contract number 718/UN6.3.1/PL/2017.

References
[1] Ben-Tal, A., Nemirovskii, A., 2002, Robust optimization methodology and applications., *Mathematical Programming* **92(3, Ser. B)** pp. 453-480.

[2] Chaerani, D., Roos, C., Handling Optimization under Uncertainty Problem Using Robust Counterpart Methodology, *Jurnal Teknik Industri* **Vol. 15, No. 2** pp. 111-118, DOI: 10.9744/jti.15.2.111-118, ISSN 1411-2485 print, ISSN 2087-7439 online., 2013.

[3] Chaerani, D., C. Roos, Modeling Robust Design Problems via Conic Optimization, Proceedings of OR 2006, Operations Research Proceedings 2004, Springer Verlag, Berlin, Heidelberg, New York, 2006.

[4] Chaerani, D., Ruchjana, B.N., Setiawan, A., Rejito, J., Dewanto, S.P., Dharmawan, I.A., Rosandi, Y., Determining Robust Counterpart of Spatial Optimization Model for Water Supply Allocation, AIP Conf. Proc. 1716, 020002 (2016); http://dx.doi.org/10.1063/1.4942985.

[5] FAO AQUASTAT. 2008. Irrigation in the Middle East region in figures. AQUASTAT Survey 2008. Iran (Islamic Republic of).

[6] Den Hertog, D, Ben-Tal, A., Brekelmans., Practical Robust Optimization, Lecture Notes LNMB Course, Spring 2015.

[7] Gorissen, B. L., Yanikoglu, I., den Hertog, D., A Practical Guide to Robust Optimization. *Omega* 53 (2015): 124-137, 2015.
[8] Karandish, F., dan Asadi, R. 2015. Applying the economic water use efficiency concept to determine the optimal cropping pattern in Rayen City. Journal of Biodiversity and Environmental Sciences (JBES), Vol. 6, No. 3, 351-358.
[9] Murray, A.T, Gober, P., Anselin, L., Rey, S. J., Sampson, D., Padegimas, P. D., Liu, Y., Spatial Optimization Models for Water Supply Allocation, *Water resources management*, Vol 26 No 8, pp 2243–2257, Springer, 2012.
[10] Peraturan Pemerintah Nomor 20 Tahun 2006 tentang Irigasi. Pasal 2.
[11] Rossiter, D. G., dan Wambeke, A. R. 1997. Automated Land Evaluation System ALES Version 4.65d Users Manual. Cornell University, Department of Soil, Crop and Atmospheric Sciences, SCAS Teaching Series No. T93-2 Revision 6.
[12] Sabouni, M. S., dan Mardani, M. 2013. Application of Robust Optimization Approach for Agricultural Water Resource Management under Uncertainty. Journal of Irrigation and Drainage Engineering, Vol. 139, No. 7, 571-581. ASCE.
[13] Undang-Undang Nomor 7 Tahun 2004 tentang Sumber Daya Air. Pasal 41.
[14] Watkins, D. W., dan McKinney, D. C. 1997. Finding robust solutions to water resources problems. Journal of Water Resource and Planning Management, Vol. 123, No.1 , 49-58.
[15] Yu, H. X., dan Jin, Li. 2012. An Brief Introduction to Robust Optimization Approach. International Journal of Pure and Applied Mathematics, Vol. 74, No.1, 121-124.

Selecting the best rayon in customer's perspective using fuzzy analytic hierarchy process

E G Sonjaya[1], E Paulus[2], A Hidayat[3]

Department of Computer Sciences, Universitas Padjadjaran, Bandung, Indonesia

Email: egsonjaya@gmail.com, erick.paulus@unpad.ac.id, akik@unpad.ac.id

Abstract: Annually, the best *Rayon* selection is conducted by the assessment team of PT.PLN (Persero) Cirebon with the goal to increase the spirit of company members in providing an improved service for customers. However, there is a problem in multiple criteria decision making in this case, which is the importance intensity of each criterion in the selection are often assessed subjectively. To solve this problem, *Fuzzy Analytical Hierarchy Process* are used to cover AHP scale deficiency in the form of *'crisp'* numbers. So, it should be considered to use *Fuzzy logic approach* to handle uncertainty. Fuzzy approach, especially triangular fuzzy number towards AHP scale, are expected to minimize the handling of subjective input, which then will make a more objective result. Thus, this research was conducted to help the management or assessment team in the selection of the best *Rayon* with a more objective selection in according to the company criteria.

Keyword: Multiple Criteria, Fuzzy Analytical Hierarchy Process, Triangular Fuzzy Number

1. Introduction

Rayon is a subunit under the Area which helps a closer customer service and electricity distribution network. The Rayon formation is based on the number of customers in the region, with a purpose to improve customer services. The quality and spirit given by the Rayon can help the progress of a company to actualize their vision and mission, based on interviews from the company.

The best Rayon selection is conducted by the assessment team, which includes the Head of Planning and Evaluation Division, and Area Manager. The best Rayon is chosen based on their achievement of targets determined by the company by various criteria. Each criterion has a different scale/intensity of urgency. The best Rayon selection is completed by manual calculation, which multiplying the urgency scale score with the Rayon score and then summed up. And then, the Rayon with the highest score will be chosen as the best Rayon.

Problems which are faced with this selection process is that the assessment team cannot select the best Rayon accurately because the score of every criterion are based on the assessment team subjectivity. Thus, the final result of the best Rayon is still uncertain *(fuzzy=blurred or unclear).* The inaccuracy of the score given by the assessment team will have an impact on the process of decision making, which will make the decisions less precise. These problems can be solved by building a Decision Support

System (DSS) to reduce the subjectivity of the assessment team in selecting the best Rayon. Therefore, the Fuzzy Analytical Hierarchy Process (F-AHP) method can be used to solve the problems.

F-AHP method solves the problem of thinking using fuzzy set theory concepts and hierarchical structure analysis [1]. Essentially, the F-AHP method is an improvement method of the AHP, which usually uses real number calculations, to the F-AHP which uses fuzzy number calculations. F-AHP is a more developed method from the AHP method as it is combined with the fuzzy concept approach. F-AHP is designed to cover the weaknesses of AHP, which lies in the problem of subjective criteria. Some fuzzy AHP implementation on college ranking [2], risk assessment [3] and the best metro station [4] is succeed and give promising results

The selection of the best Rayon which uses the F-AHP method is expected to help the Planning and Evaluation Division of PT.PLN (Persero) Cirebon Area to make the best decision that can be done fairly and precisely.

2. Research Methods

2.1. Fuzzy Analytical Hierarchy Process (FAHP)

F-AHP is a combination of the AHP method with a fuzzy concept approach [4]. Because the AHP concept basically excludes unclear personal considerations, AHP has been improved by using a fuzzy logic approach. In thc FAHP alternative criteria, comparison in pairs is indicated by linguistic variables. FAHP approach, Triangular Fuzzy Number (TFN) is used for falsification process from a crisp comparison matrix [5] [6]. Blurred data will be in the TFN. Each membership function is defined in 3 parameters which include I, M, and U, where I is the lowest possible score, M is the middle possible score, and U is the highest possible score at the decision interval for the decision makers. The score of I, M, and you can also be determined by the decision makers. In this article, the writers propose three parameters of fuzzy numbers to represent Saaty scale (1-9) according to their importance, i.e. [1]

$$\begin{aligned} \tilde{1} &\equiv (1,1,1) \\ \tilde{x} &\equiv (x-1, x, x+1)\ ;\ \forall\, x = 2,3,\dots,8 \\ \tilde{9} &\equiv (9,9,9) \end{aligned} \tag{1}$$

Triangular Fuzzy Number (TFN) can show the subjectivity of a paired computability or can show a definite degree of vagueness (uncertainty). In this case, linguistic variables can be used by the decision makers to represent the uncertainty of data, in case of any inconvenience with TFN. TFN and its linguistic variables corresponding to the Saaty scale are shown in the following table [1]:

Table 1. Triangular Fuzzy Numbers and linguistic variable

Definition	Saaty Scale	TFN
Equally Important	1	(1,1,1)
Moderately more important	3	(2,3,4)
Strongly More Important	5	(4,5,6)
Very strongly more important	7	(6,7,8)
Extremely more important	9	(9,9,9)
Intermediate Values	2,4,6,8	(1,2,3), (3,4,5), (5,6,7), and (7,8,9)

To do local priorities of the fuzzy pairwise comparison matrix, there are various methods developed by previous experts. By combining AHP procedures with arithmetic operations for fuzzy numbers, local priority can be obtained using the following equation [7]:

$$S_i = \sum_{j=i}^{m} M_{g_i}^j \otimes \left[\sum_{i=1}^{n}\sum_{j=1}^{m} M_{g_i}^j\right]^{-1} \qquad (2)$$

Where:
S_i = fuzzy synthetic extent
g_i = goal set (i = 1, 2, 3, …, n)
M_{gi}^{j} = Triangular Fuzzy Number (j = 1, 2, 3, …, m)

$$S_i = \left(\sum_{j=1}^{m} l_j ; \sum_{j=1}^{m} m_j ; \sum_{j=1}^{m} u_j\right) \otimes \left(\frac{1}{\sum_{i=1}^{n} u_i} ; \frac{1}{\sum_{i=1}^{n} m_i} ; \frac{1}{\sum_{i=1}^{n} l_i}\right) \qquad (3)$$

Where:
l = Lower bound score (Lowest possibility)
m = The most promising score (Middle possibility)
u = Upper bound score (Highest possibility)

The formula to determine the pairwise comparison score of Si:

$$V(S_i \geq S_k) = \begin{cases} 1; & m_{S_i} \geq m_{S_k} \\ 0; & l_{S_k} \geq u_{S_i} \\ \frac{(l_{S_k} - u_{S_i})}{\left((m_{S_i} - u_{S_i}) - (m_{S_k} - l_{S_k})\right)}; & lainnya \end{cases} \qquad (4)$$

Where:
$V(S_i \geq S_k)$ = Comparison score between *fuzzy synthetic extent*,
S_i = *fuzzy synthetic extent* score for i criteria,
S_k = *fuzzy synthetic extent* score for k criteria,

$$d'_i = min(V(S_i \geq S_k) \text{ ; for } k = 1, 2, 3, \ldots, n;\ k \neq i, \qquad (5)$$

$$W' = (d'_i, d'_{i+1}, d'_{i+2}, \ldots, d'_{i+n-1})^T; \qquad (6)$$

$$d_i = \frac{d'i}{\sum_{i=1}^{n} d'i} \qquad (7)$$

$$W = (d_i, d_{i+1}, d_{i+2}, \ldots, d_{i+n-1}) \qquad (8)$$

where:
d'_i = values of i criteria
W' = vector values criteria
d_i = values normalization
W = vector values normalization criteria
n = number of criteria

Arithmetic operation for fuzzy numbers can be seen from the following operation:

$$
\begin{aligned}
&1.\ \tilde{n}_1 \oplus \tilde{n}_2 = (\tilde{n}_{1l} + \tilde{n}_{2l}; \tilde{n}_{1m} + \tilde{n}_{2m}; \tilde{n}_{1u} + \tilde{n}_{2u}) \\
&2.\ \tilde{n}_1 \otimes \tilde{n}_2 = (\tilde{n}_{1l} \times \tilde{n}_{2l}; \tilde{n}_{1m} \times \tilde{n}_{2m}; \tilde{n}_{1u} \times \tilde{n}_{2u}) \\
&3.\ {}^{1}/_{\tilde{n}_1} = \left({}^{1}/_{\tilde{n}_{1u}}; {}^{1}/_{\tilde{n}_{1m}}; {}^{1}/_{\tilde{n}_{1l}}\right)
\end{aligned} \tag{9}
$$

While global priority is obtained by multiplying the normalization of each w_j criterion scale by normalizing $d(A_i)$ values and summing all the multiplication results of each criterion. The equation can be written as follows:

$$\tilde{P}_i = (\tilde{w}_1 \otimes d(A_1)) \oplus (\tilde{w}_2 \otimes d(A_2)) \oplus \cdots \oplus (\tilde{w}_j \otimes d(A_i)) \tag{10}$$

3. Result and Discussion

3.1. Data Collection Result

The interviews with the company results in several criterion and alternatives, which will be used for the best Rayon selection based on the customer perspective. It is then represented in the hierarchical structure. The problems formulated to build hierarchical structure are goal identification, criterion identification, and alternatives identification (rayons) to be assessed. The best Rayon hierarchy structure in customer perspective can be seen in the following picture:

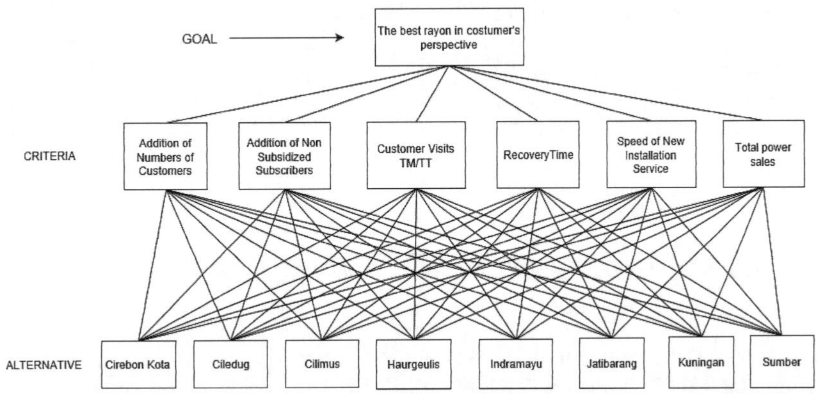

1. The result of interviews from the Head of Distribution Planning says that the best Rayon in the costumer's perspective selection consists of several criteria:
2. Addition of numbers of customers, rayon's achievement in achieving the target of increasing the number of subscribers, using units of subscribers.
3. Addition of non-subsidized subscribers, rayon's achievement in achieving the target of increasing the number of customers, especially for the level of not using subsidies, using units of subscribers.
4. Costumer visits TM/TT, rayon's achievement in achieving the target number of high voltage / high voltage customer visits, using units of subscriber numbers.
5. Recovery time, rayon's achievement in targeted customer service should not be extinguished within a certain time, using units of minutes and calculated based on target number / month.
6. Speed of new installation service, rayon's achievement in achieving the target of a new installation service time is calculated from the day of the new pairs of submissions until completion.
7. Total power sales, rayon's achievement in achieving the target of total sales of electricity.

IOP Conf. Series: Materials Science and Engineering **332** (2018) 012012 doi:10.1088/1757-899X/332/1/012012

3.2. Data Processing

3.2.1. Values Score Criteria. Before determining the matrix of pairwise comparisons between criteria, we first determine the *values* or scale of each criterion. The function of determining the importance, intensity or scale of each criterion is to avoid inconsistency. This is because when we input the score of the comparison between criteria into the matrix in pairs, it is often inconsistent. Therefore, to avoid inconsistency, we need to determine the intensity of interest for each criterion. The interview process is conducted by one person from the Head of Planning and Evaluation Division of the related company. The intensity of importance score is given by the company in the range of 1 to 9.

Table 2. Values score criteria on costumer perspective.

Criteria		Point value
K1	Addition of Numbers of Customers	7
K2	Addition of Non Subsidized Subscribers	6
K3	Customer Visits TM/TT	4
K4	Recovery Time	4
K5	Speed of New Installation Service	6
K6	Total Power Sales	8

3.2.2. Pairwise Comparison Matrix. The existence of value score can be directly calculated for the AHP pairwise comparison matrix for each criterion. The following is the result of pairwise comparison matrix calculation.

Table 3. Pairwise comparison matrix, criteria on costumer perspective

	K1	K2	K3	K4	K5	K6
K1	1	7/6	7/4	7/4	7/6	7/8
K2	6/7	1	6/4	6/4	6/6	6/8
K3	4/7	4/6	1	4/4	4/6	4/8
K4	4/7	4/6	4/4	1	4/6	4/8
K5	6/7	6/6	6/4	6/4	1	6/8
K6	8/7	8/6	8/4	8/4	8/6	1

Then, the score of AHP pairwise comparison matrix (table 4) is converted into a triangular fuzzy equation or Triangular Fuzzy Number (TFN). The F-AHP scale has three scores which include the lowest score (lower, I), the middle score (median, m), and the highest score (upper, u)

Table 4. TFN Pairwise comparison matrix criteria on costumer perspective

	K1			K2			K3			K4			K5			K6		
	L	M	U	L	M	U	L	M	U	L	M	U	L	M	U	L	M	U
K1	1	1	1	0,167	1,167	2,167	0,75	1,75	2,75	0,75	1,75	2,75	0,167	1,167	2,167	0,875	0,875	1,875
K2	0,857	0,857	1,857	1	1	1	0,5	1,5	2,5	0,5	1,5	2,5	1	1	1	0,75	0,75	1,75
K3	0,571	0,571	1,571	0,667	0,667	1,667	1	1	1	1	1	1	0,667	0,667	1,667	0,5	0,5	1,5
K4	0,571	0,571	1,571	0,667	0,667	1,667	1	1	1	1	1	1	0,667	0,667	1,667	0,5	0,5	1,5
K5	0,857	0,857	1,857	1	1	1	0,5	1,5	2,5	0,5	1,5	2,5	1	1	1	0,75	0,75	1,75
K6	0,143	1,143	2,143	0,333	1,333	2,333	1	2	3	1	2	3	0,333	1,333	2,333	1	1	1

3.2.3. F-AHP Calculation. The F-AHP calculation starts from the calculation of fuzzy synthesis score, fuzzy vector and ordinate score, F-AHP vector values, and value priority normalization which will be obtained a global value priority criterion, assisted by using software that is Microsoft excel and web-based application for supporting calculation. The result of F-AHP calculation as follows:

a. Fuzzy synthetic score (*Si*)

Table 5. Fuzzy synthetic (*Si*) on costumer perspective.

Criteria	Fuzzy synthetic score (*Si*)		
	L	*M*	*U*
K1	0,057	0,200	0,498
K2	0,071	0,171	0,415
K3	0,068	0,114	0,329
K4	0,068	0,114	0,329
K5	0,071	0,171	0,415
K6	0,059	0,229	0,541

b. F-AHP Vector Score (v) and defuzzy ordinate score (d')

This process applies a fuzzy approach which is a fuzzy minimum implication (min) function. After comparing fuzzy synthesis scores, defuzzified ordinate score will be obtained This process implements *fuzzy* approach which is a fuzzy minimum implication function with a minimum score of d'. The result of v and d' calculations on costumer perspective is as follows:

1. *d'(VsK1)* = min (1, 1, 1, 1, 0.938) = **0,938**
2. *d'(VsK2)* = min (0.925, 1, 1, 1, 0.860) = **0,860**
3. *d'(VsK3)* = min (0.760, 0.819, 1, 0.819, 0.701) = **0,701**
4. *d'(VsK4)* = min (0.760, 0.819, 1, 0.819, 0.701) = **0,701**
5. *d'(VsK5)* = min (0.925, 1, 1, 1, 0.860) = **0,860**
6. *d'(VsK6)* = min (1, 1, 1, 1, 1) = **1**

c. Fuzzy vector *values* score (w')

Based on the ordinate score of each criterion, the vector value score can be obtained as follows:

Costumer Perspective

$W' = (0.938,\ 0.860,\ 0.701,\ 0.701,\ 0.860,\ 1)^T$

$\sum W' = 5{,}068$

d. Fuzzy vector values score normalization (w)

Vector values score normalization is where each vector values elements are divided by the number of the vector values itself. It is where the number of values that has been normalized will be equal to 1. Normalization of fuzzy vector values scores criterion is equal to the global priority value score (which is the goal).

Costumer perspective

$W_{local} = (0.185,\ 0.170,\ 0.139,\ 0.139,\ 0.170,\ 0.198)$

3.2.4. Alternative Ranking and Decision Results. Alternative ranking is a step to determine the final decision. At this process, the activity that occurs is multiplying the local values (W) (Criterion *values*) with the alternative score and summed up with each alternative elements in the criterion-affected level. The sum of the score are ranked and resulted in the decision of the best Rayon in multi perspective in PT PLN (Persero) Cirebon Area. The following is the conclusion table of *values* criterion and conclusion of ranking using FAHP.

Table 6. Alternative ranking on customer perspective

	K1	K2	K3	K4	K5	K6	TOTAL	F-AHP RANKING
Cirebon Kota	17,523	17,005	13,865	13,865	17,005	18,407	97,672	4
Ciledug	15,739	17,005	13,865	13,865	17,005	19,734	97,214	5
Cilimus	17,549	17,005	13,865	13,865	17,005	19,065	98,355	2
Haurgeulis	12,599	17,005	13,865	13,865	17,005	19,734	94,074	6
Indramayu	11,173	14,908	13,865	13,865	17,005	19,734	90,550	8
Jatibarang	12,898	17,005	13,865	13,865	17,005	19,092	93,731	7
Kuningan	**18,526**	**17,005**	**13,865**	**13,865**	**17,005**	**18,682**	**98,948**	**1**
Sumber	17,475	17,005	13,865	13,865	17,005	18,575	97,791	3

It can be concluded that Kuningan Rayon has the highest score compared to other Rayons with a score of 98,948. Then the second position is Cilumus Rayon with a score of 98,355 and the third position is Sumber Rayon with a score of 97,791.

4. Conclusion

Fuzzy AHP can recommend an objective selection by recalculating the important intensity of six criteria to make a decision of the best Rayon. We have investigated nine Rayons based on six criterions as input value. The result shows that the best Rayon Selection of PT PLN (Persero) Cirebon Area on costumer perspective is the Kuningan Rayon with the highest final result scoring at 98,948. This research is also meant to be a recommendation to assist the planning and evaluation division to make decisions.

References

[1] Alias M. A., Hashim, S. Z., & Samsudin S. Int. J. Advance. Soft Comput Appl., vol. 1, no. 1. (2009).
[2] V. Agarwa, IEEE Int. Conf. Comput. Commun. Control. IC4 2015 (2016)
[3] J. Tian and Z. F. Yan, J. Appl. Res. Technol., vol. 11, no. 4, 568–577 (2013).
[4] K. Kepaptsoglou, M. G. Karlaftis, and J. Gkountis, KSCE J. Civ. Eng., vol. 17, no. 5, 1109–1116 (2013).
[5] X. Zeshui and L. Huchang, Fuzzy Syst. IEEE Trans., vol. 22, no. 4, 749–761 (2014).
[6] M. Tahri, M. Maanan, M. Maanan, H. Bouksim, and M. Hakdaoui, Prog. Phys. Geogr., vol. 41, no. 3, 268–285 (2017).
[7] Kahraman, C., Cebeci, U., & Ruan, D. Int. J. Production Economics 87, 171-184 (2004).

ECONOMIC ORDER QUANTITY (EOQ) OPTIMAL CONTROL CONSIDERING SELLING PRICE AND SALESMAN INITIATIVE COST

Elis Hertini*, NursantiAnggriani, Winda Mianna, Asep K Supriatna

Department of Mathematics Universitas Padjadjaran Bandung Indonesia
Email : *elis.hertini@unpad.ac.id

Abstract. Retailers usually offer several types of similar products. A larger number of available stock products in display space will lead consumer to buy more, as well as giving a negative impression on other types of less available products. However, the amount of display space is limited so capacity of carrying the products is limited. Competition among products to increase demand rate is influenced by stock levels available in display space, price and salesmen's initiative in promoting the products. The Economic Order Quantity (EOQ) to replenish the stock of the product is dependent on the on-hand inventory. Salesman's initiative also affects maximum profit obtained by the seller. In this paper, Potryagin's Maximal Principle is used to determine the state of the inventory levels response to control prices of products. Sensitivity analysis of capacity allocation display space is also presented numerically.

Keyword : EOQ Model, Selling Price, Salesmen's Initiative.

1. Introduction

Inventory has important meaning and role in a company. Without an inventory the company will be faced with the situation of the company's disruption and even discontinuity. Inventories are products that wait for further processing such as production activities on manufacturing systems and marketing activities in the distribution system [1]. Inventory usually occurs in the form of finished goods or goods to be used to fulfil a particular purpose [2]. The main function of inventory in a system is to ensure the smoothness of the fulfilment mechanism of product demand in accordance with the needs of consumers so that the system can achieve optimal performance. A well managed inventory also affects the decrease of production cost of a company. Managed systems can achieve optimal performance through production planning related to good inventory control over time [3].

Retail is a sale of small quantities of commodities to consumers. Retail usually sells some kind of similar product. The basic competition of the retail market is price competition, inventory and product promotion. Retail inventories are goods purchased for the purpose of resale. Retail inventory is closely related to the limitations of display space so that the amount of inventory is limited. The large amount of inventory in the showroom attracts consumers to buy more. Conversely supplying fewer than other similar products will lead to a negative assumption.

Differences in inventory levels affect each other at the level of consumer demand among similar products. The level of demand is the amount of goods consumers wanted at a given price level and time. The market balance can be defined as a situation where at the price level formed, the consumer can buy the desired quantity of product, and the seller can sell the desired quantity of product [4]. Generally, when the amount of inventory exceed the number of requests, the cost of storage will increase, while when the price is below the equilibrium price the amount of demand will exceed the inventory resulting in stock out.

In addition to price and quantity of product availability, product promotion to consumers also affects the level of consumer demand. Many companies involve salesmen to promote their products, to increase demand for these products. In general, many consumer goods inventory displayed in shopping centers will attract customers to buy more, which means that customers have the scope to choose better design or product quality. Customer satisfaction cannot be separated from the efforts of the sales team [5].

Inventory control is an important activity that needs special attention from retail companies. It determines the optimal inventory amount with minimum total cost. In addition to price and quantity of product availability, product promotion to consumers also affects the level of consumer demand. With optimal product and salesman prices, proper inventory control can maximize retail profit. The inventory control model that gives an important role is the Economic Order Quantity (EOQ) model. The EOQ model is used to determine the optimal order quantity of a product. This optimum order quantity leads to minimal inventory costs so as to provide maximum benefits. Technically, EOQ inventory model is a model of procurement or supply of raw materials in a company. EOQ model to determine the number of orders that meet the total cost so there is no shortage of inventory has been done by [6]. Most EOQ models consider only a single product. However, in reality a producer might be produces more than one similar products or a retail might be sells more than one similar products [7].

In modeling two similar products, the author in [7] assumes that the first product demand is positively affected by the joint venture and is negatively affected by the inventory stock of the two similar products. The same assumption is used by the author for two similar products but with different models [5]. In other paper the author discusses the dynamic analysis and optimal control on the supply control model of two different products with limited production capacity as well as joint sales team initiatives [8]. Research with inventory control model of two different products on limited product capacity as well as sales team initiatives with dynamic system approach has been done, e.g [3], other researches studies about optimal controls with sales team initiatives not only during equilibrium conditions, but optimal control with the sales team effort at any time [9]. The authors in [10] extend the model to consider more than two products and also consider different form of production growth model [11]. In this paper we discuss optimal controls for product price forecasts and salesman maximizing revenue. The Pontryagin Maximum Principle is used in solving this optimal control problem as in [12].

2. Model Formulation

A similar product is defined as a product of a uniform nature. Here a similar product is limited to two types of similar products. The establishment of EOQ models of two products in a retailer's showroom for limited inventory storage, product demand is influenced by the level of product inventory in the showroom. Salesman Initiative in EOQ model of two similar products takes form using salesmen promotion media. Some of the parameters that influence the formation of EOQ models with supply-dependent demand, pricing and salesmen initiatives from two similar products can be seen in Table 1.

Table 1. Parameters that Influence the Formation of the EOQ Model

Notation	Information
D_x	Level of product demand 1
D_y	Level of product demand 2
E_x	Salesman initiative product 1
E_y	Salesman initiative product 2
R_x	Replenishment rate product 1
R_y	Replenishment rate product 2
r_x	Product replenishment rate product 1
r_y	Product replenishment rate product 2
L_x	Maximum storage capacity of the product 1
L_y	Maximum storage capacity of the product 2
L	Maximum storage capacity of both products ($L = L_x + L_y$)
S_x^{max}	The upper limit of the selling price per unit of product 1 when there is no demand
S_y^{max}	The upper limit of the selling price per unit of product 2 when there is no demand
$\delta = (r - i)$	*r* and *i* represent the interest rate and inflation of currency unity
l_1	Comparison of the level of salesmen initiative with the level of product demand 1 when the maximum salesman initiative level
l_2	Comparison of the level of salesmen initiative with the level of product demand 2 when the maximum salesman initiative level
l_3	Comparison of inventory levels with product demand level 1 when the maximum salesman initiative level
l_4	Comparison of inventory levels with product demand level 2 when the maximum salesman initiative level

2.1. Mathematical Model

The level of inventory is influenced by the level of production and demand level of each product. For example, for the production level, the amount of product inventory 1 is denoted by *X* and product 2 is denoted *Y*. The EOQ model is an inventory model in which demand depends on the level of inventory at time, $(X(t), Y(t))$, the sales price of both products $(S_x(t), S_y(t))$, and product marketing effort at time *t* by salesmen initiative $(E(t))$. The product inventory level should not be zero or negative, and should not exceed the capacity of the display space provided for each product (L_x and L_y), such that

$$0 < X(t) \leq L_x \tag{1}$$

and

$$0 < Y(t) \leq L_y \tag{2}$$

The general model used to describe production is the logistic growth model,

$$D_x = \frac{C_x E_x (1 - \frac{Y}{L})}{l_1 E + l_2 X}, \tag{3}$$

and

$$D_y = \frac{C_y E_y (1 - \frac{X}{L})}{l_3 E + l_4 Y}, \tag{4}$$

where

$$C_x = (1 - \frac{S_x}{S_x^{max}}), \tag{5}$$

and

$$C_y = \left(1 - \frac{S_y}{S_y^{max}}\right), \tag{6}$$

is the potential growth rate of each product. Like the production level, the demand level is also assumed to follow the logistic equation, in which the replenishment rate of product 1 and product 2 depends on the level of inventory,

$$R_x = r_x\left(1 - \frac{X}{L_x}\right)X, \tag{7}$$

and

$$R_y = r_y\left(1 - \frac{y}{L_y}\right)y\,, \tag{8}$$

where r_x and r_y are positive constants called the intrinsic growth rate.

The change in the level of product inventory at time *t* is the difference in the inventory level of the product at time *t* with the previous inventory level. The demand level is affected by the sales initiative, then for product 1, $\frac{\partial D_x}{\partial N} = \frac{\tau_x X}{L_x{}^2}$ will be positive for every X∈ (0, ∞) and $\lim_{N\to\infty} D_x \to \tau_x X$ so that $D_x \in [0, \tau_x, X]$. This means the level of product demand under conditions of infinite sales initiative influence still results in limited demand. This applies equally to product 2, and hence the differential equations for supplies that available at time *t* for product 1 and product 2 respectively given by:

$$\dot{X} = r_x\left(1 - \frac{X}{L_x}\right)X - \frac{C_x E_x(1-\frac{Y}{L})}{l_1 E + l_2 X} = G_1(X, Y), \tag{9}$$

and

$$\dot{Y} = r_y\left(1 - \frac{Y}{L_y}\right)Y - \frac{C_y E_y\left(1-\frac{X}{L}\right)}{l_3 E + l_4 Y} = G_2(X, Y). \tag{10}$$

3. Model Analysis

3.1. Model Stability Analysis

The fixed point or balance point is obtained when the inventory level change of each product is at the zero level, i.e. the condition in which the balance between the level of consumer demand and the level of demand for the product is met. In other words when the market equilibrium is reached. The equilibrium of the system is said to be stable if all eigenvalues of the system are negative [14]. In Equilibrium, the system of dynamic equations (9) and (10) at a fixed point (x^*, y^*) meet the the following conditions:

$$r_x\left(1 - \frac{x^*}{L_x}\right)x^* = \frac{C_x E_x^*(1-\frac{y^*}{L})}{l_1 E + l_2 x^*}, \tag{11}$$

and

$$r_y\left(1 - \frac{y^*}{L_y}\right)y^* = \frac{C_y E_y^*(1-\frac{x^*}{L})}{l_3 E + l_4 y^*}. \tag{12}$$

The stability behaviour of the fixed points (x^*, y^*) of the inventory model will be analysed locally and globally as the following.

Local Stability

From the system of dynamic equations (9) and (10) we find the Jacobian matrix at the fixed point (x^*, y^*):

$$A(x^*, y^*) = \begin{bmatrix} x^*\left(-\frac{r_x}{L_x} + \frac{C_x l_2 E(1-\frac{y^*}{L})}{(l_1 E + l_2 x^*)^2}\right) & x^*\left(\frac{C_x E}{L(l_1 E + l_2 x^*)}\right) \\ y^*\left(\frac{C_y E}{L(l_3 E + l_4 y^*)}\right) & y^*\left(-\frac{r_y}{L_y} + \frac{C_y l_4 E(1-\frac{x^*}{L})}{(l_3 E + l_4 y^*)^2}\right) \end{bmatrix}. \tag{13}$$

After obtaining the Jacobian matrix from the dynamic system, we find the characteristic equation with $P(\lambda) = det(A(x^*, y^*) - \lambda I) = 0$, i.e.:

$$P(\lambda) = det\begin{bmatrix} x^*(-\frac{r_x}{L_x}+\frac{C_x l_2 E(1-\frac{y^*}{L})}{(l_1E+l_2x^*)^2}) & x^*(\frac{C_x E}{L(l_1E+l_2x^*)}) \\ y^*(\frac{C_y E}{L(l_3E+l_4y^*)}) & y^*(-\frac{r_y}{L_y}+\frac{C_y l_4 E(1-\frac{x^*}{L})}{(l_3E+l_4y^*)^2}) \end{bmatrix} = 0. \quad (14)$$

The characteristic equation of $A(x^*,y^*)$ is given by

$$P(\lambda) = \lambda^2 - \lambda\psi_1(x^*,y^*) + \psi_2(x^*,y^*) = 0, \quad (15)$$

where,

$$\psi_1(x^*,y^*) = \frac{\partial}{\partial x}(G_1(x^*,y^*)) + \frac{\partial}{\partial y}(G_2(x^*,y^*))$$

$$= x^*\left[\left(-\frac{r_x}{L_x}+\left(-\frac{r_x}{L_x}+\frac{C_x l_2 E\left(1-\frac{y^*}{L}\right)}{(l_1E+l_2x^*)^2}\right)\frac{C_x l_2 E\left(1-\frac{y^*}{L}\right)}{(l_1E+l_2x^*)^2}\right)+y^*\left(-\frac{r_y}{L_y}+\frac{C_y l_4 E\left(1-\frac{x^*}{L}\right)}{(l_3E+l_4y^*)^2}\right)\right]$$

and (16)

$$\psi_2(x^*,y^*) = \frac{\partial}{\partial x}(G_1(x^*,y^*))\frac{\partial}{\partial y}(G_2(x^*,y^*)) - \frac{\partial}{\partial y}(G_1(x^*,y^*))\frac{\partial}{\partial x}(G_2(x^*,y^*))$$

$$= \left[\left\{x^*\left(-\frac{r_x}{L_x}+\frac{C_x l_2 E\left(1-\frac{y^*}{L}\right)}{(l_1E+l_2x^*)^2}\right)\right\}\left\{y^*\left(-\frac{r_y}{L_y}+\frac{C_y l_4 E\left(1-\frac{x^*}{L}\right)}{(l_3E+l_4y^*)^2}\right)\right\} - x^*y^*\left\{\frac{C_x E}{L(l_1E+l_2x^*)}\right\}\left\{\frac{C_y E}{L(l_3E+l_4y^*)}\right\}\right]$$

The point (x^*,y^*) is stable if the eigenvalues (λ) of equations (14) and (15) are negative. We note that two eigenvalues will be negative if the terms $\psi_1(x^*,y^*) < 0$ and $\psi_2(x^*,y^*) > 0$ are fulfilled.

Global Stability

Global stability analysis is done using the Liapunov Second Method approach. This method is similar to EOQ method in investigating nonlinear system. The Liapunov method corresponding to the dynamic function of equations (4) and (5) in [9], i.e.

$$V(X,Y) = \left[(X-x^*) - x^*\ln\left(\frac{X}{x^*}\right)\right] + h\left[(Y-y^*) - y^*\ln\left(\frac{Y}{y^*}\right)\right] \quad (17)$$

where h is constant for Liapunov function. The definitive positive V function and continuous continuous derivative at R_2^+ remain asymptotically stable (x^*,y^*) where $\dot{V} = \frac{\partial V}{\partial X}\frac{dX}{dt} + \frac{\partial V}{\partial Y}$. This can be written in the form $\dot{V} = [X - x^* \;\; Y - y^*]\, P \begin{bmatrix} X & x^* \\ Y & y^* \end{bmatrix}$ with $P = \begin{bmatrix} a_{11} & a_{12} \\ a_{21} & a_{22} \end{bmatrix}$, where

$$a_{11} = -\frac{r_x}{L_x} - \frac{C_x l_2 E\left(1-\frac{y^*}{L}\right)}{(l_1E+l_2X)(l_1E+l_2x^*)} \quad (18)$$

$$a_{12} = \frac{1}{2}\left\{\frac{C_x E(l_1E+l_2x^*)}{L(l_1E+l_2X)(l_1E+l_2x^*)} + \frac{hC_y E}{L(l_3E+l_4Y)((l_3E+l_4y^*)}\right\} = a_{21} \quad (19)$$

$$a_{22} = h\left[-\frac{r_y}{L_y} - \frac{C_x l_4 E\left(1-\frac{y^*}{L}\right)}{(l_3E+l_4Y)((l_3E+l_4y^*)}\right]. \quad (20)$$

The characteristic equation of the matrix P with $det(P - \lambda I) = 0$ is $\lambda^2 - \lambda(a_{11} + a_{22}) + (a_{11}a_{22} - (a_{12})^2)$ =0. The equilibrium point (x^*,y^*) is stable if the eigen value (λ) is negative. The two eigenvalues will be negative if the conditions $a_{11} + a_{22} < 0$ and $a_{11}a_{22} - (a_{12})^2 > 0$ are fulfilled.

3.2. The Optimal Control

The main goal of a company is to get the maximum amount of profit. The profit function that can be obtained by retailers is

$$\pi(X,Y,S_x,S_y,E,t) = (S_x - p_x)\left\{\frac{C_xE_x(1-\frac{Y}{L})}{l_1E+l_2X}\right\} + (S_y - p_y)\left\{\frac{C_yE_y\left(1-\frac{X}{L}\right)}{l_3E+l_4Y}\right\} - h_xX - h_yY - \gamma E \quad (21)$$

where π is the total profit function of product 1 and product 2 minus product storage cost and product marketing cost by the salesman. The parameters p_x and p_y are the cost to buy a unit of product 1 and product 2 while h_x and h_y are the storage costs for product 1 and product 2, and γ is the marketing cost incurred by the salesman.

To maximize the net income one must optimize the inventory levels of each product by taking into account the influence of the price of each product and the salesman's initiative. Using the Maximum Pontryagin principle [12], we will look for price controls for each product and sales initiative, S_x, S_y, E, by optimizing the product 1 and product 2, according to the dynamic equation system (9) and (10). Here is the optimal control problem as the Maximum Pontryagin principle [15],

$$maks\ PV = \int_0^T \pi(X,Y,S_x,S_y,E,t)e^{-\delta t}dt. \quad (22)$$

Subject to,

$$\dot{X} = r_x\left(1-\frac{X}{L_x}\right)X - \frac{C_xE_x(1-\frac{Y}{L})}{l_1E+l_2X}$$

$$\dot{Y} = r_y\left(1-\frac{Y}{L_y}\right)Y - \frac{C_yE_y(1-\frac{X}{L})}{l_3E+l_4Y}$$

$$0 \le S_x(t) \le S_x^{max}$$

$$0 \le S_y(t) \le S_y^{max}$$

$$0 \le E(t) \le E_{max}$$

$S_x^{max}, S_x^{max}, and\ E_{max}$ are the upper limit for the marketing efforts of both salesmen and the selling price of each product. To solve the maximum problem use the Hamiltonian equation

$$H = \pi(X,Y,S_x,S_y,E,t)e^{-\delta t} + \lambda_1(t)G_1(X,Y,E) + \lambda_2(t)G_2(X,Y,E) \quad (23)$$

This is equivalent to the following system:

$$H = \left[(S_x - p_x)\left\{\frac{C_xE_x(1-\frac{Y}{L})}{l_1E+l_2X}\right\} + (S_y - p_y)\left\{\frac{C_yE_y(1-\frac{X}{L})}{l_3E+l_4Y}\right\} - h_xX - h_yY - \gamma\right]e^{-\delta t}$$

$$+\lambda_1(t)\left[r_x\left(1-\frac{X}{L_x}\right)X - \frac{C_xE_x\left(1-\frac{Y}{L}\right)}{l_1E+l_2X}\right] + \lambda_2(t)\left[r_y\left(1-\frac{Y}{L_y}\right)Y - \frac{C_yE_y(1-\frac{X}{L})}{l_3E+l_4Y}\right] \quad (24)$$

where $\lambda_1(t)$ and $\lambda_2(t)$ are co-state variables. Optimal conditions are in accordance with the Maximum Pontryagin principle [13], if

$$\frac{\partial H}{\partial E} = 0, \frac{\partial H}{\partial S_x} = 0, \frac{\partial H}{\partial S_y} = 0\ dan - \frac{d\lambda_1}{dt} = \frac{\partial H}{\partial X}, -\frac{d\lambda_2}{dt} = \frac{\partial H}{\partial Y} \quad (25)$$

The value of Lagrange multipliers $\lambda_1(t)$ and $\lambda_2(t)$ as $t \to \infty$ are:

$$\lambda_1{}^*(t) = \left(\frac{Q_1}{\delta^2-\psi_1\delta+\psi_2}e^{-\delta t}\right) \quad (26)$$

and

$$\lambda_2{}^*(t) = \left(\frac{Q_2}{\delta^2-\psi_1\delta+\psi_2}e^{-\delta t}\right) \quad (27)$$

Substituting equations (26) and (27) into equation (25), we obtain the following equation,

$$\left[(S_x^* - p_x)\left\{\frac{C_xl_2x^{*2}\left(1-\frac{y^*}{L}\right)}{(l_1E+l_2x^*)^2}\right\} + (S_y^* - p_y)\left\{\frac{C_yl_4y^{*2}\left(1-\frac{x^*}{L}\right)}{(l_3E+l_4y^*)^2}\right\} - y\right]e^{-\delta t} -$$

$$\left[\frac{Q_1}{\delta^2-\psi_1\delta+\psi_2}e^{-\delta t}\right]x\left[\frac{C_xl_2x^{*2}\left(1-\frac{y^*}{L}\right)}{(l_1E+l_2x^*)^2}\right] - \left[\frac{Q_2}{\delta^2-\psi_1\delta+\psi_2}e^{-\delta t}\right]\left[\frac{C_yl_4y^{*2}\left(1-\frac{x^*}{L}\right)}{(l_3E+l_4y^*)^2}\right] = 0 \quad (28)$$

$$S_x^* = \frac{1}{2}\left[S_x^{max} + p_x + \frac{Q_1}{\delta^2-\psi_1\delta+\psi_2}\right] \quad (29)$$

$$S_y^* = \frac{1}{2}\left[S_y^{max} + p_y + \frac{Q_2}{\delta^2-\psi_1\delta+\psi_2}\right] \quad (30)$$

We conclude that under optimal conditions, to obtain the maximum net profit from the EOQ model for two similar products, the equations (11), (12), (28), (29) and (30) should be satisfied.

4. Numerical Analysis

The system will be simulated numerically. As illustration is given the following example with the parameter value in the numerical simulation taken from [8] where $r_x = r_y = 10$, $p_x = \$20$, $p_y = \$15$, $S_x^{max} = \$40$, $S_y^{max} = \$35$, $L = 100\ unit$, $L_x = 50\ unit$, $L_y = 50\ unit$, $\gamma = \$2$, $h_x = h_y = \$0.5$, $l_1 = 0{,}4$, $l_2 = 0{,}3$, $l_3 = 0{,}5$, $l_4 = 0{,}3$. inflation rate $i = 11\%$ and interest rate r = 16%, so $\delta = r\text{-}i$ = 0,05. The value of the salesman's initiative in promoting both products is taken form [7], i.e. 37.14510186, with the selling price for product 1 is $ 30,000,000 and product 2 is $ 25,000,000. By solving the equations numerically using Maple 15, the optimal solution of the inventory levels are obtain. The best conditions of inventory levels that maximize profitability are 49 unit both for product 1 and product 2. Selling at chosen price of each product, and promoting both products simultaneously with the chosen salesman initiatives, the optimal inventory of each product will maximize profit from retailers.

The stability criterion at the equilibrium point can be illustrated by the trajectory of the product 1 and product 2 areas, in Figure. 1. The figure shows that the direction of the trajectory to a certain point. Stability point at (49.20351407; 45,191271150). Both eigenvalues at the equilibrium point remain negative, i.e. -9.6 and -9.9. The eigenvalues show that the stability point is in stable condition (Stable Node). The bottom line is the fact that the levels of product inventory and production capacity meet the consumer demand and also maximize earnings. Furthermore, the graphs in Figures 2 show the effect of controls performed by the sales initiatives with regard to the maximum net income over time. The figure shows that the initial inventory level of both products is 49.7 units. To achieve the optimum stable conditions it takes 0.5 units of time. To maximize retail profit at the inventory level of 149.120351407 units and 249,190,150 units, the optimum order quantity (EOQ) of product 1 is 7.837981333 and product 2 is 7.956476643.

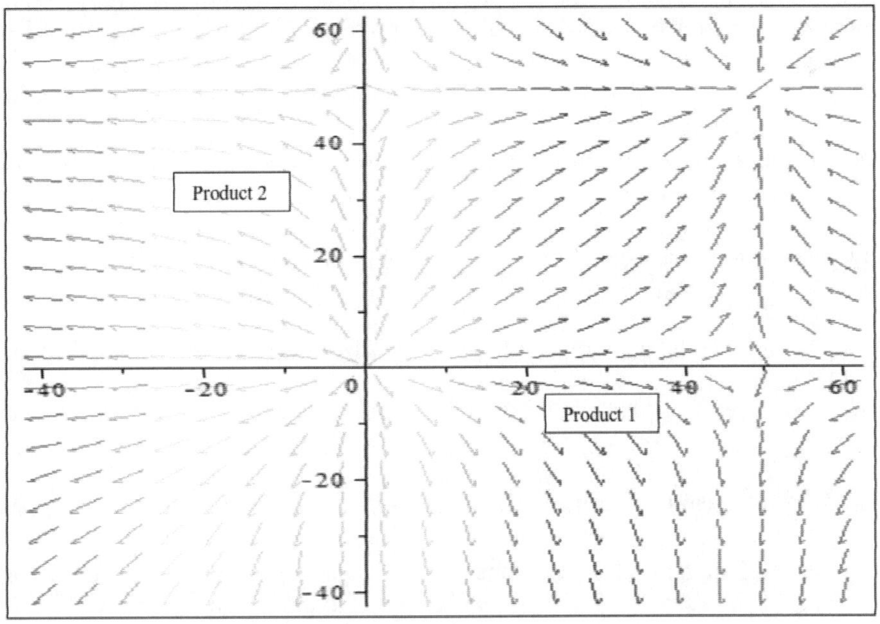

Figure 1. Phase potrait of product 1 and product 2.

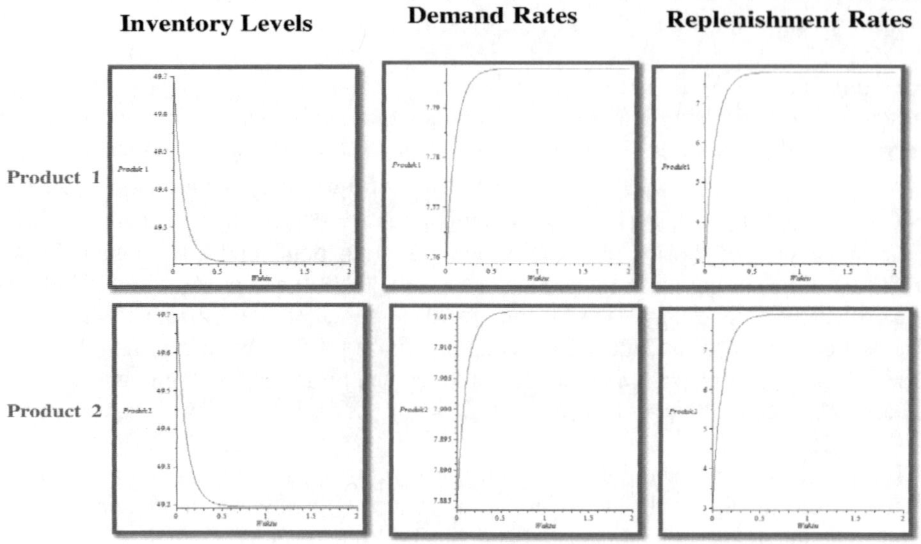

Figures 2. The Inventory Level, Demand Rates and EOQ for Product 1 and Product 2 over time

5. Conclusion

The inventory control of the EOQ model depends on inventory levels, product pricing and salesman initiatives from two similar products that use a dynamic systems approach. Using a dynamic system model, the results obtained for the inventory model of the two similar products, is the difference between the production rate limited by the production capacity, and the level of demand limited by the salesman's initiative. After obtaining a dynamic system model for fixed points (x^*, y^*) inventory levels of two similar products, a condition of equilibrium between the level of consumer demand and the level of product filling when market equilibrium is reached, local stable conditions are achieved when the condition $\psi_1(x^*, y^*) < 0$ and $\psi_2(x^*, y^*) > 0$, and stable global conditions are reached when $a_{11} + a_{22} < 0$ and $a_{11}a_{22} - (a_{12})^2 > 0$. The optimal condition of the product price and the salesman's initiative in maximizing the net profit from the inventory model of two similar products will be achieved, if they satisfy (11), (12), (28), 29, and (30).

Acknowledgments
The primery work is carried out with funding from PUPT RISTEKDIKTI 2017, with contract number 718/UN6.3.1/PL/2017.

References

[1] Rangkuti A 2013 *7 Model Riset Operasional & Aplikasinya* (cetakan ke-1) Brilian Internasional Surabaya

[2] Levin R I, McLaughlin C P, Lamone R P, Kottas J F 1972 *Productions Operations Management: Contemporary Policy for Managing Operating Systems* New York : McGraw-Hill

[3] Nursanti A 2015 Analisis Dinamik pada Model Pengendalian Dua Produk Berbeda dengan Kapasitas Produksi Terbatas serta Inisiatif Tim Sales Bersama *Jurnal Teknik Industri* Vol 17 No 1 Universitas Kristen Petra. Surabaya.

[4] Gaspersz Vincent 1999 *Ekonomi Manajerial: Pembuatan Keputusan Bisnis* Jakarta : Gramedia

Pustaka Utama.
[5] Sana S S 2011 An EOQ model of homogeneous products while demand is salesmen's initiatives and stock sensitive *Computers & Mathematics with Applications* no[62 p577–587.
[6] Kostic K 2009 Inventory Control as a Discrete System Control for the Fixed-Order Quantity System, *Applied Mathematical Modelling*, 33, pp. 4201-4212.
[7] Sana S S 2011 An EOQ model for salesmen's initiatives, stock and price sensitive demand of similar products – A dynamical system. *Applied Mathematics and Computation* no 218 p3277–3288.
[8] Sana S S 2012 The EOQ Model-A Dynamical System *Applied Mathematics and Computation* 2(18) 2012 pp 8736–8749.
[9] Al Fauzi AL, Darti I, and Suryanto A 2017 Kontrol Optimal pada Model *Economic Order Quatity* dengan Inisiatif Tim Penjualan, *Jurnal Teknik Industri* Vol 19, No 1 Universitas Kristen Petra Surabaya
[10] Husniah H and Supriatna AK 2015 Model Dinamis Persediaan Tiga Produk dengan Penjualan Bersama, *Jurnal Industrial Services* 1(1) pp. 1-6.
[11] Hennie Husniah, Nursanti Anggriani, Syifa Khairani, Islami N F, A.K. Supriatna 2015 Model Dinamis Tingkat Persediaan Dua Jenis Stock dengan Laju Produksi Sigmoid dan Penjualan Bersama, *Proceeding Seminar Sistem Produksi* XI
[12] Pontryagin, L. S., V. G. Boltyanskii., R. V. Gamkrelize., E. F. Mishchenko 1962 *The Mathematical Theory of Optimal Processes* New York : Wiley.
[13] Rao S S 1989 *Optimization: Theory and Application*.Canada: Wiley Easterm Limited.
[14] Wiggins S 1990 Introduction to Appled Nonlinier Dynamical Systems and Chaos Springer-Verlag New York
[15] Lenhart, S. and Workman J 2007 *Optimal Control Applied to Biological Models* Chapman and Hall/ CRC London

IOP Conf. Series: Materials Science and Engineering **332** (2018) 012014 doi:10.1088/1757-899X/332/1/012014

A genetic algorithm application in backcross breeding problem

E Carnia*, H Napitupulu, and A K Supriatna

Department of Mathematics, Faculty of Mathematics and Natural Sciences, Universitas Padjadjaran, Jatinangor 45363, Sumedang, Jawa-Barat, Indonesia

*Email: ema.carnia@unpad.ac.id

Abstract. In this paper we discuss a mathematical model of goat breeding strategy, i.e. the backcrossing breeding. The model is aimed to obtain a strategy in producing better variant of species. In this strategy, a female (doe) of a lesser quality goat, in terms of goat quality is bred with a male (buck) of an exotic goat which has a better goat quality. In this paper we will explore a problem on how to harvest the population optimally. A genetic algorithm (GA) approach will been devised to obtain the solution of the problem. We do several trials of the GA implementation which gives different set of solutions, but relatively close to each other in terms of the resulting total revenue, except a few. Further study need to be done to obtain GA solution that closer to the exact solution.

1. Introduction

In this paper we discuss a mathematical model of goat breeding strategy, i.e. the backcrossing breeding. The model first appeared in [1] and explored further in [2]. The backcrossing breeding strategy is aimed to obtain a better variant of species. In this strategy, a female (doe) of a lesser quality goat, in terms of goat quality such as meat and milk quality, is bred with a male (buck) of an exotic goat which has a better goat quality (a *fullblood*). It is a common practice in Indonesia to breed a doe of Kambing Kacang / *Capra aegagrus hircus*. with an exotic male goat from other variant of the same subspecies (such as Boer goat or Jamnapari goat). The resulting offspring of the first generation is called filial 1 (F1). When the F1 is a female, then she is bred with other male exotic goat of the same variant to produce an F2 female which then subsequently is bred with another male exotic goat of the same variant to produce F3. This process goes on to produce better quality of goat which is often called a *purebred* goat. It is easy to show that lim F*n* → *fullblood*, in terms of quality as *n* approaches infinity. In fact the process is faster to follow the sequence 50% (½), 75% (¾), 88% (7/8), 94% (15/16), 97% (31/32), for $n = 1,2,3,4,5$, respectively, hence the F5 is regarded as a good quality goat (a *purebred*, which the appearence is very closed to the *fullblood*). In this paper we adopt the model in [1], which has a schematic diagram as in Figure 1.

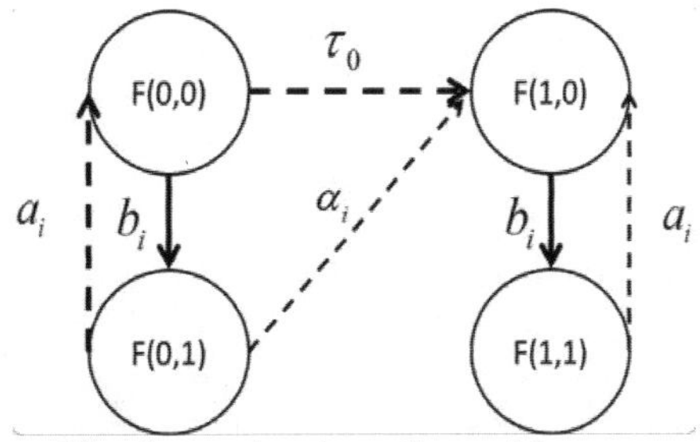

Figure 1. Schematic diagram of backcrossing breeding with two generations and two age-classes (taken from [1])

By refering to Figure 1, suppose that the number of does for each age-class is known at $t = 0$, with:

- $F_t(0,0)$ is the number of lesser quality or local does belong to the first age-class;
- $F_t(0,1)$ is the number of lesser quality or local does belong to the second age-class;
- $F_t(1,0)$ is the number of hybrid does belong to the first age-class;
- $F_t(1,1)$ is the number of hybrid does belong to the second age-class.

The authors in [1] argued that the number of does at time $t = 0$ is $F_{t=0}(G,T) = F_t(0,0) + F_t(0,1) + F_t(1,0) + F_t(1,1)$. In a more compact form, this can be written as a vector $\boldsymbol{F_{t=0}}(G,T) = [F_t(0,0), F_t(0,1), F_t(1,0), F_t(1,1)]^T$. The vector $\boldsymbol{F_t}(G,T)$ is called the initial distribution vector at time t. Further, by the same argument, $F_{t+1}(G,T) = F_{t+1}(0,0) + F_{t+1}(0,1) + F_{t+1}(1,0) + F_{t+1}(1,1)$, or more compactly can be written in the vector form $\boldsymbol{F_{t+1}}(G,T) = [F_{t+1}(0,0), F_{t+1}(0,1), F_{t+1}(1,0), F_{t+1}(1,1)]^T$, with:
$F_{t+1}(0,0) = \tau_0 F_t(0,0) + a_1 F_t(0,1)$; $F_{t+1}(1,0) = \tau_0 F_t(0,0) + \alpha_1 F_t(0,1) + \tau_0 F_t(1,0) + a_2 F_t(1,1)$;
$F_{t+1}(0,1) = b_1 F_t(0,0) + b_r F_t(0,1)$; $F_{t+1}(1,1) = b_2 F_t(1,0) + b_r F_t(1,1)$.
Hence the population growth model can be written in the matrix form as

$$\begin{bmatrix} F_{t+1}(0,0) \\ F_{t+1}(0,1) \\ F_{t+1}(1,0) \\ F_{t+1}(1,1) \end{bmatrix} = \begin{bmatrix} \tau_0 & a_1 & 0 & 0 \\ b_1 & b_r & 0 & 0 \\ \tau_0 & \alpha_1 & \tau_0 & a_2 \\ 0 & 0 & b_2 & b_r \end{bmatrix} \begin{bmatrix} F_t(0,0) \\ F_t(0,1) \\ F_t(1,0) \\ F_t(1,1) \end{bmatrix}, \quad (1)$$

which can also be written in a more compact form $\boldsymbol{F_{t+1}(G,T)} = \boldsymbol{AF_t(G,T)}$. The model is easy to solve iteratively to give the solution for the next n year is, i.e. $\boldsymbol{F_{t+n}}(G,T) = \boldsymbol{A}^n \boldsymbol{F_t}(G,T)$.

In this paper we pose a problem on how to harvest the population optimally. If Q_{1k}, Q_{2k}, Q_{3k}, and Q_{4k} kis the numbers of individual form sub population $F_k(0,0)$, $F_k(0,1)$, $F_k(1,0)$, and $F_k(1,1)$, respectively, then the problem is to maximize the profit Π along the time horizon T, with

$$\Pi = \sum_{k=0}^{T} (P_1 Q_{1k} + P_2 Q_{2k} + P_3 Q_{3k} + P_4 Q_{4k}) \rho^k,$$

and p_i is the associated unit price of the sub population i in the profit function above. The discounting factor is represented by ρ. We use genetic algorithm approach [3] to devise the solution of the problem.

2. Genetic Algorithm

Genetic algorithms (GA) was found by John Holland and developed by David Golberg. GA is numerical optimization algorithms inspired by both natural selection and natural genetics [3]. It is regarded as a heuristic approach. Some advantages of genetic algorithm compared to other optimization methods are: the direct, parallel, and powerful technique which can be used to optimize continuous or discrete variables, free derivative information, simultaneously searches from a large domain of objective function, deals with a large number of variables, optimizes variables with extremely complex objective function, provides several optimum solutions instead of one and so on [1,4,5]. This heuristic approach is applicable to many fields of optimization case, which makes GA attractive for many researchers [6]. Following is the basic algorithm of GA as suggested by Golberg in flowchart of Figure 2.

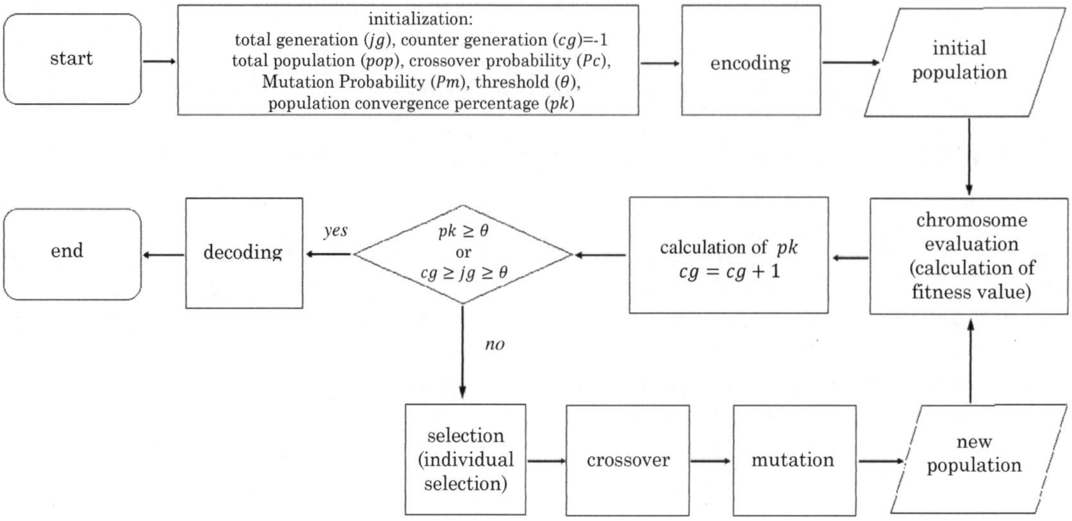

Figure 2. Genetics Algorithm flowchart developed by David Golberg

3. Genetic Algorithm implemented in Backcross Breeding Matrix

As explained earlier, in this research, our attention is directed to the population growth model given in [1]. In [1] it is explained that

a_i= Upgrading birth rate with $i = 0,1,2,\dots,n$
α_i= Non-Upgrading birth rate $i = 0,1,2,\dots,n$
b_i= Survival rate with $i = 0,1,2,\dots,n-1$
b_r= Survival rate of the last class
τ_0= Birth rate of the 1st age class

Let $\rho = \frac{1}{(1+r)}$, where r is the rate of return. Suppose that

$$\begin{aligned} Q_{ik} &= 0,1,2,3,\dots \\ Q_{1k} &\leq F_k(0,0) \\ Q_{2k} &\leq F_k(0,1) \\ Q_{3k} &\leq F_k(1,0) \\ Q_{4k} &\leq F_k(1,1) \end{aligned} \tag{2}$$

then, our consideration is to solving the following problem,

$$\max\left(\Pi = \sum_{k=0}^{T}(P_1Q_{1k} + P_2Q_{2k} + P_3Q_{3k} + P_4Q_{4k})\rho^k\right) \tag{3}$$

subject to:

$$\begin{bmatrix} F_{t+1}(0,0) \\ F_{t+1}(0,1) \\ F_{t+1}(1,0) \\ F_{t+1}(1,1) \end{bmatrix} = \begin{bmatrix} \tau_0 & a_1 & 0 & 0 \\ b_1 & b_r & 0 & 0 \\ \tau_0 & \alpha_1 & \tau_0 & a_2 \\ 0 & 0 & b_2 & b_r \end{bmatrix} \begin{bmatrix} F_t(0,0) - Q_{1t} \\ F_t(0,1) - Q_{2t} \\ F_t(1,0) - Q_{3t} \\ F_t(1,1) - Q_{4t} \end{bmatrix}. \tag{4}$$

Our purpose is to solve the maximization problem (3) with constrained $\boldsymbol{AF_t(G,T)}$ as in (4) using Genetic Algorithm method. We built the GA routine using C++ Program. Following are some remark for the GA program construction.

- The objective function is the function $\boldsymbol{\Pi}$ with maximization problem
- The calculation of fitness for maximization is using $eval(v_k) = \boldsymbol{\Pi}, k = 1,2, \ldots N$;
  N= Number of population size. We take N=10 for this case
- The total fitness calculated by: $TF = \sum_{k=1}^{N} eval(v_k), k = 1,2,..,N$
- The cumulative probability q_k for each chromosome v_k calculated by $q_k = \sum_{k=1}^{N} p_k, k = 1,2,..,N$.

Generally, the algorithm genetics processes include the following important steps: encoding, selection, crossover, and mutation ([8]). The details of the steps are explained in detail in this section.

Pre-step 1: Initialization
We begin the process with initialization as follows:

Number of generation $jg = 1000$
Counter generation $cg = -1$
Number of Population $pop = 10$
Crossover Probability $pc = 0.25$
Number of mutation per generation $mut = 4$
Threshold $\theta = 90\%$

Pre-step 2: Encoding
The objective is to find Q_{11}, Q_{12} which maximize of $\boldsymbol{\Pi}$ in the interval solution

$$0 = a_1, a_2 \leq Q_{21}, Q_{42} \leq b_1, b_2 = 0.$$

In this research we use binary encoding method to represent chromosome and the gen length (denoted by m_j). We calculate m_j ($j = 1,2$) as

$$2^{m_j - 1} < (b_j - a_j) * 2^k < 2^{m_j} - 1.$$

To convert the binary to decimal for variable Q_{11} and Q_{12}, we use formula

$$Q_{1j} = a_j + decimal(substring_j) * \frac{(b_j - a_j)}{2^{m_j - 1}}. \quad (5)$$

For this case we have $m_1 = m_2 = 4$, so that we will have length of chromosome or number of gen $m = m_1 + m_2 = 8$.

Step 1: Initial Population
The random process is done to obtain 10 initial populations. This obtained 10 random initial population are in the form of binary number with length 8. The decimal value of each initial population is calculated by substituting the decimal of each substring to formula (5)

Step 2: Chromosome Evaluation
The fitness value of chromosome, in our case is maximization problem, is the value of objective function Π. The fitness value of each chromosome is evaluated in this step, in order to see the best value of Π (maximum value).

Step 3: Calculation of Population Convergence Percentage
The Population Convergence Percentage (pk) defined as percentage of number of individual with most number of same fitness value. The formula given as follows

$$pk = \frac{n}{pop} * 100\%$$

IOP Conf. Series: Materials Science and Engineering **332** (2018) 012014 doi:10.1088/1757-899X/332/1/012014

where n is the most number of same fitness value and pop is the number of individual. The counter generation is calculated by formula,

$$cg = cg + 1.$$

Step 4: Termination Criteria
The genetic algorithm process will be terminated when the counter generation cg reach the defined number of generation ($jg = 1000$) or the convergence of population pk reach the defined threshold level ($\theta = 90\%$). After initialization of population, the value of $cg = 0$, and if there is no individual with same fitness number, $pk = 10\%$ so that the genetics algorithm process is continue.

Step 5: Chromosome Selection
Since the maximum point which are try to be found then the fitness value used are defined by

$$eval(v_k) = \Pi(Q_{11}, Q_{12})$$

then calculate the total fitness of the population

$$TF = \sum_{k=1}^{10} eval(v_k)$$

and calculate the probability selection p_i for each chromosome v_k by formula

$$p_k = \frac{eval(v_k)}{TF}.$$

Then, calculate the cumulative probability

$$q_k = \sum_{i=1}^{k} p_i.$$

Select a set of 10 chromosome using roulette wheel for 10 times, and get a random numbers $r \in [0,1]$. We apply the following criteria to addresses the position to replace the selected $k-th$ current individual,

$$if \begin{cases} q_{k-1} < r \le q_k, & select \;\; v_k \\ r < q_1, & select \;\; v_1 \end{cases}.$$

Step 6: Crossover
The crossover will be done to the new population, in this paper, the Single Point Crossover (SPX) method are used. SPX is the most simple crossover method where a random number are generated in the interval $[1,m]$ as an intersection point. According to [7] the SPX algorithm is as follows:

- Generate a number randomly in interval $[1,m]$, and use this number as the starting position of a gen in a chromosome which will encounter crossover. Mark alleles which will get crossover (from the starting position until the last gen).
- Exchange the marking allele between parent 1 and 2.

Step 7: Mutation
Number of gen in one generation is $m * N = 8 * 10 = 80$. If we want to arrange the mutation probability value equal to pm, then we should have calculate number of mutation (nm) as

$$nm = m * N * pm$$

In this simulation, we use $pm = 0.05$, then, in the mutation process we generate randomly $nm = 4$ number of integer in interval $[1, pop * m]$. This 4 number is used as locating in which gen will be mutated. The replacement process of allele is 1 and 0, as follows

$$allele \begin{cases} 0 \to 1 \\ 1 \to 0 \end{cases}.$$

After the mutation process then the calculation in one generation is finished. The next process is to repeat the process in step 2 until the termination criteria is satisfied.

IOP Conf. Series: Materials Science and Engineering **332** (2018) 012014 doi:10.1088/1757-899X/332/1/012014

Step 8: Decoding

Decoding is the process to recode the gens in a chromosome so that the value is return into the value (before encoded).

4. Numerical Results

In this section we give an example of the GA implementation discussed above. We assume that the parameters used in the example are a_1=0.56, a_2 =0.244, α_1= 0.44, b_1=0.945, b_2=0.975, b_r= 0.875, τ_0= 0, $P_1 = 650{,}000$ currency unit, $P_2 = 2{,}600{,}000$ currency unit, $P_3 = 1{,}800{,}000$ currency unit, $P_4 = 6{,}000{,}000$ currency unit, $r = 0.1$, $\rho = \frac{1}{1+r} = 0.9090$, and there is no cost associated in the model.

We run the implemented GA algorithm in C++ for 10 trials with different set of initial population of solutions. We only consider a short time horizon to go, i.e. $T = 2$. We do 10 trials just to illustrate the method. The results of one of the trial is presented in Table 1. The colums of Table 1 represent 10 different individual of initial solution. The rows represent the updated solution due to the GA algorithm. For example, initially we have a goat population of $\begin{bmatrix} F_0(0,0) \\ F_0(0,1) \\ F_0(1,0) \\ F_0(1,1) \end{bmatrix} = \begin{bmatrix} 10 \\ 0 \\ 10 \\ 0 \end{bmatrix}$ at time t=0. The population grows and at time t=1 (by ceiling each vector component) we have $\begin{bmatrix} F_1(0,0) \\ F_1(0,1) \\ F_1(1,0) \\ F_1(1,1) \end{bmatrix} = \begin{bmatrix} 0 \\ 10 \\ 0 \\ 10 \end{bmatrix}$.

According to Table 1, row 8 (the most updated solution resulting from GA algorithm) shows that at time t=1 we should take $\begin{bmatrix} Q_1 \\ Q_2 \\ Q_3 \\ Q_4 \end{bmatrix} = \begin{bmatrix} 0 \\ 0 \\ 0 \\ 7 \end{bmatrix}$ out from $\begin{bmatrix} F_1(0,0) \\ F_1(0,1) \\ F_1(1,0) \\ F_1(1,1) \end{bmatrix} = \begin{bmatrix} 0 \\ 10 \\ 0 \\ 10 \end{bmatrix}$ and sell them. The remaining unsold goat population then grows and at the final time t=2 we have $\begin{bmatrix} F_2(0,0) \\ F_2(0,1) \\ F_2(1,0) \\ F_2(1,1) \end{bmatrix} = \begin{bmatrix} 6 \\ 9 \\ 6 \\ 3 \end{bmatrix}$ which is automatically taken as the solution, i.e. $\begin{bmatrix} Q_1 \\ Q_2 \\ Q_3 \\ Q_4 \end{bmatrix} = \begin{bmatrix} 6 \\ 9 \\ 6 \\ 3 \end{bmatrix}$. The total revenue from selling these population is 84,545,454 currency unit. Figure 3 shows the plot for the total revenue from selling the goat population according to the GA results for consecutive generations, associated with the consecutive rows in Table 1, for all 10 different initial solutions. The final generation shows that at least 90% of the evolving solutions has already convergent to the same solution. Table 2 shows the results for the whole 10 trials done in the experiment. The table reveals that the GA gives different set of solutions but relatively close to each other in terms of the resulting total revenue, except a few.

Table 1. Numerical results by Genetics Algorithm of Trial 7

cg		Individual									
		1	2	3	4	5	6	7	8	9	10
1.	Q_{i1} Q_{i2} i=1,..,4	(0,0,0,0), (6,9,7,9)	(0,9,0,4), (1,1,2,6)	(0,0,0,1), (6,9,7,8)	(0,8,0,9), (2,2,2,1)	(0,10,0,0), (0,0,3,9)	(0,0,0,7), (6,9,6,3)	(0,0,0,1), (6,9,7,8)	(0,0,0,0), (6,9,7,9)	(0,1,0,8), (6,8,5,2)	(0,6,0,6), (3,4,3,4)
2.		(0,1,0,8), (6,8,5,2)	(0,0,0,7), (6,9,6,3)	(0,1,0,0), (6,8,7,9)	(0,0,0,8), (6,9,5,2)	(0,0,0,1), (6,9,7,8)	(0,0,0,7), (6,9,6,3)	(0,0,0,7), (6,9,6,3)	(0,0,0,1), (6,9,7,8)	(0,0,0,7), (6,9,6,3)	(0,0,0,7), (6,9,6,3)
3.		(0,1,0,1), (6,8,7,8)	(0,0,0,7), (6,9,6,3)	(0,0,0,7), (6,9,6,3)	(0,0,0,8), (6,9,5,2)	(0,0,0,8), (6,9,5,2)	(0,0,0,7), (6,9,6,3)	(0,0,0,7), (6,9,6,3)	(0,0,0,1), (6,9,7,8)	(0,0,0,8), (6,9,5,2)	(0,0,0,7), (6,9,6,3)

4.		(0,1,0,1), (6,8,7,8)	(0,0,0,7), (6,9,6,3)	(0,0,0,7), (6,9,6,3)	(0,0,0,10), (6,9,5,0)	(0,0,0,7), (6,9,6,3)	(0,0,0,6), (6,9,6,4)	(0,0,0,7), (6,9,6,3)	(0,0,0,1), (6,9,7,8)	(0,0,0,1), (6,9,7,8)	(0,0,0,1), (6,9,7,8)
5.		(0,1,0,1), (6,8,7,8)	(0,0,0,0), (6,9,7,9)	(0,0,0,1), (6,9,7,8)	(0,0,0,7), (6,9,6,3)	(0,0,0,1), (6,9,7,8)	(0,0,0,7), (6,9,6,3)	(0,0,0,1), (6,9,7,8)	(0,0,0,1), (6,9,7,8)	(0,0,0,1), (6,9,7,8)	(0,0,0,1), (6,9,7,8)
6.		(0,1,0,1), (6,8,7,8)	(0,0,0,1), (6,9,7,8)	(0,1,0,1), (6,8,7,8)	(0,0,0,7), (6,9,6,3)	(0,0,0,1), (6,9,7,8)	(0,0,0,7), (6,9,6,3)	(0,1,0,1), (6,8,7,8)	(0,0,0,1), (6,9,7,8)	(0,1,0,1), (6,8,7,8)	(0,0,0,7), (6,9,6,3)
7.		(0,0,0,7), (6,9,6,3)	(0,0,0,7), (6,9,6,3)	(0,1,0,0), (6,8,7,9)	(0,1,0,1), (6,8,7,8)	(0,1,0,1), (6,8,7,8)	(0,0,0,7), (6,9,6,3)	(0,0,0,7), (6,9,6,3)	(0,0,0,1), (6,9,7,8)	(0,1,0,1), (6,8,7,8)	(0,0,0,7), (6,9,6,3)
8.		(0,0,0,7), (6,9,6,3)	(0,0,0,7), (6,9,6,3)	(0,0,0,7), (6,9,6,3)	(0,0,0,7), (6,9,6,3)	(0,0,0,7), (6,9,6,3)	(0,0,0,7), (6,9,6,3)	(0,0,0,7), (6,9,6,3)	(0,0,0,7), (6,9,6,3)	(0,0,0,6), (6,9,6,4)	(0,0,0,7), (6,9,6,3)

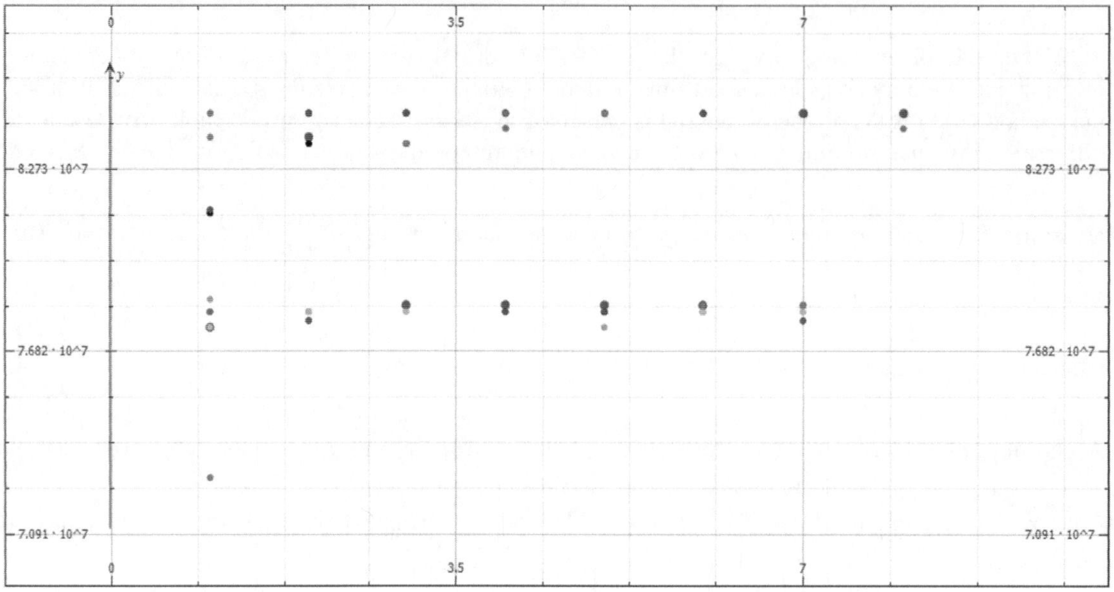

Figure 3. Plot of Numerical values by Genetics Algorithm of Trial 7

Table 2. Numerical results by Genetics Algorithm of 10 Trials

Trial	cg	Q_{i1}	Q_{i2}	$\Pi(Q_{ij})$
1.	9	(0,0,0,4)	(6,9,6,6)	83,057,851
2.	3	(0,1,0,1)	(6,8,7,8)	78,314,049
3.	21	(0,5,0,8)	(3,5,3,2)	82,190,082
4.	10	(0,0,0,4)	(6,9,6,6)	83,057,851
5.	21	(0,0,0,1)	(6,9,7,8)	78,099,173
6.	21	(0,5,0,5)	(3,5,4,5)	82,190,082
7.	7	(0,0,0,7)	(6,9,6,3)	84,545,454
8.	13	(0,6,0,8)	(3,4,3,2)	82,404,958
9.	11	(0,0,0,5)	(6,9,6,5)	83,553,719
10.	12	(0,2,0,2)	(5,7,6,7)	77,000,000

5. Conclusion

In this paper we have discussed a mathematical model of goat breeding strategy, i.e. the backcrossing breeding. The model is aimed to obtain a strategy in producing better variant of species. In this strategy, a female (doe) of a lesser quality goat, in terms of goat quality is bred with a male (buck) of an exotic goat which has a better goat quality. In this paper we posed a problem on how to harvest the population optimally. If $Q_{1k}, Q_{2k}, Q_{3k},$ and Q_{4k} is the numbers of individual form sub population

IOP Conf. Series: Materials Science and Engineering **332** (2018) 012014 doi:10.1088/1757-899X/332/1/012014

$F_k(0,0)$, $F_k(0,1)$, $F_k(1,0)$, and $F_k(1,1)$, respectively, then the problem is to maximize the profit Π along the time horizon *T*. A genetic algorithm approach has been devised to obtain the solution of the problem. The trials we did in the experiment reveals that the GA gives different set of solutions but relatively close to each other in terms of the resulting total revenue, except a few. Further study need to be done to obtain GA solution that closer to the exact solution.

Acknowledgment
The authors thank to the Universitas Padjadjaran, due to the funding of this work through the scheme of Academic Leadership Grant (ALG) with contract number 855/UN6.3.1/PL/2017.

References
[1] Raihan A U, Supriatna A K and Anggriani N 2016 *Pros. Sem. Nasional Matematika Unpar 2016* MS pp 86-92
[2] Supriatna A K, Anggriani N, Carnia E and Raihan A 2017 *AIP Conference Proceedings* 1868, 040012 (doi: 10.1063/1.4995127)
[3] Haupt R L and Haupt S E 2004 *Practical Genetic Algorithms (2nd ed.)* (New Jersey: A John Wiley & Sons, Inc., Publication)
[4] Bimlendu S, Sujata D, Abhinav M, Vinay K, Prasanna K C 2016 A Review over Genetic Algorithm and Application of Wireless Network System *Procedia Computer Science* **78** pp 431-438
[5] Wagner F S, Claudio M N A, Pereira, Pius P M S, Roberto S 2002 *Genetic algorithms applied to turbine extraction optimization of a pressurized-water reactor.* Applied Energy **773** pp 217-222
[6] Kramer O 2017 *Studies in Computational Intelligence Volume 679: Genetic Algorithm Essensials* (Oldenburg: Springer)
[7] Weise T 2009 *Global Optimization Algorithms Theory and Application* (http://www.it-weise.de/, downloaded 10/10/2017)
[8] Adriana F and Fadly R L 2015 *Algoritma Genetika dan Penerapannya* (Yogyakarta: Teknosain)

IOP Conf. Series: Materials Science and Engineering **332** (2018) 012015 doi:10.1088/1757-899X/332/1/012015

Analysis of forecasting and inventory control of raw material supplies in PT INDAC INT'L

E Lesmana *, **B Subartini, Riaman, D A Jabar**

Departemen Matematika FMIPA Universitas Padjadjaran, Jl. Raya Bandung-Sumedang km 21 Jatinangor, Indonesia

*Email: man.msie@gmail.com, riaman@unpad.ac.id

Abstract. This study discusses the data forecasting sales of carbon electrodes at PT. INDAC INT L uses winters and double moving average methods, while for predicting the amount of inventory and cost required in ordering raw material of carbon electrode next period using Economic Order Quantity (EOQ) model. The result of error analysis shows that winters method for next period gives result of MAE, MSE, and MAPE, the winters method is a better forecasting method for forecasting sales of carbon electrode products. So that PT. INDAC INT L is advised to provide products that will be sold following the sales amount by the winters method.

1. Introduction

This study discusses the data forecasting sales of carbon electrodes at PT. INDAC INT L uses winters and double moving average methods, while for predicting the amount of inventory and cost required in ordering raw material of carbon electrode next period using Economic Order Quantity (EOQ) model. The result of error analysis shows that winters method for next period gives result of MAE, MSE, and MAPE, the winters method is a better forecasting method for forecasting sales of carbon electrode products. So that PT. INDAC INT L is advised to provide products that will be sold following the sales amount by the winters method.

The rapid development of science and technology increasingly sophisticated push the rate of economic growth. So the competition between companies is getting tighter, therefore every company should estimate the estimated amount of sales and establish the control of raw material inventory effectively and efficiently, so that the company can achieve the desired goals. Raw materials are an initial process of production transformation which will then be processed into finished products [2]. Every company should have a goal to make a profit, but to gain profit is not easy, because it is influenced by several factors to be overcome. One of the factors that influence it is the production problem. If the production process smoothly then the company will be able to achieve its objectives, but if the production process is not smooth, then the purpose of the company will not be achieved. Usually one of the factors causing the failure of production is due to the absence of raw materials that can be processed in the production process.

Errors in the determination of raw material purchases will also affect the profits of the company. Purchase of raw materials that are too large will affect the amount of storage costs, which include maintenance costs, warehouse rental costs and costs incurred in connection with material damage. But if the purchase of raw materials are too low, will affect the profits of the company, because there will be a stock out which resulted in a company failed production.

Previous research related to this issue was carried out by Womack et al in 1990 on the principle of lean production to solve inventory problems that consume too much physical space, create financial burdens, and increase the likelihood of damage to the goods. Wacker et al in 1996 on inventory management and control. Tatoglu and Zaim in 2007 on the concepts and techniques of inventory implementation in the business world and its impact on company performance. In 2008 Koumanakos in the management of inventory system with EOQ, MRP and ERP method, while in 2012 Pamungkas and Sutanto test analysis of raw material control, the method used is the method of forecasting is a trend projection method where this technique adjusts with the trend line series The historical data point of a company which is then projected with forecasts for the coming period, and the EOQ method for determining the optimal number of orders or purchases. Lwiki et al in 2013 examines the impact of inventory management on financial performance. In 2014 Katiandagho et al studied the wood supply control system for the furniture industry, the method used was the EOQ method to analyze the optimal volume of wood raw materials required for production, as well as calculate the total cost of inventory to be incurred by the company.

2. Forecasting

Forecasting is an attempt to predict future circumstances through testing of the past. The essence of forecasting is the approximate future events on the basis of patterns in the past, and the use of policies against projections with patterns in the past [7]. The Time Series method is a quantitative method used to analyze data sets that are a function of time. This method assumes that some pattern or combination of patterns always repeats over time, and the archetype can be identified on the basis of historical data of the time series.

1. According [3] pattern of a series of data in time series can be grouped into 4 kinds namely: 1. Horizontal (constant), i.e. when the data fluctuates around the average stably. The pattern is a straight horizontal line.
2. Trend, that if the data has a tendency, either the direction increases or decreases over time.
3. Seasonal (seasonal) is when the pattern is a repetitive movement regularly in any given period.
4. Cyclical i.e. when data is influenced by long-term economic fluctuations, such as business life cycle.

2.1 Triple Exponential Smoothing Winters

The Winters method is based on three smoothing equations, one for stationary elements, one for trends, and one for seasonality. The basic equations for the winters method are as follows:

Over all smoothing $$\boldsymbol{S_t = \alpha \frac{X_t}{I_{t-L}} + (1-\alpha)(S_{t-1} + b_{t-1})} \quad (1)$$

Trend slicing $$b_t = \gamma(S_t - S_{t-1}) + (1-\gamma)b_{t-1} \quad (2)$$

Seasonal Smoothing $$I_t = \beta \frac{X_t}{S_t} + (1-\beta)I_{t-L} \quad (3)$$

Forecast $$F_{t+m} = (S_t + b_t m)I_{t-L+m} \quad (4)$$

with :

X_t = actual data at time t
S_t = smoothing factor at time t
L = seasional length
b = trend smoothinh factor
I = seasional adjustment factor
F_{t+m} = predicted value for m next periode

Initialization

Preliminary estimate S_{L+1}, b_{L+1}, I_L can be obtained as follows :

1. Initial smoothing using :
$$S_{L+1} = X_{L+1} \tag{5}$$
2. Estimate trend using :
$$b_{L+1} = \frac{1}{L}\left[\frac{X_{L+1}-X_1}{L} + \frac{X_{L+2}-X_2}{L} + \cdots + \frac{X_{L+L}-X_L}{L}\right] \tag{6}$$
3. Seasional factor use :
$$I_L = \frac{X_L}{\bar{X}} \tag{7}$$
Where :
$$\bar{X} = \frac{1}{L}\sum_{i=1}^{L} X_i \tag{8}$$

The Winters method is based on three smoothing equations, one for stationary elements, one for trends, and one for seasonality. Forecasting with this method is based on the serial data projection that is passed by moving average. The estimated value for a period is the average of the last observed N value [4].

2.2 Double Moving Average

According to [3] forecasting with this method is based on the serial data projection that is passed by the moving average. The estimated value for a period is the average of the last observed N value . There are several steps used to determine the forecast by double moving averages method, which is as follows:

1. Calculate the first moving average (S_t') , obtained from existing historical data so that it can be written as follows:
$$S_t' = \frac{X_t + X_{t-1} + X_{t-2} + \cdots + X_{t-N+1}}{N}. \tag{9}$$
2. Calculate the second moving average (S_t'') , obtained from the first moving average so that it can be written as follows:
$$S_t'' = \frac{S_t' + S_{t-1}' + S_{t-2}' + \cdots + S_{t-N+1}'}{N} \tag{10}$$
3. Determine the a_t constant value:
$$a_t = S_t' + (S_t' - S_t'') \tag{11}$$
4. Determine slope value b_t :
$$b_t = \frac{2}{N-1}(S_t' - S_t'') \tag{12}$$
5. Determine the forecast value F_{t+m} :
$$F_{t+m} = a_t + b_t m \tag{13}$$

with m is the period ahead of time t.

2.3 Standard Statistics Size

X_t is the actual data for period t and F_t is the forecast for the same period then the error value can be defined as:
$$e_t = X_t - F_t \tag{14}$$

If there is actual data and forecasting for time period n, then there are a number of n errors and standard statistical measures can be defined as follows:

1. Mean Absolute Error (MAE)
$$MAE = \frac{1}{n}\sum_{t=1}^{n} |e_t| \tag{15}$$
2. *Mean Squared Error (MSE)*

$$MSE = \frac{1}{n}\sum_{t=1}^{n} e_t{}^2 \quad (16)$$

3. *Percentage Error (PE_t)*

$$PE_t = \frac{X_t - F_t}{X_t} \cdot 100\% \quad (17)$$

4. *Mean Absolute Percentage Error (MAPE)*

$$MAPE = \frac{1}{n}\sum_{t=1}^{n} |PE_t| \quad (18)$$

with :

e_t = error value
X_t = actual data at time t
n = number of time periods
F_t = forecast value in perio t

3. Inventory

According to [1] inventory is an asset that includes goods owned by the company for the purpose of sale in a normal period of business. So the inventory is the number of goods provided to meet the demand from customers.

According to [5] EOQ is the most economical volume or amount of purchases to run on every purchase. To meet that need, it can be calculated to meet the most economical needs of a number of items that will be obtained by purchase using a minimal cost. The raw materials available in ensuring the smoothness of the production process and the costs incurred by the company with respect to the company to a minimum, then the action that needs to be done is to determine the amount of raw material inventory, the number of buffer stock and reorder point.

According to [3] EOQ is the most economical volume or amount of purchases to run on every purchase. To meet that requirement, it can be calculated to fulfill the most economical needs of a number of goods that will be obtained by purchase using a minimal cost.

The raw materials available in ensuring the smoothness of the production process and the costs incurred by the company in relation to the company to a minimum, then the action that needs to be done is to determine the amount of raw material inventory, the amount of buffer stock and reorder point. Terms of data using the EOQ method are:

a. demand level is known and is constant (deterministic)
b. lead times are known and are constant
c. the ordered goods are assumed to be available soon or the production rate of goods ordered abundantly
d. every order is received in one shipment and can be used immediately
e. there is no back order because it is out of stock
f. purchase price or manufacturing cost is unchanged
g. no discounts (quantity discount)
h. Large carrying costs depend on a straight line with the average inventory amount.
i. The amount of ordering cost or set up cost ordered is fixed and does not depend on the number of items in each lot.
j. Items are a one-of-a-kind product and have no relationship with other products.

3.1 *EOQ Model with More Capacity*

If the production capacity is denoted by P and the demand is denoted by D, where $P > D$, then in every production cycle there will always be an over-production of (P - D) units. The main features in the EOQ model with more capacity are:

a. Excess stock due to market demand is smaller than production capacity.
b. Stock will incur more storage costs continuously during the production process takes place.

Total Optimum Cost of EOQ Model More Capacity

$$TC = \frac{1}{2}.\left(\frac{Q_0}{P}\right).(P-D).C_c + \left(\frac{D}{Q_0}\right).C_s \quad (19)$$

$$\frac{dTC}{dQ_0} = \frac{P-D}{2P} \cdot C_c - \frac{D}{{Q_0}^2} \cdot C_s = 0$$

$$\frac{P-D}{2P} \cdot C_c = \frac{D}{{Q_0}^2} \cdot C_s$$

$${Q_0}^2 = \frac{2 \cdot P \cdot D \cdot C_s}{(P-D) \cdot C_c}$$

$$Q_0 = \sqrt{\frac{2.P.D.C_s}{(P-D).C_c}} \quad (20)$$

4. Result And Discussion

Data taken from PT. INDAC INT L includes sales data of carbon electrodes from 2014 to 2016 seen in Table 1.

Table 1. Sales Data of Carbon Electrode PT. INDAC INT L
Period January 2014 - December 2016

2014		2015		2016	
Month	Sales	Month	Sales	Month	Sales
Jan	382.414	Jan	374.994	Jan	336.254
Feb	344.456	Feb	355.069	Feb	400.880
Mar	218.447	Mar	378.202	Mar	350.878
Apr	403.949	Apr	521.911	Apr	469.887
May	387.080	May	511.122	May	458.165
Jun	525.646	Jun	454.074	Jun	604.351
Jul	270.406	Jul	334.385	Jul	234.281
Aug	439.912	Aug	477.189	Aug	369.692
Sep	505.750	Sep	515.231	Sep	405.017
Oct	390.415	Oct	454.074	Oct	416.751
Nov	463.436	Nov	466.970	Nov	364.620
Dec	435.005	Dec	342.842	Dec	387.380

The first step of data processing is to analyze the shape of the data pattern through the data pattern graph by plot the data in Table 1, making it easier in determining the method of forecasting that will be used. The plot of data the number of sales of carbon electrodes at PT. INDAC INT L can be seen in Figure 1.

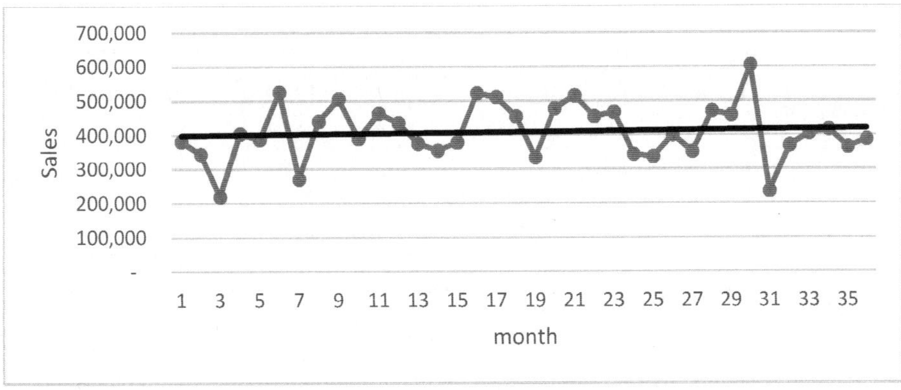

Figure 1. Plot Data Total Sales of Carbon Electrodes at PT. INDAC INT L

Seasonal factors are identified in data on the number of sales of carbon electrodes at PT. INDAC INT L. In Figure 1 it is shown that the pattern of changes is repeated automatically from year to year. In January to December the data increases and decreases annually. This indicates that the data on the number of sales of carbon electrodes is data that has a trend and seasonal patterns with seasonal variations that fluctuate.

4.1 *Forecasting Using Winters Method*

The Winters method is based on three equations, one for stationers, one for trend and one for seasonality. In predicting the data using the Winters method it is necessary to determine the optimal α, β, and agar in order to obtain the forecast result which has the smallest error. From Table 1, data processing is performed to determine the MAE (Mean Absolute Error), MSE (Mean Squared Error) and MAPE (Mean Absolute Percentage Error) with the smallest value according to the calculation in, to obtain the values α, β, and γ in Table 2 as follows.

Table 2. Determination Process α, β, and γ in Winters Method

Alpha	Beta	Gamma	MAE	MSE	MAPE
0,65	0,1	0,1	53881,39	7250589173,63	13,1844
0,7	0,1	0,1	53850,01	7243649546,78	13,1505
0,75	0,1	0,1	54067,13	7288454335,98	13,1910
0,7	0,55	0,1	51803,83	6954395820,63	12,9208
0,7	0,6	0,1	51751,46	6943710012,76	12,9271
0,7	0,65	0,1	51801,39	6937339821,57	12,9511
0,7	0,6	0,1	51751,46	6943710012,76	12,9271
0,7	0,6	0,15	51299,98	6944105781,19	12,8048
0,7	0,6	0,2	51475,99	7039913176,73	12,8779

Based on Table 2 it can be seen that the most α, β, and nilai values are $\alpha = 0.7$, $\beta = 0.6$ and $\gamma = 0.15$. Furthermore, the calculation process using winters method, which is assumed $S_{13} = X_{13}$ based on equation (1) obtained value $S_{13} = 374.994$. Next will be calculated value b_{13} based on the equation (2) obtained. After the initialization process is complete, then the calculation of single, seasonal and seasonal smoothing for the period from January 2014 to December 2016, with $\alpha = 0{,}7$, $\beta = 0{,}6$, and $\gamma = 0{,}15$ (parameter obtained by trial method and error).

Furthermore, for the forecast of sales of carbon electrodes in the next year is obtained in Table 3.

IOP Conf. Series: Materials Science and Engineering **332** (2018) 012015 doi:10.1088/1757-899X/332/1/012015

Table 3. Data Processing Using Winters Method($\alpha = 0.7$, $\beta = 0.6$, and $\gamma = 0.15$)

Period (t)	Actual Value	Single exp smoothing	seasional smoothing	Trend smoothing	Forecast	$\lvert e_t \rvert$
1	382414		0,96			
2	344456		0,87			
3	218447		0,55			
4	403949		1,02			
5	387080		0,97			
6	525646		1,32			
7	270406		0,68			
8	439912		1,11			
9	505750		1,27			
10	390415		0,98			
11	463436		1,17			
12	435005		1,10			
13	374994	374994	0,99	34929		
14	355069	409614	0,87	34883	355451	382,4
15	378202	614777	0,59	60425	244432	133769,8
16	521911	561834	0,96	43419	686600	164688,8
17	511122	548755	0,95	34945	589769	78646,5
18	454074	415318	1,19	9688	772372	318297,6
19	334385	471364	0,70	16641	289305	45080,7
20	477189	448034	1,08	10646	540424	63234,9
21	515231	420887	1,24	4977	583968	68737,0
22	454074	451170	1,00	8773	418544	35530,2
23	466970	418174	1,14	2507	536583	69612,7
24	342842	345360	1,03	-8791	460671	117829,4
25	336254	339917	0,99	-8289	331544	4710,2
26	400880	423171	0,92	5443	287505	113375,6
27	350878	545534	0,62	22981	252485	98393,1
28	469887	511717	0,94	14461	548116	78228,6
29	458165	495940	0,93	9925	499143	40977,5
30	604351	508675	1,19	10347	599594	4756,9
31	234281	390685	0,64	-8904	362237	127955,7
32	369692	353705	1,06	-13115	413091	43399,3
33	405017	330125	1,23	-14685	423609	18592,5
34	416751	387240	1,04	-3915	314490	102261,3
35	364620	339543	1,10	-10482	435712	71092,6
36	387380	361057	1,06	-5683	340134	47245,8
37					350954	
38					320029	
39					213815	
40					316876	
41					310604	

Period (t)	Actual Value	Single exp smoothing	seasional smoothing	Trend smoothing	Forecast	$\|e_t\|$
42					388090	
43					205286	
44					334505	
45					382311	
46					317773	
47					328095	
48					309615	

Data forecast of the sales of carbon electrodes at PT. INDAC INT L uses the Winters method with $\alpha = 0.7$, $\beta = 0.6$ and $\gamma = 0.15$ can be seen in Figure 2.

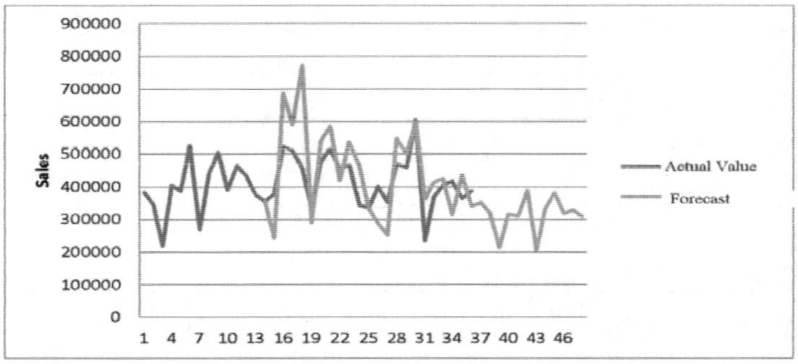

Figure 2. Plot Comparison of Actual Data with Sales Forecast Data with Winters Method

In predicting the data using the double moving average method, it is necessary to determine the optimal number of N periods to obtain the forecast result which has the smallest error. The statistical measures used to determine the optimal N period are MAE (Mean Absolute Error), MSE (Mean Squared Error) and MAPE (Mean Absolute Percentage Error), the process of selecting the N period can be seen in Table 4.

Table 4. Selection Process Period *N* on Double Moving Average Method

Comparison MAE, MSE, MAPE			
MA(NxN)	MAE	MSE	MAPE
2x2	117530,7	22821305711	31,20734
3x3	102046,5	18668955972	27,5938
4x4	84464,04	12432935095	22,47225
5x5	75811,95	10852098796	20,20566
6x6	74878,28	9422501714	19,3164
7x7	73941,68	8671545666	18,96529
8x8	68671,93	8539855893	17,84794
9x9	68247,29	8335163182	18,46958
10x10	70346,97	8331330579	19,03834
11x11	68946,46	9068398796	19,33037
12x12	72237,58	9188822106	20,43182

Based on Table 4, from the minimum MAE, MSE and MAPE values, the period N = 9 is used. The results of the calculation process can be seen in Table 5.

Table 5. Data Processing Using Double Moving Average Method (N = 9)

Period (t)	Actual Value	S'_t	S''_t	a_t	b_t	Forecast	$\lvert e_t \rvert$
1	382414						
2	344456						
3	218447						
4	403949						
5	387080						
6	525646						
7	270406						
8	439912						
9	505750	386451,2					
10	390415	387340,3					
11	463436	400560,2					
12	435005	424622,2					
13	374994	421405,0					
14	355069	417848,1					
15	378202	401465,5					
16	521911	429410,5					
17	511122	437322,7	411825,1	462820,3	6374,4		
18	454074	431580,9	416839,5	446322,4	3685,4	469195	15120,6
19	334385	425355,4	421063,4	429647,4	1073,0	450008	115622,4
20	477189	426883,5	423988,2	429778,9	723,8	430720	46468,5
21	515231	435797,5	425229,9	446365,1	2641,9	430503	84728,7
22	454074	444584,2	427805,4	461363,1	4194,7	449007	5067,1
23	466970	457017,7	432157,5	481877,9	6215,0	465558	1412,4
24	342842	453088,8	437893,5	468284,2	3798,8	488093	145251,1
25	336254	432460,2	438232,3	426688,1	-1443,0	472083	135829,0
26	400880	420211,1	436331,1	404091,2	-4030,0	425245	24364,8
27	350878	408744,9	433793,7	383696,1	-6262,2	400061	49183,1
28	469887	423800,6	433621,0	413980,3	-2455,1	377434	92453,2
29	458165	421686,9	433043,6	410330,3	-2839,2	411525	46640,3
30	604351	431589,1	432576,0	430602,3	-246,7	407491	196859,9
31	234281	407167,7	428418,6	385916,8	-5312,7	430356	196074,2
32	369692	396359,0	421678,7	371039,2	-6329,9	380604	10912,4
33	405017	403267,3	416143,0	390391,6	-3218,9	364709	40307,4
34	416751	412211,4	413893,1	410529,8	-420,4	387173	29578,6
35	364620	408182,5	412556,6	403808,4	-1093,5	410109	45489,8
36	387380	412238,2	412944,8	411531,7	-176,6	402715	15335,2

Period (t)	Actual Value	S'_t	S''_t	a_t	b_t	Forecast	$\lvert e_t \rvert$
37						411355	
38						411178	
39						411002	
40						410825	
41						410649	
42						410472	
43						410295	
44						410119	
45						409942	
46						409765	
47						409589	
48						409412	

Plot data forecast of the sales of carbon electrodes at PT. INDAC INT L by using the method of double moving average for period N = 9 can be seen in Figure 3

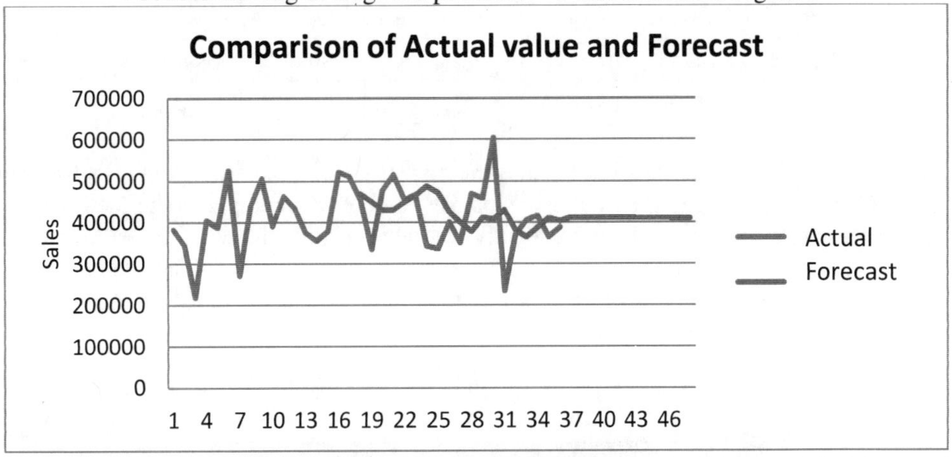

Figure 3. Plot Comparison of Actual Data with Sales Forecasting Data Using Double Moving Average Method For Period N = 9

From the sales prediction value obtained by using the winters and double moving average methods, the real data comparison with the forecast results is as follows.

Table 6. Comparison of Sales Data and Forecast Results

Actual Data 2016		Forecast with *Winters method* 2017		Forecast with *Double Moving Average* 2017	
Month	Sales	Month	Sales	Month	Sales
Jan-16	336.254	Jan-17	350.954	Jan-17	411.355
Feb-16	400.880	Feb-17	320.029	Feb-17	411.178
Mar-16	350.878	Mar-17	213.815	Mar-17	411.002
Apr-16	469.887	Apr-17	316.876	Apr-17	410.825
May-16	458.165	May-17	310.604	May-17	410.649
Jun-16	604.351	Jun-17	388.090	Jun-17	410.472
Jul-16	234.281	Jul-17	205.286	Jul-17	410.295
Aug-16	369.692	Aug-17	334.505	Aug-17	410.119
Sep-16	405.017	Sep-17	382.311	Sep-17	409.942
Oct-16	416.751	Oct-17	317.773	Oct-17	409.765
Nov-16	364.620	Nov-17	328.095	Nov-17	409.589
Dec-16	387.380	Dec-17	309.615	Dec-17	409.412

From Table 6 it can be seen that by using the method of double moving average get higher sales forecast value than Winters method. However, by using the double moving average method, sales will continue to decrease over each period due to negative b_t value. Unlike the winters method, the number of sales will continue to increase and decrease with the increasing number of sales smooth.
In addition, to determine the correct forecasting method, an error analysis of MAE, MSE and MAPE can be calculated. The calculation results can be seen in Table 7.

Table 7. Comparison of Error Analysis Winters Method and Double Moving Average Method

No.	Error Analysis	*Winters*	*Double Moving Average*
1	MAE	51299,9775	68247,2914
2	MSE	6944105781	8335163182
3	MAPE	12,8048356	18,46958009

In Table 7 can be seen comparison of error analysis results between winters and double moving average method. Based on the comparison of error analysis results winters method has a smaller error value than the method of double moving average. So that from Table 6 and Table 7 is directly proportional, which shows that the winters method is a better method to maximize sales at PT. INDAC INT L.

Implementation of EOQ method more capacity on inventory data obtained from PT. INDAC INT L is known as follows.

D = average number of carbon electrode demand in one year 3,777,953

P = average production capacity of 5,000,000 carbon electrodes in one year

C_c = carbon storage electrode cost Rp 1200 per unit
C_s = carbon electrode order cost Rp 150181,4 per unit
Based on the amount of known inventory data, it can be obtained the optimal order quantity as follows:

$$Q_0 = \sqrt{\frac{2 \cdot P \cdot D \cdot C_s}{(P-D) \cdot C_c}}$$
$$= \sqrt{\frac{2 \cdot 5.000.000 \cdot 3.777.953 \cdot 150181,4}{(5.000.000 - 3.777.953) \cdot 100}}$$
$$= 215.472 \text{ unit per period}$$

$$TC = \frac{1}{2} \cdot \left(\frac{Q_0}{P}\right).(P - D).C_c + \left(\frac{D}{Q_0}\right).C_s$$
$$= \frac{1}{2} \cdot \left(\frac{150.181}{5.000.000}\right).(5.000.000 - 3.777.953).100 + \left(\frac{3.777.953}{150.181}\right).150181,4$$
$$= 63.196.285 \text{ per period}$$

From the calculation results obtained, in order not to accumulate the production results, because the supply is greater than the demand, then PT. INDAC INT L is advised to purchase raw materials of 215,472 kg per period, with total expenses of Rp 63,196,285.00

5. Conclusion

Based on the results of data processing in the previous chapter, it can be drawn conclusion as follows:

Comparison of data patterns of sales amount of carbon electrodes at PT. INDAC INT L from January 2014 to December 2016 with the results of the next forecast, using the Winters method will provide better results for the company.

From the calculation results obtained, in order not to accumulate the production, because the inventory is greater than the demand, then PT. INDAC INT L is advised to purchase raw materials of 215,472 kg per period, with total expenses of Rp 63.196.28

References

[1] Assauri, S. 1980. *Management Produksi*. Jakarta : Lembaga Penerbit Fakultas Ekonomi Universitas Indonesia.
[2] Bahagia, S. N. 2006. *Sistem Inventory*. Bandung : Penerbit ITB.
[3] Herjanto, E. 2007. *Manajemen Operasi Edisi Ketiga*. Jakarta: PT Grasindo.
[4] Lesmana, E. 2016.Lecture note *Inventory Control*.
[5] Lwiki, T.; Boniface P.; Galo, N.; and Kirigo, V. 2013. The impact of inventory management practices on financial performance of sugar manufacturing firms in kenya. International Journal of Business, Humanities and Technology. Vol.3 No.5
[6] Makridakis, S., Spyros, G. 1983. *Forecasting:Methods and Applications 2nd Edition*. Singapore: John Wiley & Sons.
[7] Prasetya, H., Lukiastuti, F. 2009. *Manajemen Operasi*. Yogyakarta : MedPress.
[8] Siagian, P. 1987. *Penelitian Operasional: Teori dan Praktek*. Jakarta: Penerbit Universitas Indonesia.
[9] Simbar, M.; Katiandagho, T.; Lolowang, T.; dan Baroleh, J. 2014. Analisis pengendalian persediaan bahan baku kayu cempaka pada industri mebel dengan menggunakan metode *EOQ*. J. Ilmiah.
[10] Wahyudi, R. 2015. Analisis pengendalian persediaan barang berdasarkan metode *EOQ* di toko era baru samarinda. Ejournal ilmu administrasi bisnis, 2(1): 162-173.

IOP Conf. Series: Materials Science and Engineering **332** (2018) 012016 doi:10.1088/1757-899X/332/1/012016

Determining the Optimal Solution for Quadratically Constrained Quadratic Programming (QCQP) on Energy-Saving Generation Dispatch Problem

E. Lesmana, D. Chaerani, H. N Khansa
Department of Mathematics
Faculty of Mathematics and Natural Sciences Universitas Padjadjaran
Jalan Raya Bandung Sumedang KM 21 Jatinangor Sumedang Indonesia
E-mail: eman.lesmana@unpad.ac.id, d.chaerani@unpad.ac.id, snabilakhansa@gmail.com

Abstract. Energy-Saving Generation Dispatch (ESGD) is a scheme made by Chinese Government in attempt to minimize CO2 emission produced by power plant. This scheme is made related to global warming which is primarily caused by too much CO2 in earth's atmosphere, and while the need of electricity is something absolute, the power plants producing it are mostly thermal-power plant which produced many CO2. Many approach to fulfill this scheme has been made, one of them came through Minimum Cost Flow in which resulted in a Quadratically Constrained Quadratic Programming (QCQP) form. In this paper, ESGD problem with Minimum Cost Flow in QCQP form will be solved using Lagrange's Multiplier Method

1. Introduction

Energy needs in the world increasing as the times and has become something crucial, one of them electrical energy. The fullfillment of electrical energy is achieved by the existence of power plants with various power sources; such as water, steam, gas, geothermal power, waves, solar, nuclear, and others. Steam power plants with coal fuel are the most widely operated in Indonesia because of the cheap and abundant coal, but unfortunately the coal-fired power plant is not environmentally friendly. Given the increasingly polluted environment and global warming experienced by the earth, environmental pollution, one of which is from carbon emissions of power plants, must be stopped. The transition from conventional power generation to renewable power is done gradually, but for a long period of time coal and other conventional steam power plant will still play a major role as the fullfillment of electrical energy needs.

Refers to [1], China as a country that also relies on coal as its main fuel power plant creates a policy scheme that is *Energy-Saving Generation Dispatch* (ESGD) to minimize carbon emissions generated by power plants while still optimizing its performance. This scheme developed into many models of both mathematical and non-mathematical models, some of which are: environmental and economic coherence on ESGD by D Jun in 2009, the impact of ESGD on the feasibility of developing energy utilization by L Tianqi in 2010, ESGD analysis related to carbon emissions limits and renewable energy in 2012 by Nan He, and ESGD with *Minimum Cost Flow* by 2015 by Zhan'an Zhang and Xingguo Cai (see [2]). ESGD with *Minimum Cost Flow* by Zhan'an Zhang and Xingguo Cai looks at the power generation system as a network which best describes the topology of the power system and takes into account the power losses occurring in the transmission line. The resulting mathematical model is in the form of Quadratic Optimization Model with Quadratic Constraint or also known as Quadratic Constrained Quadratic Programming (QCQP).

QCQP is one of the foundations of Non-Linear Programming (NLP) (Lu *et al.*, 2011 [4]). QCQP covers issues like *Trust Region Method*, Max-Cut Problems, Quadratic Programming, and more. In determining the optimal global solution of QCQP various approaches such as the Semidefinite Programming approach for QCQP with two quadratic constraints by Ye and Zhang (2003) [7] have been done, and the decomposition of the matrix for the completion of the QCQP subclass in polynomial time by Sturm and Zhang (2003) [6]. To obtain the optimal global conditions of QCQP itself has been widely published in one of the writings in Gao's Canonical Theory of Duality in 2000. In this dual canonical approach, a dual canonical function is created so that each critical point corresponds to the Karush-Kuhn Tucker primal solution. Lu *et al* (2011) [4] developed an algorithm to produce a Karush-Kuhn Tucker solution that satisfies optimal global conditions for QCQP.

In this paper, ESGD completion is done by modeling *Minimum Cost Flow* in the form of QCQP Problem with Lagrange Multiplier Method. This problem includes Non-Conventional Non-Linear Programming so it is important to ensure an optimal and optimal global local solution.

2. Methodology: Quadratically Constrained Quadratic Programming

According to Cheng Lu *et al.* (2011) ([4]) ,a Quadratic Programming problem with Quadratic Constraints otherwise known as QCQP (Quadratically Constrained Quadratic Programming) is a Quadratic Programming problem with constraints of quadratic function as well. QCQP is one of the foundations of Non-Linear Programming with theoretical interest and application interest. Some subclasses of QCQP for example are Trust Region Method, problem Max-Cut, Quadratic Programming 0-1, and Box Constrained Quadratic Programming. QCQP includes the NP-Hard problem (Non-Linear Programming Difficult) which means it can not be solved in polynomial time unless P = NP. The general form of the QCQP Model is as follows,

$$\min \quad f(x) = \frac{1}{2}x^T Dx + c^T x \tag{1}$$

$$\text{s.t.} \quad \frac{1}{2}x^T D_i x + c_i^T x \leq b_i, i = 1, 2, ..., m \tag{2}$$

$$x \geq 0 \tag{3}$$

with D and D_i is a symmetric matrix of $n \times n$ whereas c and c_i are real vectors in $\mathbb{R}^n$. Defined $b = [b_1, ..., b_m]^T$. If the D matrix, $D_1, ..., D_m$ is a semidefinite positive matrix, then this QCQP problem becomes a matter of Convex Optimization. These subclasses can be completed in polynomial time via the Second Order Conical Programming, but to get the optimal solution to common QCQP problems is even more difficult.

To obtain an optimal global solution to the QCQP problem can be found in various literatures, such as: Non-Convent QCQP with two quadratic constraints and identification of subclass sets that can be solved by approach via Semidefinite Programming by Ye and Zhang (2003) [7], QCQP non- convex to Konrad Linear Programming by Sturm and Zhang (2003) [6], approximation algorithms for solving QP (*Quadratic Programming* problems) with some ellipsodial constraints by Fu, Luo, and Ye (1998) [3], or an approximation algorithm for QCQP with criteria special with Semidefinite Programming relaxation. In a dual canonical approach, a dual canonical function is generated such that any critical point of this function corresponds to Karush-Kuhn Tucker's primal solution. Furthermore, the Karush-Kuhn Tucker (KKT) solution corresponds to the optimal solution of the initial problem.

The Lagrange function for QCQP can be seen as follows.

$$\begin{aligned} L(x,\lambda) &= \frac{1}{2}x^T Dx + c^T x + \sum_{i=1}^{m} \lambda_i (\frac{1}{2}x^T D_i x + c_i^T x - b_i) \\ &= \frac{1}{2}x^T (D + \sum_{i=1}^{m} \lambda_i D_i)x + (c + \sum_{i=1}^{m} \lambda_i c_i)^T x - b^T \lambda \end{aligned} \tag{4}$$

with $\lambda \in \mathbb{R}^m_+$.

For a non-convex QCQP problem, a necessary requirement for an optimal local feasible solution x^* from the condition of the KKT is the existence of a vector Lagrange $lambda^* \in \mathbb{R}^m$ in such so,

$$\begin{aligned} \nabla_x L(x,\lambda^*)|_{x=x^*} = (D + \sum_{i=1}^{m} \lambda_i^* D_i)x^* + (c + \sum_{i=1}^{m} \lambda_i^* c_i) = 0, \\ \frac{1}{2}x^{*T} D_i x^* + c_i^T x^* \leq b_i, \qquad \lambda_i^* \geq 0, \\ \lambda_i^* (\frac{1}{2}x^{*T} D_i x^* + c_i^T x^* - b_i) = 0, \qquad i = 1, 2, ..., m \end{aligned} \tag{5}$$

and for the x^* corresponding point of the Lagrange vector λ^* it can be shown that the optimal value of the $f(x)$ objective function in System (1) $f(x^*) = L(x^*,\ lambda^*)$.

3. Problem Description: Energy Saving Generation Dispatch dengan Minimum Cost Flow

Sources of electrical energy in China still rely on Steam Power Plant with coal fuel, which is also a major contributor to carbon emissions. Energy-Saving Generation Dispatch is a scheme drafted by the Chinese government in 2007 in order to overcome the challenge of balancing energy supply fullfillment and sustaining the environmental balance.

Initially, the Chinese government compiled ESGD on the basis of 'average hours of work and generation' also known as 'percentage-based release'. This means the level of quality (in which case the carbon emissions it produces a little is good to say) a power plant viewed from the working hours of the power plant. At this stage, the working hours of the plant in a year for the same type of plant are almost the same so the annual generation quota can be said to be proportional to the unit capacity. However, it is not effective to see carbon emissions generated because the rate of fuel consumption (in other words the rate of coal consumption) is in fact not always proportional to the capacity of the unit. For example, a 1,000-MW power plant with a capacity of 1,000-MW removes 270 g of standard coal per kWh while a 50-MW power plant will issue 450 g of standard coal per kWh (Ding and Yang, 2013 [5]).

In 2007, the Chinese government developed ESGD. In this new version, power plants are ranked according to the fuel consumption rate from low to high; if the fuel consumption rate is the same then the power plant will be sorted based on the intensity of its carbon emission from low to high. The rankings of these power plants are more accurate, although there are other obstacles such as the reallocation of distribution of financial interest or investment attraction, where high efficiency power generation attracts more investors, or technical constraints such as power losses on transmission lines. This other constraint seems unavoidable, but can be minimized, therefore there is always the development of ESGD implementation from all sides. ESGDs with sequential coordination for example to generate better power plant clock scheduling to attract investors, or ESGDs by paying attention to lost power on the transmission line, and so on. But of all the emerging ESGD versions, the objective function or objective remains the same: minimizing the operating costs of power generation systems that are considered to represent the fuel costs that are directly proportional to the carbon emissions generated so that both seek to comply with government ESGD schemes.

IOP Conf. Series: Materials Science and Engineering **332** (2018) 012016 doi:10.1088/1757-899X/332/1/012016

The general model for Energy-Saving Generation Dispatch is the following.

$$\min \quad \sum_{t=1}^{T}\sum_{i=1}^{N} F_{it}(P_{it}) = \sum_{t=1}^{T}\sum_{i=1}^{N}(\alpha_i P_{it}^2 + \beta_i P_{it} + \gamma_i) \tag{6}$$

$$s.t \quad \sum_{i=1}^{N} P_{i,t} = P_{Dt} + P_{L,t}, \qquad i = 1,2,...,N, \qquad t = 1,2,...,T, \tag{7}$$

$$P_{i\min} \leqslant P_{i,t} \leqslant P_{i\max}, \qquad i = 1,2,...,N, \qquad t = 1,2,...,T, \tag{8}$$

$$U_{i\min} \leq U_{i,t} \leq U_{i\max} \qquad i = 1,2,...,N, \tag{9}$$

$$\begin{cases} P_{i,t} - P_{i,t-1} \leq UR_i, & \text{jika pembangkitan naik,} \\ P_{i,t-1} - P_{i,t} \leq DR_i, & \text{jika pembangkitan turun,} \end{cases} \tag{10}$$

$$\sum_{i=1}^{N}(P_i\max \times \lambda_{i,t}) \geq (P_{Dt} + R_t), \tag{11}$$

$$\begin{cases} P_i^{\min} \leq P_i \leq P_{i,l}^{l}, \\ P_{i,j-1}^{u} \leq P_i \leq P_{i,j}^{l}, \\ P_{i,z_i}^{u} \leq P_i \leq P_i^{\max}, & j = 2,3,...,z_i, \end{cases} \tag{12}$$

$$S_i \leq S_i^{\max}, \qquad i = 1,2,...,NL. \tag{13}$$

The notations can be seen in Table 1.

Table 1. Decision Variables

Variables	Details
F_{it}	Objective function for $i-$th generators
λ_i	Coefficient of objective function for $i-$th generators
β_i	Coefficient of objective function for $i-$th generators
γ_i	Coefficient of objective function for $i-$th generators
P_{it}	power for $i-$th generators at time t
N	Total number of generator which works at the operating system
$d_{it}(P_{it})$	Fungsi *electrical carbon-characteristic* (carbon emision CO_2 at time$-t$ where the output of the $-i$ unit is P_{it})
e	Emision coeffisient CO_2
q	The heating value of the fuel unit (= 8.14 kWh/kg from standard coal)
η	Power generator efficiency

The objective function intends to minimize the fuel costs that indirectly minimize the carbon emissions generated by the power generation system. Furthermore, this problem is viewed through Minimum Cost Flow which takes into account the shape of the network.

In the analysis of an electrical power system, there are three points to consider: (1) generating system, (2) transmission and distribution system (distribution), and (3) electricity user load. These three components are equally to be considered because they are interrelated and affect each other. Power generation establishes a network of electrical components, with the generator acting as a source component of the power source. The flow that flows in this network is the power flow. System optimization is done to deliver optimal power from generator to demand points using minimum cost. The cost here is the cost of using fuel (in this case coal) which is directly proportional to the amount of carbon emissions produced by the system, so minimize the cost of proportional to minimize carbon emissions. The term losses refers to the electrical energy lost to the transformer core (transformer) and the conductor or transmission cable in the network. This losses factor can look well on the representation of power generation networks by network flow diagram because a system of power generation itself is essentially a network bus power system.

4. Result and Discussion

4.1. Electrical Diagram Conversion (Bus Power System) into Network Flow Diagram

A network diagram flow is composed by dots and lines (Zhang and Cai, 2015 [2]. In the selection of dots that will be node in network flow diagram here are three, ie generator, point demand, and transformer. These three components are chosen because of the increase and loss of power. Generator as power producer becomes point source and point demand becomes point sink. The transmission cable as the power flow path becomes edge and the transformer becomes the point between the point source and demand following the network bus power system.

Suppose forming a network flow diagram as a representation of electrical diagrams. Optimal flow will be searched in each edge so that the existing flow is optimal with minimal system cost. To flow from point i to point j denoted stream f_{ij}. Suppose u_{ij} is capacity in existing edge and c_{ij} is the cost coefficient per unit flowing from point i to point j, then a stream f_{ij} in edge (i, j) requires $c_{ij}f_{ij}$ and f_{ij} can not be more than edge u_{ij}. Network is a directed network and is a weighted graph because every edge which represents the transmission cable in it has a non-negative load so that it can qualify Minimum Cost Flow. Each edge also has a c and u cost whereas each point has a number of supply or demand of b_j with value b_j positive for point supply and negative for point demand.

4.2. Solving ESGD via Minimum Cost Flow Model

The main purpose of this model is to maximize the power flow from the generator (point textit source) to the point textit demand, so the problem is in the form of textit Maximum Flow, but because every textit edge in textit network is associated with cost, the more general the form of this problem is textit Minimum Cost Flow. The form of the problem in textit Minimum Cost Flow can be expressed as follows:

$$\begin{aligned} \min \quad & \sum_{(i,j)\in E} c_{ij} f_{ij} \\ s.t \quad & \sum_{j=1}^{n} f_{ij} - \omega_{ji} \sum_{j=1}^{n} f_{ji} = b_i \qquad \forall i \\ & 0 \leq f_{ij} \leq u_{ij} \end{aligned} \tag{14}$$

The cost function c_{ij} is an economic function that can be a constant, linear, quadratic, or other. Here, the cost function c_{ij} is a quadratic function (see (6)) which is then downgraded to power to get a minimal cost,

$$c_{ij} = \frac{df(P_{it})}{dP} = \frac{\alpha_i P_{it}^2 + \beta_i P_{it} + \gamma_i}{dP} = 2\alpha_i P_{it} + \beta_i \tag{15}$$

where f_{ij} represents $F(P_i)$ or P_L which shows a lot of flow of power flowing, so c_{ij} can be written as,

$$c_{ij} = 2\alpha_i f_{ij} + \beta_i \tag{16}$$

In addition to fuel consumption (which is represented by the cost function), loss of power on the transmission line is also a power-related function f_{ij} on a network system. The loss of power occurring at transformer points and other points where there is an increase or decrease in power is represented by ω_{ji} associated with f_{ij}, so that finally the model is a Quadratic Programming with Quadratic Constraints.

Suppose the increment or power reduction coefficient is represented by κ_{ji}, in this case the value is between 0 and 1 because all the existing lines are *loss arc* or there is a reduction in power. In considering this power loss, the number of lost power units is converted into fuel consumption standards (coal) where one Mega Watt equals 0.123 ton/h fuel consumption. So ω_{ji} as the cost coefficient of missing power calculated in objective function can be written as,

$$\omega_{ji} = 0.123 \times \kappa_{ji} \times f_{ij}$$

maka secara lengkap permasalahan ESGD dalam bentuk *Minimum Cost Flow* dapat dituliskan sebagai,

$$\begin{aligned} \min \quad & \sum_{(i,j)\in E} (2\alpha_i f_{ij} + \beta_i) f_{ij} + (0.123\kappa_{ji} f_{ij}) f_{ij} \\ \text{s.t.} \quad & \sum_{j=1}^{n} f_{ij} - \kappa_{ji} f_{ij} \sum_{j=1}^{n} f_{ji} = b_i \qquad \forall i \\ & 0 \leq f_{ij} \leq u_{ij} \end{aligned} \tag{17}$$

and subsequently rewritten as,

$$\begin{aligned} \min \quad & \sum_{(i,j)\in E} 2\alpha_i f_{ij}^2 + \beta_i f_{ij} + (0.123\kappa_{ji} f_{ij}^2) \\ \text{s.t.} \quad & \sum_{j=1}^{n} f_{ij} - \kappa_{ji} \sum_{j=1}^{n} f_{ji}^2 = b_i \qquad \forall i \\ & 0 \le f_{ij} \le u_{ij} \end{aligned} \tag{18}$$

4.3. Energy-Saving Generation Dispatch Model in Quadratic Programming Form with Quadratic Constraints

The general form of QCQP has been shown in Equation (1), System (18) will be written in that general form.

$$\begin{aligned} \min \quad & f(x) = \frac{1}{2} f_{ij}^T D f_{ij} + c^T f_{ij} \\ \text{s.t.} \quad & \frac{1}{2} f_{ij}^T D_i f_{ij} + c_i^T f_{ij} \le b_i, \qquad i = 1, 2, ..., m \end{aligned} \tag{19}$$

The D matrix is the matrix with the $\{d_i | d_i = \alpha_i, i = 1, ..., m\}$ and vector c containing $\{\delta_i | \delta_i = \beta_i, i = 1, ..., m\}$. The same is true of the D_i and vector c_i matrices. Suppose λ Lagrange Multiplier for inequality and μ Lagrange Multiplier for the equation, Lagrange equation is formed as follows;

$$\begin{aligned} L(x, \lambda, \mu) &= \frac{1}{2} x^T D x + c^T x + \sum_{i=1}^{m} \lambda_i (\frac{1}{2} x^T D_i x + c_i^T x - b_i) \\ &= \frac{1}{2} x^T (D + \sum_{i=1}^{m} \lambda_i D_i) x + (c + \sum_{i}^{m} \lambda_i c_i)^T x - b^T \lambda \end{aligned} \tag{20}$$

We can then use solver to solve Eq. (20).

The following is the example for numerical calculations for Energy-Saving Generation Dispatch Problems in Quadratic Programming form with Quadratic Constraints. Suppose a bus power system as follows,

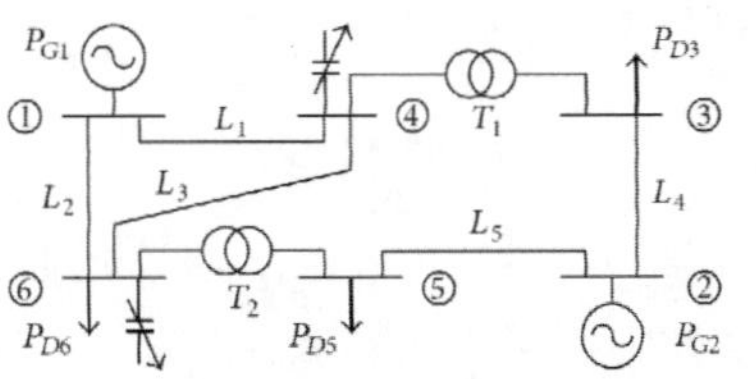

Figure 1. Contoh Diagram *Bus Power System*

and system parameters as follows,

IOP Conf. Series: Materials Science and Engineering **332** (2018) 012016 doi:10.1088/1757-899X/332/1/012016

Table 2. Transitional Cable Parameters

Route	V (kV)	Model	r_0 (Ω/km)	L (km)	R (Ω)
L_1	110	LGJ-185	0.17	71.2	12.1
L_2	110	LGJ-185	0.17	71.2	12.1
L_3	110	LGJ-185	0.17	71.2	12.1
L_4	10	LGJ-120	0.242	10	2.42
L_5	10	LGJ-120	0.242	10	2.42

Table 3. Transformer Parameters

Transformator	V (kV)	Model	$\triangle P_k$ (kw)	R (Ω)
T_1	110/11	SFPL1-50000/110	250	1.21
T_2	110/11	SFPL1-50000/110	250	1.21

Table 4. Generator Parameter

Generator	Capacity (MW)	Output	α_i	β_i	γ_i
G_1	100	$50 \leq P_{G1} \leq 100$	0.0014	0.25	2.5
G_2	100	$50 \leq P_{G2} \leq 100$	0.0018	0.18	5.0

To change the diagram in Figure 1 to a *network diagram flow*, then select the points that become the source point first, in this case there are two points, namely point 1 and point 2 representing the generator as power producer. Then select the point where the destination point, here there are three destination point or point demand power that is point 3, 5, and 6. The last step is to assemble the network on network flow diagram by following the flow in bus power system, other points in the network are points where there is a reduction of power during delivery, ie the point at which there is a transformer or capacitor. Network diagram flow generated from the diagram bus power systemin Figure 1 is as follows. The objective function here aims to minimize the

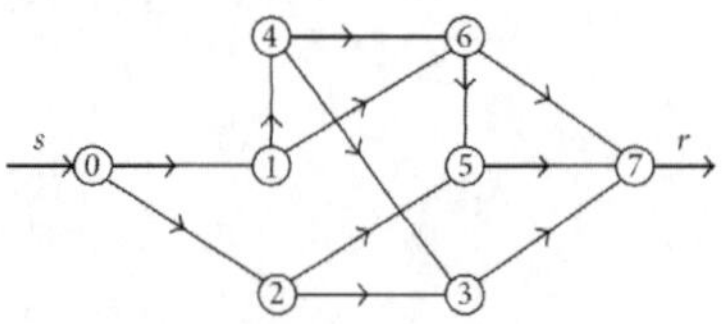

Figure 2. *Network Diagram Flow* from Figure 1

cost of operation, so that the line that is considered is the line where the power, ie lines associated with generators, transformers, and capacitors. For the coefficient of increment or reduction of power κ_{ji} is determined from the existing system device specifications. The constraint function is seen from the form of the system by following the concept of Minimum Cost Flow, and in each line corresponding to the transformer and capacitor, in which there is a loss of power, a corresponding reduction of power loss coefficients like that of the system (18). Here form the model is formed,

$$\begin{aligned} \min \quad F(x) = {} & (0.25x_{01} + 0.0028x_{01}^2) + (0.18x_{02} + 0.0036x_{02}^2) \\ & + 0.0002 \times 0.123 \times (x_{43}^2 + x_{65}^2) \\ & + 0.0004 \times 0.123 \times (x_{25}^2 + x_{23}^2) \\ & + 0.002 \times 0.123 \times (x_{14}^2 + x_{46}^2 + x_{16}^2) \end{aligned} \tag{21}$$

s.t.

$$x_{01} - x_{14} - x_{16} = 0 \tag{22}$$

$$x_{02} - x_{25} - x_{23} = 0 \tag{23}$$

$$x_{23} - 0.0002x_{23}^2 + x_{43} - 0.0001x_{43}^2 - x_{37} = 0 \tag{24}$$

$$x_{14} - 0.001x_{14}^2 - x_{46} - x_{43} = 0 \tag{25}$$

$$x_{65} - 0.0001x_{65}^2 + x_{25} - 0.0002x_{25}^2 - x_{57} = 0 \tag{26}$$

$$x_{46} - 0.001x_{46}^2 + x_{16} - 0.001x_{16}^2 - x_{65} - x_{67} = 0 \tag{27}$$

$$x_{37} + x_{57} + x_{67} = P_{D3} + P_{D5} + P_{D6} \tag{28}$$

$x_{01} \leq 100$	(29)	$x_{23} \leq 100$	(37)	$x_{16} \geq 0$	(45)
$x_{02} \leq 100$	(30)	$x_{67} \leq 100$	(38)	$x_{43} \geq 0$	(46)
$x_{14} \leq 100$	(31)	$x_{57} \leq 100$	(39)	$x_{25} \geq 0$	(47)
$x_{46} \leq 100$	(32)	$x_{37} \leq 100$	(40)	$x_{65} \geq 0$	(48)
$x_{16} \leq 100$	(33)	$x_{01} \geq 50$	(41)	$x_{23} \geq 0$	(49)
$x_{43} \leq 50$	(34)	$x_{02} \geq 50$	(42)	$x_{67} \geq 0$	(50)
$x_{25} \leq 100$	(35)	$x_{14} \geq 0$	(43)	$x_{57} \geq 0$	(51)
$x_{65} \leq 50$	(36)	$x_{46} \geq 0$	(44)	$x_{37} \geq 0$	(52)

There are 31 constraints with 7 equations and 24 inequalities and 4 of which are quadratic. The inequality constraints on this system will be seen as a variable limit.

IOP Conf. Series: Materials Science and Engineering **332** (2018) 012016 doi:10.1088/1757-899X/332/1/012016

The input textit demand in this calculation or $(P_{D3} + P_{D5} + P_{D6})$ is 193.8.

$$
\begin{aligned}
L(x,\mu) =& 0.25x_1 + 0.0028x_1^2 + 0.18x_2 + 0.0036x_2^2 \\
&+ 0.0000246x_3^2 + 0.0000246x_4^2 + 0.0000492x_5^2 + 0.0000492x_6^2 \\
&+ 0.000246x_7^2 + 0.000246x_8^2 + 0.000246x_9^2 \\
&+ \mu_1(x_1 - x_7 - x_9) \\
&+ \mu_2(x_2 - x_5 - x_6) \\
&+ \mu_3(x_6 - 0.0002\,{x_6}^2 + x_3 - 0.0001\,{x_3}^2 - x_{10}) \\
&+ \mu_4(x_6 - 0.0002\,{x_6}^2 + x_3 - 0.0001\,{x_3}^2 - x_{10}) \\
&+ \mu_5(x_4 - 0.0001\,{x_4}^2 + x_5 - 0.0002\,{x_5}^2 - x_{11}) \\
&+ \mu_6(x_8 - 0.001\,{x_8}^2 + x_9 - 0.001\,{x_9}^2 - x_4 - x_{12}) \\
&+ \mu_7(x_{10} + x_{11} + x_{12} - 193.8) \\
=& \mu_1x_1 - \mu_1x_7 - \mu_1x_9 + \mu_2x_2 \\
&- \mu_2x_5 - \mu_2x_6 + \mu_3x_6 - 0.0002\mu_3{x_6}^2 \\
&+ \mu_3x_3 - 0.0001\mu_3{x_3}^2 - \mu_3x_{10} + \mu_4x_7 \\
&- 0.001\,\mu_4{x_7}^2 - \mu_4x_8 - \mu_4x_3 + \mu_5x_4 \\
&- 0.0001\,\mu_5{x_4}^2 + \mu_5x_5 - 0.0002\,\mu_5{x_5}^2 - \mu_5x_{11} \\
&+ \mu_6x_8 - 0.001\,\mu_6{x_8}^2 + \mu_6x_9 - 0.001\,\mu_6{x_9}^2 \\
&- \mu_6x_4 - \mu_6x_{12} + \mu_7x_{10} + \mu_7x_{11} \\
&+ \mu_7x_{12} - 193.8\,\mu_7 + 0.000246\,{x_9}^2 + 0.000246\,{x_8}^2 + 0.000246\,{x_7}^2 \\
&+ 0.0000492\,{x_5}^2 + 0.0000492\,{x_6}^2 + 0.0000246\,{x_3}^2 + 0.0000246\,{x_4}^2 \\
&+ 0.0036\,{x_2}^2 + 0.18\,x_2 + 0.25\,x_1 + 0.0028\,{x_1}^2
\end{aligned}
\tag{53}
$$

IOP Conf. Series: Materials Science and Engineering **332** (2018) 012016 doi:10.1088/1757-899X/332/1/012016

4.4. Kondisi Karush-Kuhn Tucker untuk Contoh Numerik

Equation (53) will be downgraded to the x_i with $i = 1, \ldots, 12$ and μ_j variables with $j = 1, \ldots, 7$.

$$\frac{\partial L}{\partial x_1} = 0.25 + 0.0056\, x_1 + \mu_1 = 0 \tag{54}$$

$$\frac{\partial L}{\partial x_2} = 0.18 + 0.0072\, x_2 + \mu_2 = 0 \tag{55}$$

$$\frac{\partial L}{\partial x_3} = 0.0000492\, x_3 + \mu_3\,(1 - 0.0002\, x_3) - \mu_4 = 0 \tag{56}$$

$$\frac{\partial L}{\partial x_4} = 0.0000492\, x_4 + \mu_5\,(1 - 0.0002\, x_4) - \mu_6 = 0 \tag{57}$$

$$\frac{\partial L}{\partial x_5} = 0.0000984\, x_5 - \mu_2 + \mu_5\,(1 - 0.0004\, x_5) = 0 \tag{58}$$

$$\frac{\partial L}{\partial x_6} = 0.0000984\, x_6 - \mu_2 + \mu_3\,(1 - 0.0004\, x_6) = 0 \tag{59}$$

$$\frac{\partial L}{\partial x_7} = 0.000492\, x_7 - \mu_1 + \mu_4\,(1 - 0.002\, x_7) = 0 \tag{60}$$

$$\frac{\partial L}{\partial x_8} = 0.000492\, x_8 - \mu_4 + \mu_6\,(1 - 0.002\, x_8) = 0 \tag{61}$$

$$\frac{\partial L}{\partial x_9} = 0.000492\, x_9 - \mu_1 + \mu_6\,(1 - 0.002\, x_9) = 0 \tag{62}$$

$$\frac{\partial L}{\partial x_{10}} = -\mu_3 + \mu_7 = 0 \tag{63}$$

$$\frac{\partial L}{\partial x_{11}} = -\mu_5 + \mu_7 = 0 \tag{64}$$

$$\frac{\partial L}{\partial x_{12}} = -\mu_6 + \mu_7 = 0 \tag{65}$$

$$\frac{\partial L}{\partial \mu_1} = x_1 - x_7 - x_9 = 0 \tag{66}$$

$$\frac{\partial L}{\partial \mu_2} = x_2 - x_5 - x_6 = 0 \tag{67}$$

$$\frac{\partial L}{\partial \mu_3} = x_6 - 0.0002\, {x_6}^2 + x_3 - 0.0001\, {x_3}^2 - x_{10} = 0 \tag{68}$$

$$\frac{\partial L}{\partial \mu_4} = x_7 - 0.001\, {x_7}^2 - x_8 - x_3 = 0 \tag{69}$$

$$\frac{\partial L}{\partial \mu_5} = x_4 - 0.0001\, {x_4}^2 + x_5 - 0.0002\, {x_5}^2 - x_{11} = 0 \tag{70}$$

$$\frac{\partial L}{\partial \mu_6} = x_8 - 0.001\, {x_8}^2 + x_9 - 0.001\, {x_9}^2 - x_4 - x_{12} = 0 \tag{71}$$

$$\frac{\partial L}{\partial \mu_7} = x_{10} + x_{11} + x_{12} - 193.8 = 0 \tag{72}$$

A Quadratic Programming with Quadratic Constraints is a subclass of Non-Linear Programming, used *solver NLPSolve* to solve equations (53). The solution of this numerical example (with rounding) is

Table 5. Numerical Calculation Result

Variables	Variable in Maple	Value
x_{01}	x_1	100
x_{02}	x_2	100
x_{14}	x_7	48.1
x_{16}	x_9	51.9
x_{23}	x_6	50
x_{25}	x_5	50
x_{43}	x_3	41.6
x_{46}	x_8	4.2
x_{65}	x_4	0
x_{67}	x_{12}	53.3
x_{57}	x_{11}	49.5
x_{37}	x_{10}	91

and the cost required for the operation of this power plant system is 108.5 unit cost. When viewed again in the diagram network that has been created then the flow of power that occurs on each line is as follows.

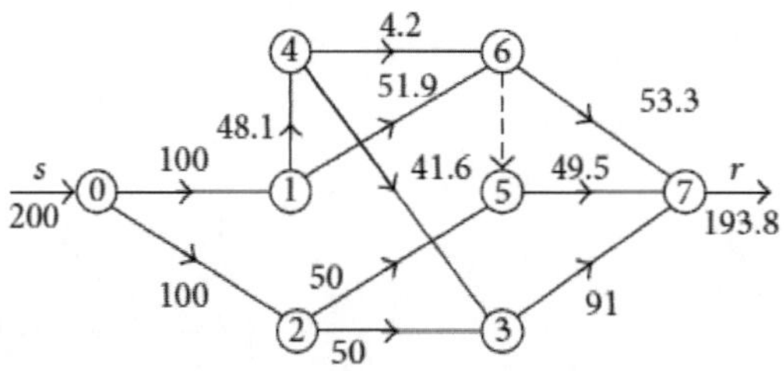

Figure 3. The Flow of Power that occurs on each line

This calculation is done with the help of textit Software Maple, takes 0.09 seconds and 38.37M memory on Intel (R) Core i5-4210U CPU system @ 1.70GHz, 4.00 GB RAM and 64-bit operating system. Furthermore, as the load changes occur, recalculation of the power flowing in each path adjusts to the need and a new combination of lines in each line will be obtained for the planning of the operation of the power generation system again.

Acknowledgement
This research is funded by Competency Based Research or *Penelitian Berbasis Kompetensi* from Indonesian Ministry of Research Technology and Higher Education for the year of 2017 contract number 718/UN6.3.1/PL/2017.

References

[1] Zhong, H., Xia, Q., Chen, Y. dan Kang, C. 2015. Energy-saving generation dispatch toward a sustainable electric power industry in China. *Energy Policy*, vol. 83, pp. 14-25.
[2] Zhang, Z. dan Cai X. 2015. Energy-Saving Generation Dispatch Using Minimum Cost Flow. *Mathematical Problems in Engineering*, Article ID 562462, 9 pages.
[3] Fu, M., Luo, Z., and Ye, Y. 1998. Approximation algorithms for quadratic programming, *J. Comb. Optim.*, vol. 2, pp. 29-50.
[4] Lu, C., Fang, Shu-Cheng., and Jin, Xingwei. 2011. KKT Solution and Conic Relaxation for Solving Quadratically Constrained Quadratic Programming Problems. *SIAM Journal on Optimization*, vol. 21, pp. 1475-1490.
[5] Ding, Y. dan Yang, H. 2013. Promoting Energy-Saving and Environmentally Friendly Generation Dispatching Model in China: Phase Development and Case Studies. *Energy Policy*, vol. 57,109118.
[6] Sturm, J.F and Zhang, S. 2003. On cones of nonnegative quadratic functions. *Math. Oper. Res.*, vol. 28, pp. 246-267.
[7] Ye, Y. and Zhang, S. 2003. New results on quadratic minimization. *SIAM J. Optim.*, vol. 14, pp. 245-267.

A comparative study of three pillars system and banking methods in accounting long-term purposes of retiree in Indonesian saving account

E S Hasbullah[1], M Suyudi[1], N A Halim[2], Sukono[2*], F Gustaf, A S Putra[3]

[1]Department of Mathematics, Faculty of Mathematics and Natuaral Sciences, Universitas Padjadjaran, INDONESIA
[2]Department of Mathematics, Center of Analysis for Informatics and Mathematics, Universiti Malaysia Terengganu, MALAYSIA
[3]Department of Geophysics, Faculty of Mathematics and Natuaral Sciences, Universitas Padjadjaran, INDONESIA

*Corresponding author's e-mail: sukono@unpad.ac.id;

Abstract. Human productivity is the main capital in economic activity. This main factor leads to the fact that the continuity of human resources in economic sector depends on the limited productivity age. In other word, once the economic agents has reach the limit of the productivity age. Hence they enter the pension state. In this case, the preparation of 'old-age' fund become crucial and should be initiated before the pension state to avoid the destitute condition of retiree. Two most simple and familiar methods in preparing the pension fund are The Three Pillars system and banking methods. Here we simulate the both of the methods for the synthetic data of investment program and analyse the result. The result gives the idea that the Three Pillar System has effective prospect in Long-term scheme. However, the banking method is likely adapted to the short-term plan.

Keywords: Pension Fund, Normal Fee, Three Pillars System, Banking Methods, Saving

1. Introduction

The human age will increase as well as people will grow older. This affects human productivity in finance and economic needs. There are even a moment that humans can no longer meet their need for finances altogether. At that time, the routine of life should be kept as usual. The cost of living is the same or even greater than usual activities will be very burdensome due to reduced finances. According to this background, It is necessary to make a strategy to solve this problem. One solution is to prepare the pension funds prepared during the productive age.

In Indonesia there are many methods of calculating pension funds. Even some agencies offer services to prepare the pension fund. We have to be able to select a best method that will give best result for preparation of pension fund. The Three Pillar System ([1]) and Banking Methods are the methods used to perform calculation of pension funds in several countries. The Three Pillar System has the advantage of being able to show advantages and disadvantages in the calculation of pension funds. While the Banking Method has an advantage because of its exact calculation. Basically both these systems have different advantages and disadvantages.

The objective of this research is to analyze the use of Three Pillar System with Banking Calculation Method in Indonesia and to find advantages and disadvantages in its application in Indonesia.

2. Basic Concept

The Three Pillar System is a retirement format established by the World Bank in 1998 and has been widely adopted by other countries in Central and Eastern Europe. The purpose of the Three Pillar System is to separate the focus of the object on the calculation of pension funds into three main focuses, namely: (1) Pillar 1 - State Pension, (2) Pillar 2 - Employee Retirement, (3) Pillar 3 - Individual Retirement.

Due to GOV.UK, State pensions are regular government pension payments allowing residents of the country to claim pension funds when they reach the state of pension age. While Retirement Worker is a pension plan that is produced by a company or organization for the benefit of its employees [2], [3]. In this fee scheme, employers and employees contribute to fund the investment that grows during the savings period. In non-contribution schemes, only the employer contributes. The amounts paid to employees on retirement will depend on the type of the scheme and reflect the contribution entered into the employee's last salary. This Scheme is also known as a defined contribution scheme and is one of two employment pension schemes (the other is the final salary scheme). The money-purchase scheme is a scheme where employees effectively have their own pension funds in working pension funds.

According to [3], Individual Retirement is defined as a contribution arrangement between an individual and a pension provider, usually a financial organization such as a community equivalent, a bank, or an insurance company where the employer may also contribute. Individual Retirement can be arranged directly between individuals and providers. In many cases, it is usually fixed by the employer as a means provided to all employees to be able to access pension programs run by pension providers. These are often referred to as private group pensions or in other words employers who sponsor pensions.

2.1 Actuarial Basic Function

The actuarial basic function is a function used for the calculation of normal contributions and actuarial liabilities. Actuarial basic functions used include composite survival function, interest rate function, salary function, benefit function and annuity function. The following is an explanation of each of these functions ([4]).

***Composite Survival Function* ([4])**

Composite survival function is a function that indicates the chance of participants until the time allowed to retire. For example, if a participant is x, the chances of staying active during the active period for the next n year are formulated as follows:

$$ {}_nP_x^{(T)} = \frac{l_{x+n}^{(T)}}{l_x^{(T)}} \tag{1} $$

where ${}_nP_x^{(T)}$ is the probability of an x-old participant will remain in service until n next year. $l_{x+n}^{(T)}$ Is the total active employers in age of $x+n$. While, $l_x^{(T)}$ is the active employers in age of x.

Interest Rate Function **([4])**

The interest rate function is used to discount future payments for now. If i is the interest rate assumed to be constant for n years, then the discount factor for n years is calculated as follows:

$$v^n = \frac{1}{(1+i)^n} \tag{2}$$

with v^n is the discontinue factor within n years.

Salary Function **([4])**

The participant's salary for one year at age x is denoted by s_x, S_x represents the cumulative salary of participants from the age of entry y to age x-1 whereas the cumulative salary of participants from the age of admission y to r-1, which is formulated as follows:

$$S_r = \sum_{t=y}^{r-1} s_t \tag{3}$$

with s_t is salary of participants for a years when age t year.

Benefit Function **([4])**

Benefit functions are used to determine the amount of benefits paid to participants at retirement, accelerated retirement, disability, or death. There are three types of benefit functions used to obtain cumulative benefits, namely *flat dollar unit benefits*, *career average*, and *final average*. In this study the function of benefits used for normal retirement age is a career average with the following formula:

$$B_r = KS_r \tag{4}$$

with B_r is the cumulative benefit achieved by participants from ages y years to r-1, K is the proportion of the average salary provided per year of service.

Annuity Function **([4])**

An annuity is a payment made by the Pension Fund to a certain number of participants, conducted over a certain period of time or duration, on an ongoing basis. Annuities can be obtained by the following formula:

$$\ddot{a}_x = \sum_{t=0}^{\infty} {}_tP_x^{(m)}.v^t \tag{5}$$

with is $\ddot{a}_x$ the present value of the lifetime annuity at the age of x, ${}_tP_x^{(m)}$ is an opportunity x-age participants will remain alive for t years later.

3. Methodology

The data for this research is defined as the Table of *Mortality* and the *Morbidity*. The Table of Mortality indicates the estimated mortality rate occurs each year in each age group. The amount of pure premium to be paid is determined by mortality rate. While the table of Morbidity indicates the percentile of events each age group. In this research, the *dependent variable* is the amount of the participant's pension benefit (rupiah), the contribution to be paid by the pension participant.

3.1 Pay as You Go Method

According to [4], the Pay as You Go Method is a Methodology used to establish an actuarial balance in pension calculations. In the Pay as You Go Method or called the Current Cost Method, the employer only finances an employee's or an employee's retirement benefit as needed outside of the last salary.

In a simplified form, actuarial balance (AB) can be expressed as:

$$AB = \left[\frac{TF_0 + y_0 \sum_{t=0}^{74} \theta_t N_t \prod_{h=1}^{t} \frac{(1+g_h)}{(1+r_h)}}{y_0 \sum_{t=0}^{74} N_t \prod_{h=1}^{t} \frac{(1+g_h)}{(1+r_h)}} \right] - \left[\frac{B_0 \sum_{t=0}^{74} R_t \prod_{h=1}^{t} \frac{(1+\lambda_h)}{(1+r_h)} + \prod_{h=1}^{74} \frac{(TF_{74})}{(1+r_h)}}{y_0 \sum_{t=0}^{74} N_t \prod_{h=1}^{t} \frac{(1+g_h)}{(1+r_h)}} \right] \approx 0 \qquad (6)$$

whereas:

TF_0	: Asset value at the beginning of the appraisal period
θ_t	: Tax payroll (contribution) rate in year t
Y_0	: Average base contribution in year 0
N_t	: Number of contributors in year t
g	: Annual real wage growth rate
r	: Projected outcome on trust fund assets
BO	: Average retirement (benefit) in year 0
R_t	: Number of pensioners in year t
λ	: The real yearly growth rate benefits
$TF74$	: Asset value at the end of the valuation period

From the actuarial equilibrium equation we get the normal fee equation based on PAYG method

$$NC_x = y_0 \sum_{t=0}^{74} \theta_t N_t \prod_{h=1}^{t} \frac{(1+g_h)}{(1+r_h)} \qquad (7)$$

while the the retirement benefits of participants at age r based on PAYG are formulated by equation

$$B_r = B_0 \sum_{t=0}^{74} R_t \prod_{h=1}^{t} \frac{(1+\lambda_h)}{(1+r_h)} \qquad (8)$$

3.2 Constant Percent of Salary Method ([5])

Constant Percent of Salary is a method of pension funding that calculates pension benefits based on employee salaries from the first employment. This method is used to calculate the amount of actuarial obligation that the company must incur to the employee at the time of normal retirement. The calculation of actuarial liabilities with Constant Percent type is derived from the multiplication of the present value of future benefits with an annuity comparison based on the participant's salary. Then, the normal contribution is obtained from the multiplication of the salary of participants with the proportion of "K" wages of Constant Percent type.

Actuarial Liabilities ([5])

Actuarial liability is an obligation to provide benefits to pension plan participants at either retirement due to death, disability, early retirement or resignation. The actuarial obligations of the Constant Percent type can be written as follows:

$$^{CPr}(AL)_x = \frac{^s\ddot{a}^{(T)}_{(\overline{yx}-\overline{y})}}{^s\ddot{a}^{(T)}_{(\overline{yr}-\overline{y})}}(B_r \cdot {}_{r-x}P^{(T)}_x . v^{r-x} . \ddot{a}_r) \tag{9}$$

with information as follows:

$^{CPr}(AL)_x$: actuarial liabilities for active participants at age x years with normal retirement age using Constant Percent.

$^s\ddot{a}^{(T)}_{(\overline{yx}-\overline{y})}$: cumulative current value of lifetime annuity of y years until the age of r-y years based on participant's salary

$^s\ddot{a}^{(T)}_{(\overline{yr}-\overline{y})}$: cumulative current value of lifetime annuity of y years up to the age of x-y years based on participant's salary.

B_r : cumulative benefits achieved by participants from ages y years to r-1.

${}_{r-x}P^{(T)}_x$: the chances of age participants will continue to work until r-x years upcoming.

v^{r-x} : discount factor from age to age.

$\ddot{a}_r$: the present value of a lifetime annuity at retirement age.

The function $B_r \cdot {}_{r-x}P^{(T)}_x . v^{r-x} . \ddot{a}_r$ is a function to obtain the value $^r(PVFB)_x$, ie the present value of the normal pension benefit at age *x* years. So the equation can be written as follows:

$$^{CPr}(AL)_x = \frac{^s\ddot{a}^{(T)}_{(\overline{yx}-\overline{y})}}{^s\ddot{a}^{(T)}_{(\overline{yr}-\overline{y})}}\,{}^r(PVBV)_x \tag{10}$$

Normal Cost **([5])**

The normal fee is the fee that must be fulfilled or paid annually by the participant during the active period of work. In the Constant Percent method, the contribution is normally derived from the percentage of the participant's salary. The normal contribution to the Constant Percent method is derived from multiplying the proportion of "*K*" values with the participant's salary:

$$^{CPr}(NC)_x = Ks_x \tag{11}$$

with $^{CPr}(NC)_x$ is the normal contribution for active participants aged *x* to age *r* years using Constant Percent. *K* is the proportion of participant's salary based on Constant Percent type, where the value of "*K*" is obtained with the following formula:

$$K = \frac{^r(PVFB)_x}{s_y\, {}^s\ddot{a}^{(T)}_{(\overline{yr}-\overline{y})}} \tag{12}$$

with s_{yy} is the salary of the participant for a year at the age of *y* years $^r(PVFB)_x$ is the present value of the normal pension benefit at the age of y years.

Traditional Credit Unit

The Traditional Credit Unit is an actuarial method whose actuarial obligations are the value at the date of valuation, the pension benefit from the date of entry into the plan, by the date of valuation [5]. This method is also called the Cost Unit Credit Method by Erisa and Anderson, and the Cost Benefit Method by McGill and Grubbs.

If the participant is a pensioner at the age of r with the annual retirement of B_r, therefore $B_r\ddot{a}_r^{(12)}$ is sufficient to finance this pension at the age of r. B_r shows the benefits gained during the active years of service participants, from age e to age r.

$$b_x = \frac{B_r}{r-e} \tag{13}$$

The normal cost at the beginning of each year is the cost of the pension benefits received in that year. Given

$$NC_x = b_x.\frac{D_r^{(\tau)}}{D_x^{(\tau)}}.\ddot{a}_r^{(x)} \tag{14}$$

The normal cost for younger participants is lower due to greater effect than in discount rates $D_r^{(\tau)} / D_x^{(\tau)}$. The normal total cost for a plan each year is the sum of the normal costs for all participants receiving the benefits that should be received. The normal total cost in the coming years will be affected by *aging, the size of unit benefits, withdrawals, pensions, deaths,* and *newcomers.*

$$NC_x = b_x.\frac{D_r^{(\tau)}}{D_x^{(\tau)}} = b_x.v^t t.P_x.\ddot{a}_r^{(12)} \tag{15}$$

in the state of $t=r-x$, hence:

$$AL_x = NC_x.(x-e) \tag{16}$$

Saving Account ([6])

Saving Account is a deposit account held in a bank or other financial institution that provides basic security and a simple interest rate. Depending on the type of savings account, account holders cannot write checks from accounts (without adding additional fees or fees) and accounts tend to have a number of transfers / free transactions.

In a simplified form, the calculation of a deposit account can be expressed as:

$$B_n = deposit.\frac{\left(1+\frac{j}{m}\right)^{mn}-1}{\left(1+\frac{j}{m}\right)^{\frac{m}{p}}-1} \tag{17}$$

with information as follows:

B_n : Future value of saved funds
$deposit$: Large initial storage fund
j : Nominal percent interest
m : interest rate conversion
n : storage period
p : additional deposits per year

4. Result and Discussion

4.1 Comparison of the First Pillar

In this section, the calculation of contributions and retirement benefits for a person aged $x = 25$. With income of Rp5,000,000 per month and retirement at age $r = 56$ years, percentage of salary increase $S = 10\%$ and percentage of normal pension benefit equal to $k = 2\%$. The calculation for obtaining pension benefits for normal pensions using *Pay-as-You-GO method* is done using equation (8). Then it will be calculated the value of contributions and pension benefits using the Saving Account Method with equation (17).

A similar calculation is made to calculate the pension benefit and the normal contribution for the age of 25 to 56 years. The results are summarized in Table 1.

Table 1. Comparison Results on the First Pillar

	First Pillar System		Saving Account	
Age	Contribution	Benefit	Contribution	Benefit
25	6,000,000	6,000,000	6,000,000	6,000,000
26	13,200,000	12,240,000	12,000,000	12,120,000
27	21,780,000	18,727,200	19,200,000	18,362,400
28	31,944,000	25,468,992	27,120,000	24,729,648
29	43,923,000	32,472,965	35,832,000	31,224,241
30	57,978,360	39,746,909	45,415,200	37,848,726
31	74,405,562	47,298,822	55,956,720	44,605,700
32	93,538,421	55,136,912	67,552,392	51,497,814
33	115,753,796	63,269,607	80,307,631	58,527,771
34	141,476,861	71,705,554	94,338,394	65,698,326
35	171,187,002	80,453,632	109,772,234	73,012,293
36	205,424,403	89,522,950	126,749,457	80,472,538
37	244,797,413	98,922,860	145,424,403	88,081,989
38	289,990,782	108,662,957	165,966,843	95,843,629
39	341,774,850	118,753,089	188,563,527	103,760,501
40	401,015,824	129,203,360	213,419,880	111,835,712
41	468,687,245	140,024,142	240,761,868	120,072,426
42	545,882,791	151,226,073	270,838,055	128,473,874
43	633,830,574	162,820,072	303,921,861	137,043,352
44	733,909,085	174,817,341	340,314,047	145,784,219
45	847,664,994	187,229,372	380,345,451	154,699,903
46	976,832,993	200,067,957	424,379,996	163,793,901
47	1,123,357,942	213,345,195	472,817,996	173,069,779
48	1,289,419,550	227,073,494	526,099,796	182,531,175
49	1,477,459,901	241,265,587	584,709,775	192,181,798
50	1,690,214,127	255,934,535	649,180,753	202,025,434
51	1,930,744,599	271,093,735	720,098,828	212,065,943
52	2,202,479,024	286,756,928	798,108,711	222,307,262
53	2,509,252,888	302,938,212	883,919,582	232,753,407
54	2,855,356,735	319,652,044	978,311,540	243,408,475
55	3,245,588,822	336,913,255	1,082,142,694	254,276,645
56	3,685,313,759	354,737,053	1,196,356,963	265,362,178

And we can see in Graph comparing contributions and pension benefits from First Pillar and Banking Calculations.

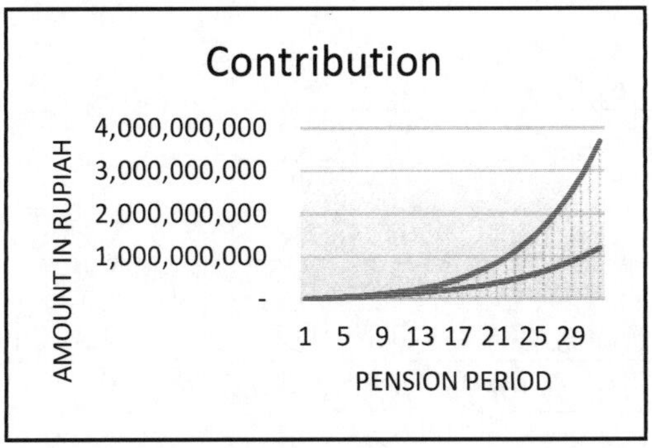

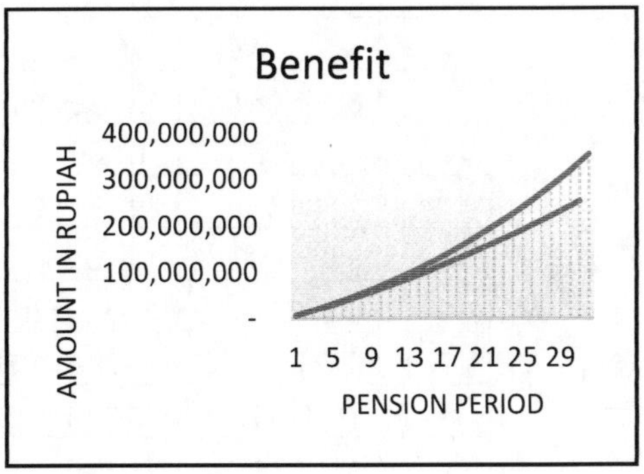

Figure 1. The graph comparing the first pillar (red line) and the saving account (blue line) for (a). Normal Contribution and (b). Benefit

4.2 Comparison on the Second Pillar

In this section a retirement benefit calculation is calculated for a person aged $x = 25$. With revenues of Rp5,000,000 per month and retirement at age $r = 56$ years, the percentage of salary increase $S = 10\%$ and the proportion of participant's salary equal to $K = 2\%$. The calculation for obtaining pension benefits for normal retirement using the Constant Percent of Salary method is done using equation (4).

Furthermore, calculated the normal contribution paid by the participant pension using equation (5). Then it will be calculated the value of contributions and benefits of pension using Saving Account Method with equation (12). A similar calculation is made to calculate the pension benefit and the normal contribution for the age of 25 to 56 years. The results are summarized in Table 2.

Table 2. Comparison Results on the Second Pillar

	First Pillar System		Saving Account	
Age	Contribution	Benefit	Contribution	Benefit
25	1,200,000	1,200,000	1,200,000	1,200,000
26	1,320,000	2,904,000	1,320,000	2,666,400
27	1,452,000	5,270,760	1,452,000	4,443,701
28	1,597,200	8,503,493	1,597,200	6,583,032
29	1,756,920	12,861,533	1,756,920	9,143,082
30	1,932,612	18,674,946	1,932,612	12,191,150
31	2,125,873	26,362,798	2,125,873	15,804,344
32	2,338,461	36,455,984	2,338,461	20,070,934
33	2,572,307	49,625,708	2,572,307	25,091,895
34	2,829,537	66,719,008	2,829,537	30,982,643
35	3,112,491	88,802,999	3,112,491	37,875,017
36	3,423,740	117,219,959	3,423,740	45,919,509
37	3,766,114	153,655,830	3,766,114	55,287,803
38	4,142,725	200,225,366	4,142,725	66,175,640
39	4,556,998	259,577,885	4,556,998	78,806,066

	First Pillar System		Saving Account	
40	5,012,698	335,028,524	5,012,698	93,433,104
41	5,513,968	430,721,046	5,513,968	110,345,911
42	6,065,364	551,829,669	6,065,364	129,873,476
43	6,671,901	704,809,116	6,671,901	152,389,941
44	7,339,091	897,704,243	7,339,091	178,320,604
45	8,073,000	1,140,533,240	8,073,000	208,148,718
46	8,880,300	1,445,761,660	8,880,300	242,423,162
47	9,768,330	1,828,888,500	9,768,330	281,767,117
48	10,745,163	2,309,170,523	10,745,163	326,887,869
49	11,819,679	2,910,517,014	11,819,679	378,587,868
50	13,001,647	3,662,594,610	13,001,647	437,777,235
51	14,301,812	4,602,190,996	14,301,812	505,487,869
52	15,731,993	5,774,897,443	15,731,993	582,889,382
53	17,305,192	7,237,183,974	17,305,192	671,307,079
54	19,035,712	9,058,957,871	19,035,712	772,242,255
55	20,939,283	11,326,716,991	20,939,283	887,395,092
56	23,033,211	14,147,434,900	23,033,211	1,018,690,505

Based on the data in Table 2. We can make a graph as Figure 2.

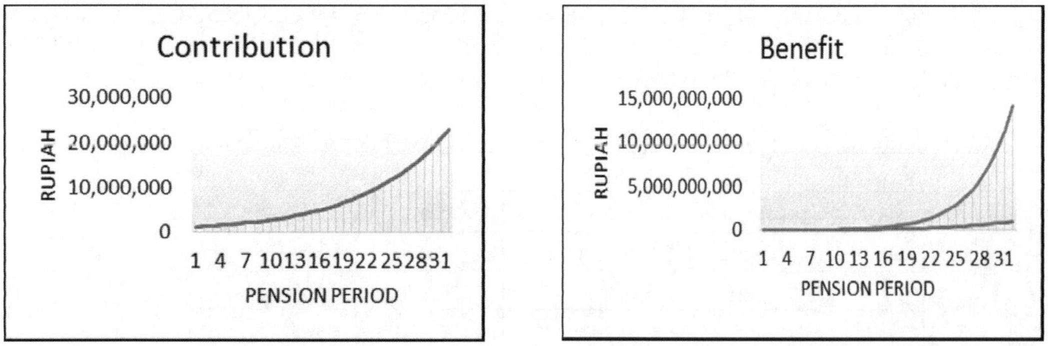

(a) (b)

Figure 2. The graph comparing the second pillar (red line) and the saving account (blue line) for (a). Normal Contribution and (b). Benefit

4.3 Comparison on the Third Pillar

In this section, the calculation of retirement benefit for a person aged $x = 35$ years old at the time of evaluation. It is assumed that a person is calculated to be a retirement participant at the age of $e = 25$ years and retired at age $r = 56$ years. The calculation to obtain pension benefit for normal pension using Traditional Credit Unit Method is done by using the formula (13).

Furthermore, the calculated normal contribution paid by the participant pension using equation (14). Then it will be calculated the value of contributions and pension benefits using the Saving Account Method with equation (18), namely: The same calculations performed to calculate the benefits of pensions. The normal fee and actuarial liability for the age of 25 to 56 years. The results are summarized in Table 3.

Table 3. Comparison Results on the Third Pillar

	First Pillar System		Saving Account	
Age	Contribution	Benefit	Contribution	Benefit
25	2,667,493	0	2,667,493	2,667,493
26	2,667,493	6,000,000	2,667,493	5,388,335
27	2,667,493	12,000,000	2,667,493	8,163,594
28	2,667,493	18,000,000	2,667,493	10,994,359
29	2,667,493	24,000,000	2,667,493	13,881,739
30	2,667,493	30,000,000	2,667,493	16,826,866
31	2,667,493	36,000,000	2,667,493	19,830,896
32	2,667,493	42,000,000	2,667,493	22,895,007
33	2,667,493	48,000,000	2,667,493	26,020,399
34	2,667,493	54,000,000	2,667,493	29,208,300
35	2,667,493	60,000,000	2,667,493	32,459,958
36	2,667,493	66,000,000	2,667,493	35,776,650
37	2,667,493	72,000,000	2,667,493	39,159,676
38	2,667,493	78,000,000	2,667,493	42,610,362
39	2,667,493	84,000,000	2,667,493	46,130,062
40	2,667,493	90,000,000	2,667,493	49,720,156
41	2,667,493	96,000,000	2,667,493	53,382,051
42	2,667,493	102,000,000	2,667,493	57,117,185
43	2,667,493	108,000,000	2,667,493	60,927,021
44	2,667,493	114,000,000	2,667,493	64,813,054
45	2,667,493	120,000,000	2,667,493	68,776,808
46	2,667,493	126,000,000	2,667,493	72,819,837
47	2,667,493	132,000,000	2,667,493	76,943,726
48	2,667,493	138,000,000	2,667,493	81,150,093
49	2,667,493	144,000,000	2,667,493	85,440,588
50	2,667,493	150,000,000	2,667,493	89,816,892
51	2,667,493	156,000,000	2,667,493	94,280,723
52	2,667,493	162,000,000	2,667,493	98,833,830
53	2,667,493	168,000,000	2,667,493	103,477,999
54	2,667,493	174,000,000	2,667,493	108,215,052
55	2,667,493	180,000,000	2,667,493	113,046,845
56	2,667,493	186,000,000	2,667,493	117,975,275

And the comparison between the methods is in Figure 3.

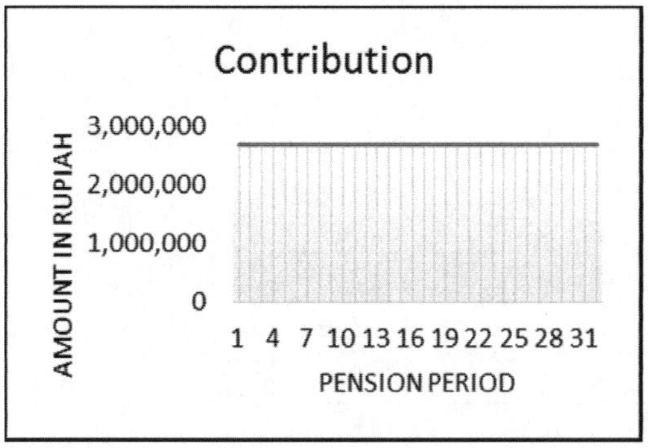

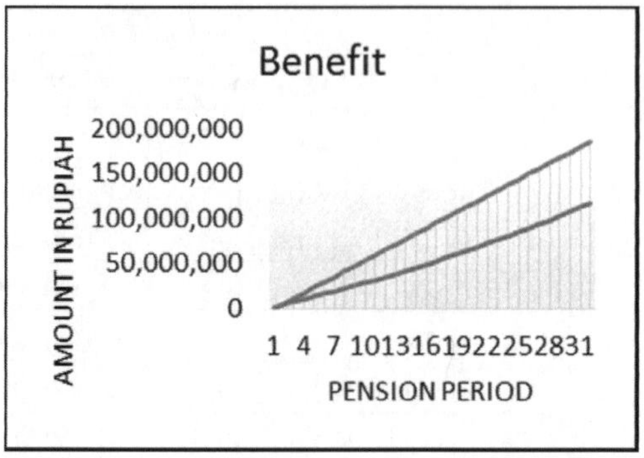

(a) (b)

Figure 3. The graph comparing the Third pillar (red line) and the saving account (blue line) for (a). Normal Contribution and (b). Benefit

From the results, we can see that the three pillar system provides more efficient benefits than conventional bank calculation system. With three different treatments, almost all calculations with the three pillar system give greater benefit value as the pension length increases. In the first pillar system, has not seen the advantages of the three pillar system. This is because the value of contributions in the first pillar system is greater than the calculation of the bank. Nevertheless, the benefits gained are greater. In the second pillar system, the value of the contribution is the same. Here it appears that the three pillar system can provide greater benefits. Similarly, the third pillar system that provides greater benefits with the same value contribution between the three pillar systems with saving account method.

Due to the calculations, it give us a new idea that the saving conventional account is too old to be used since the Three Pillars system may have more efficient calculation on giving benefit.

5. Conclusion

From the previous discussion, the authors conclude that; in the First Pillar the premium per age is greater and its benefits are greater as well compared to premium by the Saving Account Method. In the second Pillar, the obligatory premium is relatively constant but relatively larger than the method of Saving Account, while the benefits obtained are greater than the Saving Account Method. In the third Pillar, the premium is equal but their benefit is greater than the calculation by using Saving Account Method. So it can be concluded that the Three Pillar System is better than the Banking Calculation Method.

6. Acknowledgement

Further thanks to the Rector, Director of DRPMI, and Dean of FMIPA, Universitas Padjadjaran, which has a grant program Riset Fundamental Unpad (RFU) under the coordination of Mr. Endang S. Hasbullah, which is a means to increase research activities and publications to researchers at the Universitas Padjadjaran.

References

[1] http://www.investopedia.com/terms/s/savingsaccount.asp (accessed on 29 December 2016)
[2] Calculation of Retirement Benefits 2009 *The Eligibility Requirements and the Method of Calculating Retirement Benefits for Minnesota Public Employee Pension Plans*

[3] Crone S 2013 *Understanding the Basics of Actuarial Methods.* Pension review board. April No. 13-001: pp 1-24

[4] Indriya R S and Satyahadewi N 2014 Penerapan Metode Cost Prorate Tipe Constant Percent untuk Perhitungan Nilai Kewajiban Aktuaria Dana Pensiun. *Buletin Ilmiah Mat. Stat dan Terapannya (Bimaster)* Vol 03 No. 3 pp 223 - 226

[5] Pugh C 2006 Funding Rules and Actuarial Methods *OECD Working Papers On Insurance and Private /Pensions*, No. 1, OECD Publishing. Shepard, Mark. 2009. Minnesota Public Pensions. Research Department. St. Paul, MN 55155. Legislative Analyst/651-296-5051

[6] Yuli A, Des A Z, and Ensiwi M 2009 Perhitungan Dana Pensiun untuk Pensiun Normal Berdasarkan Metode Constant Dollar; Studi Kasus: PT. Taspen Palembang *Jurnal Penelitian Sains* **12** 2(A) 12202

IOP Conf. Series: Materials Science and Engineering **332** (2018) 012018 doi:10.1088/1757-899X/332/1/012018

Application of decision rules for empowering of Indonesian telematics services SMEs

E T Tosida*, O Hairlangga, F Amirudin, M Ridwanah

Computer Science Department, Pakuan University, Jl. Pakuan PO Box 452, Bogor, Indonesia

*Email: enengtitatosida@unpak.ac.id, oktahairlangga.062@gmail.com, fadlanamirudin043@gmail.com, Mridwanah@gmail.com

Abstract. The independence of the field of telematics became one of Indonesia's vision in 2024. One effort to achieve it can be done by empowering SMEs in the field of telematics. Empowerment carried out need a practical mechanism by utilizing data centered, including through the National Economic Census database (Susenas). Based on the Susenas can be formulated the decision rules of determining the provision of assistance for SMEs in the field of telematics. The way it did by generating the rule base through the classification technique. The CART algorithm-based decision rule model performs better than C45 and ID3 models. The high level of performance model is also in line with the regulations applied by the government. This becomes one of the strengths of research, because the resulting model is consistent with the existing conditions in Indonesia. The rules base generated from the three classification techniques show different rules. The CART technique has pattern matching with the realization of activities in The Ministry of Cooperatives and SMEs. So far, the government has difficulty in referring data related to the empowerment of SMEs telematics services. Therefore, the findings resulting from this research can be used as an alternative decision support system related to the program of empowerment of SMEs in telematics.

1. Introduction

The Indonesian government is supporting business independence for *technopreneur* or entrepreneurs who are into the core businesses involving technology-based. Moreover, the start-up of Information and Communication Technology (ICT) or also known as telematics, is currently experiencing a significant increase. Self-reliance of start-ups based on telematics is needed to be one of the supporters of development in Indonesia. Self-reliance in the field of telematics became one of Indonesia's vision in 2024. One of Indonesia's vision in telematics is to realize an independent, competitive, dignified and ethical personality with function as an efficient meta-infrastructure, and ICT that provides opportunities and resources justly in achieving prosperity [1]. The increasing number of SME telematics is a potential that must be supported to have a competitive value, especially in the face of the ASEAN Economic Community (AEC).

Achieving that vision requires synergism from many parties. Particularly central or local governments responsible for implementing relevant national policies encourage independence for ICT-based technopreneurs. But there are many difficulties in deciding to develop technopreuneur through the provision of assistance. A practical mechanism is needed in determining the provision of assistance to SMEs in desperate need of support. One of the causes of such difficulties is the amount of data that is out of sync. Another cause is the varying qualification standards in determining the feasibility of SMEs for assistance [2]. In fact, the Indonesian government has a centralized and consistent source of data related to SMEs, including SMEs in telematics. The data in question comes from the National Economic Census (Susenas) owned by the Central Bureau of Statistics (BPS). But so far data related to the condition of SMEs in Susenas precisely has not been utilized optimally for the activities of empowerment of SMEs.

Susenas data that holds data about SMEs of telematics services includes many attributes appropriate to the empowerment process [3]. Therefore, this research is proposed to give one way of empowering Indonesian SMEs telematics services through study of feasibility of rule-based aid based on Susenas data. Although the data used in this study is still using Susenas 2006 data, but the principles, ideas and the concept of empowerment through the utilization of Susenas data is relatively new. The Susenas data can be generated to form the decision rule of eligibility through the Data Mining approach, using the classification method. The classification technique used is a decision tree technique that is formed through three techniques namely J48 or C45 algorithm, Classification and Regression Tree (CART) and Iterative Dichotomies Tree (ID3).

Based on the trials conducted with the three classification techniques are expected to find a rule of thumb that is practical. The decision rule is expected to be utilized by the government and other stakeholders in an objective but comprehensive assistance process.

2. Method

The method applied in this research uses the stages of data mining or also known as Knowledge Discovery and Data Mining (KDD) [4]. As a series of processes, data mining can be divided into the several phases presented in Figure 1. These stages are interactive, the user is directly involved or through the knowledgebase mediation.

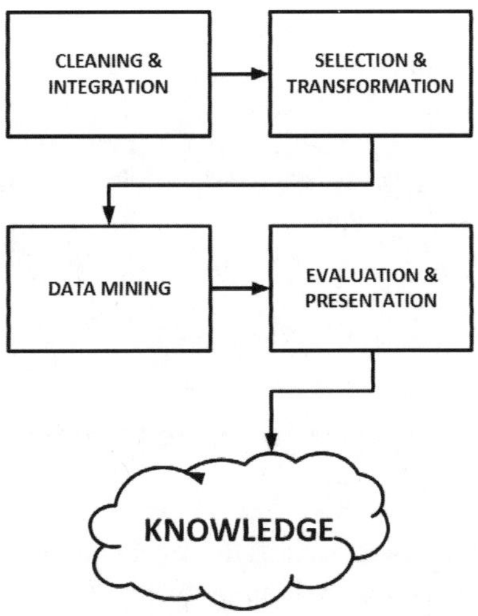

Figure 1. Steps of Data Mining [4]

Explanation of the steps undertaken in this study are as follows:

IOP Conf. Series: Materials Science and Engineering **332** (2018) 012018 doi:10.1088/1757-899X/332/1/012018

2.1. Data Cleaning and Integration
Data cleansing is the process of eliminating noise and inconsistent data or data irrelevant. As well as data integration is the combination of data from different databases into the new database. Not infrequently the data needed for data mining is not only derived from the database but also from multiple databases or text files, but this stage was done in previous studies by [2].

2.2. Data Selection and Transformation
The existing data in the database is often not all used, therefore only the appropriate data to be analyzed to be retrieved from the database. As well as the transformation of the data is modified or merged data into a format suitable for processing in data mining. For this study using entropy-based method for selection of the data. Selection of these attributes also according [5]. function to sort the attributes of the most influential in the process of classification so that it becomes more compact and optimized.

2.3. Mining Process
This stage is a major process when the method is applied to find valuable and hidden knowledge from data. To process the data mining of this study using Decision Tree algorithm, J48, CART and ID3. The performance of the three algorithms are analyzed through a high degree of accuracy and analytical relevance to SMEs competence condition Indonesian telematics services. The end result of this mining process is used to examine the decision-making rules of eligibility of assistance.

2.4. Pattern Evaluation
The process of identifying unique patterns and engaging into knowledgebased found done through the stages of evaluation pattern. In this stage the results of data mining techniques in the form of a distinctive pattern and prediction models were evaluated to assess whether the hypothesis that there is indeed reached. If the result is not in accordance with the hypothesis there are several alternatives that can be taken such as making feedback to improve the process of data mining, or accept this result as an out-of-expectation that may be useful. Evaluation of this study using the confusion matrix and proceed with the analysis of the relevance of SMEs competence condition Indonesian telematics services until 2006.

2.5. Knowledge Presentation
This stage is the visualization and presentation of knowledge about the methods used to obtain the knowledge acquired. The last stage of the data mining process is how to formulate a decision or action of the analytical results obtained. There are times when it must involve people who do not understand the data mining. Therefore presenting the results of data mining in the form of knowledge that can be understood by everyone is a necessary stage in the process of data mining. In this presentation, the results in the form of meaningful charts decision tree that contains multiple rules or rule is implemented in the form of a web-based system.

3. Result and Discussion
The result of this research is the classification system of SMEs business assistance of telematics service using J48, CART and ID3 algorithm then the algorithm is executed using R studio application using 80% training data and 20% test data from 8798 total data, thus generating decision tree to determine the feasibility decision a business entity to receive assistance. After obtaining the output of decision tree, rule or rule of the decision tree is implemented using adobe dream weaver then applied in the framework and database designed and created in the MySQL application.

This study is constituted by [2] and is a continuation of [6]. Data cleaning and integration were performed in accordance with the results of the study [6]. The selection of attributes is done gradually through a specific strategy, resulting from the resulting decision tree indicating a rule that is too large.

The first stage attribute selection is done by summarizing some attribute classes. Among them are the provincial attribute class which originally amounted to 31 provinces to 5 islands (Sumatra, Java, Kalimantan, Papua and Bali-Nusa Tenggara). The attributes of the owner's education which originally amounted to seven classes were summarized into four classes, as well as business group attributes (initially seven classes into four classes). This attribute class summary is intended as one of the rule base summary strategies that will be generated by the inference engine. The second attribute selection is done by trimming many attributes into 10 attributes with entropy-based techniques, which are implemented using R studio software. The result of attribute selection is presented in Table 1.

Table 1. Attributes Selection Result

No	Attribute	Value			
1	The form of legal entity	1.	CV	5.	Individual
		2.	Firm	6.	Limited Firm
		3.	Special permission from the relevant agencies	7.	Foundation
		4.	Cooperative		
2	Business group	1.	Telecomunication Services	3.	Software consultation
		2.	Hardware consultation	4.	others
3	Sales	1.	Mikro	3.	Medium
		2.	Small		
4	Difficulty	1.	Raw material / merchandise	6.	Capitals
		2.	Fuel oil /energy	7.	No difficulty
		3.	Workmanship	8.	Transportation
		4.	Others	9.	Wage laborers
		5.	Marketing		
5	Cooperative	1.	Yes	2.	No
6	Partnership	1.	Yes	2.	No
7	Receive training	1.	Yes	2.	No
8	Marketing	1.	Eksport	3.	Outside of the district
		2.	District	4.	Province
9	Business prospects	1.	Better	4.	Just as bad
		2.	Worse	5.	Not comparable
		3.	Sama baik		
10	Plan	1.	No	4.	Yes, improve skills
		2.	Yes, expand business premises	5.	Yes, others
		3.	Yes, opening branch		

The selected classification techniques J48, CART and ID3 have equality in the inference process, and each has deficiencies and advantages [5]. The purpose class for the inference process of SMEs assistance in telematics services is to convert from the attribute "never received help or never received help". Based on the results of this conversion is expected to obtain the base rules for determining eligible SMEs to be given assistance. Recapitulation accuracy of all three classification techniques using a variety of model scenario are shown in Table 2. Base the rules resulting from the three classification techniques showed different rules. The resulting rule base for J48 algorithm, CART and ID3 shown successively in Figures 2, 3 and 4.

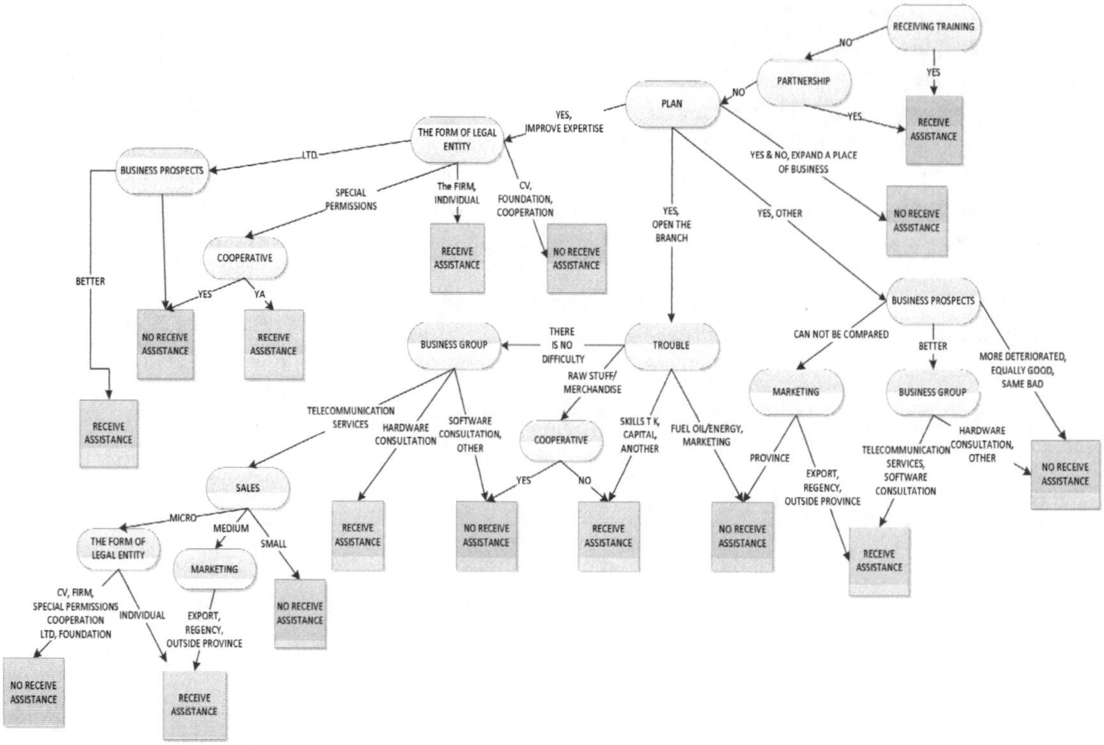

Figure 2. The rule base of SMEs assistance of Indonesian telematics services using J48 Algorithm

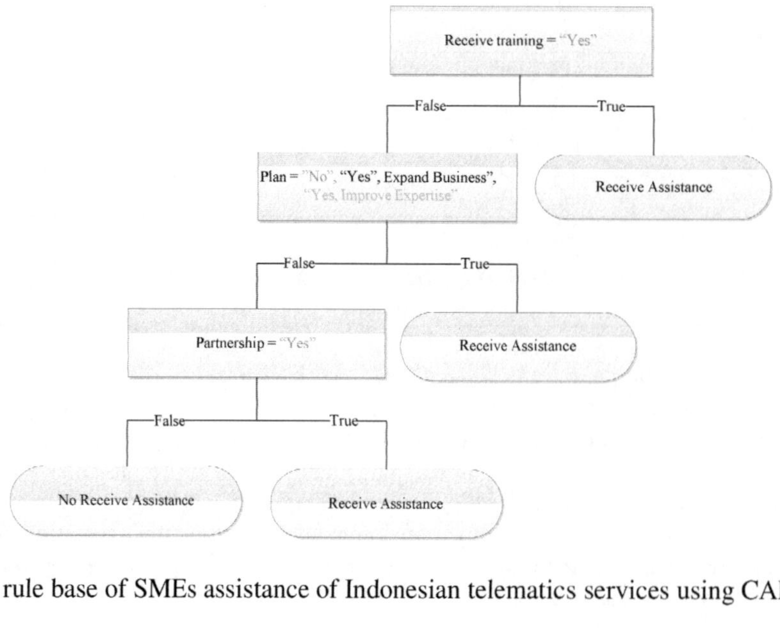

Figure 3. The rule base of SMEs assistance of Indonesian telematics services using CART Algorithm

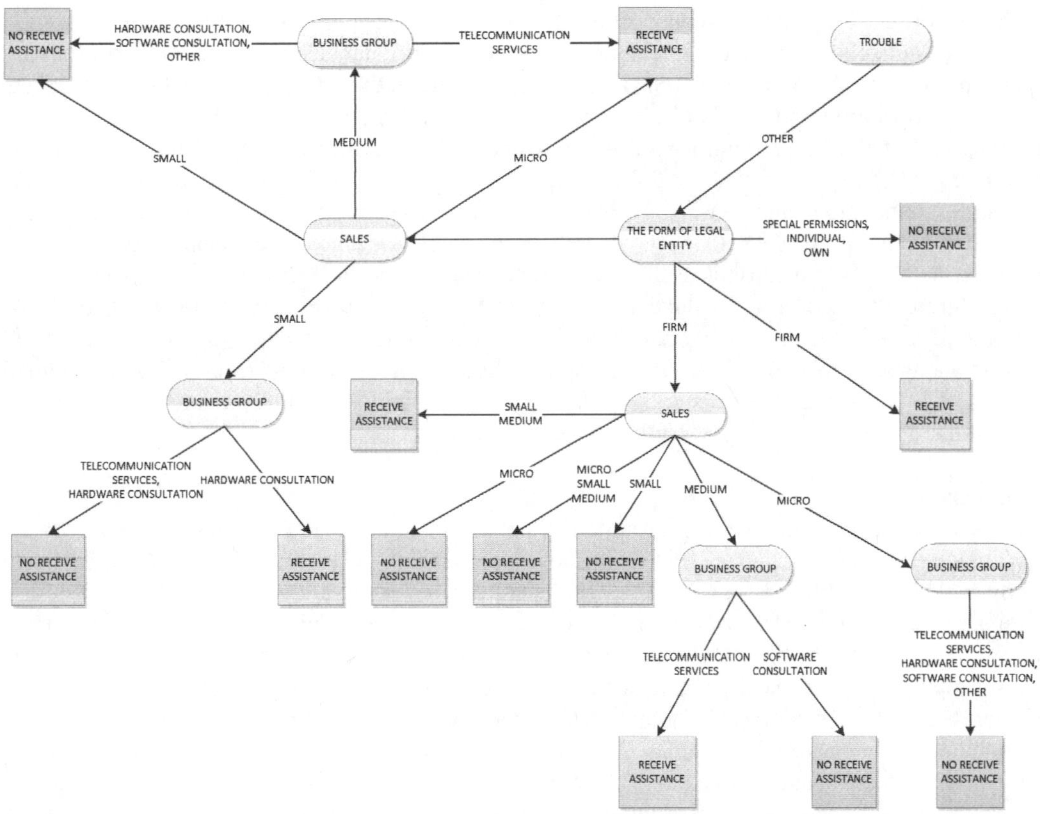

Figure 4. The rule base of SMEs assistance of Indonesian telematics services using ID3 Algorithm

The built-in model scenario is intended to improve model accuracy. Different model scenarios are used for each algorithm. In J48 algorithm the scenarios used are decision pruning process (scenario 1) and without pruning (scenario 2). In the CART and ID3 algorithms use different scenarios ie without data reduction (scenario1) and data reduction (scenario 2). Based on the information in Table 1 shows that each algorithm has a different performance scheme but produces almost the same decision rule. The performance of the J48 and ID3 algorithm-based decision model has better accuracy in the train data than the test data, for both model scenarios. Other conditions are shown in the performance of CART algorithm-based decision model. In the CART model the test data results in a higher accuracy for both model scenarios. This suggests that the CART algorithm has a better ability to classify the process of providing assistance to MSMEs Indonesian telematics services. This is an interesting finding, especially considering the data source used is Susenas data covering the condition of SMEs of Indonesian telematics services that represent all provinces.

Table 2. Model Performance

No	Algorithm	Accuracy (Scenario 1)		Accuracy (Scenario 2)	
		Training data	Testing data	Training data	Testing data
1	J48	64.88%	57.04%	79.71%	55.00%
2	CART	62.21%	73.97%	61.18%	63.92%
3	ID3	99.00%	24.00%	82.00%	26.00%

The performance results of this decision rule model are also in line with the policies adopted by regulators. Regulators in question in this case include Indonesian Communication & Information Ministry, Indonesian Cooperation and SMEs Ministry. Based on the referral of activities undertaken in Indonesian Cooperation and SMEs Ministry related to the selection for assistance for SMEs is based on training activities [7]. Training for start-ups of telematics has been facilitated through the techno park scheme [8]. This shows that CART technique has pattern match with realization of activity in Indonesian Cooperation and SMEs Ministry. But the rule base generated from these three classification techniques needs to be analyzed more deeply because it indicates the existence of unbalanced data on some attributes. Therefore obtaining the resulting accuracy is expected to be improved through the process of balancing data [9]. This research gives a very important result to be used as a reference in the process of providing assistance for SMEs telematics services. Because so far the government has difficulty in referring data related to the empowerment of SMEs telematics services. Therefore, the findings resulting from this research can be used as an alternative supporting decision related to the SMEs empowerment program in general.

4. Conclusion

The process of SMEs empowering through the provision of assistance for Indonesian telematics services SMEs can be modeled with Data Mining approach using J48, CART and ID3 classification techniques. This research is a new study, as far as we examine the various studies, especially for the determination of the rules of assistance for SMEs in the field of telematics services, one of them [10]. Although the approach is still using the classic decision rule approach but the strength of this research lies in the optimization of Susenas data which has been very rarely used for related research needs. Moreover, Susenas 2006 data is not equipped with the output of determinant class of assistance. The determination of the output class is done through the optimization of one of the attributes assumed as the output class (the attribute in question is 'ever get help'). Model performance is demonstrated by the accuracy of various schemes and data mining techniques used. The CART algorithm-based decision rule model performs better than C45 and ID3 models. The high level of performance model is also in line with the regulations applied by the government. This becomes one of the strengths of research, because the resulting model is consistent with the existing conditions in Indonesia. The weakness of this research is Susenas data which used is very old data (2006). This happens because the Susenas process in Indonesia is still done every ten years. In 2016 CBS has implemented the process of preparing the list of instruments and respondents and supporting components. Implementation of the census to the SMEs of Indonesian telematics services is only implemented in 2017, so that when the implementation of this study the data cannot be accessed and published to the public.

Acknowlegments

1. DRPM Ristek Dikti, as the main sponsor, which gives us Competitive Grants Scheme
2. Computer Science Department, Mathematics and Natural Science Faculty, Pakuan University,and Research Institute Pakuan University, for supporting, coordinating and facilitating to achieve this grants.
3. Indonesian Communication & Information Ministry, Indonesian Cooperation and SMEs Ministry and Bandung Technopark for active participation in the activities of interviews anduser requirement.

References

[1] Indonesian Communication & Information Ministry. Preparation of a long-term binding roadmap for development of the ICT sector. 2045 Towards 100 years of independent Indonesia. Research and Development Resources, Tools and Implementation of Post and Information Technology, Research and Human Resources. (2016).

[2] Tosida ET, Maryana S, Thaheer H, Damin FA. Visualization model of small and medium enterprises (SMEs) telematics services potentiality map in Indonesia. InInformation &

Communication Technology and Systems (ICTS), 2015 International Conference on 2015 Sep 16 (pp. 151-156). IEEE (2015).

[3] Tosida, E.T., Seminar, K.B. and Herdiyeni, Y. Attribut Selection of Indonesian telematic services MSMEs assisstance feasibility, using AHP. *Kursor*, *8*(2), 97-106 (2016).

[4] Han, J., Kamber M, Pei J. Data mining : Concepts and techniques. Third Edition. Morgan Kaufmann is an imprint of Elsevier, 225Wyman Street,Waltham, MA 02451, USA. (2012).

[5] Sohn SY, Kim JW. Decision tree-based technology credit scoring for start-up firms : Korean case. *J. Expert Systems with Applications* 39 (4) : 4007-40112. Elsevier. doi:10.1016/j.eswa.2011.09.075.(2012).

[6] Tosida, E. T., S. Maryana, and H. Thaheer. Implementation of Self Organizing Map (SOM) as decision support: Indonesian telematics services MSMEs empowerment. In *IOP Conference Series: Materials Science and Engineering*, vol. 166, no. 1, p. 012017. IOP Publishing (2017).

[7] McGuirk H, Lenihan H, Hart M. Measuring the impact of innovation human capital on small firms propensity to innovative. *J. Research Policy, Vol.44 (4)* : 965-976. Elsevier. doi:http://dx.doi.org/10.1016/j.respol.2014.11.008. (2015).

[8] Soenarso Wisnu S, Nugraha D, Listyaningrum E. Development of science and technology park (stp) in indonesia to support innovation-based regional economy: Concept and early stage development. *J.World Technopolis Review (WTR).* 2 (1) : 32-42.World Technopolis Association. dioi:10.7165/wtr2013.2.1.32. (2013).

[9] Sun Z, Song Q, Zhu X, Sun H, Xu, Zhou Y. A novel ensemble method for classifying imbalanced data. Patter Recognition 48(5) : 1623-1637. http://dx.doi.org/10.1016/j.patcog.2014.11.014. (2015).

[10] Sadatrasoul, SM, Gholamian MR, Siami M, HajimaohammadiZ. Credit scoring in banks and financial institutions via data mining techniques : A literatur review. *Journal of AI and Data Mining*. Vol. 1, No. 2, 2013, 119-129. [Jan 5th, 2015]. Link : http://jad.shahroodut.ac.ir/article_124_0.html. (2013).

IOP Conf. Series: Materials Science and Engineering **332** (2018) 012019 doi:10.1088/1757-899X/332/1/012019

Eigenvectors determination of the ribosome dynamics model during mRNA translation using the Kleene Star algorithm

Ernawati[1*], E Carnia[2], A K Supriatna[2]

[1]Master Student in Mathematics Study Program, Faculty of Mathematics and Natural Sciences, Universitas Padjadjaran, INDONESIA
[2]Department of Mathematics, Faculty of Mathematics and Natural Sciences, Universitas Padjadjaran, INDONESIA

*Corresponding author: ernawati.nizib@gmail.com

Abstract. Eigenvalues and eigenvectors in max-plus algebra have the same important role as eigenvalues and eigenvectors in conventional algebra. In max-plus algebra, eigenvalues and eigenvectors are useful for knowing dynamics of the system such as in train system scheduling, scheduling production systems and scheduling learning activities in moving classes. In the translation of proteins in which the ribosome move uni-directionally along the mRNA strand to recruit the amino acids that make up the protein, eigenvalues and eigenvectors are used to calculate protein production rates and density of ribosomes on the mRNA. Based on this, it is important to examine the eigenvalues and eigenvectors in the process of protein translation. In this paper an eigenvector formula is given for a ribosome dynamics during mRNA translation by using the Kleene star algorithm in which the resulting eigenvector formula is simpler and easier to apply to the system than that introduced elsewhere. This paper also discusses the properties of the matrix $B_{\lambda}^{\otimes n}$ of model. Among the important properties, it always has the same elements in the first column for $n = 1, 2, \ldots$ if the eigenvalue is the time of initiation, $\lambda = \tau_{in}$, and the column is the eigenvector of the model corresponding to λ.

1. Introduction

Max-plus algebra is widely used to model time events discrete system. A typical application of discrete events system are production lines, where every machine must wait with a starting a new operation until the operations on other machines are completed as in the scheduling, transporting, manufacturing, queuing, and protein production in cell. One problem that often appears in the max-plus algebra is the eigen problem. The eigen problem for max-plus matrices describes the steady state of the system. A more details examples of the eigen problems can be seen in [4], [5], [7], [9], [10], [11].

The eigenvalues and eigenvectors in a max-plus model can be used to get an overview of the system dynamics such as in train system scheduling, scheduling production systems, and scheduling lesson activities on moving classes. In the translation stage of cellular protein production, in which the ribosomes moves uni-directionally along an mRNA strand to building amino acid chains, the eigenvalues and eigenvectors are useful to calculate the protein production rate and density of ribosomes on the mRNA. Based on this, it is important to examine the eigenvalues and eigenvectors of the model.

In the previous work, the authors in [3] provided a formula for determining the eigenvalues and eigenvectors of their model [3]. In this paper we will discuss a simpler formula that is easier to understand the process of determining the eigenvectors of their model. This formula is obtained by a different approach compared to the work in [3].

2. Model and Method

The details derivation of the model are given in [3]. In this section we only present a brief theoretical background required in the subsequent sections, such as mRNA translation, max-plus algebra, eigen problem in max-plus algebra and matrices of the ribosome dynamics during mRNA translation.

2.1. Messenger RNA translation

mRNA translation is one of the steps in protein production in cells [1]. The mRNAs are single strands of nucleotides, grouped in triplets called codons, holds the code for a specific chain of amino acids that makes up a protein. The translation is performed by molecular machines called ribosomes, which scan along the mRNA adding amino acids to a growing chain which will become the protein. In this paper we will examine the eigenvalues and eigenvectors of the matrix B in (1) which is a max-plus model of ribosome dynamics during mRNA translation introduced by [3]. See [3] for the details.

$$B = \begin{bmatrix} \tau_{in} & 0 & \varepsilon & \varepsilon & \cdots & \varepsilon \\ \tau_{in}\otimes\tau_1 & \tau_1 & 0 & \varepsilon & \cdots & \varepsilon \\ \tau_{in}\otimes\tau_1\otimes\tau_2 & \tau_1\otimes\tau_2 & \tau_2 & 0 & \ddots & \vdots \\ \tau_{in}\otimes\tau_1\otimes\tau_2\otimes\tau_3 & \tau_1\otimes\tau_2\otimes\tau_3 & \tau_2\otimes\tau_3 & \ddots & \ddots & \varepsilon \\ \vdots & \vdots & \vdots & \ddots & \tau_{n-1} & 0 \\ \tau_{in}\otimes\tau_1\otimes\tau_2\otimes\tau_3\otimes\cdots\otimes\tau_n & \tau_1\otimes\tau_2\otimes\tau_3\otimes\cdots\otimes\tau_n & \tau_2\otimes\tau_3\otimes\cdots\otimes\tau_n & \cdots & \tau_{n-1}\otimes\tau_n & \tau_n\oplus\tau_{out} \end{bmatrix} \qquad (1)$$

2.2. Max-plus algebra

Bacelli *et al.* in [2] defines the max-plus algebra and denoted by $\mathcal{R}_{max}$ which is defined as the set $\mathbb{R}_{max} = \mathbb{R} \cup -\infty$, with $\mathbb{R}$ is real numbers extended by an infinite element $\varepsilon = -\infty$ and operation $\oplus$, $\otimes$ are defined by

$$a \oplus b = \max\{a, b\} \text{ and } a \otimes b = a + b \qquad (2)$$

where $a, b \in \mathbb{R}_{\max}$. It is easy to show that both operation $\oplus$ and $\otimes$ are associative and commutative. The zero and unit element in max-plus algebra are $\varepsilon = -\infty$ and $e = 0$, respectively.
We define max-plus power in the natural way

$$a^{\otimes k} = a \otimes a \otimes a \otimes a \otimes \ldots \otimes a \text{ , } k \text{ times} \qquad (3)$$

We can write $a^{\otimes k}$ in conventional algebra as $k \times a$. We also note that the $\otimes$ operator has an inverse which can be expressed as a negative power

$$a \otimes b^{\otimes(-1)} = a - b, \qquad (4)$$

but there is no inverse of the $\oplus$ operator.

Vectors and matrices can also be constructed. We denoted by a_{ij} the *i-j*th component of the matrix A. For matrices $A \in \mathbb{R}_{max}^{n\times m}$, $B \in \mathbb{R}_{max}^{n\times m}$, $C \in \mathbb{R}_{max}^{m\times l}$ the sum and product of matrices are then defined as

$$[A \oplus B]_{ij} = a_{ij} \oplus b_{ij}, \qquad (5)$$

$$[A \otimes C]_{ij} = \bigoplus_{k=1}^{m} a_{ik} \otimes c_{kj}, \qquad (6)$$

where $a_{ij} = [A]_{ij}$, $b_{ij} = [B]_{ij}$ and $c_{ij} = [C]_{ij}$, and have used the following convenient notation for summation over indices:

$\overset{n}{\underset{i=1}{\oplus}} a_i = \max\{a_i \,|\, 1 \le i \le n\}$.

IOP Conf. Series: Materials Science and Engineering **332** (2018) 012019 doi:10.1088/1757-899X/332/1/012019

For square matrix $A \in \mathbb{R}_{max}^{n\times n}$, similar to scalar in max-plus algebra, we denote $A^{\otimes k} = \underbrace{A\otimes A\otimes ...\otimes A}_{k} = A\otimes A^{k-1}$ as k^{th} power of A, while $A^{+} = A\oplus A^{2}\oplus A^{3}\oplus ... = \oplus_{k=1}^{\infty} A^{k}$ and $A^{*} = E_n \oplus A^{+}$ with E_n is the identity matrix in $\mathbb{R}_{max}^{n\times n}$. Furthermore, those relation can be written in the form of $A^{*} = \oplus_{k=0}^{\infty} A^{\otimes k}$ which is called *Kleene star*. Therefore,

$$A^{+} = A\otimes A^{*} \tag{7}$$

Eigen problems are common problem in mathematics especially in linear algebra. In max-plus algebra, similar to linear algebra, eigen problems are formulated as $A\otimes \mathbf{u} = \lambda \otimes \mathbf{u}$ for given matrix $A \in \mathbb{R}_{max}^{n\times n}$, where λ is called *eigenvalue* and $\mathbf{u} \in \mathbb{R}_{max}^{n}$ is called *eigenvector*. This problem is treated by several authors who can be such as in [4], [5], [10], [11]. The graph associated with matrix $A \in \mathbb{R}_{max}^{n\times n}$, is denoted by $G(A)$. It consists of a set of nodes and directed weighted arcs, where if there is a matrix element $[A]_{ij}$ with value $a_{ij} \neq \varepsilon$, then there is an arc $j \to i$ (from *j* to *i*) with weight a_{ij}. A series of one or more arcs between two nodes *i* and *j* is called a *path* $i \to j$. A path $i \to i$ is called a *circuit* and denoted γ. A graph is said to be *strongly connected* if for any two different nodes there is a path between them and a matrix A is said to be *irreducible* if the corresponding graph $G(A)$ is strongly connected. The *circuit weight* w_γ of a circuit γ is defined as the sum of the weight of all arcs in that circuit and the *circuit length* l_γ as the number of arcs in the circuit. The *mean circuit weight* is then defined as $\overline{w_\gamma} = \frac{w_\gamma}{l_\gamma}$. If the maximum mean circuit weight in a graph is λ, then a circuit with a mean circuit weight equal to λ is called a *critical circuit*. The *critical graph* corresponding to matrix A is denoted by $G^{c}(A)$. It is defined as the subgraph of $G(A)$ containing only the nodes and arcs which are in the critical circuits and the set of nodes in $G(A)$ denoted by $U^{c}(A)$. Furthermore, in this paper we define $\left(A_{\lambda}\right)_{ij} \in \mathbb{R}_{max}^{n\times n}$ as

$$(A_\lambda)_{ij} = (-\lambda\otimes A)_{ij} \tag{8}$$

for any λ. For the analysis of the subsequent section we need the following theorems:

Theorem 1. [6] *Let $A \in \mathbb{R}_{max}^{n\times n}$ be a matrix that has a maximal mean circuit weight of circuits in $G(A)$ equal to e then A^{+} exists and is defined by*

$$A^{+} = A\oplus A^{2}\oplus ...\oplus A^{n} = \oplus_{k=1}^{n} A^{\otimes k}.$$

Based on Theorem 1 it is given that $A^{+} = A\oplus A^{2}\oplus ...\oplus A^{n} = A\otimes\left(E_n \oplus A\oplus A^{2}\oplus ...\oplus A^{n-1}\right)$, and according to (7) it is obtained that

$$A^{*} = E_n \oplus A\oplus A^{2}\oplus ...\oplus A^{n-1}. \tag{9}$$

Theorem 2. *Let $A \in \mathbb{R}_{max}^{n\times n}$ has finite number of the maximal mean circuit weight λ , then λ is the eigenvalue of A and for any $\mathbf{u}\in U^{c}$, the column of $\left(A_{\lambda}^{*}\right)_{\mathbf{u}}$ is the eigenvector of A corresponding to the λ.*

Theorem 3. [8] *Any irreducible matrix $A \in \mathbb{R}_{max}^{n\times n}$ posseses one and only one eigenvalue λ, which is a finite number equal to the maximal mean circuit weight of circuits in the graph $G(A)$.*

Theorem 4. *Let $B \in \mathbb{R}_{max}^{(n+1)\times(n+1)}$ is given by* (1), *then* $\lambda = \max\{\tau_{in}, \tau_{out}, \tau_i \,|\, 1\le i\le n\}$.

We do not prove the theorems. Interested readers can consult the respective references.

3. Result and Discussion

In this section, we will examine how to determine the eigenvector of the matrix B in (1). The eigenvalue of the matrix B can be found from the correspomding graph $G(B)$. Fig. 1 show the example of graph $G(B)$ with $n=3$.

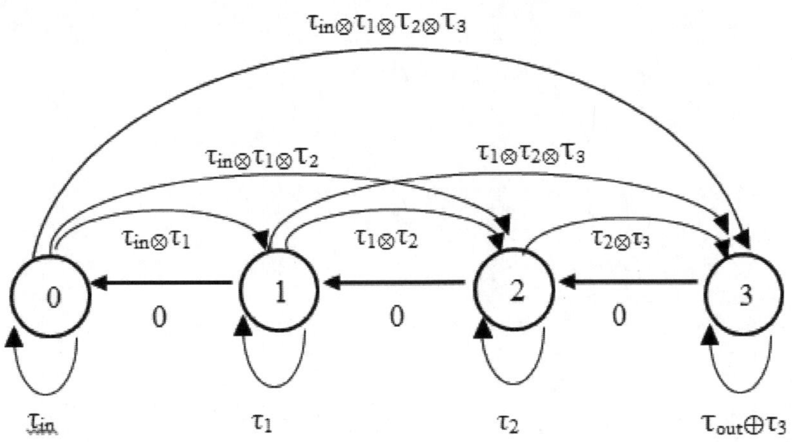

Figure 1. Diagram showing the weighted, directed graph $G(B)$ corresponding to the matrix B (Eq. (1)) with $n=3$.

The matrix B is irreducible and has exactly one eigenvalue because the graph $G(B)$ is strongly connected. Since the eigenvalue of B is the maximum mean circuit weight, then the eigenvalue of B denotes the largest waiting time of the model. In [3] the solution of determining the eigenvector depends on the corresponding eigenvalues of Theorem 4 as follows:

a. If $\lambda = \tau_{in}$ then the final expression for the eigenvector is

$$\mathbf{u} = \begin{bmatrix} \tau_1^{\otimes(-1)} \otimes \cdots \otimes \tau_n^{\otimes(-1)} \otimes u_n \\ \tau_2^{\otimes(-1)} \otimes \cdots \otimes \tau_n^{\otimes(-1)} \otimes u_n \\ \vdots \\ \tau_n^{\otimes(-1)} \otimes u_n \\ u_n \end{bmatrix}. \quad (10)$$

b. If $\lambda = \tau_{out}$ then the final expression for the eigenvector is

$$\mathbf{u} = \begin{bmatrix} u_0 \\ \tau_{out} \otimes u_0 \\ \tau_{out}^{\otimes 2} \otimes u_0 \\ \vdots \\ \tau_{out}^{\otimes n} \otimes u_0 \end{bmatrix}. \quad (11)$$

c. If $\lambda = \max\{\tau_i \,|\, 1 \le i \le n\}$ then the final expression for the eigenvector is

$$\mathbf{u} = \begin{bmatrix} u_0 \\ \tau \otimes u_0 \\ \tau^{\otimes 2} \otimes u_0 \\ \vdots \\ \tau^{\otimes q} \otimes u_0 \\ \tau_{q+1} \otimes \tau^{\otimes q} \otimes u_0 \\ \tau_{q+1} \otimes \tau_{q+2} \otimes \tau^{\otimes q} \otimes u_0 \\ \vdots \\ \tau_{q+1} \otimes \cdots \otimes \tau_n \otimes \tau^{\otimes q} \otimes u_0 \end{bmatrix}, \qquad (12)$$

and

$$\mathbf{u} = \begin{bmatrix} u_0 \\ \tau \otimes u_0 \\ \tau^{\otimes 2} \otimes u_0 \\ \vdots \\ \tau^{\otimes p} \otimes u_0 \\ \tau_{p+1} \otimes \tau^{\otimes p} \otimes u_0 \\ \tau_{p+1} \otimes \tau_{p+2} \otimes \tau^{\otimes p} \otimes u_0 \\ \vdots \\ \tau_{p+1} \otimes \cdots \otimes \tau_n \otimes \tau^{\otimes p} \otimes u_0 \end{bmatrix}. \qquad (13)$$

The next discussion is purported to calculate the eigenvectors which different to the method in [3]. It is done by using the Kleene star algorithm. We identify the eigenvectors into three cases: $\lambda = \tau_{in}$, $\lambda = \tau_{out}$, $\lambda = \max\{\tau_i \,|1 \le i \le n\}$.The steps of determining the eigenvector formula of the matrix B in (1) by the Kleene star method is undertaken with the following steps:

i. Determine the eigenvalues of B by the formula $\lambda = \max\{\tau_{in}, \tau_{out}, \tau_i \,|1 \le i \le n\}$
ii. Calculate $B_\lambda = -\lambda \otimes B$
iii. Calculate $B_\lambda^* = E_n \oplus B_\lambda \oplus B_\lambda^{\otimes 2} \oplus B_\lambda^{\otimes 3} \oplus ... \oplus B_\lambda^{\otimes n}$
iv. Choose the eigenvector of matrix B of the column vectors in B_λ^* corresponding to the location of the eigenvalues in matrix B.

Several other authors have reviewed this method such as [8]. By using this method we find the following theorem regarding the properties of the matrix $B_\lambda^{\otimes n}$ of (1) as follows:

Theorem 5. Let $B_{\max}^{(n+1)\times(n+1)}$ satisfy (1) with $\lambda = \tau_{in}$, then

i. $B_\lambda^{\otimes n}$ is the matrix with the first column always the same for every n, where $n = 1, 2, 3, ...$
ii. b_{1n} is the first column of $B_\lambda^{\otimes n}$ and it is the eigenvector of B.

Proof:

i. For $n = 1, 2, 3, ...$ we obtain

$$B_\lambda = \begin{bmatrix} 0 & -\tau_{in} & \varepsilon & \cdots & \varepsilon \\ \tau_1 & -\tau_{in} \otimes \tau_1 & -\tau_{in} & \cdots & \varepsilon \\ \tau_1 \otimes \tau_2 & -\tau_{in} \otimes \tau_1 \otimes \tau_2 & -\tau_{in} \otimes \tau_2 & \ddots & \vdots \\ \vdots & \vdots & \vdots & \vdots & -\tau_{in} \\ \tau_1 \otimes \tau_2 \otimes \cdots \otimes \tau_n & -\tau_{in} \otimes \tau_1 \otimes \cdots \otimes \tau_n & -\tau_{in} \otimes \tau_2 \otimes \cdots \otimes \tau_n & \cdots & -\tau_{in} \otimes (\tau_n \oplus \tau_{out}) \end{bmatrix}$$

$$B_\lambda^{\otimes 2} = \begin{bmatrix} 0 & -\tau_{in} & -2\tau_{in} & \varepsilon & \cdots & \varepsilon \\ \tau_1 & -\tau_{in} \otimes \tau_1 & -2\tau_{in} \otimes (\tau_1 \oplus \tau_2) & -2\tau_{in} & \cdots & \varepsilon \\ \tau_1 \otimes \tau_2 & -\tau_{in} \otimes \tau_1 \otimes \tau_2 & \cdots & -2\tau_{in} \otimes (\tau_2 \oplus \tau_3) & \ddots & \vdots \\ \tau_1 \otimes \tau_2 \otimes \tau_3 & -\tau_{in} \otimes \tau_1 \otimes \tau_2 \otimes \tau_3 & \cdots & -2\tau_{in} \otimes \tau_3 \otimes (\tau_2 \oplus \tau_3) & \ddots & \varepsilon \\ \vdots & \vdots & \vdots & \vdots & \vdots & -2\tau_{in} \\ \tau_1 \otimes \tau_2 \otimes \tau_3 \otimes \cdots \otimes \tau_n & -\tau_{in} \otimes \tau_1 \otimes \cdots \otimes \tau_n & \cdots & \cdots & \cdots & \cdots \end{bmatrix}$$

$$B_\lambda^{\otimes 3} = \begin{bmatrix} 0 & \cdots & \cdots & \cdots & \cdots & \varepsilon \\ \tau_1 & \cdots & \cdots & \cdots & \cdots & \varepsilon \\ \tau_1 \otimes \tau_2 & \cdots & \cdots & \cdots & \ddots & \vdots \\ \tau_1 \otimes \tau_2 \otimes \tau_3 & \cdots & \cdots & \cdots & \ddots & \ddots \\ \vdots & \vdots & \vdots & \vdots & \vdots & \ddots \\ \tau_1 \otimes \tau_2 \otimes \tau_3 \otimes \cdots \otimes \tau_n & \cdots & \cdots & \cdots & \cdots & \cdots \end{bmatrix}; \quad B_\lambda^{\otimes 4} = \begin{bmatrix} 0 & \cdots & \cdots & \cdots & \cdots & \varepsilon \\ \tau_1 & \cdots & \cdots & \cdots & \cdots & \varepsilon \\ \tau_1 \otimes \tau_2 & \cdots & \cdots & \cdots & \ddots & \vdots \\ \tau_1 \otimes \tau_2 \otimes \tau_3 & \cdots & \cdots & \cdots & \ddots & \ddots \\ \vdots & \vdots & \vdots & \vdots & \vdots & \ddots \\ \tau_1 \otimes \tau_2 \otimes \tau_3 \otimes \cdots \otimes \tau_n & \cdots & \cdots & \cdots & \cdots & \cdots \end{bmatrix}$$

$$B_\lambda^{\otimes 5} = \begin{bmatrix} 0 & \cdots & \cdots & \cdots & \cdots & \varepsilon \\ \tau_1 & \cdots & \cdots & \cdots & \cdots & \varepsilon \\ \tau_1 \otimes \tau_2 & \cdots & \cdots & \cdots & \ddots & \vdots \\ \tau_1 \otimes \tau_2 \otimes \tau_3 & \cdots & \cdots & \cdots & \ddots & \ddots \\ \vdots & \vdots & \vdots & \vdots & \vdots & \ddots \\ \tau_1 \otimes \tau_2 \otimes \tau_3 \otimes \cdots \otimes \tau_n & \cdots & \cdots & \cdots & \cdots & \cdots \end{bmatrix}; \quad B_\lambda^{\otimes 6} = \begin{bmatrix} 0 & \cdots & \cdots & \cdots & \cdots & \varepsilon \\ \tau_1 & \cdots & \cdots & \cdots & \cdots & \varepsilon \\ \tau_1 \otimes \tau_2 & \cdots & \cdots & \cdots & \ddots & \vdots \\ \tau_1 \otimes \tau_2 \otimes \tau_3 & \cdots & \cdots & \cdots & \ddots & \ddots \\ \vdots & \vdots & \vdots & \vdots & \vdots & \ddots \\ \tau_1 \otimes \tau_2 \otimes \tau_3 \otimes \cdots \otimes \tau_n & \cdots & \cdots & \cdots & \cdots & \cdots \end{bmatrix}$$

and so on. It can be seen that all the first columns of the matrix $B_\lambda^{\otimes n}$ have elements that are always the same for $n = 1,2,3, \ldots$ ■

ii. $B_\lambda^* = E_n \oplus B_\lambda \oplus B_\lambda^{\otimes 2} \oplus \cdots$

$$B_\lambda^* = \begin{bmatrix} 0 & \varepsilon & \varepsilon & \cdots & \varepsilon \\ \varepsilon & 0 & \varepsilon & \cdots & \varepsilon \\ \varepsilon & \varepsilon & 0 & \ddots & \vdots \\ \vdots & \vdots & \vdots & 0 & \varepsilon \\ \varepsilon & \varepsilon & \varepsilon & \cdots & 0 \end{bmatrix} \oplus \begin{bmatrix} 0 & \cdots & \cdots & \cdots & \cdots \\ \tau_1 & \cdots & \cdots & \cdots & \cdots \\ \tau_1 \otimes \tau_2 & \cdots & \cdots & \cdots & \cdots \\ \vdots & \vdots & \vdots & \vdots & \vdots \\ \tau_1 \otimes \cdots \otimes \tau_n & \cdots & \cdots & \cdots & \cdots \end{bmatrix} \oplus \begin{bmatrix} 0 & \cdots & \cdots & \cdots & \cdots \\ \tau_1 & \cdots & \cdots & \cdots & \cdots \\ \tau_1 \otimes \tau_2 & \cdots & \cdots & \cdots & \cdots \\ \vdots & \vdots & \vdots & \vdots & \vdots \\ \tau_1 \otimes \cdots \otimes \tau_n & \cdots & \cdots & \cdots & \cdots \end{bmatrix} \oplus \cdots$$

Since $\lambda = \tau_{in}$ in column 1 of matrix B corresponds to Theorem 2, the eigenvector is the first column of the matrix B_λ^* as follows

$$b_{1n} = \begin{bmatrix} 0 \\ \tau_1 \\ \tau_1 \otimes \tau_2 \\ \vdots \\ \tau_1 \otimes \tau_2 \otimes \cdots \otimes \tau_n \end{bmatrix}$$

■

In the next section will examine the eigenvectors of the matrix (1) by the Kleene star algorithm .

3.1. Case (i): $\lambda = \tau_{in}$

In this case we show that the waiting time for initiation of translations is longer that for tRNA capture and termination, i.e. $\tau_{in} > \tau_{out}, \tau_i$ for $1 \le i \le n$.

a. $\lambda = \tau_{in}$

b. By using Theorem 5 the resulting eigenvector of matrix (1) is

$$\mathbf{u} = \begin{bmatrix} 0 \\ \tau_1 \\ \tau_1 \otimes \tau_2 \\ \tau_1 \otimes \tau_2 \otimes \tau_3 \\ \vdots \\ \tau_1 \otimes \tau_2 \otimes \tau_3 \otimes \ldots \otimes \tau_n \end{bmatrix} \tag{14}$$

3.2. Case (ii): $\lambda = \tau_{out}$

In this case we show that the waiting for termination of translation is longer that for tRNA capture and initiation, i.e. $\tau_{out} > \tau_{in}, \tau_i$ for $1 \le i \le n$.

a. $\lambda = \tau_{out}$

b. $$B_\lambda = \begin{bmatrix} -\tau_{out} \otimes \tau_{in} & -\tau_{out} & \varepsilon & \varepsilon \\ -\tau_{out} \otimes \tau_{in} \otimes \tau_1 & -\tau_{out} \otimes \tau_1 & -\tau_{out} & \varepsilon \\ -\tau_{out} \otimes \tau_{in} \otimes \tau_1 \otimes \tau_2 & -\tau_{out} \otimes \tau_1 \otimes \tau_2 & -\tau_{out} \otimes \tau_2 & -\tau_{out} \\ -\tau_{out} \otimes \tau_{in} \otimes \tau_1 \otimes \tau_2 \otimes \tau_3 & -\tau_{out} \otimes \tau_1 \otimes \tau_2 \otimes \tau_3 & -\tau_{out} \otimes \tau_2 \otimes \tau_3 & 0 \end{bmatrix}$$

$$B_\lambda^{\otimes 2} = \begin{bmatrix} -2\tau_{out} \otimes \tau_{in} \otimes (\tau_{in} \oplus \tau_1) & -2\tau_{out} \otimes (\tau_{in} \oplus \tau_1) & -2\tau_{out} & \varepsilon \\ -2\tau_{out} \otimes \tau_{in} \otimes \tau_1 \otimes (\tau_{in} \oplus \tau_1 \oplus \tau_2) & -2\tau_{out} \otimes \tau_1 \otimes (\tau_{in} \oplus \tau_1 \oplus \tau_2) & -2\tau_{out} \otimes (\tau_1 \oplus \tau_2) & -2\tau_{out} \\ -2\tau_{out} \otimes \tau_{in} \otimes \tau_1 \otimes \tau_2 (\tau_{in} \oplus \tau_1 \oplus \tau_2 \oplus \tau_3) & -2\tau_{out} \otimes \tau_1 \otimes \tau_2 (\tau_{in} \oplus \tau_1 \oplus \tau_2 \oplus \tau_3) & -2\tau_{out} \otimes \tau_2 \otimes (\tau_1 \oplus \tau_2 \oplus \tau_3) & -\tau_{out} \\ -\tau_{out} \otimes \tau_{in} \otimes \tau_1 \otimes \tau_2 \otimes \tau_3 & -\tau_{out} \otimes \tau_1 \otimes \tau_2 \otimes \tau_3 & -\tau_{out} \otimes \tau_2 \otimes \tau_3 & 0 \end{bmatrix}$$

$$B_\lambda^{\otimes 3} = \begin{bmatrix} \cdots & \cdots & \cdots & -3\tau_{out} \\ \cdots & \cdots & \cdots & -2\tau_{out} \\ \cdots & \cdots & \cdots & -\tau_{out} \\ \cdots & \cdots & \cdots & 0 \end{bmatrix}$$

c. $B_\lambda^* = E_n \oplus B_\lambda \oplus B_\lambda^{\otimes 2} \oplus B_\lambda^{\otimes 3}$

Since $\lambda = \tau_{out}$ in column 4 of matrix B corresponds to theorem 2, the eigenvector is the fourth column of the matrix B_λ^*. The resulting eigenvector is

$$\mathbf{u} = \begin{bmatrix} -3\tau_{out} \\ -2\tau_{out} \\ -\tau_{out} \\ 0 \end{bmatrix} = \begin{bmatrix} \tau_{out}^{\otimes(-3)} \\ \tau_{out}^{\otimes(-(3-1))} \\ \tau_{out}^{\otimes(-(3-2))} \\ 0 \end{bmatrix}.$$

By continuing this step for the matrix $B_{\max}^{(n+1)\times(n+1)}$ we obtain a general formula for eigenvector associated with $\lambda = \tau_{out}$ as follows

$$\mathbf{u} = \begin{bmatrix} \tau_{out}^{\otimes(-n)} \\ \tau_{out}^{\otimes(-(n-1))} \\ \tau_{out}^{\otimes(-(n-2))} \\ \tau_{out}^{\otimes(-(n-3))} \\ \vdots \\ 0 \end{bmatrix}. \qquad (15)$$

3.3. Case (iii): $\lambda = \max\{\tau_i \mid 1 \le i \le n\}$

In this case we assume that there are two places in the bulk of the mRNA, with waiting times equals to τ. That is, there exist integers p and q such that $\tau_p, \tau_q = \tau$, where $1 < p < q < n$.

a. $\lambda = \max\{\tau_i \mid 1 \le i \le n\}$

b. For simplicity we use $n = 4$ so

$$B_\lambda = \begin{bmatrix} -\tau\otimes\tau_{in} & -\tau & \varepsilon & \varepsilon & \varepsilon \\ -\tau\otimes\tau_{in}\otimes\tau_1 & -\tau\otimes\tau_1 & -\tau & \varepsilon & \varepsilon \\ \tau_{in}\otimes\tau_1 & \tau_1 & 0 & -\tau & \varepsilon \\ \tau_{in}\otimes\tau_1\otimes\tau & \tau_1\otimes\tau & \tau & 0 & -\tau \\ \tau_{in}\otimes\tau_1\otimes\tau\otimes\tau_4 & \tau_1\otimes\tau\otimes\tau_4 & \tau\otimes\tau_4 & \tau_4 & -\tau\otimes(\tau_4\oplus\tau_{out}) \end{bmatrix}$$

$$B_\lambda^{\otimes 2} = \begin{bmatrix} -2\tau\otimes\tau_{in}\otimes(\tau_{in}\oplus\tau_1) & -2\tau\otimes(\tau_{in}\oplus\tau_1) & -2\tau & \varepsilon & \varepsilon \\ -\tau\otimes\tau_{in}\otimes\tau_1 & -\tau\otimes\tau_1 & -\tau & -2\tau & \varepsilon \\ \tau_{in}\otimes\tau_1 & \tau_1 & 0 & -\tau & -2\tau \\ \tau_{in}\otimes\tau_1\otimes\tau & \tau_1\otimes\tau & \tau & 0 & -\tau \\ \tau_{in}\otimes\tau_1\otimes\tau\otimes\tau_4 & \tau_1\otimes\tau\otimes\tau_4 & \tau\otimes\tau_4 & \tau_4 & -\tau\otimes\tau_4 \end{bmatrix}$$

$$B_\lambda^{\otimes 3} = \begin{bmatrix} -2\tau\otimes\tau_{in}\otimes\tau_1 & -2\tau\otimes\tau_1 & -2\tau & -3\tau & \varepsilon \\ -\tau\otimes\tau_{in}\otimes\tau_1 & -\tau\otimes\tau_1 & -\tau & -2\tau & -3\tau \\ \tau_{in}\otimes\tau_1 & \tau_1 & 0 & -\tau & -2\tau \\ \tau_{in}\otimes\tau_1\otimes\tau & \tau_1\otimes\tau & \tau & 0 & -\tau \\ \tau_{in}\otimes\tau_1\otimes\tau\otimes\tau_4 & \tau_1\otimes\tau\otimes\tau_4 & \tau\otimes\tau_4 & \tau_4 & -\tau\otimes\tau_4 \end{bmatrix}$$

$$B_\lambda^{\otimes 4} = \begin{bmatrix} -2\tau\otimes\tau_{in}\otimes\tau_1 & -2\tau\otimes\tau_1 & -2\tau & -3\tau & -4\tau \\ -\tau\otimes\tau_{in}\otimes\tau_1 & -\tau\otimes\tau_1 & -\tau & -2\tau & -3\tau \\ \tau_{in}\otimes\tau_1 & \tau_1 & 0 & -\tau & -2\tau \\ \tau_{in}\otimes\tau_1\otimes\tau & \tau_1\otimes\tau & \tau & 0 & -\tau \\ \tau_{in}\otimes\tau_1\otimes\tau\otimes\tau_4 & \tau_1\otimes\tau\otimes\tau_4 & \tau\otimes\tau_4 & \tau_4 & -\tau\otimes\tau_4 \end{bmatrix}.$$

c. $B_\lambda^* = E_n \oplus B_\lambda \oplus B_\lambda^{\otimes 2} \oplus B_\lambda^{\otimes 3} \oplus B_\lambda^{\otimes 4}$

Now, focus on the third and fourth columns of the matrix B_λ^* corresponds to Theorem 2 about the existence of its eigenvalues in matrix B. The resulting eigenvector is

$$\mathbf{u}_{\tau_p} = \begin{bmatrix} -2\tau \\ -\tau \\ 0 \\ \tau \\ \tau_4 \otimes \tau \end{bmatrix} = \begin{bmatrix} \tau^{\otimes 0} \otimes \tau^{\otimes(-2)} \\ \tau^{\otimes 1} \otimes \tau^{\otimes(-2)} \\ \tau^{\otimes 2} \otimes \tau^{\otimes(-2)} \\ \tau_{2+1} \otimes \tau^{\otimes 2} \otimes \tau^{\otimes(-2)} \\ \tau_{2+2} \otimes \tau_{2+1} \otimes \tau^{\otimes 2} \otimes \tau^{\otimes(-2)} \end{bmatrix} = \begin{bmatrix} \tau^{\otimes 0} \otimes \tau^{\otimes(-p)} \\ \tau^{\otimes 1} \otimes \tau^{\otimes(-p)} \\ \tau^{\otimes p} \otimes \tau^{\otimes(-p)} \\ \tau_{p+1} \otimes \tau^{\otimes p} \otimes \tau^{\otimes(-p)} \\ \tau_{p+2} \otimes \tau_{p+1} \otimes \tau^{\otimes p} \otimes \tau^{\otimes(-p)} \end{bmatrix}$$

and $\mathbf{u}_{\tau_q} = \begin{bmatrix} -3\tau \\ -2\tau \\ -\tau \\ 0 \\ \tau_4 \end{bmatrix} = \begin{bmatrix} \tau^{\otimes 0} \otimes \tau^{\otimes(-3)} \\ \tau^{\otimes 1} \otimes \tau^{\otimes(-3)} \\ \tau^{\otimes 2} \otimes \tau^{\otimes(-3)} \\ \tau^{\otimes 3} \otimes \tau^{\otimes(-3)} \\ \tau_{3+1} \otimes \tau^{\otimes 3} \otimes \tau^{\otimes(-3)} \end{bmatrix} = \begin{bmatrix} \tau^{\otimes 0} \otimes \tau^{\otimes(-q)} \\ \tau^{\otimes 1} \otimes \tau^{\otimes(-q)} \\ \tau^{\otimes 2} \otimes \tau^{\otimes(-q)} \\ \tau^{\otimes q} \otimes \tau^{\otimes(-q)} \\ \tau_{q+1} \otimes \tau^{\otimes q} \otimes \tau^{\otimes(-q)} \end{bmatrix}.$

By continuing this step for the matrix $B_{\max}^{(n+1)\times(n+1)}$ we obtain a general formula for eigenvector associated with $\lambda = \max\{\tau_i \,|\, 1 \le i \le n\}$ as follows

$$\mathbf{u}_{\tau_p} = \begin{bmatrix} \tau^{\otimes 0} \otimes \tau^{\otimes(-p)} \\ \tau^{\otimes 1} \otimes \tau^{\otimes(-p)} \\ \vdots \\ \tau^{\otimes p} \otimes \tau^{\otimes(-p)} \\ \tau_{p+1} \otimes \tau^{\otimes p} \otimes \tau^{\otimes(-p)} \\ \vdots \\ \tau_{p+1} \otimes \ldots \otimes \tau_n \otimes \tau^{\otimes p} \otimes \tau^{\otimes(-p)} \end{bmatrix}, \tag{16}$$

and

$$\mathbf{u}_{\tau_q} = \begin{bmatrix} \tau^{\otimes 0} \otimes \tau^{\otimes(-q)} \\ \tau^{\otimes 1} \otimes \tau^{\otimes(-q)} \\ \vdots \\ \tau^{\otimes q} \otimes \tau^{\otimes(-q)} \\ \tau_{q+1} \otimes \tau^{\otimes q} \otimes \tau^{\otimes(-q)} \\ \vdots \\ \tau_{q+1} \otimes \cdots \otimes \tau_n \otimes \tau^{\otimes q} \otimes \tau^{\otimes(-q)} \end{bmatrix}. \tag{17}$$

4. Conclusion

This paper shows the formula for determining the eigenvectors of the system introduced in [3] using the Kleene star algorithm, where the eigenvector formula produced in this study is simpler and easier to use in finding eigenvectors of the matrix B. In this paper we also show the properties of the matrix $B_{\lambda}^{\otimes n}$ that if the eigenvalues of the matrix B is $\lambda = \tau_{in}$ then the first column of the matrix $B_{\lambda}^{\otimes n}$ is always the same for every $n = 1, 2, \ldots$ and they becomes the eigenvector of the matrix B.

Acknowledgment
The authors acknowledge support from Universitas Padjadjaran, due to the funding of this work through the scheme of Academic Leadership Grant (ALG) with contract number 855/UN6.3.1/PL/2017.

References

[1] Alberts B, Johnson A., Walter P and Lewis J 2008 *Molecular Biology of the Cell, 5th edn* Garland Pub Inc
[2] Bacelli F, Cohen G, Olsder G J and Quadrat J P 1992 *Synchronization and Linearity in Algebra for Discrete Event Systems* (New York: Wiley)
[3] Brackley C A, Broomhead D S, Romano M C and Thiel M 2009 A max-plus model of ribosome dynamics during mRNA translation *J. Theor Biol.* **303** 128-140
[4] Cuninghame-Green R A 1979 *Minimax Algebra* (Lecture Notes in Economics and Mathematical System 166) (Berlin: Springer-Verlag)
[5] Cuninghame-Green R A 1995 *Minimax Algebra and Application* In: Advances in Imaging and Electron Physics 90 (P W Hawkes, ed) (New York: Academic Press)
[6] Farlow K G 2009 *Max-Plus Algebra* (Virginia Polytechnic Institute and State University: Master's Thesis Virginia Polytechnic Institute and State University)
[7] Gavalec M and Plavka J 2007 Eigenproblem in extremal algebra *In: Proc Of the 9th Int. Symp. Oper. Res. '07* Nova Gorica Slovenia 2007 15-21
[8] Heidergott B, Olsder G J and van der Woude J 2006 *Max Plus at Work: Modeling and Analysis of Synchronized Systems: A Course on Max-Plus Algebra and Its Applications.* (New Jersey: Princeton University Press)
[9] Plavka J 2001 Eigenproblem for circulant matrices in max-algebra *Optimization* 50(2001)477-483
[10] Zimmermann K 1976 Extremal Algebra in (in Czech) *Ekon ústav ČSAV Praha*
[11] Zimmermann U 1981 *Linear and Combinatorial Optimization in Ordered Algebraic Structure* Ann Discrete Math No 10 (Amsterdam: North Holland)

IOP Conf. Series: Materials Science and Engineering **332** (2018) 012020 doi:10.1088/1757-899X/332/1/012020

Performance comparison of genetic algorithms and particle swarm optimization for model integer programming bus timetabling problem

F D Wihartiko[1*], H Wijayanti[2], F Virgantari[2]

[1]Departement of Computer Science, Pakuan University, Indonesia.
[2]Departement of Mathematic, Pakuan University, Indonesia

*E-mail: fajardelli@unpak.ac.id

Abstract. Genetic Algorithm (GA) is a common algorithm used to solve optimization problems with artificial intelligence approach. Similarly, the Particle Swarm Optimization (PSO) algorithm. Both algorithms have different advantages and disadvantages when applied to the case of optimization of the Model Integer Programming for Bus Timetabling Problem (MIPBTP), where in the case of MIPBTP will be found the optimal number of trips confronted with various constraints. The comparison results show that the PSO algorithm is superior in terms of complexity, accuracy, iteration and program simplicity in finding the optimal solution.

1. Introduction

The problem of optimal bus timetabling (or scheduling the trip of bus) is a matter of determining the number of bus departures by taking into account various bus operational constraints. Various bus scheduling models has been developed, one of them is through the integer programming model [1] which in the model has considered the constraints on the elements of manager and passengers. In the constraint managers elements include the number and capacity of the bus, the number and capacity of employees and the cost and travel time of the bus. In the passenger element of the model has paid attention to the number of passengers and minimum service standards that must be given to the passenger management.

The problem of integer programming has grown over the past 50 years. The problem can be categorized as NP-hard optimization problem. IP can be resolved precisely and also through an artificial intelligence based algorithm approximation. The exact problem of IP can be solved by the branch and bound method (B & B) which is the development algorithm of the simplex algorithm [2].

Artificial intelligence-based IP settlement can be solved with genetic algorithm and PSO. GA is a common algorithm used to solve cases of numerical optimization. This algorithm is based on artificial intelligence and uses the theory of life evolution as the basic idea of programming [3]. Particle Swarm Optimization is one of alternative algorithm to solve optimization problem. This algorithm puts forward the idea of colonizing from a bunch of feeding animals and can be used to solve integer programming problems [4]

The application of genetic algorithm has been done to solve the optimization problem of bus rapid transit. Among them can be found in [1], [5], [6], [7], [8] and [9]. The majority of these studies do not show the accuracy of GA with respect to the optimal solution. Research [1] reveals that the accuracy value reaches 99% with an opportunity to obtain a 100% accuracy value of 0.17. This raises new challenges to how to solve the problem of *Model Integer Programming for Bus Timetabling Problem*

(***MIPBTP***) which will produce 100% accuracy with the chance of reaching an optimal solution approaching 1. So this study aims to find a better method to find MIPBTP solution. The PSO algorithm itself is a rarely optimized optimization algorithm for solving optimization in the case of Bus Timetabling Problem [10] & [11]. Both of the above approximation algorithms have different advantages and disadvantages in their application in the integer programming model in the case of bus time tables. This paper will discuss the performance of GA and PSO in MIPBTP. This paper provides a contribution to better algorithm recommendations of solving MIPBTP.

2. Research Method

The model used is linear programming integer model [1]. The data used in this experiment are real to existing BRT service in Bogor (Indonesia) and generated data to see various cases in progress using GA and PSO. The GA concept used is according to the study [1] with the flowchart as follows:

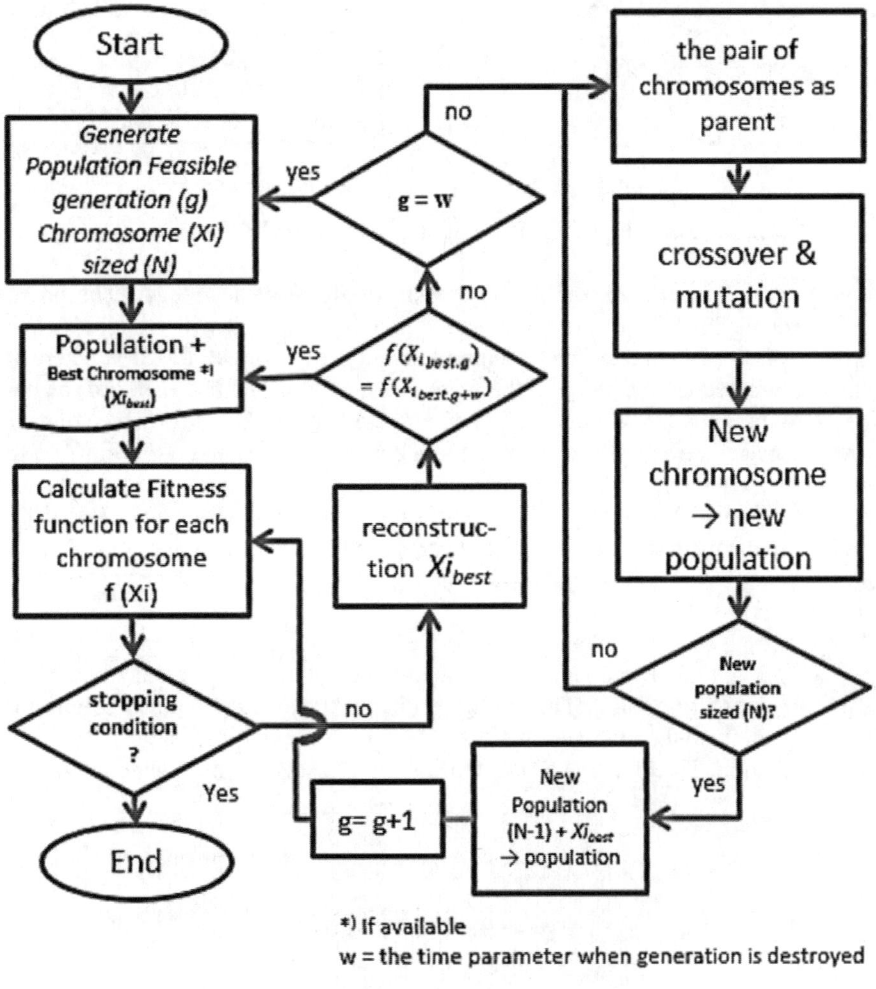

Figure 1. The GA Concept for MIPBTP

The concept of the PSO used is the development of the concept [4] modified according to the needs of model [1] with the result according to the flowchart in Figure 2:

IOP Conf. Series: Materials Science and Engineering **332** (2018) 012020 doi:10.1088/1757-899X/332/1/012020

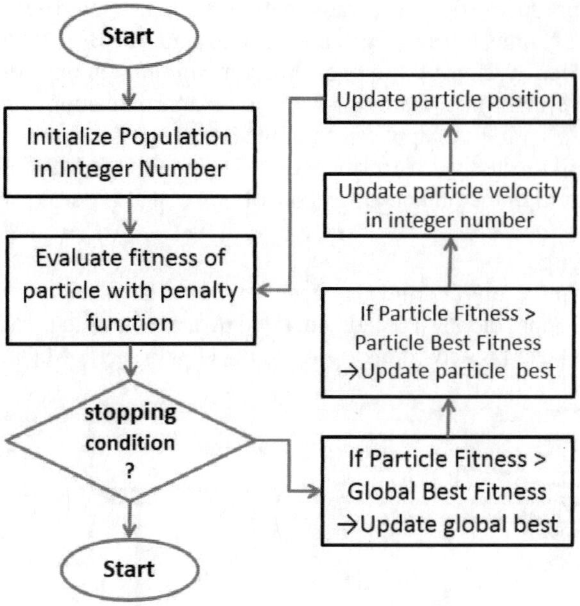

Figure 2. The PSO Concept for MIPBTP

In the initialization phase the vector is generated in the form of an integer. The process of particle movement is a rounded integer of the result rounding down so that it will produce a change of particle position which is also an integer. The evaluation process uses a penalty function where any constraints that are violated will reduce the value of the objective function. The parameter of the penalty function used is similar to the study [1]. Comparison is done by looking at the performance of each algorithm in reaching the optimal solution to the case of the number of variables and constraints on the category of small, medium (real data) and large. Trials in small and large categories use generated data for MIPBTP. In the small category use 2 variables and 6 constraints. While on a large scale used 100 variables and 202 constraints. In the medium category used real data onto 18 variables and 38 constraints.

3. Results and Analysis

The graph of the result of GA implementation in the case of the best solution according to the research [1] can be seen in Figure 3. The image shows the optimal solution improvement to reach the optimum point in each generation (iteration). The red line indicates the average optimal solution found for a set of generation (iterations) when the red line reaches the bottom point indicates that there is a process of population annihilation. The blue line shows the best fitness value of each generation.

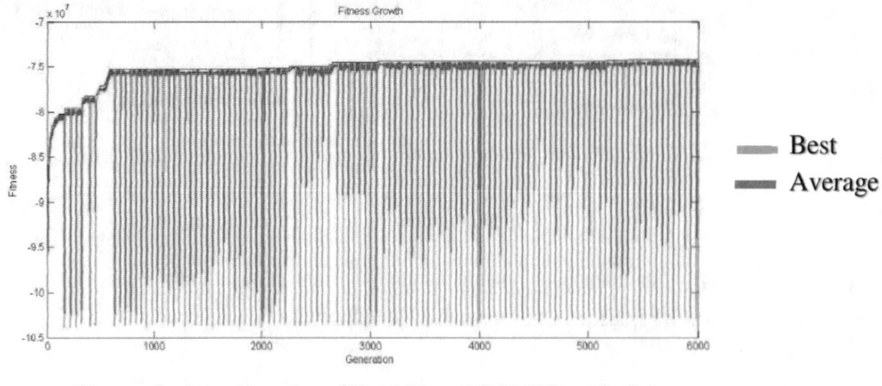

Figure 3. Best Results of Real Data MIPBTP with GA

The result of swarm movement towards 4000 iterations in finding the optimal solution can be seen in the Figure 4. This illustration uses the transformation function $f: x^{18} \rightarrow x^2$.

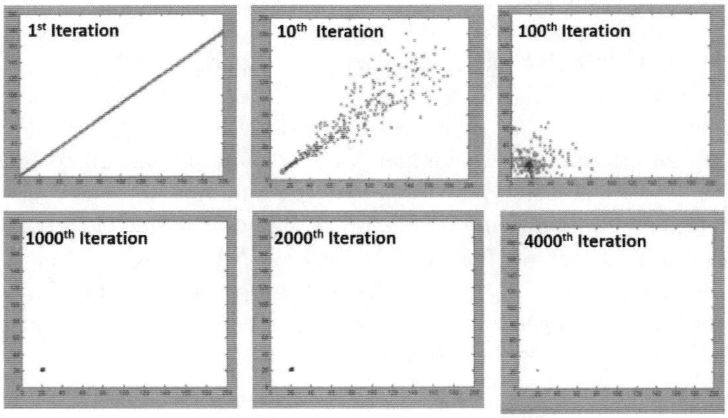

Figure 4. Particle Movement in Order To Find Optimum Solution of Real Data MIPBTP

The comparison performance of GA and PSO based on the number of iterations and the accuracy to get optimal solution in real data of MIPBTP can be see in the following graphics:

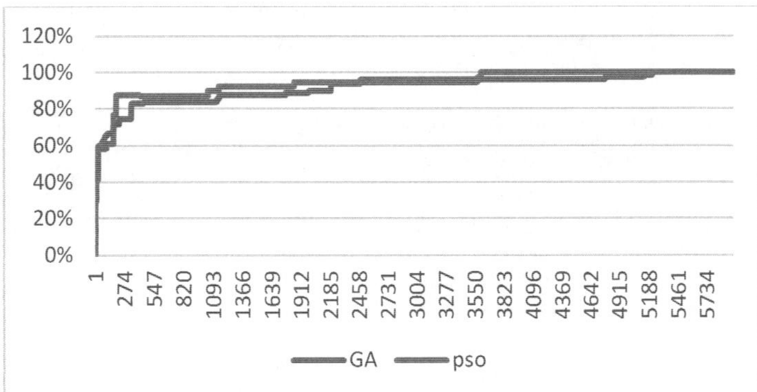

Figure 5. Comparison of GA and PSO Accuracy Charts

From the graph it is seen that the PSO algorithm is superior in terms of speed to reach the optimal solution. The difference between iteration in terms of achieving an optimal solution to 1567 iterations. Both algorithms have achieved the optimal solution. However, the GA probability of obtaining an optimal solution to 0.17% with an average accuracy of 99% [1] is different from the PSO which always got the solution to 100% accuracy. The trial results for MIPBTP case resolution from several categories can be seen to the following table.

Table 1. Comparison of GA and PSO Performance

	Small	Medium	Big
Variable	2	18	100
Constrain	6	38	202
Data	Generate	Real	Generate
Iteration	600	6000	60000

	Small		Medium		Big	
Result	**Average Accuracy**	**Best Iteration**	**Average Accuracy**	**Best Iteration**	**Average Accuracy**	**Best Iteration**
GA	100%	49	99%	5221	93%	56743
PSO	100%	52	100%	3654	100%	44711

In Table 1 it is seen that there is no significant difference when using GA and PSO for cases with relatively small data. The decrease in accuracy of GA is seen when the number of variables increases. Some chromosome extermination techniques (or population destruction techniques) are required to keep accuracy high. This technique is done when the best fitness value of one generation does not change until the specified time. In this case the fitness value of one generation reaches the saturation point, so that chromosomal destruction of that generation is done, except the best chromosome. then the best chromosomes are added to the new N-1 chromosome population. This process continues in GA until it reaches the termination criteria. This technique is able to improve accuracy even though GA is not always able to find the optimal solution to each iteration on a medium and large scale. In contrast to GA, PSO has a stable condition in finding the optimal solution. No need to add population destruction techniques (as in genetic algorithms that have to destruct populations that have saturated at certain iterations to improve their accuracy), PSO is able to get optimal solutions in various conditions. In summary the results of the comparison of the two algorithms in terms of complexity, iteration and accuracy on MIPBTP can be seen in the following table.

Table 2. Summary of Comparative Results

Indicator	GA	PSO
Complexity	$O(n^2)$	$O(n)$
Accuracy	Solutions in large variables and constraints result in feasible solutions that are near optimal	Resolution produces the optimal solution
Iterate	The more variables and constraints, the more iterations required. In general, it takes much more than the PSO	The more variables and constraints, the more iterations required. Generally, it takes fewer iterations of the GA
Additional techniques used	It takes the addition of chromosome extermination techniques in certain generations in order to approach the optimum solution	No additional techniques

4. Conclusion

The results of GA and PSO implementation in the bus timetabling problem show that the PSO algorithm is superior in finding the optimal solution in terms of accuracy and iteration. In addition the PSO algorithm is also superior to the simplicity of the techniques used. For small scale there is no significant difference between the two methods. Differences are seen in medium and large scale where genetic algorithms can only produce feasible solutions that are near optimal. PSO algorithm has ease of implementation and also has high calculation accuracy.

References

[1] Wihartiko FD, Buono A and Silalahi BP, *Integer programming model for optimizing bus timetable using genetic algorithm.* IOP Conference Series: Materials Science and Engineering. 166(1) 1-9 (2017)

[2] Genova K and Guliashki V, *Linear Integer Programming Methods and Approaches – A Survey*, Cybernetics and Information Technologies 11(1) 3-25 (2011)

[3] Huy PNA, San CTB and Triantaphyllou E. *Solving Integer Programming Problems Using Genetic Algorithms*. ICEIC. Ha-Noi. (2004).

[4] E.C. Laskari, K.E. Parsopoulos and M.N. Vrahatis, *Particle Swarm Optimization for Integer Programming*. IEEE. (2002).

[5] Chuanjiao S, Wei Z, Yuanqing W. Scheduling Combination and Headway Optimization of Bus

Rapid Transit. *Journal of Transportation Systems Engineering and Information Technology*. 2008; 8(5): 61–67

[6] Kidwai FA. *A Genetic Algorithm Based Bus Schedulling Model for Transit Network*. Eastern Asia Society for Transportation Studies. Bangkok. 2005; 05: 477–489

[7] Zhang M. *The Research on Multi-objective School Bus Vehicle Routing Problems Based On Bi-level Programming*. Chengdu, China: Southwest Jiaotong University; 2008.

[8] Ben Sghaier S, Ben Guedria N, Mraihi R, editors. *Solving School Bus Routing Problem with genetic algorithm*. Advanced Logistics and Transport (ICALT), 2013 International Conference on; 2013: IEEE. pp. 7–12.

[9] Minocha B, Tripathi S, editors. *Solving School Bus Routing Problem Using Hybrid Genetic Algorithm: A Case Study*. Proceedings of the Second International Conference on Soft Computing for Problem Solving (SocProS 2012), December 28–30, 2012; 2014: Springer. pp. 93–103.

[10] Zhang Y, Wang S, and Ji1 G, A *Comprehensive Survey on Particle Swarm Optimization Algorithm and Its Applications*. Hindawi Publishing Corporation Volume 2015, Article ID 931256, 38 page

[11] Ab Wahab MN, Nefti-Meziani S, Atyabi A (2015) A Comprehensive Review of Swarm Optimization Algorithms. PLoS ONE 10(5): e0122827. doi:10.1371/journal.pone.0122827

IOP Conf. Series: Materials Science and Engineering **332** (2018) 012021 doi:10.1088/1757-899X/332/1/012021

Optimization of land use of agricultural farms in Sumedang regency by using linear programming models

F M Zenis[1]**, S Supian**[2]**, E Lesmana**[2]

[1]Master Students of Mathematic's study program, Faculty of Mathematics and Natural Sciences, Universitas Padjadjaran, INDONESIA
[2]Department of Mathematics, Faculty of Mathematics and Natural Sciences, Universitas Padjadjaran, INDONESIA

Corresponding author: fhani.mulyani@gmail.com

Abstract. Land is one of the most important assets for farmers in Sumedang Regency. Therefore, agricultural land should be used optimally. This study aims to obtain the optimal land use composition in order to obtain maximum income. The optimization method used in this research is Linear Programming Models. Based on the results of the analysis, the composition of land use for rice area of 135.314 hectares, corn area of 11.798 hectares, soy area of 2.290 hectares, and peanuts of 2.818 hectares with the value of farmers income of IDR 2.682.020.000.000,-/year. The results of this analysis can be used as a consideration in decisions making about cropping patterns by farmers.

1. Introduction

Land is one of the most important assets for farmers. If a land has a good quality, it will have good potential also for the welfare of the surrounding population. If a land is used for non-agricultural development, it will cause some problems such as food self-sufficiency, environment, and employment. For example, declining agricultural yields, pollution from waste, floods, and increasing unemployment. Also bad for the health of local people, due to the pollution. In West Java, there are several types of land, namely Wetland, Non-Rice Farming Land, and Non-Agricultural Land. The paddy field consists of irrigated rice fields, which are not planted with rice, and while not cultivated. While non irrigated rice field consists of planted with rice, which is not planted with rice, and which temporarily not cultivated. A broad understanding of land is a surface area of the Earth whose characteristics encompass all the identifiers of both a sufficiently steady and predictable nature of the biosphere, atmosphere, soil, geology, hydrology and plant and animal populations, and the results of the activity human beings in the past and the present, as far as the identification marks have an influence over the human use of land in the present and future. [1]

Productive land use is crucial to the productivity of agricultural commodities, especially food crops, as a source of carbohydrates to ensure the food security of people in West Java. Food crops consist of grains and crops. Rice is one of the most important cultivation plants in civilization. Rice production in the world ranks third of all cereals, after maize and wheat. However, rice is a major source of carbohydrates for the majority of the world's population. [2]

A problem will be solved to obtain optimum results in accordance with the given constraints. If the problem is formulated appropriately, it can give the optimum decision variable value. Once the optimum solution is obtained, the problem is often re-evaluated under different conditions to obtain a new settlement. [3]

Previous research that discusses the optimization of land use linear programming is Bisschop. Discusses a farmer who owns a plot of land and must make a decision about what type of plant to plant in the field. Limitations used are the area of land, the amount of labor comprised of family labor, wage labor, and temporary labor, and also constraints of water limitations. The ultimate goal is to maximize profits.

In this paper discuss about how to formulate optimization model of agricultural land use in order to obtain optimal results.

2. The basic concept of optimization

Optimization is the act of getting the best results under certain circumstances. In the design, construction, and maintenance of engineering systems, engineers must take many technological and managerial decisions at several stages. The ultimate goal of all decisions is to minimize the effort required or to maximize the desired benefits. [4] In everyday life, whether consciously or unconsciously, people always do the optimization to meet their needs. The optimization done by ordinary people is more based on intuition than optimization theory.

Linear programming is a way to solve the problem of allocation of limited resources such as labor, raw materials, machine work hours and the like in the best way possible to obtain optimal results. [5]

Linear programming has become the basis of manufacturing decisions, marketing, bank loans, stock quotes, taxi fares, coordination transport, telephone charges, computer access, optimization of land and other practical issues. [6]

The general form of Linear programming model can be described as follows:

Objective function: max/min $Z = \sum_{j=1}^{n} C_j x_j$ (1)

Subject to

$$\sum_{i=1}^{m} a_{ij} x_j \{\leq, =, \geq\} b_i$$

$$x_j \geq 0$$

The solution to meet these linear constraints is called a feasible solution. Three elements in linear programming are decision variables, constraints and objective functions.

3. Model of Land Use Optimization

The model of land use optimization is implemented on Sumedang Regency Farm where the parameters are:

3.1. Parameterization

Index

c plants / commodities

t month

Parameter of the plant

p_c c plant price [Rp / ton]

Parameter of land

L available land area [ha]

l_{tc} land area in month t [ha]

Labor parameters

V available labor [people]

v_c labor needed c plant [person]

V^P female labor [person]

V^L male labor [person]

r^P annual wage rate for female labor [Rp / person]

r^L annual wage rate for male labor [Rp / person]

Parameter of fertilizer

F annual amount of fertilizer available [kg]

f_c fertilizer needed for plant c [kg]

p^F fertilizer price [Rp / kg]

Decision Variables

x_c the number of each c plant grown [ha]

V^P number of female workers [people]

V^L number of men [man]

s_c sale from c plant [ton]

p_c the price of each plant c [Rp / ton]

3.2. Numerical Calculation

Sumedang has a land area of 152.220 hectares consisting of rice fields and non-rice fields. While, the data used in the agricultural land use model of Sumedang Regency are as follows:

Table 1. Planting area (ha)

Rice	80.261
Corn	11.798
Soy	2.290
Peanuts	2.818
Total	**97.167**

Table 2. Cropping area (ha)

Rice	74.933
Corn	10.207
Soy	2.288
Peanuts	2.915
Total	**90.343**

Table 3. Analysis of Farming.

Commodities	Result (kg)	Price (Rp/Kg)	Value of Results / Production (Rp)	Total Production Cost (Rp)	Farm Income (Rp)	R/C
Rice	5.510	4.800	26.448.000	16.910.000	9.538.000	1,56
Corn	6.500	1.800	11.700.000	8.215.000	3.485.000	1,42
Soy	1.571	3.800	5.969.800	5.570.000	399.800	1,07
Peanuts	1.230	8.300	10.209.000	9.480.000	729.000	1,08

Table 4. Total labor.

	Man			Woman			
	Hired worker	Family	Total	Hired worker	Family	Total	Total
Rice	118	16	134	85	6	91	225
Corn	70	13	83	58	4	62	145
Soy	54	12	66	-	1	1	67
Peanuts	120	22	142	144	6	150	292
Total	362	63	425	287	17	304	**729**

Table 5. Volume Fertilizer of each Commodity.

	Volume (Kg)				Total
	Urea	SP -36	KCl	NPK Phonska	
Rice	200	100	75	50	425
Corn	150	50	50	50	300
Soy	50	100	50	50	250
Peanuts	100	200	100	-	400
Total					**1.375**

Table 6. Price of Fertilizer.

Type of Fertilizer	**Price (Rp)**
Urea	1.800
SP-36	2.000
KCl	2.000
NPK	2.000

The purpose of this study is to maximize the profits of farmers for a year, with revenue from sales minus the costs incurred. While the obstacles used in this study are limited land, labor needs, and fertilizer needs.

Based on (1), and referring to [7], the above problem can be formulated as follows:

$$\text{Max} \quad \sum_c p_c s_c - r^L V^L - r^P V^P - p^F f_c x_c$$

subject to

$$\begin{aligned} &\sum_c x_c \le L \\ &\sum_c l_{tc} x_c \le L && \forall t \\ &\sum_c v_c x_c \le L\left(V^L + V^P\right) && \forall t \\ &\sum_c f_c x_c \le LF \\ &x_c \ge 0 && \forall c \\ &V^P \ge 0 \\ &s_c \ge 0 && \forall c \end{aligned} \tag{2}$$

Based on data obtained from Dinas Pertanian, Peternakan, dan Perikanan Kabupaten Sumedang in 2016 which has been described above, it can be obtained a formula for Linear Programming optimization model as follows:

Objective function: Max $z = 19128000x_1 + 6985000x_2 + 3149800x_3 + 1459000x_4$

subject to:

$$
\begin{aligned}
&x_1 + x_2 + x_3 + x_4 \le 152220 \\
&x_1 \ge 80261 \\
&x_2 \ge 11798 \\
&x_3 \ge 2290 \\
&x_4 \ge 2818 \\
&225x_1 + 145x_2 + 67x_3 + 292x_4 \le 110968380 \\
&425x_1 + 300x_2 + 250x_3 + 400x_4 \le 209302500 \\
&x_1, x_2, x_3, x_4 \ge 0
\end{aligned}
\tag{3}
$$

where
x_1 = Rice land area
x_2 = Corn land area
x_3 = Soy land area
x_4 = Peanuts land area

The optimal solution of equation (3) is obtained by using the simplex table is to arrange the equations in a table, the simplex table. The initial simplex table for this problem is as follows:

Table 7. First Simplex's Table.

VB	z	x_1	x_2	x_3	x_4	s_1	s_2	s_3	s_4	s_5	s_6	s_7	NK
z	1	-19128000	-6985000	-3149800	-1459000	0	0	0	0	0	0	0	0
s_1	0	1	1	1	1	1	0	0	0	0	0	0	152220
s_2	0	1	0	0	0	0	1	0	0	0	0	0	80261
s_3	0	0	1	0	0	0	0	1	0	0	0	0	11798
s_4	0	0	0	1	0	0	0	0	1	0	0	0	2290
s_5	0	0	0	0	1	0	0	0	0	1	0	0	2818
s_6	0	225	145	67	292	0	0	0	0	0	1	0	110968380
s_7	0	425	300	250	400	0	0	0	0	0	0	1	209302500

To get the optimal result, the above problem is solved by 4 iteration process. Thus, using the help of LINDO's software, obtained optimal results of the area of rice area of 135.314 ha, corn area of 11.798 ha, soy area of 2.290 ha, and peanuts area of 2.818 ha. The optimal income obtained from the pattern of land area is IDR 2.682.020.000.000,-/year.

4. Conclusion

From the description above can be concluded that the income earned by farmers of IDR 2.682.020.000.000,-/year, if farmers grow rice, corn, soy, and peanuts with the following land size: 135.314 ha of rice area, 11.798 ha of corn area, 2.290 ha of soy area, and 2.818 ha of peanuts. Thus, it can be used as a consideration of the farmers for the planning of land use patterns.

References

[1] FAO. 1977. A framework for land evaluation. ILRI Publ. 22. Wageningen.Viii+87h
[2] Hatta, Muhammad. 2016. *Mukjizat Herbal dan Khasiatnya dalam Alquran Vol 3*. Mirqat: Jakarta

IOP Conf. Series: Materials Science and Engineering **332** (2018) 012021 doi:10.1088/1757-899X/332/1/012021

[3] Rio Armindo. 2006. *Penentuan Kapasitas Optimal Produksi CPO (Crude Palm Oil) Dipabrik Kelapa Sawit PT. Andira Argo Dengan Menggunakan Goal Programming.* ITB
[4] Rao, S. S. 1996. Engineering optimization: theory and practice
[5] Dimyati, Tjuju T. dan Dimyati, A. 2002. *Operation Research : Model-model Pengambilan Keputusan*. Sinar Baru Algesindo, Bandung
[6] Wu, J., & GE, X. 2012. Optimization Research of Generation Investment Based on Linear Programming Model. *International Conference on Applied Physics and Industrial Engineering*, 1400-1405
[7] Bisschop, J. 2017. *Optimization Modelling*. Netherlands: AIMMS

IOP Conf. Series: Materials Science and Engineering **332** (2018) 012022 doi:10.1088/1757-899X/332/1/012022

Markov chain model for demersal fish catch analysis in Indonesia

Firdaniza*, N Gusriani

Department of Mathematics, Faculty of Mathematics and Natural Sciences, Padjadjaran University. Jl.Raya Bandung-Sumedang km 21 Jatinangor, 45363 Indonesia. Tel./Fax. +62-22-7794696

*Corresponding author: firdaniza@unpad.ac.id

Abstract. As an archipelagic country, Indonesia has considerable potential fishery resources. One of the fish resources that has high economic value is demersal fish. Demersal fish is a fish with a habitat in the muddy seabed. Demersal fish scattered throughout the Indonesian seas. Demersal fish production in each Indonesia's Fisheries Management Area (FMA) varies each year. In this paper we have discussed the Markov chain model for demersal fish yield analysis throughout all Indonesia's Fisheries Management Area. Data of demersal fish catch in every FMA in 2005-2014 was obtained from Directorate of Capture Fisheries. From this data a transition probability matrix is determined by the number of transitions from the catch that lie below the median or above the median. The Markov chain model of demersal fish catch data was an ergodic Markov chain model, so that the limiting probability of the Markov chain model can be determined. The predictive value of demersal fishing yields was obtained by calculating the combination of limiting probability with average catch results below the median and above the median. The results showed that for 2018 and long-term demersal fishing results in most of FMA were below the median value.

1. Introduction

As an archipelagic country, Indonesia has the potential of marine and fishery resources that can improve the Indonesian economy. The Government of Indonesia has been divided the Fisheries Management Area (FMA) into 11 regions. Each FMA has provided fish catches that vary according to geographical location.

One of the fish resources that boost Indonesia's economy is demersal fish. Demersal fish is a fish with a habitat in the muddy seabed. Although demersal fish does not become the target of Indonesian exports, but this fish is very popular with the community because it tastes good and the price is quite affordable. Some types of demersal fish are black pomfret, silver pomfret, halibut, baramundi, red snapper, jack travelie, hairtail, giant cat fish and others. During the period of 2005-2014 demersal fish production was quite varied and contributed an average of 130 thousand tons each year. Most demersal fish were caught in the Java sea and Karimata straits, Natuna sea and South China Sea [1]. The study conducted related to the results of fish catch, namely the prediction of potential areas of pelagic fishing in Mamuju district [5]. To predict a value in the future, various methods can be used, such as the Markov chain or times series. Markov chain method has been used in stock price prediction [2], in predicted fishing results in India [3], predicted dominant fish catch areas in India [6].

In addition to the study of catch areas, research on the prediction of fish catch is also necessary because it will have an impact on Indonesian exports. In this study the Markov chain method will be used to analyze the demersal fishing results in Indonesia. The objective is to know the predictions of demersal fishing results across Indonesia's FMA. This is expected to be useful for the government (Marine Affairs and Fisheries) in taking policy for future anticipation if the catch is decreased.

2. Markov Chain Model

The Discrete-time Markov chain is a stochastic process with the nature that the future state depends only on the present state of being free from the past.

Definition 1. Let $\{X(n), n = 0,1,2, ...\}$ is a discrete time stochastic process with state space i = 0,1,2, ..., if

$$P\{X(n+1) = j|X(0) = i_0, X(1) = i_1, ..., X(n-1) = i_{n-1}, X(n) = i\}$$
$$= P\{X(n+1) = j|X(n) = i\} = p_{ij}$$

For all $i_0, i_1, ..., i_{n-1}, i, j$ and n, then the process is called the discrete-time Markov chain of, and p_{ij} is called transition probability [4].

Definition 2.

The one-step transition probability matrix of a Markov chain is defined as $\mathbf{P} = [p_{ij}] = \begin{bmatrix} p_{00} & p_{01} & p_{02} & \cdots \\ p_{10} & p_{11} & p_{12} & \cdots \\ p_{20} & p_{21} & p_{22} & \cdots \\ \vdots & \vdots & \vdots & \cdots \end{bmatrix}$

where $p_{ij} \geq 0$ and $\sum_{j=0}^{\infty} p_{ij} = 1$, $i, j = 0, 1, 2, ...$

The n-step transition probability can be calculated using the Chapman-Kolmogorov equation. In matrix form can be written as $\mathbf{P^n} = \mathbf{P^r}.\boldsymbol{P^{n-r}}$ [4].

2.1. State Classification

For a Markov chain, if there is only one class of communication, the Markov chain is said to be irreducible. This means that all the existing state communicate with each other.

State *i* and *j* are said to communicate with each other, if there is n ≥ 0 such that $p^n{}_{ij} > 0$ *and* $p^n{}_{ji} > 0$.

State *i* is said recurrent if and only if $\sum_{n=1}^{\infty} p^n{}_{ii} = \infty$.

State *i* is called periodic with period $d(i)$, if $d(i) = greatest\ common\ divisor\ \{n|\ p^n{}_{ii} > 0, n \geq 1\}$. If $d(i) = 1$, *then* state *i* is called aperiodic.

2.2. Stationary Distribution

A Markov chain is ergodic if irreducible, positive recurrent, and aperiodic. Ergodic's Markov chains have limiting probabilities, $\lim_{n\to\infty} p_{ij}^n = \pi_j$, $i, j = 0, 1, 2, ...$ which is free from the initial state *i*, $[\pi_j]$ is called the stationary distribution (steady state) of the Markov chain [4].

3. Methodology

Demersal fishing results has changed every period and its catching area. Discrete time markov chain can model problems with data that changes with time.

Demersal fishery data obtained from the Directorate of Fisheries Fishing Ministry of Marine Affairs and Fisheries Indonesia, the data is compiled by year and FMA. Assume that the number of Markov chain states were 2, i.e., state 1; the number of catches below the median value, and state 2; the number of catches above the median value. For each FMA it has been done the same stepsfigure:

a. Determine the state "1" and state "2" of each demersal fish catch data based on the value of the captured median.
b. Calculate the number of transitions from state "1" to state "2" and vice versa

c. find the transition probability matrix
d. Check if Markov chain is ergodic
e. Determine the stationary distribution of the markov chain

4. Result and Discussion

According to Minister of Marine Affairs and Fisheries Regulation no. PER.18 / Men / 14 and PER.01 / Men / 09 on fisheries management area (FMA), it is mentioned that FMA in Indonesia has been divided into 11 FMA, that is; (1) Malacca Strait and Andaman Sea, (2) Hinda Ocean in West Sumatera and Sunda Strait, (3) Indian Ocean to Southern Java South to Nusa Tenggara, Savu Sea and West Timor Sea, (4) Karimata Strait, Sea (7) Tolo Bay and the Banda Sea, (8) Tomini Bay, Maluku Sea, Halmahera Sea, (5) Seram Sea, and Berau Bay, (9) Sulawesi Sea and North Halmahera Island, (10) Cendrawasih Bay and Pacific Ocean, (11) Aru Gulf, Arafuru Sea, East Timor Sea. The map of Indonesia's FMA is shown in Figure 1.

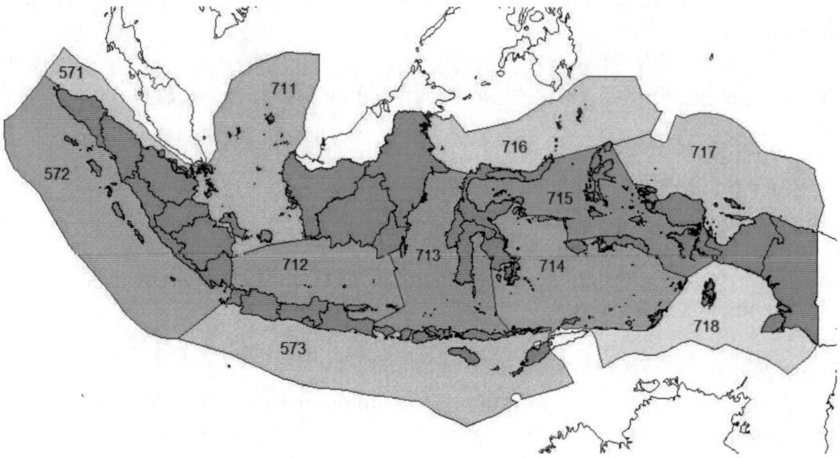

Figure 1. The map of Indonesia's FMA

Data of demersal fish catch were shown in table 1 below:

Table 1. Production of demersal fish catch at 11 Indonesia's FMA in 2005-2014 (ton)

FMA	2005	2006	2007	2008	2009	2010	2011	2012	2013	2014
FMA571	106440	107166	122221	113179	109064	105279	139161	140692	149551	148410
FMA572	115057	116891	138983	142455	154072	145982	134324	167895	178400	167446
FMA573	78383	71697	51252	63698	60373	54719	65199	56753	58438	60565
FMA711	155331	161485	177191	206442	202364	213977	203843	199866	202706	222175
FMA712	261739	277106	259945	304202	259808	266271	266545	276757	307715	314485
FMA713	129183	123830	130337	145305	162579	154833	162167	155768	156477	164766
FMA714	150546	134766	106686	64029	62800	83607	103324	101453	97105	143658
FMA715	46931	37227	55020	40516	38818	46944	44851	38037	50485	72145
FMA716	17058	22763	19799	17653	15773	23571	24165	26984	27540	26955
FMA717	26014	21071	28015	29233	33193	36724	38236	37823	45573	42474
FMA718	210705	211048	210449	179286	178796	296883	246255	221103	224974	194498

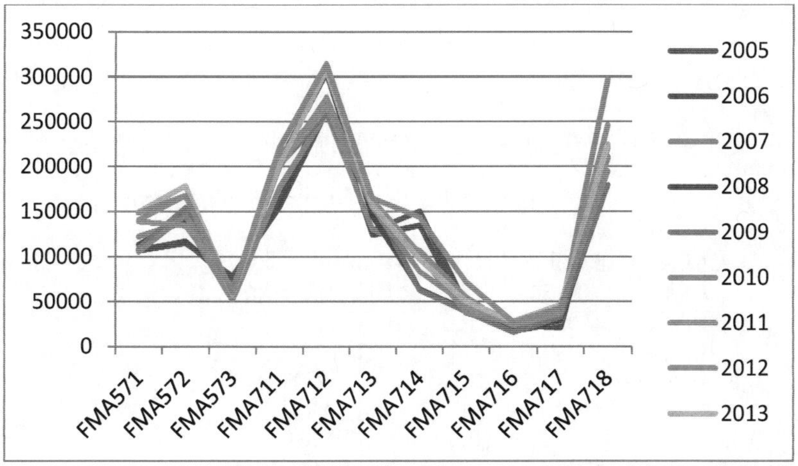

Figure 2. Graph of demersal fish catch in 11 FMA 2005-2014

From the data was determined median for each FMA. The median value of each FMA was presented in Table 2.

Table 2. The median value of the demersal fish catch for each FMA

FMA	Median
FMA571	117700
FMA572	144218.5
FMA573	60469
FMA711	202535
FMA712	271651
FMA713	155300.5
FMA714	102388.5
FMA715	45891
FMA716	23167
FMA717	34958.5
FMA718	210876.5

After calculating the amount of displacement of each state, an transition probability matrix was obtained, $\mathbf{P} = [p_{ij}]$ with $p_{ij} = \frac{n_{ij}}{n_i}$, n_{ij} = number of transitions from state i to state j, the number of transitions from state i.

From the available data we have obtained a transition probability matrix from the Markov chain for each FMA.

FMA571, FMA572, FMA713 : $\mathbf{P} = \begin{bmatrix} 0{,}6 & 0{,}4 \\ 0{,}25 & 0{,}75 \end{bmatrix}$

FMA573, FMA715 : $\mathbf{A} = \begin{bmatrix} 0{,}4 & 0{,}6 \\ 0{,}75 & 0{,}25 \end{bmatrix}$

FMA711, FMA712 : $\mathbf{B} = \begin{bmatrix} 0{,}4 & 0{,}6 \\ 0{,}5 & 0{,}5 \end{bmatrix}$

FMA714 : $\mathbf{C} = \begin{bmatrix} 0{,}6 & 0{,}4 \\ 0{,}5 & 0{,}5 \end{bmatrix}$

FMA716, FMA717 : $\mathbf{D} = \begin{bmatrix} 0{,}8 & 0{,}2 \\ 0 & 1 \end{bmatrix}$

FMA718, : $\mathbf{E} = \begin{bmatrix} 0{,}5 & 0{,}5 \\ 0{,}4 & 0{,}6 \end{bmatrix}$

From all the transition probability matrix it is seen that all states communicate with each other, so the Markov chain is said to be irreducible.

The n step transition probability matrix can be calculated using the Chapman Kolmogorov equation, we got

$$\mathbf{P}^2 = \begin{bmatrix} 0{,}46 & 0{,}54 \\ 0{,}3375 & 0{,}6625 \end{bmatrix}, \quad \mathbf{P}^3 = \begin{bmatrix} 0{,}411 & 0{,}589 \\ 0{,}3681 & 0{,}6319 \end{bmatrix}, \quad \mathbf{P}^4 = \begin{bmatrix} 0{,}3938 & 0{,}6061 \\ 0{,}3788 & 0{,}6212 \end{bmatrix}$$

$$\mathbf{A}^2 = \begin{bmatrix} 0{,}61 & 0{,}39 \\ 0{,}4875 & 0{,}5125 \end{bmatrix}, \quad \mathbf{A}^3 = \begin{bmatrix} 0{,}5365 & 0{,}4635 \\ 0{,}5792 & 0{,}4206 \end{bmatrix}, \quad \mathbf{A}^4 = \begin{bmatrix} 0{,}5622 & 0{,}4378 \\ 0{,}5472 & 0{,}4528 \end{bmatrix}$$

$$\mathbf{B}^2 = \begin{bmatrix} 0{,}46 & 0{,}54 \\ 0{,}45 & 0{,}55 \end{bmatrix}, \quad \mathbf{B}^3 = \begin{bmatrix} 0{,}454 & 0{,}546 \\ 0{,}455 & 0{,}545 \end{bmatrix}, \quad \mathbf{B}^4 = \begin{bmatrix} 0{,}4546 & 0{,}5454 \\ 0{,}4545 & 0{,}5455 \end{bmatrix}$$

$$\mathbf{C}^2 = \begin{bmatrix} 0{,}56 & 0{,}44 \\ 0{,}55 & 0{,}45 \end{bmatrix}, \quad \mathbf{C}^3 = \begin{bmatrix} 0{,}556 & 0{,}444 \\ 0{,}555 & 0{,}445 \end{bmatrix}, \quad \mathbf{C}^4 = \begin{bmatrix} 0{,}5556 & 0{,}4444 \\ 0{,}5555 & 0{,}4445 \end{bmatrix}$$

$$\mathbf{D}^2 = \begin{bmatrix} 0{,}64 & 0{,}36 \\ 0 & 1 \end{bmatrix}, \quad \mathbf{D}^3 = \begin{bmatrix} 0{,}512 & 0{,}488 \\ 0 & 1 \end{bmatrix}, \quad \mathbf{D}^4 = \begin{bmatrix} 0{,}4096 & 0{,}5904 \\ 0 & 1 \end{bmatrix}$$

$$\mathbf{E}^2 = \begin{bmatrix} 0{,}45 & 0{,}55 \\ 0{,}44 & 0{,}56 \end{bmatrix}, \quad \mathbf{E}^3 = \begin{bmatrix} 0{,}445 & 0{,}555 \\ 0{,}444 & 0{,}556 \end{bmatrix}, \quad \mathbf{E}^4 = \begin{bmatrix} 0{,}4445 & 0{,}5555 \\ 0{,}4444 & 0{,}5556 \end{bmatrix}$$

From P, P^2, P^3, P^4 we obtained

$\sum_{n=1}^{\infty} p_{00}^n = p_{00}^1 + p_{00}^2 + p_{00}^3 + p_{00}^4 + p_{00}^5 + \cdots = 0{,}46 + 0{,}411 + 0{,}3938 + \cdots = \infty.$

$$\sum_{n=1}^{\infty} \boldsymbol{p_{11}^n} = \boldsymbol{p_{11}^1} + \boldsymbol{p_{11}^2} + \boldsymbol{p_{11}^3} + \boldsymbol{p_{11}^4} + \boldsymbol{p_{11}^5} + \cdots = \mathbf{0{,}6625} + \mathbf{0{,}6319} + \mathbf{0{,}6212} + \cdots = \infty.$$

Because $\sum_{n=1}^{\infty} p^n{}_{ij} = \infty$, then state "1" and state "2" were recurrent state. So it was said the Markov chain is irreducible and recurrent. The same is true for the matrices A, B, C, D, and E.

The next step, determine the periodicity of the Markov chain with the transition matrix P, A, B, C, D, E, we obtained

$d(0) = gcd\{1,2,3, \ldots\} = 1$, then state "0" was aperiodic

$d(1) = gcd\{1,2,3, \ldots\} = 1$, the state "1" was aperiodic.

it could be concluded that the Markov chain was aperiodic. For $n \to \infty$, we get

$\boldsymbol{P^n} = \begin{bmatrix} 0{,}3846 & 0{,}6154 \\ 0{,}3846 & 0{,}6154 \end{bmatrix}$. Because every state of Markov chain was recurrent and aperiodic, then $p^n{}_{ij} \to \frac{1}{\mu_j}$. So we get $\boldsymbol{\mu} = [\mu_0, \mu_1] = [2{,}6 \quad 1{,}63]$.

Because $\mu_0 = 2{,}6 < \infty$ and $\mu_1 = 1{,}63 < \infty$, then the Markov chain is said to be positive recurrent.

Since the Markov chains were irreducible, positively recurrent and aperiodic, it was said to be ergodic Markov chains.

Since the Markov chain was ergodic, then a stationary distribution was $\boldsymbol{\pi} = [0{,}3846 \quad 0{,}6154]$. This means for a Markov chain with a transition probability matrix P, the probability of demersal fish quantities above the median value was 0,6154.

In the same way, a steady state distribution for the transition probability matrix A, B, C, D, and E are:

For matrix A, $\boldsymbol{\pi} = [0{,}5556 \; 0{,}4444]$

For matrix **B**, $\boldsymbol{\pi} = [0{,}4545 \; 0{,}5455]$

For matrix **C**, $\boldsymbol{\pi} = [0{,}5556 \; 0{,}4444]$

For matrix **D**, $\boldsymbol{\pi} = [0 \; 1]$

For matrix **E**, $\boldsymbol{\pi} = [0{,}4444 \; 0{,}5556]$

Long-term predictions of demersal fishing results for each FMA

$y = a_0\pi_0 + a_1\pi_1$

where

a_0 = average catch below median

a_1 = average catch above median
π_0 = the probability of catch is below median
π_1 = the probability of catch is above median

The predictions of demersal fish yields throughout FMA can be seen in the table 3.

Table 3. Predicted number of long-term demersal fish catches

FMA	Predicted catch results (ton)
571	127783,9
572	149983,7
573	61462,66
711	195929,4
712	280967,5
713	151254,1
714	102240
715	46295,91
716	25843
717	40166
718	219918,7

5. Conclusion

From the Markov model analysis, it was concluded that the demersal fish catch in some FMA was below the median value. This should be anticipated by the government with various efforts to make demersal fish production is always increasing.

Acknowledgments

Thanks to Padjadjaran University due to the funding of this work through the scheme of Fundamental Grant 2017.

References

[1] Directorate of Capture Fisheries 2015 *Statistic of Marine Capture Fisheries by Fisheries Management Area (FMA) 2005-2014.*

[2] K J Doubleday and J N Esunge 2011 *Application of Markov Chains to Stock Trends* Journal of Mathematics and Statistics 7 (2) p 103-106

[3] Mini K G and Somy Kuriakose 2004 *Longterm projection of marine fish landing along Soutwest coast of India using Markov chain model* J. Mar. Biol. Ass. India 46(2) p 209-214

[4] Osaki Shunji 1992 *Applied Stochastic System Modelling* Springer-Verlag Berlin Heidelberg

[5] Safruddin, M Zainuddin, C Rani 2014 *Prediksi Derah Potensial Penangkapan Ikan Pelagis Besar di Perairan Kabupaten Mamuju* Jurnal IPTEKS PSP Vol 1 (2)

[6] Somy Kuriakose & Mini K G. 2006 *A Stochastic model to analyse pelagic fishery resource dominance along the Karnataka coast (west coast of India).* Indian Journal of Marine Science. Vol. 35(3) p 257-262

A heuristic approach to handle capacitated facility location problem evaluated using clustering internal evaluation

G R Sutanto[1,2*], S Kim[1], D Kim[1], H Sutanto[3]

[1]Department of Industrial and Management Engineering, Myongji University, Republic of Korea
[2]Department of Industrial Engineering, University of Buddhi Dharma, Indonesia
[3]Department of Manufacturing Engineering, Atma Jaya Catholic University, Indonesia

Corresponding author: gerardrandi@yahoo.com

Abstract. One of the problems in dealing with capacitated facility location problem (CFLP) is occurred because of the difference between the capacity numbers of facilities and the number of customers that needs to be served. A facility with small capacity may result in uncovered customers. These customers need to be re-allocated to another facility that still has available capacity. Therefore, an approach is proposed to handle CFLP by using *k*-means clustering algorithm to handle customers' allocation. And then, if customers' re-allocation is needed, is decided by the overall average distance between customers and the facilities. This new approach is benchmarked to the existing approach by Liao and Guo which also use *k*-means clustering algorithm as a base idea to decide the facilities location and customers' allocation. Both of these approaches are benchmarked by using three clustering evaluation methods with connectedness, compactness, and separations factors.

1. Introduction

The facility location problem (FLP) is a research topic as a branch of Operations Research and Computational Geometry, which has been a research topic for a long time. One of the main issues in this topic is about space equity and capacity limitation of suppliers (Nauss, 1978; Chudak and Williamson, 2005; Liao and Guo, 2008). In FLP, facilities need to be placed to cover a certain number of customers. Finding the most optimal placements for facilities is the main issues in this topic. Various approaches from various fields of study had been invented to find the most optimal FLP solution. Most common approaches are from Cluster Analysis (Clustering). One example of clustering application to solve FLP is the solution to solve Logistics Center Location Problem (Qingfang Ruan et al., 2010). However, many approaches are invented to solve FLP with stochastic environment or constraint. One example is FLP under small sample or no-sample cases (Wena et al., 2015). Or another FLP with capacity limitation as constraint, which is defined as Capacitated Facility Location Problem (CFLP). CFLP is a branch of Facility Location Problem (FLP), which is concerned with the optimal placement location of each facility, to cover a certain number of customers. In addition, each facility also has limited number of capacity as a constraint.

To solve CFLP, results must cover both the location and the coverage of each facility. Allocating and re-allocating customers is one way to solve the CFLP, because the existing facilities are used into their full extent, without necessities to make a large amount of investment. One of the existing approaches using customers' allocation and re-allocation to solve CFLP is based on the *k*-

means clustering algorithm (Liao and Guo, 2008). The *k*-means clustering algorithm is an algorithm which aims to partition *n* objects into *k* clusters. In Liao's Approach clusters are the coverage area of the facilities, with each cluster centers act as the facility location.

In this research, another approach is proposed to solve CFLP. This approach also use *k*-mean clustering algorithm to allocate customers into their designated facilities. However, if customers' re-allocation is needed, it'll be done based on the average distance between customers and available facilities. This approach is benchmarked to Liao's Approach, using clustering evaluation method. Without any benchmark dataset to be compared with, the only suitable evaluation method is the Internal Evaluation.

An evaluation package for R language is done by taking Internal Evaluation as one of its evaluation method (Brock et al., 2008). Other factors which selected for the clustering Internal Evaluation are compactness, connectedness, and separations. Therefore, in this research, Liao's approach is evaluated with evaluation methods based on those factors.

2. Liao's Approach

This existing approach is based on paper "A Clustering-Based Approach to the Capacitated Facility Location Problem" published in 2008 by Liao and Guo. This particular approach is adapted from *k*-means clustering algorithm (figure 1). Similar with *k*-means clustering algorithm, centroids (facility locations), are randomly initialized. Capacity constraints are considered when allocating customers into facility. Customers are allocated into facility based on their distance. These steps will be repeated until the new facility location is same as the previous facility location.

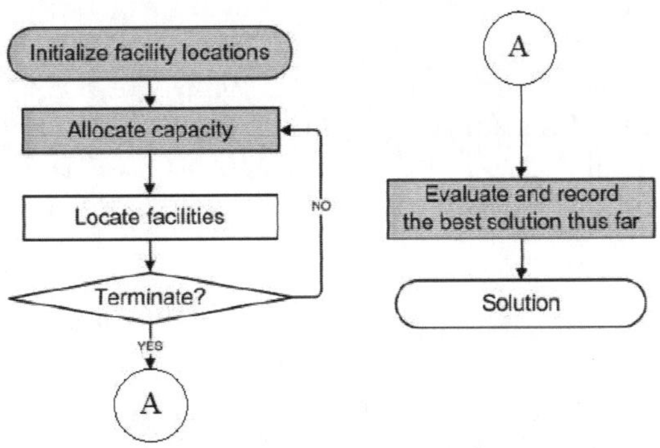

Figure 1. Flowchart for Liao's Approach

In the Initialize Facility Location step, facility locations are initialized by generating random value of *x* and *y* coordinates. Total number of facility locations is also given in this step. Next, in the Allocate Capacity step, a capacity constraint is given to each facility. In this research, all facilities have an equal number of capacities. Then, locations for each facility are determined in the Locate Facilities step. Locations for each facility are located by calculating the means value of all customers which each facility covered. Facilities only covering a certain number of customers based on their capacity, and priority is given to the closest customer from their locations. Those processes will always be repeated until the locations for each facility are same with their locations from previous iteration. The solution is the facility locations after the whole steps are terminated.

3. Proposed Approach

This proposed approach also use *k*-means clustering algorithm to allocate customers into facility. In addition, *k*-means also used to determine the facility locations. However, if there is one or more facility serving more customers than its capacity, re-allocation will happen. Re-allocation will be done only for the un-served customers, based on average distance method between the customers and the available facilities.

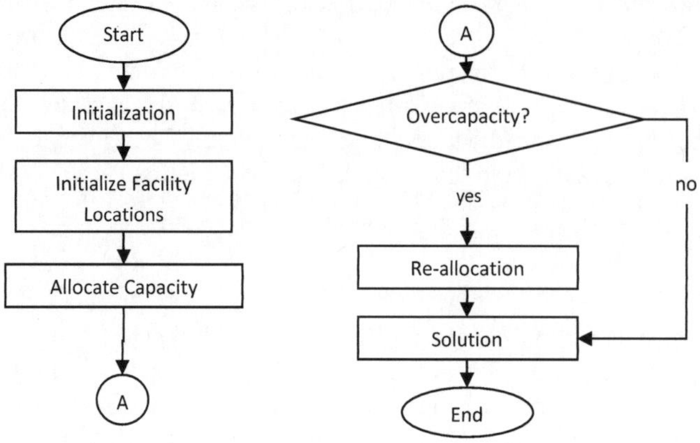

Figure 2. Flowchart for the proposed approach

Initialization is the first step to assign the desired number of facility location, and its capacity (figure 2). Capacity should be uniform for each facility. This step also read customer locations data, and put it into a matrix. Customer locations should be consists of *x* and *y* coordinates. In Initialize Facility Location step, *k*-means Clustering Algorithm is used to determine facility location. The *k*-means Clustering Algorithm will run to process the customer locations data, with desired number of facility as a constraint. The centroids from *k*-means Clustering Algorithm's result are the initial facility locations, as can be seen on figure 3.

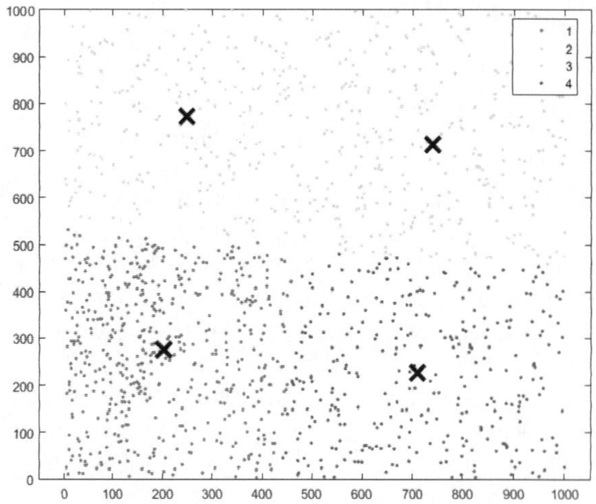

Figure 3. An example if a company decided to build 4 facilities (k=4)

Then, for each facility location, capacity constraints are applied. After applying the capacity constraint, each facility needs to be checked for overcapacity. If there's any facility containing more than its capacity, some customers from that particular capacity will be omitted (in figure 4). These customers need to be designated into another facility which still has available capacity. Omitted customers will be re-allocated based on average distance of its neighboring customers within a certain range.

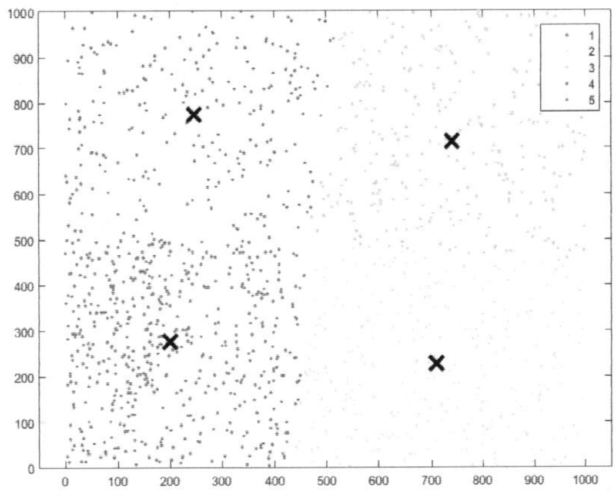

Figure 4. An example of omitted customers (purple points), from the initial data in Figure 3

3.1. Re-allocation step. Priorities will be given for each un-served customer. Priorities will be based on distance between every un-served customer with another customer in a certain range. How wide the range for each un-served customer is decided using Elbow Method. Ranges are the *k*, and average distances of all customers in that particular range are the cost functions. The value of range in "elbow" position of the graph would be used for deciding priorities. In a circle where the un-served customer act as the center and the value of range act as the borderline, every customer inside the circle will act as the deciding factor for re-allocation.

For example, for re-allocating customer X, a certain *k* range is selected by using Elbow Method, and containing another customer A, B, and C. Customer A and B are served by Facility 1 with average distance to customer X is 10. C is served by Facility 2 with average distance to customer X is 20. So customer X would be served by Facility 1, because the average distance value is the smallest one compared to another average distance value in that particular range. This example is shown in figure 5, with A and B are the triangles which means from the same facility (Facility 1), and C is square which means C is from another facility (Facility 2). In addition, re-allocation only will happen if the designated facility still have empty slot of capacity.

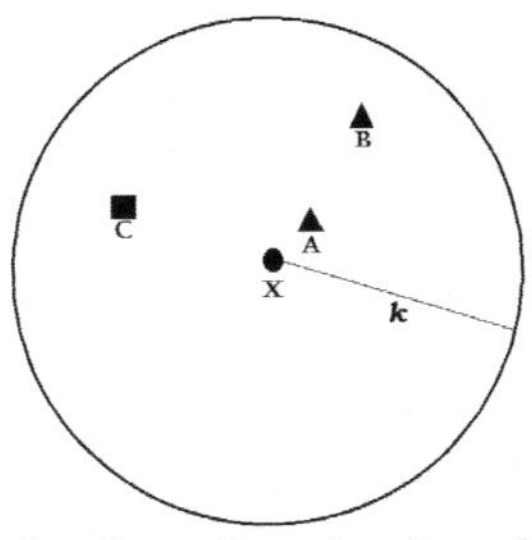

Figure 5. An example of re-allocation based on average distance

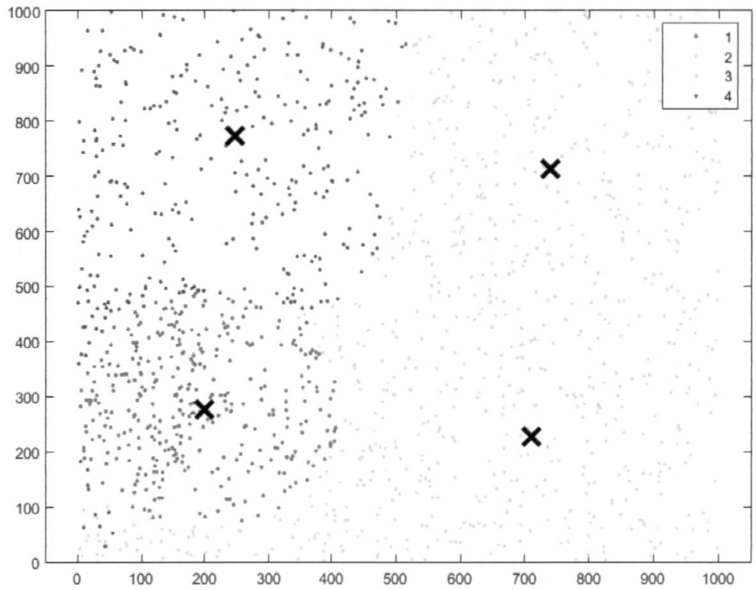

Figure 6. Solution after re-allocation for data in Figure 4

4. Evaluation Method

CFLP problems are replicated by randomly generating a number of datasets. These datasets are consisting of potential locations for facilities and a set of customer locations. Customer locations are consisting of x and y coordinates. Two types of synthetic datasets are used. First dataset is evenly distributed, and the other is clustered. And then, For evaluating the approach's solutions, a variety of evaluation methods are selected. These evaluation methods are aimed to determining results from a cluster analysis. All selected evaluation methods are categorized into the Internal Evaluation, which means evaluation for evaluating the results based on the results condition only. There is no comparison with another or existing result.

In Internal Evaluation, three factors were taken into consideration: (1) Compactness, (2) Separation, and (3) Connectedness. Connectedness is a factor that shows how well each observation connected to another observation in the same. Compactness is a factor that relates into homogeneity between observations in the same cluster. Separation is a factor that relates with how far are the clusters from each other. Compactness and separations are combined into single evaluation method, Silhouette (Rousseuw, 1987). Another evaluation method which also combines compactness and separation in one evaluation method is Davies-Bouldin Index (Davies and Bouldin, 1979). On the other hand, evaluation method for Connectedness is Connectivity cluster (Handl et al., 2005). All of these methods are common to determining how good a cluster analysis result. Silhouette is a method of evaluation based on how well each observation clustered into its cluster. Silhouette (*ASW*) can be calculated by:

$$s(i) = \frac{b(i)-a(i)}{max\{a(i),b(i)\}} \quad (1)$$

$$ASW = \frac{1}{n}\sum_{i=1}^{n} s(i) \quad (2)$$

where:

n = the number of observations
$a(i)$ = average distance of observation i with all other observations in the same cluster (which cover the compactness factor)
$b(i)$ = average distance of i with all other observations in the nearest another cluster (which cover the separations factor)
ASW = average silhouette width

The result of Silhouette evaluation method (*ASW*) is between -1 and 1, with the result closer to 1 is meaning the better result. Davies-Bouldin Index is a method of evaluation based on the average distance of all observations in a cluster to its center, and also based on distance between clusters centers. This index can be calculated by:

$$DB = \frac{1}{m}\sum_{i=1\ j=i+1}^{m} max(\frac{\sigma_i+\sigma_j}{d(c_i,c_j)}) \qquad (3)$$

where:
m = the number of clusters
c_i = centroid of cluster *i*
c_j = centroid of cluster *j*
σ_i = average distance of all observations in cluster *i* to centroid c_i (which cover the compactness factor)
σ_j = average distance of all observations in cluster *j* to centroid c_j (which cover the compactness factor)
$d(c_i, c_j)$ = distance between centroid c_i and c_j (which cover the separations factor)

Smaller number in the result of Davies-Bouldin Index evaluation method (*DB*) is meaning the better result. Connectivity is a criterion of evaluation based on the value of certain number of the closest neighbors from a certain observation. This index can be calculated by:

$$Conn(C) = \sum_{i=1}^{n}\sum_{j=1}^{l} f(x_i, z_{i(j)}) \qquad (4)$$

where:
n = the number of observations
$z_{i(j)}$ = the *j*-th nearest neighbor of observation *i*
$f(x_i, z_{i(j)})$ = 0, if *i* and $z_{i(j)}$ are in the same cluster, or
$f(x_i, z_{i(j)}) = {}^{1}/_{j}$, if *i* and $z_{i(j)}$ are in different clusters
l = number of neighbors that contribute to the connectivity calculation

Smaller number in the result of Connectivity evaluation method (*C*) is meaning the better result. Each approach is iterated 5 times for each dataset. Then, for each dataset result, the average of 5 time iteration results is taken to compare performances as shown in tables below. For Connectivity, extra iterations are done to accommodate its variations. Extra iterations are done for l = 1, l = 5, and l = 10. l = 1 means only 1 neighbor is contributing to Connectivity calculation, so l = 5 means 5 neighbors are contributing, and l = 10 means 10 neighbors are contributing.

5. Analysis Results

The results are evaluated using Silhouette, BI Index, and Davies-Bouldin Index evaluation methods. Each approach is tested into 2 randomly generated datasets. First dataset is normally distributed with 1400 customers to serve, and sufficient capacity is given in this test. Number of facility available is 4, and each has 350 capacities, so it's enough to serve all 1400 customers. Table 1 contains comparison in evenly distributed dataset.

Table 1. Results comparison in evenly distributed dataset

	Liao's Approach			Proposed Approach		
Silhouette	0.2337			0.4559		
BI Index	L = 1	L = 5	L = 10	L = 1	L = 5	L = 10
	120	537.2667	894.0032	164	607.2	911.1063
Davies-Bouldin	2.0818			1.1045		

From results in table 1, in terms of Silhouette evaluation, the result from Liao's approach is worse than the proposed approach, because in Silhouette evaluation, bigger value result is the better result. Same thing happened in Davies-Bouldin Index evaluation, the result from Liao's approach is also worse, because smaller value result is the better result in this evaluation. But in BI Index evaluation, the result from Liao's approach is better than the proposed approach, because in BI Index evaluation smaller value result is the better result.

The second dataset is a clustered dataset with 800 customers to serve, and sufficient capacity is given in this test. The number of facility available is 4, and each has 200 capacities. Table 2 contains comparison in clustered dataset.

Table 2. Results comparison in clustered dataset

	Liao's Approach			Proposed Approach		
Silhouette	0.4328			0.4219		
BI Index	L = 1	L = 5	L = 10	L = 1	L = 5	L = 10
	52	185.4	287.0397	158	372.8	509.1063
Davies-Bouldin	1.5878			1.435		

Table 2 shows based on Silhouette evaluation, the result from Liao's approach is better than the proposed approach, because the result's value is biggest one, but only by slight difference. In evaluation based on BI Index also shows that Liao's Approach have better result than the proposed approach, because the result's value is smaller than the proposed approach. On the contrary, based on Davies-Bouldin Index evaluation, the proposed approach is better than Liao's Approach, because the result's value is smaller compared to others results.

6. Conclusions

A clustering approach to solve CFLP also needs to be evaluated as a clustering method. In Internal Evaluation, three factors were taken into consideration: (1) Compactness, (2) Separation, and (3) Connectedness. Compactness and Separation are able to be evaluated using Silhouette and Davies-Bouldin Index Evaluation Method. And Connectedness is able to be evaluated using Connectivity Evaluation Method. Two heuristic approaches are evaluated and compared. One is an existing approach based on *k*-means clustering algorithm (Liao's Approach), and another is a new approach with different method in customers re-allocation step. Both approaches are evaluated in CFLP with evenly distributed customer locations and clustered customer locations.

In terms of CFLP with evenly distributed customer locations, the new proposed approach had good results in general. The proposed approach was better in Silhouette and Davies-Bouldin Index evaluation methods. In other hand, in terms of CFLP with clustered customer locations, all results from the Liao's approach are better than the first proposed approach.

References

[1] Brock, G., Pihur, V., Datta, S. and Datta, S. (2008), clValid, an R package for cluster validation, *Journal of Statistical Software*, vol. 25, pp. 1-22.
[2] Chaves, A. A. and Lorena, L. A. N. (2010), Clustering search algorithm for the capacitated centered clustering problem, *Computers & Operations Research*, vol. 37(3), pp. 552-558.
[3] Chudak, Fabian A., Williamson, David P. (2005). Improved Approximation Algorithms for Capacitated Facility Location Problems, *Journal Mathematical Programming: Series A and B*, vol. 102, Issue 2, pp. 207 – 222.
[4] Davies, David L.,Bouldin, Donald W. (1979). A Cluster Separation Measure, *IEEE Transactions on Pattern Analysis and Machine Intelligence*. PAMI-1 vol. (2), pp. 224–227.
[5] Hongtao Yu, Liqun Gao, Yanhua Lei. (2012), Model and solution for capacitated facility location problem, *In*: *Control and Decision Conference (CCDC)*, 2012 24th Chinese, pp. 1773-1776, 23-25 May 2012.
[6] Julia Handl, Joshua Knowles, Douglas B. Kell. (2005), Computational cluster validation in post-genomic data analysis, *Bioinformatics*, vol. 21 (15), pp. 3201-3212.
[7] Kaufman, L. and P. J. Rousseeuw (1990). *Finding Groups in Data: An Introduction to Cluster Analysis*, Wiley, New York, USA.
[8] Liao, K. and Guo, D. (2008), A Clustering-Based Approach to the Capacitated Facility Location Problem, *Transactions in GIS*, vol. 12, pp. 323–339
[9] Mahmoodi Darani, N., Ahmadi, V., Saadati Eskandari, Z., & Yousefikhoshbakht, M. (2013). Solving Capacitated Clustering Problem by a Combined Meta-Heuristic Algorithm, *Journal of in Computer Research*, vol. 4(1), pp. 89-100.
[10] Qingfang Ruan, Lixin Miao, Zhijun Zheng. (2010). A novel clustering-based approach for the location of multi-logistics centers, *In: 8th International Conference on Supply Chain Management and Information Systems (SCMIS)*, 2010, 1-5, 6-9 Oct. 2010.
[11] Robert M. Nauss. (1978). The 0–1 knapsack problem with multiple choice constraints, *European Journal of Operational Research*, vol. 2, Issue 2, pp.125-131.
[12] Seyed Mohsen Mousavi and Seyed Taghi Akhavan Niaki. (2013). Capacitated location allocation problem with stochastic location and fuzzy demand: A hybrid algorithm, *Applied Mathematical*, vol. 37, Issue 7, pp. 5109-5119.
[13] Silva, F. and Serra, D. (2007). A Capacitated Facility Location Problem with Constrained Backlogging Probabilities, *International Journal of Production Research*, vol 45, Issue 21.
[14] Verter Verdat. (2011). *Foundation of Location Analysis*, Springer, USA.
[15] Wena, M., Qinc, Z., Kangb, R. (2015). The Capacitated Facility Location-Allocation Problem under Uncertain Environment, *Journal of Intelligent & Fuzzy Systems*, vol. 29, no. 5, pp. 2217-2226.

IOP Conf. Series: Materials Science and Engineering **332** (2018) 012024 doi:10.1088/1757-899X/332/1/012024

Steepest descent method implementation on unconstrained optimization problem using C++ program

H Napitupulu[1*], Sukono[1], I Bin Mohd[2], Y Hidayat[3], S Supian[1]

[1]Department of Mathematics, Faculty of Mathematics and Natural Sciences, Universitas Padjadjaran, Indonesia
[2]Institute for Mathematical Research (INSPEM), Universiti Putera Malaysia, Selangor, Malaysia
[3]Department of Statistics, Faculty of Mathematics and Natural Sciences, Universitas Padjadjaran, Indonesia

*Email: napitupuluherlina@gmail.com

Abstract. Steepest Descent is known as the simplest gradient method. Recently, many researches are done to obtain the appropriate step size in order to reduce the objective function value progressively. In this paper, the properties of steepest descent method from literatures are reviewed together with advantages and disadvantages of each step size procedure. The development of steepest descent method due to its step size procedure is discussed. In order to test the performance of each step size, we run a steepest descent procedure in C++ program. We implemented it to unconstrained optimization test problem with two variables, then we compare the numerical results of each step size procedure. Based on the numerical experiment, we conclude the general computational features and weaknesses of each procedure in each case of problem.

1. Introduction

Consider the following nonlinear unconstrained minimization problem

$$\text{find } x^* \in R^n \text{ such that } f\left(x^*\right) = \min_{x \in R^n} f\left(x\right) \tag{1}$$

where $f : R^n \to R$. Steepest descent method is the simplest gradient method for solving unconstrained optimization which is designed by Cauchy in 1847. This gradient method search along the negative gradient function, can ensure a reduction of the objective function as long as the current iterate point is not a stationary point ([1]). Many researches have been done which is concern in finding the better and more appropriate step size procedure which affecting to the significance of minimizing the objective functions. In this paper we give some reviews the properties of steepest descent method, its development, and compare the numerical results of steepest descent step size from some literatures on solving global optimization problem. We run a steepest descent in C++ program to test the performance of tested step size procedure, then compare the numerical results obtained of program execution. Based on the numerical experiment, the features as well as the weaknesses of each step size procedure in each case of problem are concluded.

2. Steepest Descent Method

Steepest descent method is the simple minimizing gradient method for solving nonlinear equations since it is based on the linear approximation of Taylor series

$$f(x+\delta) \approx \hat{f}(x+\delta) = f(x) + \nabla f(x)^T \delta \quad (1)$$

The term $\nabla f(x)^T \delta$ is the directional derivatives of f at x. The step δ is a descent direction if the directional derivative is negative, which guarantees that the function f could be reduced along this direction. Furthermore, the step δ is need to be chosen to make the linear directional derivative $\nabla f(x)^T \delta$ as negative as possible, which give the maximum reduction of f but no consuming too much time in make a choice.

Assume that a function $f(x)$ is continuous in the neighborhood of point x, $d = -g$ is the steepest descent direction at point x and change δ in x given by $\delta = \alpha d$.

A small positive constant α is called step size, which its selection make an effect to the reduction value of $f(x)$. By solving one dimensional global optimization problem

$$\min_{\alpha>0} \phi(\alpha) = f(x+\alpha d) \quad (2)$$

the maximum reduction in $f(x)$ can be obtained.

The procedure of steepest descent iteration is performed by formula

$$x_{k+1} = x_k + \alpha_k d_k \quad (3)$$

Starting with an initial point x_0, a direction $d_0 = -g_0$ and the step size α_0 that minimizes $f(x_0 + \alpha d_0)$ can be determined so that the next point $x_1 = x_0 + \alpha_0 d_0$ is obtained. The procedure in (1) is repeated until the convergence is achieved or the value $\|\alpha_k d_k\|$ is sufficiently small. The algorithm of steepest descent is presented as follows.

```
ALGORITHM STEEPEST DESCENT
given :
  starting point x_0 ∈ domain f
  set k:=0
repeat
1. Compute steepest descent direction d_k = -g_k
2. Update step size α_k = min_{α>0} φ(α) = f(x_k + α d_k)
3. Update x_{k+1} := x_k + α_k d_k
until convergence or stopping criterion is satisfied
```

3. Steepest Descent Step Size

The step size in steepest descent method plays the important role in obtaining the minimizing the objective function. The step size of each iteration α_k, must be effectively chosen to give a significant reduction of function value, while the efficiently timing of choosing it also be optimized.

The early step size procedure of steepest descent is considered as step size which obtained by exact line search,

$$\alpha_k = \arg\min_{\alpha} \{ f(x_k + \alpha(-g_k)) \} \quad (4)$$

But if the steepest descent direction is used with exact line search step size would bring to the zigzag behavior which make the convergence is very slowly. Akaike ([2]) analyzing that by this search procedure the converges is linearly and badly affected by ill-conditioning, by shown that the two asymptotic directions are in a two dimensional sub space spanned by two eigen vector of the Hessian matrix. Greenstadt ([3]) studied the efficiency of steepest descent method and shown the bound of the ratio between the reductions obtained by Cauchy step and by the Newton's method, and the rate of convergence of steepest descent was proved by Forsythe ([4]).

Besides using exact line search, step size α_k could be computed by type of inexact line search methods, for instance Armijo condition ([5]), Goldstein condition ([6]), Wolfe condition ([7]), Powell ([8]), Dennis and Schnabel ([9]), Fletcher ([10]), Potra and Shi ([11]), Lemaréchal ([12]), Moré and Thuente ([13]) and the others. Steepest descent method with mentioned conditions is always convergent theoretically, which is will not terminate unless the stationary point is found. However, step size computed by those inexact line search procedure also appearing the bad zigzag behavior as well as in exact line search procedure.

Considering the weakness of existing step sizes, Barzilai and Borwein ([10]) proposed the the step size along negative gradient direction from a two point approximation to the secant equation from quasi-Newton's method so called BB method. If the dimension is two, Barzilai and Borwein established R-superlinear convergence result for the method and their analyses indicated that the convergence rate is faster. For the general $n-$dimensional strictly convex quadratic function, Raydan ([14]) proved that the two-point step size gradient method is globally convergent, and the convergence rate is $R-$linear (Dai and Liao ([15]). For the non-quadratic case, Raydan ([14]) incorporated the globalization scheme of the two point step size gradient method using the traditional technique of non-monotone line search by Grippo et al ([16]). The resulted algorithm is competitive and sometimes preferable to several famous conjugate gradient algorithm for large scale unconstrained optimization. Due it simplicity and numerical efficiency the two-point step size gradient method has received many studies. They have been successfully applied to obtaining the local minimizers of large scale real problems ([17]). The algorithm of Raydan ([15]) was further generalized by Birgin et al ([18]) for the minimization of differentiable function on closed convex sets, yielding an efficient projected gradient method. Efficient projected algorithm based on BB -like methods have also been designed (Serafini et al ([19]) and Dai and Fletcher([20]) for special quadratic programs arising from training support vector machine, that has a singly linear constraint in addition to box constraints. The BB method has also received much attention in finding sparse approximation solution to large undetermined linear systems of equations from signal/ image processing and statics for example in Wright et al ([21]). However Fletcher ([22]) shown that for some problems of non-quadratic function this method may very slow.

There are many existing step sizes modification, some of them are easy to be applied but some are satisfied the complicated algorithm. Following are some step size procedure which is simple to be applied to steepest descent method.

Table 1. List of Some Recent Simple Step Size

No	Step Size	Remark
1	$\alpha_k = \dfrac{g_k^T g_k}{g_k^T H_k g_k}$	Step size method by Cauchy (1847) which is computed by exact line search (**C** step size).
2	Given $s > 0, \beta, \sigma \in (0,1)$, $\alpha_k = \max\left\{s, s\beta, s\beta^2, ...\right\}$ such that $f(x_k + \alpha_k d_k) \le f(x_k) + \sigma \alpha_k g_k^T d_k$	This procedure of step size using inexact line search so called Armijo's line search (**A** step size).
3	Given $\beta, \sigma \in (0,1)$, $\hat{\alpha}_0 = 1$ $\alpha_k = \beta \hat{\alpha}_k$	This procedure is so called Backtracking line search (**B** step size), this inexact line search is use just the sufficient decrease

	such that $f(x_k+\alpha_k d_k) \le f(x_k)+\sigma\alpha_k g_k^T d_k$	condition to terminate the line search procedure.
4	$\alpha_k = \dfrac{s_{k-1}^T y_{k-1}}{\lVert y_{k-1}\rVert_2^2}$ **(BB1)** and $\alpha_k = \dfrac{\lVert s_{k-1}\rVert_2^2}{s_{k-1}^T y_{k-1}}$ **(BB2)** where $s_{k-1}=x_k-x_{k-1}$ $y_{k-1}=g_k-g_{k-1}$	These two point step sizes are method of non-line search procedure which is namely after the inventors, Barzilai and Borwein's Formula. These formula reduce the value of objective function and gradient and improved the convergence from linear to $R-$ superlinear.
5	$\alpha_k = \dfrac{g_k^T g_k \alpha_{k-1}^2}{2\left(f(x_k+\alpha_{k-1}d_k)-f(x_k)+\alpha_{k-1}g_k^T g_k\right)}$	This step size is so called elimination line search (**EL** step size), which is estimate the step size without compute the Hessian.

4. Numerical Experiment

In this section, several step size procedure of the steepest descent method shown in Table 1 are tested computationally. The programming is using Visual C++ 6.0, which is applied to 14 testing function of global optimization given in the following.

1. Beale Function

$$\min f(x_1,x_2)=(1.5-x_1+x_1x_2)^2+(2.25-x_1+x_1x_2^2)^2+(2.625-x_1+x_1x_2^3)^2$$
$$s.t. -4.5 \le x_1,x_2 \le 4.5,\ x_0=(1,0)$$

2. Bohachevsky I Function

$$\min f(x_1,x_2)=x_1^2+2x_2^2-0.3\cos(3\pi x_1)-0.4\cos(4\pi x_2)+0.7$$
$$s.t. -50 \le x_1,x_2 \le 50,\ x_0=(0.3,0.3)$$

3. Bohachevsky II Function

$$\min f(x_1,x_2)=x_1^2+2x_2^2-0.3\cos(3\pi x_1)\cos(4\pi x_2)+0.3$$
$$s.t. -50 \le x_1,x_2 \le 50,\ x_0=(0.1,0.1)$$

4. Booth Function

$$\min f(x_1,x_2)=(x_1+2x_2-7)^2+(2x_1+x_2-5)^2$$
$$s.t. -10 \le x_1,x_2 \le 10, x_0=(-4,4)$$

5. Matyas Function

$$\min f(x_1,x_2)=0.26\left(x_1^2+x_2^2\right)-0.48x_1x_2$$
$$s.t. -3 \le x_1,x_2 \le 3,\ x_0=(2,2)$$

6. Rosenbrock Simplified Function

$$\min f(x_1,x_2)=0.5\left(x_1^2-x_2\right)^2+\left(x_1-1\right)^2$$
$$s.t. -3 \le x_1,x_2 \le 3, x_0=(-1.6,0.9)$$

7. Six Hump Back Camel Function

$$\min\ f(x_1,x_2)=4x_1^2-2.1x_1^4+\frac{1}{3}x_1^6-x_1x_2-4x_2^2+4x_2^4$$
$$s.t.\ -3\le x_1,x_2\le 3,\ x_0=(2,-1)$$

8. Rastrigin Function

$$\min\ f(x_1,x_2)=x_1^2+x_2^2-\cos(18x_1)-\cos(18x_2)$$
$$s.t.\ -2\le x_1,x_2\le 2,\ x_0=(-1,1)$$

9. Three Hump Back Camel Function

$$\min\ f(x_1,x_2)=2x_1^2-1.05x_1^4+\frac{x_1^6}{6}-x_1x_2+x_2^2$$
$$s.t.\ -3\le x_1,x_2\le 3,\ x_0=(-2,-1)$$

10. Treccani Function

$$\min\ f(x_1,x_2)=x_1^4+4x_1^3+4x_1^2+x_2^2$$
$$s.t.\ -3\le x_1,x_2\le 3,\ x_0=(4,2)$$

11. Two Dimensional Function c= 0.05

$$\min\ f(x_1,x_2)=\left(1-2x_2+0.05\sin\left(4\pi x_2\right)-x_1\right)^2+\left(x_2-0.5\sin\left(2\pi x_1\right)\right)^2$$
$$s.t.\ -10\le x_1,x_2\le 10,\ x_0=(0.5,0.5)$$

12. Two Dimensional Function c=0.2

$$\min\ f(x_1,x_2)=\left(1-2x_2+0.2\sin\left(4\pi x_2\right)-x_1\right)^2+\left(x_2-0.5\sin\left(2\pi x_1\right)\right)^2$$
$$s.t.\ -10\le x_1,x_2\le 10,\ x_0=(2,-2)$$

13. Two Dimensional Function c=0.5

$$\min\ f(x_1,x_2)=\left(1-2x_2+0.5\sin\left(4\pi x_2\right)-x_1\right)^2+\left(x_2-0.5\sin\left(2\pi x_1\right)\right)^2$$
$$s.t.\ -10\le x_1,x_2\le 10,\ x_0=(0,0)$$

14. Goldstein and Price Function

$$\min f(x_1,x_2)=\left(1+(x_1+x_2+1)^2\left(19-14x_1+3x_1^2-14x_2+6x_1x_2+3x_2^2\right)\right)\cdot$$
$$\left(30+(2x_1-3x_2)^2\left(18-32x_1+12x_1^2+48x_2-36x_1x_2+27x_2^2\right)\right)$$
$$s.t.\ -3\le x_1,x_2\le 3,\ x_0=(-1,-1)$$

By referring to [23] for Armijo procedure (**A**), $s:=1$, $\beta:=0.75$, $\sigma:=0.38$, and for Backtracking procedure $\sigma:=0.0001$. The initial step size for Barzilai-Borwein step size 1 (**BB1**) and Barzilai-Borwein step size 2 (**BB2**) is 0.01, initial step size for elimination line search is $\alpha_0:=0.1$. The comparison of Numerical result will be based on: time execution, number of total iteration, total percentage of function, gradient and Hessian evaluation and the most decreased value of objective function obtained. Stopping criteria are $\left\|\alpha_k d_k\right\|\le 10^{-6}$ or $\left\|d_k\right\|\le 10^{-6}$. Following are the abbreviations used in of numerical results of Table 2.

IOP Conf. Series: Materials Science and Engineering **332** (2018) 012024 doi:10.1088/1757-899X/332/1/012024

Abbreviation in column one:
C = Cauchy step size
A = Armijo step size
B = Backtracking step size
BB1 = Barzilai & Borwein step size 1
BB2 = Barzilai & Borwein step size 2
EL = Elimination line search step size

Abbreviation in row one:
Iter = percentage of the total number of iteration until termination
Time = percentage of the total running time until the termination in second
Ft = percentage of the total number of function evaluation until termination
Gt = percentage of the total number of gradient evaluation until termination
Ht = percentage of the total number of Hessian evaluation until termination

Table 2. Numerical Results

No.	Step	x	f	Time	Iter %	Ft %	Gt %	Ht %
1.	C	(2.99995, 0.499987)	4.73368e-10	0.009	22.06	0	18.80	100.0
	A	(2.9999, 0.499975)	1.6441e-9	0.005	24.33	36.35	22.54	0
	B	(2.9999, 0.499974)	1.69506e-9	0.004	29.79	41.21	25.55	0
	BB1	(2.99985, 0.499963)	3.44039e-9	0	2.97	0	7.59	0
	BB2	(2.99987, 0.499968)	2.73212e-9	0.002	4.53	0	11.58	0
	EL	(2.9999, 0.499974)	1.74086e-9	0.004	16.29	22.42	13.90	0
2.	Step	x	f	Time	Iter %	Ft %	Gt %	Ht %
	C	(0.360538, 0.266918)	1.65345	0	15.78	0	7.51	100.0
	A	(1.16965e-6, 0.469363)	0.469528	0.001	19.73	62.74	28.32	0
	B	(1.94405e-6, 0.469363)	0.469528	0	13.15	20.91	9.82	0
	BB1	(-6.15555e-7, 1.6234e-7)	6.31717e-12	0	18.42	0	24.27	0
	BB2	(9.47959e-7, 0.469363)	0.469528	0	17.10	0	22.54	0
	EL	(6.95877e-7, 0.469363)	0.469528	0	15.78	16.33	7.51	0
3.	Step	x	f	Time	Iter %	Ft %	Gt %	Ht %
	C	(0.618401, 2.71507e-7)	0.412619	0	15.71	0	8.00	100.0
	A	(1.258e-6, 9.10494e-8)	2.28987e-11	0	14.28	55.21	30.66	0
	B	(4.89841e-23, -5.45896e-7)	7.66054e-12	0	32.85	34.35	19.33	0
	BB1	(8.44811e-7, 1.9562e-7)	1.12145e-11	0.001	12.85	0	19.33	0
	BB2	(2.70968e-7, -5.34494e-8)	1.12599e-12	0	12.85	0	18.00	0
	EL	(-5.56201e-7, 4.30077e-7)	9.18932e-12	0	11.42	10.42	6.00	0
4.	Step	x	f	Time	Iter %	Ft %	Gt %	Ht %
	C	(0.999994, 3.00001)	7.93254e-11	0	27.33	0	19.92	100.0
	A	(0.999994, 3.00001)	6.89661e-11	0.001	27.09	49.89	21.53	0
	B	(-2.82681e-10, 0.459324)	0.459962	0.001	24.38	42.82	18.50	0
	BB1	(0.999999, 3)	6.12972e-13	0	2.21	0	4.80	0
	BB2	(0.999997, 3)	1.31356e-11	0	14.77	0	32.02	0
	EL	(0.999999, 3)	2.04087e-12	0	4.18	7.27	3.20	0
5.	Step	x	f	Time	Iter %	Ft %	Gt %	Ht %
	C	(2.44249e-15, 2.44249e-15)	2.3863e-31	0.001	0.17	0	0.34	100.0
	A	(1.7006e-5, 1.7006e-5)	1.15682e-11	0.001	49.65	49.86	49.14	0
	B	(1.7006e-5, 1.7006e-5)	1.15682e-11	0.001	49.65	49.86	49.14	0
	BB1	(2.21911e-12, 2.21911e-12)	1.96979e-25	0	0.17	0	0.51	0
	BB2	(2.21889e-12, 2.21889e-12)	1.96939e-25	0	0.17	0	0.51	0

	EL	(3.08067e-11, 3.08067e-11)	3.79621e-23	0	0.17	0.26	0.34	0
6.	Step	*x*	*f*	Time	Iter %	Ft %	Gt %	Ht %
	C	(-1.04076, 1.04076)	0.179775	0	11.11	0	4.12	100.0
	A	(-0.693844, 0.693844)	-1.03121	0	18.51	70.34	53.60	0
	B	(-1.04076, 1.04076)	0.179775	0	29.62	22.06	17.52	0
	BB1	(1.04076, -1.04076)	0.179775	0	11.11	0	9.27	0
	BB2	(-1.04076, 1.04076)	0.179775	0	11.11	0	9.27	0
	EL	(-1.04076, 1.04076)	0.179775	0	18.51	7.58	6.18	0
7.	Step	*x*	*f*	Time	Iter %	Ft %	Gt %	Ht %
	C	(1, 1)	8.87118e-13	0.001	28.07	0	21.30	100.0
	A	(0.999992, 0.999977)	8.98588e-11	0.001	29.43	52.50	22.97	0
	B	(1, 1.00001)	2.48894e-11	0.001	16.25	28.60	12.55	0
	BB1	(0.999995, 0.999985)	3.9723e-11	0	8.74	0	19.81	0
	BB2	(0.999994, 0.999983)	4.87009e-11	0	6.65	0	15.06	0
	EL	(0.999996, 0.999989)	2.10143e-11	0.001	10.83	18.89	8.27	0
8.	Step	*x*	*f*	Time	Iter %	Ft %	Gt %	Ht %
	C	(1.6071, -0.568651)	2.10425	0	9.89	0	7.19	100.0
	A	(1.6071, -0.568655)	2.10425	0.001	43.95	61.53	38.12	0
	B	(0.0898408, 0.712656)	-1.03163	0	19.78	26.03	16.54	0
	BB1	(1.6071, -0.568652)	2.10425	0	7.69	0	15.10	0
	BB2	(1.6071, -0.568652)	2.10425	0	7.69	0	15.10	0
	EL	(1.6071, -0.568652)	2.10425	0	10.98	12.42	7.91	0
9.	Step	*x*	*f*	Time	Iter %	Ft %	Gt %	Ht %
	C	(-1.74755, -0.873778)	0.298638	0.001	8.78	0	6.09	100.0
	A	(-1.74755, -0.873788)	0.298638	0.001	46.86	68.14	34.34	0
	B	(-1.74755, -0.873772)	0.298638	0	12.97	18.83	9.69	0
	BB1	(-1.74755, -0.87378)	0.298638	0	9.20	0	18.28	0
	BB2	(-1.74755, -0.873778)	0.298638	0	12.55	0	24.93	0
	EL	(-1.74755, -0.87378)	0.298638	0	9.62	13.01	6.64	0
10.	Step	*x*	*f*	Time	Iter %	Ft %	Gt %	Ht %
	C	(1.10892e-8, 3.925e-7)	1.54548e-13	0	1.07	0	0.32	100.0
	A	(7.64024e-19, 8.86385e-5)	7.85679e-9	0.003	67.89	70.55	19.62	0
	B	(-2, 3.18205e-5)	1.01254e-9	0.002	26.72	27.68	77.33	0
	BB1	(-6.26718e-8, 1.34651e-6)	1.82881e-12	0	1.30	0	1.10	0
	BB2	(-6.26718e-8, 1.34651e-6)	1.82881e-12	0	1.30	0	1.10	0
	EL	(3.55761e-7, 2.78967e-6)	8.2885e-12	0	1.68	1.75	0.49	0
11.	Step	*x*	*f*	Time	Iter %	Ft %	Gt %	Ht %
	C	(0.402355, 0.287455)	4.41419e-11	0	30.00	0	18.18	100.0
	A	(0.402355, 0.287455)	4.30738e-11	0	31.76	76.92	23.07	0
	B	(0.148658, 0.402127)	3.7532e-13	0	5.29	14.20	4.54	0
	BB1	(0.402359, 0.287451)	6.32753e-12	0	10.00	0	17.83	0
	BB2	(0.402356, 0.287453)	1.02648e-11	0	18.82	0	33.56	0
	EL	(0.148657, 0.402127)	1.42512e-14	0	4.11	8.87	2.79	0
12.	Step	*x*	*f*	Time	Iter %	Ft %	Gt %	Ht %
	C	(2.73691, -0.787967)	0.0884257	0.001	7.76	0	4.44	100.0
	A	(1.59016, -0.270037)	2.91343e-11	0.001	11.80	29.37	12.27	0
	B	(1.87779, -0.345587)	2.2265e-11	0.001	40.99	59.17	24.55	0
	BB1	(1.87779, -0.345586)	3.68925e-13	0.001	19.56	0	33.62	0
	BB2	(1.87779, -0.345586)	5.19563e-12	0.001	11.80	0	20.28	0
	EL	(1.87779, -0.345586)	7.07142e-13	0	8.07	11.44	4.80	0
13.	Step	*x*	*f*	Time	Iter %	Ft %	Gt %	Ht %

	C	fail	fail	fail	fail	fail	fail	fail
	A	(0.102535, 0.300418)	2.8179e-10	0	37.16	66.66	33.88	0
	B	(0.0419747, -0.094749)	0.517355	0	16.89	20.46	10.63	0
	BB1	(0.0419732, -0.0947486)	0.517355	0	14.86	0	21.92	0
	BB2	(0.0419751, -0.0947489)	0.517355	0.001	18.24	0	26.91	0
	EL	(0.0419794, -0.0947485)	0.517355	0	12.83	12.87	6.64	0
14.	Step	x	f	Time	Iter %	Ft %	Gt %	Ht %
	C	(-0.599993, -0.400006)	30	0.002	23.05	0	20.97	100.0
	A	(-0.599984, -0.400015)	30	0.005	50.11	66.03	52.11	0
	B	(-9.04948e-6, -1)	3	0.005	26.82	33.96	26.90	0
	BB1	(-1,-1)	2100 fail	fail	fail	fail	fail	fail
	BB2	(-1,-1)	2100 fail	fail	fail	fail	fail	fail
	EL	(-1,-1)	2100 fail	fail	fail	fail	fail	fail

From the numerical results of Table 2 we summarize the number of successful (***ns***), number of local minimizer obtained (***nl***), and number of failure (***nf***) presented in Table 3.

Table 3. Summary of Numerical Results

Step Size	***ns***	***nl***	***nf***
C	7	6	1
A	10	4	0
B	9	5	0
BB1	10	3	1
BB2	9	4	1
EL	9	4	1

From the numerical result obtained, the **A** method and **BB1** method are the method with the most successful among others for the case of two dimensional global optimization given. **A** method never failed in solving the testing problems but have 4 times go into another local minimizer. The **C** method is the least in obtaining the minimizer successfully with one failure. The **B**, **BB2** and **EL** method reach the same level in successfully obtain the minimizer. The **B** method reach without any fail, but **BB2** and **EL** have 1 fail. From this numerical experiment we can conclude that **A** is better than others based on the number of minimizer obtained in the testing example given in given initial condition.

5. Conclusion

The advantages and disadvantages of steepest descent method are, this method is sensitive to the initial point; this method has descent property and it is a logical starting procedure for all gradient based methods. Since $\|x_{k+1} - x_k\| = |\alpha_k| \|\nabla f(x_k)\| \to 0$ as $\nabla f(x^*) = 0$, so x_k approaches the minimizer rather slowly, in fact in a zigzag way. It is not economical to do linear searches thoroughly. All that is necessary to do is to obtain reduction in function values at successive iterates/

According to the numerical experiment of the given test function, we can conclude that in the case of two dimensional unconstrained global optimization problem given, the **A** and **BB1** is considered as the better method among others and **C** method is good enough, for the case of testing example given with given initial condition.

Acknowledgments
Authors wishing to the Rector, Director of DRPMI, and Dean of FMIPA, Universitas Padjadjaran, which provides a grant program of the Academic Leadership Grant (ALG) and a grant program of Competence Research of Lecturer of Unpad (Riset Kompetensi Dosen Unpad/RKDU).

References

[1] Yuan Y 2008 Step-sizes for the gradient method *In Third International Congress of Chinese Mathematicians. Part 1, 2 AMS/IP Stud. Adv. Math., 42, Pt. 1 2 785–796. Amer. Math. Soc.,* Providence, RI
[2] Akaike H 1959 On a successive transformation of probability distribution and its application to the analysis of the optimum gradient method *Ann, Inst. Statist. Math.* Tokyo 11 pp 1-7
[3] Greenstadt J 1967 On the relative efficiencies of gradient methods *Math. Comp.* **21** pp 360-367
[4] Forshyte G E 1968 On the Asymptotic directions of the s-dimensional optimum gradient method *Numerische Mathematik* **11** pp 57-76
[5] Armijo L 1966 Minimization of function having Lipschitz continuous first partial derivatives *Pacific J. of Mathematics* **6** pp 1-3
[6] Goldstein A A 1965 On steepest descent. SIAM Journal on Control, vol. 3, 147-151.
[7] Wolfe P 1968 Convergence conditions for ascent methods *SIAM Review* **11** pp 226-235
[8] Powell M J D 1976 Some global convergence properties of a variable-metric algorithm for minimization without exact line searches *SIAM-AMS Proceedings*, Philadelphia, **9** pp 53-72
[9] Dennis J E and Schnabel R B 1983 Numerical Methods for Unconstrained Optimization and Nonlinear Equations *Prentice-Hall* Englewood Cliffs New Jersey
[10] Barzilai J and Borwein J 1988 Two-point step size gradient methods *IMA J. of Numer. Anal.* 8 141-148
[11] Potra F A and Shi Y 1995 Efficient line search algorithm for unconstrained optimization *JOTA* **85**, no.3 pp 677-704
[12] Lemaréchal C 1981 A view of line search in A. Auslander, W. Oettli and J. Stoer (Eds.). *Optimization and Optimal Control* Springer Verlag, pp 59-78
[13] Moré J and Thuente D J On line search algorithms with guaranteed sufficient decrease. Mathematics and Computer Science Division Preprint MCS-P153-0590 1990 Argonne National Laboratory Argonne.
[14] Raydan M 1993 On the Barzilai and Borwein choice of steplength for the gradient method *IMA J. Numer. Anal.* **13** pp 321-326
[15] Dai Y H, and Liao L Z. 2002. R-linear convergence of the Barzilai and Borwein gradient method. IMA. J. Numer. Anal. 26, 1-10
[16] Grippo L, Lampariello F, and Lucidi S 1986. A nonmonotone line search technique for Newton's Method *SIAM J. Numer. Anal.* **23** pp707-716
[17] Luengo F and Raydan M 2003 Gradient method with dynamical retards for large scale Optimization problems *Electronic Transactions on Numer. Anal.* vol 16 Kentstate University
[18] Birgin E G, Floudas C A, Martínez J M *Math. Programming*; Heidelberg **125** Iss. 1 pp 139-162
[19] Serafini T G, Zanghirati and Zanni L 2005 Gradient projection methods for quadratic programs and application in training support vector machine. *Optim. Meth. Software* **20** pp 347-372
[20] Dai Y H and Fletcher 2006 New algorithms for singly linearly constrained quadratic subject to lower and upper bound *Math Program Ser A.* **106**, pp 403-421
[21] Wright J S, Nowak R D, and Figueiredo M A T 2009 Sparse Reconstruction by separable approximation. IEEE Transaction on Signal processing. 57, 2479-2493
[22] Fletcher R. 2001 On the Barzilai-Borwein Method *Numer. Anal. Report NA/20*
[23] Wen G K, Mamat M, Mohd B I, and. Dasril Y 2012 A novel Step Size Selection Procedures for Steepest Descent Method *Appl. Math. Sci.* **6** No. 51, pp 2507-2518

IOP Conf. Series: Materials Science and Engineering **332** (2018) 012025 doi:10.1088/1757-899X/332/1/012025

Study of atmospheric condition during the heavy rain event in Bojonegoro using weather research and forecasting (WRF) model: case study 9 February 2017

I J A Saragih*, A G Meygatama, F M Sugihartati, M Sidauruk, A Mulsandi

School of Meteorology Climatology and Geophysics (STMKG), Jakarta, Indonesia

*Corresponding author: immanuel.saragih@bmkg.go.id

Abstract. During 2016, there are frequent heavy rains in the Bojonegoro region, one of which is rain on 9 February 2016. The occurrence of heavy rainfall can cause the floods that inundate the settlements, rice fields, roads, and public facilities. This makes it important to analyze the atmospheric conditions during the heavy rainfall events in Bojonegoro. One of the analytical methods that can be used is using WRF-Advanced Research WRF (WRF-ARW) model. This study was conducted by comparing the rain analysis from WRF-ARW model with the Himawari-8 satellite imagery. The data used are Final Analysis (FNL) data for the WRF-ARW model and infrared (IR) channel for Himawari-8 satellite imagery. The data are processed into the time-series images and then analyzed descriptively. The meteorological parameters selected to be analyzed are relative humidity, vortices, divergences, air stability index, and precipitation. These parameters are expected to indicate the existence of a convective activity in Bojonegoro during the heavy rainfall event. The Himawari-8 satellite imagery shows that there is a cluster of convective clouds in Bojonegoro during the heavy rainfall event. The lowest value of the cloud top temperature indicates that the cluster of convective clouds is a cluster of Cumulonimbus cloud (CB).

Keywords: heavy rain, WRF-ARW, Himawari-8, convective cloud

1. Introduction

Indonesia is a tropical region that receives enough heat energy from the solar radiation. There are many weather patterns are formed caused by of its convective influence triggered by the heat energy from the solar radiation (Zakir et al., 2010). Indonesia consisted of a larger ocean area than the land. It makes the atmospheric condition over Indonesia region be moist because of the high relative humidity value.
This condition potentially causes the convective cloud growth frequently due to high evaporation in the Indonesian ocean.

Cloud is the atmospheric object that consists of the water drops in the atmosphere. Clouds grow from water vapour condensed by the hygroscopic particles such as salt or dust particles, usually called condensation nuclei. The clouds will grow if the air in each atmospheric level reaches the moist condition in a range of relative humidity value about 60-100%. Convective is a local-scale factor that important in cloud formation process. Convective clouds are clouds that occur due to lifted air over relatively hot surfaces, such as Cumulonimbus clouds and Cumulus clouds (Tjasyono, 2007).

IOP Conf. Series: Materials Science and Engineering **332** (2018) 012025 doi:10.1088/1757-899X/332/1/012025

One of the extreme phenomena that frequently occurred in Indonesia region is heavy rain. Bojonegoro is one of the district in East Java Province that located in -7.3175°S and 111.7615°E. Administratively, Bojonegoro location is adjacent to Tuban regency in the north, Lamongan regency in the east, Madiun, Nganjuk, and Jombang regencies in the south, Ngawi and Blora regencies (Central Java Province) in the west (Bojonegoro District Government, 2017).

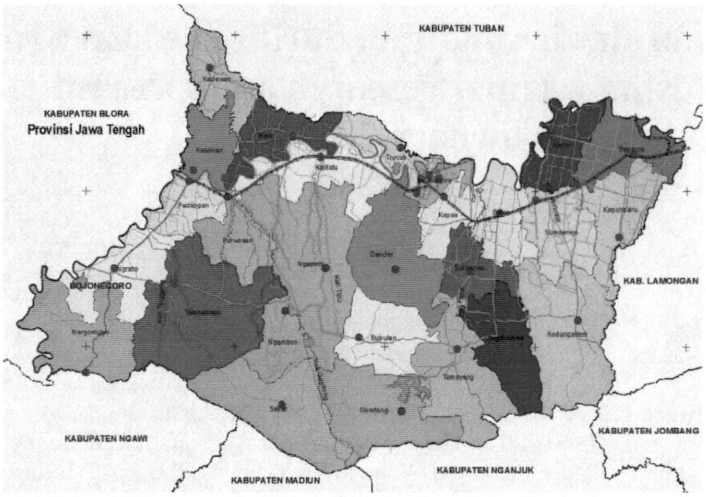

Figure 1. Bojonegoro administrative map (Bojonegoro Disctrict Government, 2017).

Analysing a weather phenomenon or event such as heavy rain requires data of meteorological parameters in the occurrence location or in the area adjacent to the occurrence location. But usually, the necessary data is difficult to obtain considering that there is no meteorological station in the region so the resulting data is not representative. To solve this problem, weather satellite data can serve as an alternative solution (Endarwin, 2012).

Rain can cause a disaster depending on the rain intensity and topography of the affected region. A topographic condition in Bojonegoro is dominated by land with a high slope of the land. So that the occurrence of the heavy rain continuously will cause the disaster such as landslide, floods, etc.

Table 1. Criteria of rainfall intensity over Indonesia region (BMKG, 2010)

Categories	Measured rainfall	
	mm/hour	mm/day
Slight rain	1-5	5-20
Moderate rain	5-10	20-50
Heavy rain	10-20	50-100
Very heavy rain	>20	>100

2. Material and methods

2.1. Research location

The research location is Bojonegoro and its surrounding region. This research location is based on the occurrence of heavy rain in Bojonegoro on 9 February 2016.

2.2. Research data

2.2.1. Final Analysis (FNL) data. FNL consisted of a global meteorological data in every 6 hours. FNL data is downloaded from https://rda.ucar.edu. The FNL data used in this study are 24-hours FNL data consisted from 00 UTC on 9 February 2016 to 00 UTC on 10 February 2016.

2.2.2. Infrared (IR) channel of Himawari-8 satellite data. This study used the 24-hours Himawari-8 satellite data from 00 UTC to 24 UTC on 9 February 2016. The Himawari-8 satellite data used in this research accessed from Agency of Meteorological, Climatological, and Geophysics (BMKG), Jakarta.

The FNL data processed by using the WRF-ARW model to mapping the local-scale atmospheric condition during heavy rainfall events. The meteorological parameters as the output of WRF-ARW model are plotted by using The Grid Analysis and Display System (GrADS). The meteorological parameters from WRF-ARW model are used to interpret the atmospheric condition that supports the occurrence of heavy rain in Bojonegoro, such as the values of relative humidity, vortices, divergence. Meanwhile, the Himawari-8 satellite data processed by using the Satellite Animation and Interactive Diagnosis (SATAID) GMSLPD. The infrared channel data are used to show the time series and distribution of the value of cloud top temperature during the heavy rain event in Bojonegoro. The Himawari-8 satellite data used to identify the cloud type that caused the heavy rain in Bojonegoro on 9 February 2016.

2.3. Research method

The research method used in this study is the descriptive analysis method of atmospheric condition from WRF-ARW model supported by satellite image output from Himawari-8 satellite imagery. This study used the WRF-ARW model version 3.8.1 that the centre of the model scheme is in -7.3175°S and 111.7615°E as the centre of Bojonegoro region. The WRF-ARW model parameterization scheme used in this study detailed in Table 1 below.

Table 2. WRF-ARW model parameterization scheme used in this study

Parameterization	1st Domain	2nd Domain
Centre longitude	111.7615°E	
Centre latitude	-7.3175°S	
Horizontal dimension X	100	88
Vertical dimension Y	40	28
Grid-Point Resolution	10km	3.3km
Geographic data resolution	30s	30s
Time step	60	
Microphysics option	WRF Single-Moment 3 scheme	
Longwave radiation option	RRTM scheme	
Shortwave radiation option	Dudhia scheme	
Surface layer option	Monin-Okbuhov Similarity scheme	
Surface option	Noah Land Surface model	
Planetary Boundary Layer option	YSU scheme	
Cumulus option	Kain-Fritsch scheme	

3. Result and discussion

The heavy rain that occurred in Bojonegoro on 9 February 2016 indicates as a phenomenon caused by convective clouds. The convective cloud growth in Bojonegoro supported by the humid atmospheric condition from the surface layer to the upper air layer. The relative humidity value from the WRF-ARW

IOP Conf. Series: Materials Science and Engineering **332** (2018) 012025 doi:10.1088/1757-899X/332/1/012025

model at 850hPa layer is about 78-84%, the 700hPa layer is about 87-93%, and the 500hPa layer is about 84-90%. The time series of relative humidity values in Bojonegoro on 9 February 2016 shows that the atmospheric condition is relatively wet so that there is much water vapour as the raw material of forming and growth processes of a convective cloud.

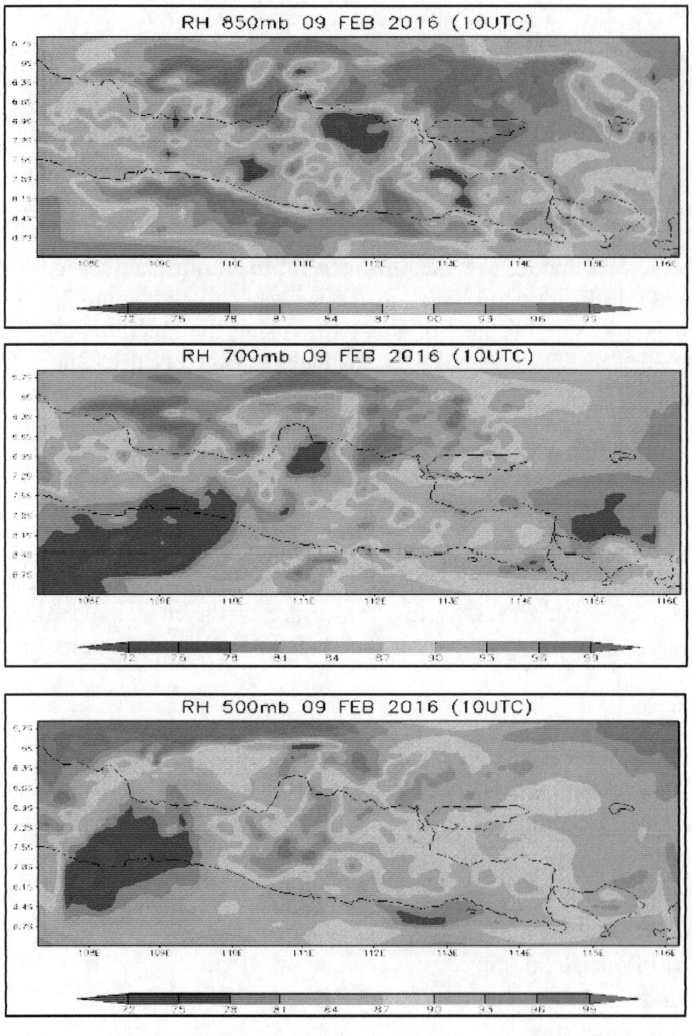

Figure 2. Relative humidity (RH) of 850hPa (top), 700hPa (middle), and 500hPa (bottom) at 10 UTC on 9 February 2016 in Bojonegoro region which obtained from the output of WRF-ARW model

Vorticity is a movement of air mass cyclonically. To support the convective activities, the vorticity value should be negative in the southern hemisphere and positive in the northern hemisphere. The negative vorticity indicates there are the convergence patterns in the atmosphere while the positive vorticity indicates the divergence patterns. The vorticity values from the WRF-ARW model during the heavy rain event in Bojonegoro shows the negative vorticity. It means that there are the convergence patterns in the atmosphere over Bojonegoro as the northern hemisphere. The vorticity in 850hPa layer is about -4/s to -1/s, in 700hPa layer is about -3/s to -2/s, and in 500hPa layer is about -3/s to -1/s. The convergence pattern means that the air masses move to the same point, so generally, in the affected areas of this convergence there are a lot of cloud growth and this condition cause the occurrence of the heavy rain in this region.

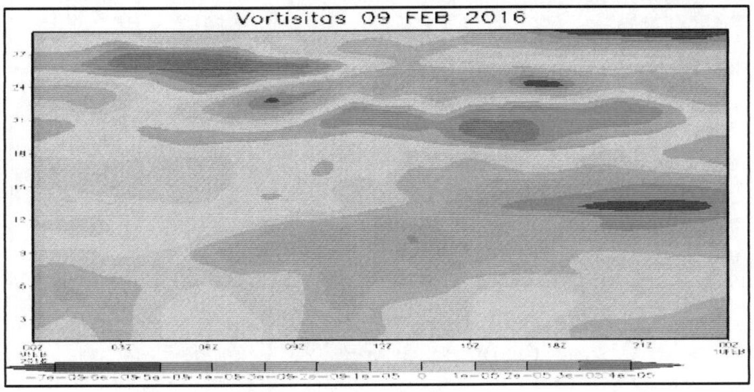

Figure 3. Time series of vorticity values in Bojonegoro on 9 February 2016 which obtained from the output of WRF-ARW model

Divergence is the parameter that indicates the spread of air masses. The WRF-ARW model output show divergence value over Bojonegoro during the heavy rain event is negative. The negative divergence means that the spread pattern of air masses in the atmosphere is convergent. The divergence value in 850hPa layer is about -3/s to -4/s, in 700hPa layers is about -2/s to -1/s, and in 500hPa layer is about -1/s to 0/s. The convergent pattern makes the air mass gathered in an area and there will grow a convective cloud over that region. This shows that in this region there is a convergent movement of air mass gathered into one so that the formation of convective clouds in the region.

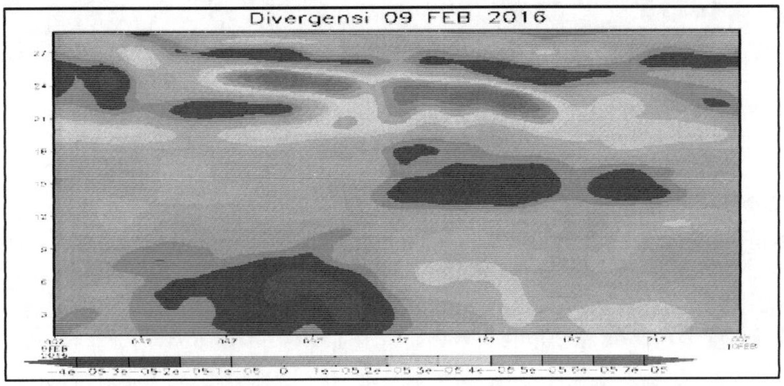

Figure 4. Time series of divergence values in Bojonegoro on 9 February 2016 which obtained from the output of WRF-ARW model

Wind is the movement of the air period in the horizontal direction caused by the difference in air pressure from one place to another. The wind moves from high-pressure area to low-pressure area. Wind speed and direction have an effect on the air mass and cloud movement. The results of WRF-ARW model output for wind at 925hPa layer indicates that the wind moves from west to east at a speed of 2-4kt. Generally, if the wind is blowing hard, then the cloud will move in the direction of the movement of the wind so it is unlikely the rain will fall in the region, but here the wind speed is categorized calm so that will not influence to the cloud movement. This condition causes the convective cloud which formed and growth over Bojonegoro region will stay and potentially caused heavy rain in Bojonegoro.

IOP Conf. Series: Materials Science and Engineering **332** (2018) 012025 doi:10.1088/1757-899X/332/1/012025

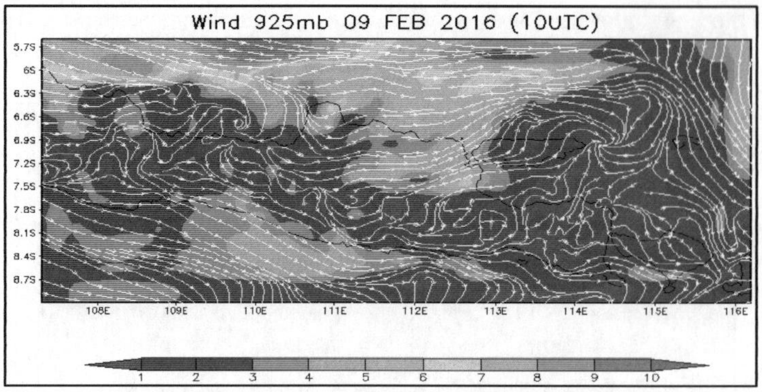

Figure 5. The wind condition in Bojonegoro on 9 February 2016 which obtained from the output of WRF-ARW model

According to BMKG as shown in Table 1, the rain categorized as a very heavy rain is if the measured rainfall more than 20mm/hour or 100mm/day. The rainfall analysis in the Bojonegoro region based on the WRF-ARW model output from 00 UTC to 01 UTC shows that there is a rainfall occurrence exceed 26 mm/hours in that area which indicated by the dark-black area. The rainfall map at 03 UTC to 04 UTC and 06 UTC to 07 UTC shows that the rainfall reach more than 26mm/hour. The heavy rainfall also measured at 09 UTC to 10 UTC. At 12 UTC to 13 UTC, the rain clouds have begun to shift away from Bojonegoro but rainfall continues to exceed 26mm/hour.

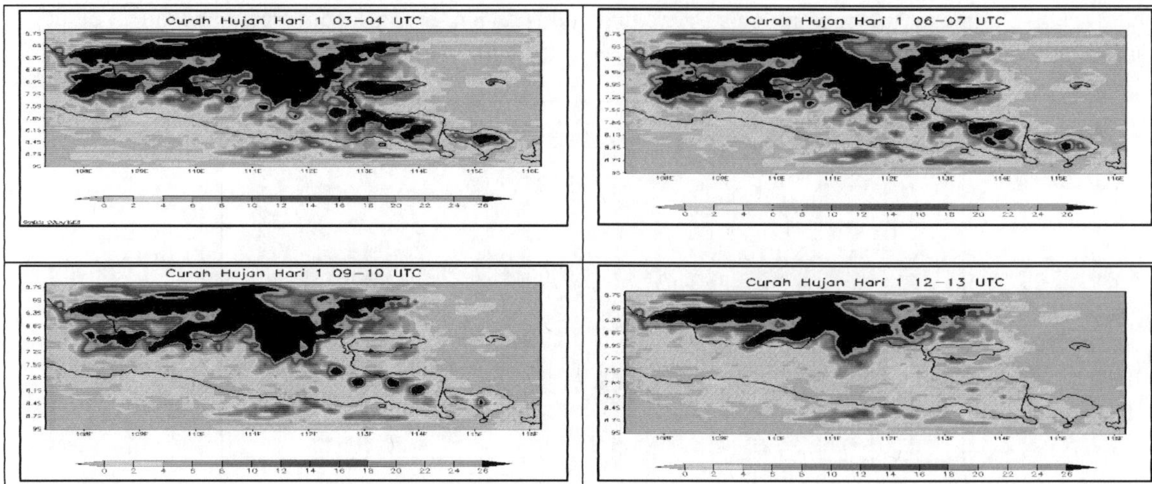

Figure 6. The rainfall map in Bojonegoro on 9 February 2016 which obtained from the output of WRF-ARW model

Cloud is a component that can't be separated from human life. Cloud is one of the important components in the rain occurrence, especially the convective cloud. Cloud imagery from satellite data can be processed and analysed from GMSLPD. The recent studies usually use the infrared channel of Himawari-8 satellite imagery to identify the cloud type over a region. The heavy rain event that occurred in Bojonegoro can be analysed using Himawari-8 satellite imagery. The result of analyses from satellite image shows that convective cloud growth (cumulus stage) begins at 08 UTC and reach the mature stage at 10 UTC. The satellite imagery shows the bright-white clouds over the Bojonegoro region as the cluster of convective clouds.

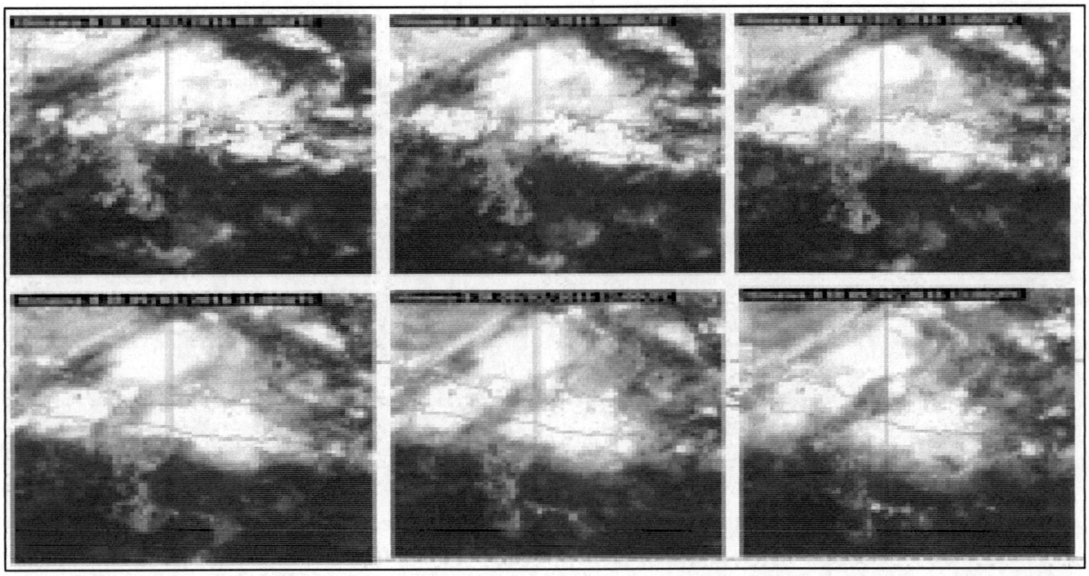

Figure 7. Imageries of infrared channel of Himawari-8 satellite over Bojonegoro region on 9 February 2016 which obtained from SATAID GMSLPD

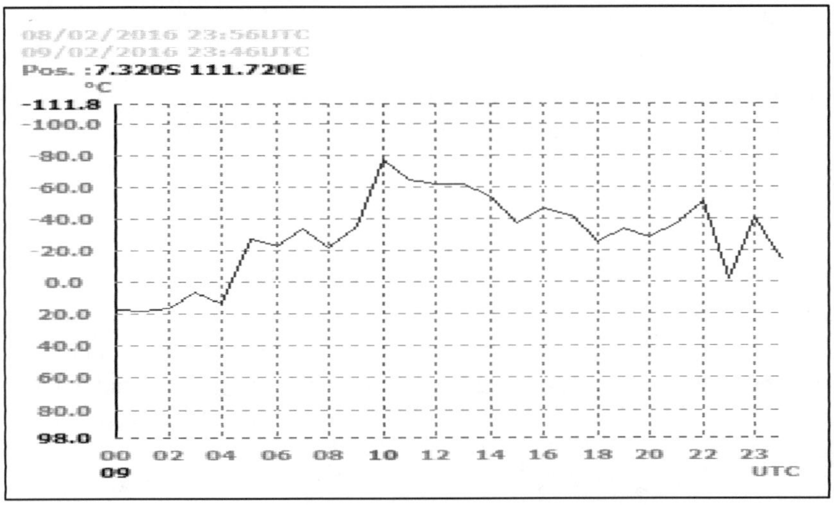

Figure 8. Time series of cloud top temperature over Bojonegoro region (7.320°S 111.720°E) on 9 February 2016 which obtained from the output of SATAID GMSLPD

The SATAID GMSLPD not only used to show the images of the clouds but also to show the time series of cloud top temperature over a region. In the previous study, the time series of cloud top temperature used to analyse the change of cloud top temperature and identify the cloud type. The time series of cloud top temperature shows that at 10 UTC the cloud over Bojonegoro region reaches the minimum temperature of -80°C. This values of cloud top temperature indicate that the cloud is classified as a convective cloud, the Cumulonimbus cloud (CB), and the heavy rain that occurred in Bojonegoro are from this CB. At 12 UTC, the cloud top temperature begins to decrease to -60°C indicating that the CB has begun to experience to the dissipation stage.

IOP Conf. Series: Materials Science and Engineering **332** (2018) 012025 doi:10.1088/1757-899X/332/1/012025

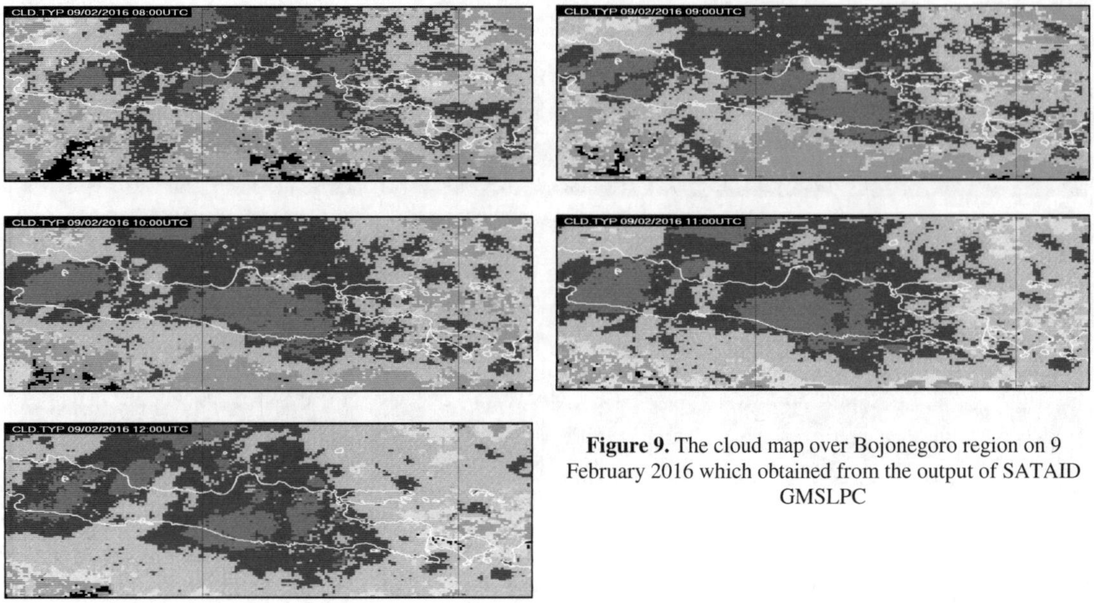

Figure 9. The cloud map over Bojonegoro region on 9 February 2016 which obtained from the output of SATAID GMSLPC

The SATAID GMSLPC is one of the SATAID software used to generate/show the cloud image processed from the Himawari-8 satellite data. The GMSLPC is also able to show the colour of the different cloud type. Based on cloud type identification using GMSLPC output, there is a cluster of convective clouds, categorized as the CB, which move from north to south as the result of wind movement effect. Observed at 10 UTC the cluster of CB entered the Bojonegoro region, indicated by the red colour, and continuously decreased until 12 UTC.

4. Conclusions

The results of this study indicate that there was heavy rain in Bojonegoro on February 9, 2016. The Himawari-8 satellite imagery shows that the heavy rain event came from the Cumulonimbus cloud (CB) over Bojonegoro region. The WRF-ARW model that using the parameter schemes according to the parameterization used in this study relatively can describe the atmospheric condition during that heavy rain event. The selected meteorological parameter such as the relative humidity, vorticity, divergence, and wind indicated the atmospheric condition that supports the convective activities over Bojonegoro region. But, to improve the accuracy in analysing the heavy rain events, a complete WRF-ARW parameterization scheme and the addition of the number of meteorological parameters analysed in some heavy rain events are needed.

References

[1] Endarwin, 2012, Analisis Objektif Terhadap Kejadian Cuaca Ekstrim Di Indonesia Memanfaatkan Data Satelit Cuaca, *Prosiding Cuaca Ekstrim,* No. 19, Vol.1 Nov 2012, Jakarta

[2] Fadholi, A., Sari, F. P., Aji, P., and Dewi, R., 2014, Pemanfaatan model Weather Research and Forecasting (WRF) dalam analisis cuaca terkait hujan lebat Batam 30-31 Januari 2011, *Jurnal Fisika dan Aplikasinya*, Vol. 10(1), page 24-30.

[3] Tjasyono, B., HK., 2007, *GM-322 Meteorologi Fisis*, Institut Teknologi Bandung, Bandung

IOP Conf. Series: Materials Science and Engineering **332** (2018) 012025 doi:10.1088/1757-899X/332/1/012025

[4] Zakir, A., et al., 2010, *Perspektif Operasional Cuaca Tropis,* Badan Meteorologi Klimatologi dan Geofisika, Jakarta
[5] BMKG, 2010, *Press Release Kondisi Cuaca Ekstrem dan Iklim Tahun 2010-2011*, 12 October 2010, Jakarta
[6] Bojonegoro District Government, 2017, *Geographical Conditions Bojonegoro,* official website of Bojonegoro district at http://www.bojonegorokab.go.id/geografi accessed on 01 December 2017

Representation mutations from standard genetic codes

I Aisah*, M Suyudi, E Carnia, Suhendi, A K Supriatna

Department of Mathematics, Faculty of Mathematics and Natural Sciences, Universitas Padjadjaran Bandung, Indonesia

*Corresponding author: isah.aisah@unpad.ac.id

Abstract. Graph is widely used in everyday life especially to describe model problem and describe it concretely and clearly. In addition graph is also used to facilitate solve various kinds of problems that are difficult to be solved by calculation. In Biology, graph can be used to describe the process of protein synthesis in *DNA*. Protein has an important role for *DNA* (*deoxyribonucleic acid*) or *RNA (ribonucleic acid*). Proteins are composed of amino acids. In this study, amino acids are related to genetics, especially the genetic code. The genetic code is also known as the triplet or codon code which is a three-letter arrangement of *DNA* nitrogen base. The bases are adenine (*A*), thymine (*T*), guanine (*G*) and cytosine (*C*). While on *RNA* thymine (*T*) is replaced with Urasil *(U*). The set of all Nitrogen bases in *RNA* is denoted by $N -\{C\ U, A, G\}$. This codon works at the time of protein synthesis inside the cell. This codon also encodes the stop signal as a sign of the stop of protein synthesis process. This paper will examine the process of protein synthesis through mathematical studies and present it in three-dimensional space or graph. The study begins by analysing the set of all codons denoted by *NNN* such that to obtain geometric representations. At this stage there is a matching between the sets of all nitrogen bases *N* with $Z_2 \times Z_2$; $C = (\bar{0}, \bar{0}), \mathrm{U} = (\bar{0}, \bar{1}), \mathrm{A} = (\bar{1}, \bar{0}), \mathrm{G} = (\bar{1}, \bar{1}\ )$ .By matching the algebraic structure will be obtained such as group, group Klein-4,Quotien group etc. With the help of Geogebra software, the set of all codons denoted by *NNN* can be presented in a three-dimensional space as a multicube *NNN,* and also can be represented as a graph, so that can easily see relationship between the codon.

Keyword: *Codon, Graph , DNA, RNA*

1. Introduction

Genes are made up of nucleic acids called deoxyribonucleic acids (*DNA*). *DNA* forms *RNA* (ribonucleic acid) through the transcription process. *DNA*-forming molecules are pentose sugars (deoxyribose), phosphates, and nitrogenous bases comprising purines (guanine (*G*) and adenine (*A*)) and pyrimidine (thymine (*T*) and cytosine (*C*)). One of the differences between *DNA* and *RNA* lies in its pyrimidine base. *RNA* contains uracid pyrimidine (*U*) and cytosine (*C*).

The genetic code is the composition of three nitrogenous bases of nucleic acids (*DNA* and *RNA*) called codons, or standard genetic codes. The number of codons there are 64 pieces. Of these 64 codons, 61 codons encode amino acids while the remaining three (*UAA*, *UAG*, and *UGA*) encode a stop signal or as a signal indicating the end of the protein formation process.

The nitrogen base of *RNA* can be represented as the set $N = \{C, U, A, G\}$. The set of all these nitrogen bases can be matched with $\mathbb{Z}_2 \times \mathbb{Z}_2 = \{(\bar{0}, \bar{0}), (\bar{0}, \bar{1}), (\bar{1}, \bar{0}), (\bar{1}, \bar{1})\}$ [6]. In addition, the set of all nitrogen bases can also be matched with a polynomial ring over $\mathbb{Z}_2$. The matching of *N* with the polynomial ring over $\mathbb{Z}_2$causes N to form a four-element galois field denoted by $GF(4)$ and NNN which is the set of all codons can be presented into a three-dimensional space as multicube NNN . Based on the corresponding transformation it is found that 24 representations of NNN [6]. Through this presentation it can be seen mutations that occur in the standard genetic code. Besides through multicube, its representation can be seen more easily through graph.

2. Theoretical Model

2.1. Graph

Definition 2.1: A graph G is defined as $G = (V, E)$ Where V is a set of all vertices and E is a set of all edges in the graph.

Each side of the graph is associated with one or two vertices. These vertexes are called end vertices. The side associated with only one point is called a loop. Two different sides connecting the same vertices are called parallel sides. Two vertices are said to be connected or adjacent if there is a side connecting the two. If all the sides in a directed graph then the graph is called directed graph (directed graph, abbreviated digraph). Conversely, if all the sides in the graph are not directional, then the graph is called undirected graph.

2.2. Genetic

Deoxyribonucleic acids (*DNA*) are very important chemical compounds for living things. *DNA* consists of three kinds of molecules, namely:
1) The pentose sugar, known as deoxyribose
2) Posterior acid
3) Nitrogen base, distinguished from primidin consisting of cytosine (*C*) and thymine (*T*) and purine consisting of adenine (*A*) and guanine (*G*)

Beside *DNA*, there are other important nucleic acids, namely ribonucleic acid (*RNA*). Like *DNA*, *RNA* consists of three kinds of molecules, namely pentose sugars, posropicacid, and nitrogenous bases. Differences of *DNA* and *RNA* one of them lies in its primidin base. The pyrimidine content in *DNA* consists of cytosine and thymine while the pyrimidine base in *RNA* consists of cytosine and uracil (*U*). Based on location and function, *RNA* is divided into three kinds, namely:

1) *RNA* ambassadors (*RNA-d*)
RNA ambassadors or *RNA-d* are located within the nucleus and are made (printed) by *DNA* in a process called transcription. *RNA-d* serves to carry the genetic information it receives from *DNA*, which is information to form proteins.
2) *RNA* transfer (*RNA-t*)
Transfer R*NA* or *RNA*-t is printed in the nucleus, before placing itself in the cytoplasm. *RNA-t* has the task of binding amino acids in the cytoplasm.
3) *RNA* ribosome (*RNA-r*)
RNA-r is made in the nucleus and located within the ribosome. *RNA-r* is the meeting place between *RNA-d* and *RNA-t* at the time of protein synthesis [10].

2.3. Codon

Codon is nucleotide sequences in nucleic acids (*DNA* and *RNA*) that encodes the amino acids in the protein chain. *DNA* and *RNA* bases' built by the nucleotides that will specify the 20 amino acids. Thus, if each nucleotide is translated into amino acids, then there would be only 4 of the 20 amino acids that will be specified. If the nucleotide sorted by 2 pieces (example: *AG*, *GT*), then there would be 16 amino acids to be specified. This amount is not enough to specify the 20 amino acids. Therefore, there must be combination of at least three nucleotides (a triplet of nucleotides) to determine each particular amino acid. This nucleotide triplet code shall be referred to the standard genetic code or codon. Codons will provide $4^3 = 64$ amino acids to be specified, this amount is more than enough so that it will no amino acids are specified by more than one codon [1]. Nucleotida at codon basic component is written in the set of nucleotide bases found in *RNA* is {*U, G, C, A*} List Codon arranged at the *RNA* can be seen in Figure 1 as follows:

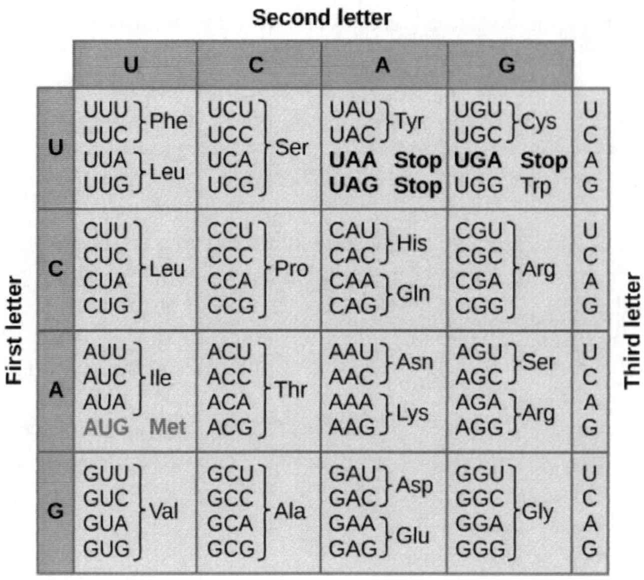

Figure 1. Code, Amino Acid and Encoded Stop Signal
(Source: [11])

2.4. Genetic Mutation

Mutation of genes is a change that occurs in nitrogen bases [3]. Based on the size of the number of altered bases, gene mutations are divided into two types namely point mutations and large (gross) mutations. Point mutations are mutations that involve only one nitrogen change while large mutations are mutations involving the change of more than one nitrogen base. An example of a point mutation is a *CCC* mutation into *CAC* and an example of a large mutation is *CCC* mutation into an *ACA*.

Gene mutations are also grouped based on their effects on the amino acids and *stopsignal* formed. Mutation of genes is generally differentiated into three types of mutations [3], i.e.,

1. Mute mutation (silent) are gene mutations that do not alter the amino acids and stopsignal formed. An example of this type of mutation is a mutation from CCC to CCA which both produce proline amino acids.
2. Nonsense mutation is a mutation of the gene that causes the amino acids to be formed to become stopsignal so that the protein synthesis process stops prematurely and the resulting amino acid chain becomes shorter than it should be .
3. Mutations of mis-meaning (missense) is a mutation that causes amino acids formed different from amino acids that should be formed.

2.5. Algebra Structure

Definition 2 (Group) [4]

Let *G* a nonempty set together with a binary operation that assigns to each ordered pair (*a, b*) of elements of *G* an element in *G* denoted by *ab*. We say *G* is a group under this operation if the following three properties are satisfied.

1. Associativity. The operation is associative;that is $((ab)c = a(bc)$ for all *a,b,c*in*G*.
2. Identity. There is an element e (called the identity) in G such that $ae = ea = a$for all *a* in *G*.

3. Inverse. For each element a in G,there is an element b in G (called an inverse of a) such that $ab = ba = e$

If a group has $ab = ba$ for every a and b in G, then the group is called abelian group.

Definition 3 (order of a group) [4]
The number of elements of a group (finite or infinite) is called is *orde*r. We will use $|G|$ to denote the order of G.

Ring Theory
Definition 4 (Ring) [4]
A ring R is a nonempty set with two binary operation, addition (denoted by $a + b$) and multiplication(denoted by ab), such that for all *a,b, c* in R :

1. $a + b = b + a$
2. $(a + b) + c = a + (b + c)$
3. There is an additive identity 0. That is , there is an element 0 in R such that $a + 0 = a$ for all $a \in R$.
4. There is an element $-a \in R$ such that $a + (-a) = 0$
5. $a(bc) = (ab)c$
6. $(b + c) = a \cdot b + ac$and $(b + c)a = ba + ca$.

Vector Spaces
Definition 5 : A set V is said to be a vector spaces over a field F if V is an abelian group under addition (denoted by +) and, if for each $\alpha \in F$ and $v \in V$, there is an element αv in V such that the following conditions hold for all α, β in F and all u, v in V.

1. $\alpha(u + v) = \alpha u + \alpha v$
2. $(\alpha + \beta) u = \alpha u + \beta u$
3. $\alpha(\beta u) = (\alpha\beta)u$
4. $1 u = u$

Finite Field (Galois Field)
Definiion 5 : A set V is said to be a vector spaces over a field F if V is an abelian group under addition (denoted by +) and, if for each $\alpha \in F$ and $v \in V$, there is an element αv in V such that the following conditions hold for all α, β in F and all u, v in V

1. $\alpha(u + v) = \alpha u + \alpha v$
2. $(\alpha + \beta) u = \alpha u + \beta$
3. $\alpha(\beta u) = (\alpha\beta)u$
4. $1 u = u$

3. Result

3.1. Representation of Genetic Code in a Three- Dimensional Space

To see the representation of the standard Genetic code, we investigate the structure of Algebra from N. Let N = *{C, U, A, G}* can be matched with $\mathbb{Z}_2 \times \mathbb{Z}_2 = \{(\bar{0}, \bar{0}), (\bar{0}, \bar{1}), (\bar{1}, \bar{0}), (\bar{1}, \bar{1})\}$ such that $\boldsymbol{C} = (\bar{0}, \bar{0})$, $\boldsymbol{U} = (\bar{0}, \bar{1})$, $\boldsymbol{A} = (\bar{1}, \bar{0})$, and$\boldsymbol{G} = (\bar{1}, \bar{1})$.

Based on Table 1 it appears that *(N, +)* is closed and associative, there is an identity element at *N*ie*C*, each element in N has an inverse ie$(C)^{-1} = C$, $(A)^{-1} = A$, $(U)^{-1} = U$, $(G)^{-1} = G$, so *(N, +)* abelian groups. Furthermore that the set of all codons or *NNN* forms a vector space over the field $\mathbb{Z}_2$.If the matching between N = *{C, U, A, G}* with the polynomial ring over $\mathbb{Z}_2$ i.e $\mathbb{Z}_2[x]/\langle x^2 + x + 1\rangle = \{0 + \langle x^2 + x + 1\rangle,\ 1 + \langle x^2 + x + 1\rangle,\ x + \langle x^2 + x + 1\ \rangle,\ x + 1 + \ \langle x^2 + x + 1\rangle\}$,this matching resulted in N having a four-element Galois field structure (GF(2^2)).

IOP Conf. Series: Materials Science and Engineering **332** (2018) 012026 doi:10.1088/1757-899X/332/1/012026

Table 1. Cayley table of (*N* ,+)

+	*C*	*U*	*A*	*G*
C	*C*	*U*	*A*	*G*
U	*U*	*C*	*G*	*A*
A	*A*	*G*	*C*	*U*
G	*G*	*A*	*U*	*C*

To facilitate the presentation of *NNN* in a three-dimensional space we use matching between $N = \{C, U, A, G\}$ with $\{0,1,2,3\}$ and with polynomial ring over $\mathbb{Z}_2$ so $C = 0 = 0$, $U = 1 = 1$, $A = x = 2$, dan $G = x + 1 = 3$. Thus it is concluded that DNA forms a vector space over a four-element Galois field ($\mathrm{GF}(2^2)$). [6].

Thus, a representation of *NNN* can be formed in a three-dimensional space called multicube*NNN*. The representation of *NNN* in a three-dimensional space which can be formed as many as 24 *NNN* multicubes relies on matching between *N* with $\{0,1,2,3\}$ and $\mathbb{Z}_2 \times \mathbb{Z}_2$ (Joséet al, 2012: 125 and 141). For example, if *C = 0, U = 1, A = 2,* and *G = 3* or if $C = (\bar{0}, \bar{0})$, $U = (\bar{0}, \bar{1})$, $A = (\bar{1}, \bar{0})$, dan $G = (\bar{1}, \bar{1})$ then the multycube*NNN* formed is called multicube *NNN* with the sequencing (*C, U, A, G*) . Figure 2 shows the multicube *NNN* by sequencing (*C, U, A, G*).

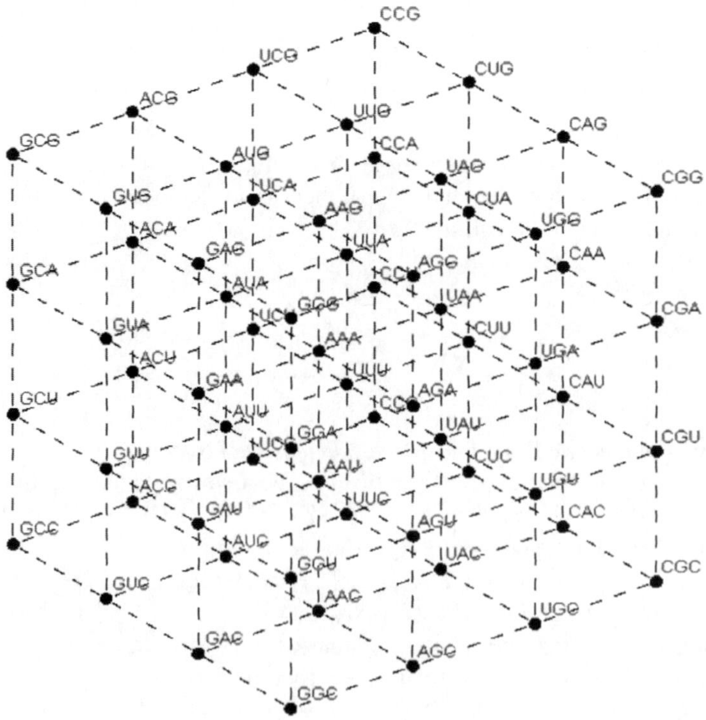

Figure 2. Multicube *NNN* by sequencing (*C, U, A, G*).

To see the mutations occurring in *NNN* will be used translations using vectors $(\bar{1},\bar{0}) \in \mathbb{Z}_2 \times \mathbb{Z}_2$. Translation of $(X_1, X_2, X_3) \in NNN$ with a vector of elements $\mathbb{Z}_2 \times \mathbb{Z}_2$ is defined as the translation of each nitrogen base (X_1, X_2, X_3) [6].

Suppose selected *(C, C, U)* ∈*NNN*, then the translations *(C, C, U)* are obtained as follows:
For *C* on *(C, C, U)*

$$T_{\binom{\bar{1}}{\bar{0}}}(C) = T_{\binom{\bar{1}}{\bar{0}}}\binom{\bar{0}}{\bar{0}} = \binom{\bar{0}}{\bar{0}} + \binom{\bar{1}}{\bar{0}} = \binom{\bar{1}}{\bar{0}} = A$$

For *A* on *(C, C, U)*

$$T_{\binom{\bar{1}}{\bar{0}}}(U) = T_{\binom{\bar{1}}{\bar{0}}}\binom{\bar{0}}{\bar{1}} = \binom{\bar{0}}{\bar{1}} + \binom{\bar{1}}{\bar{0}} = \binom{\bar{1}}{\bar{1}} = G$$

Thus, the translational result of *(C, C, U)* with vectors $(\bar{1},\bar{0}) \in \mathbb{Z}_2 \times \mathbb{Z}_2$ is *(A, A, G)*.

3.2. Graph of The Set Of Amino Acids And Stopsignal

Representations of *NNN* can be formed as many as 24 multicube *NNN* with different ordering [6]. Based on Manhattan distance, two codons connected in multicube *NNN* have distance one, while biologically, two codons connected in multicube *NNN* indicate the occurrence of a single point mutation from one codon to another codon as shown in Figure 2 above.

Examples of two codons connected in the *NNN* multicube are codons *(C, U, G)* and *(C, A, G)* where the distance between them is :

$$\begin{aligned} d((C,U,G),(C,A,G)) &= d((0,1,3),(0,2,3)) \\ &= |0-0| + |1-2| + |3-3| \\ &= 0+1+0 = 1 \end{aligned}$$

While the distance between the codons *(A, U, G)* and *(C, A, G)* is

$$\begin{aligned} d((A,U,G),(C,A,G)) &= d((2,1,3),(0,2,3)) \\ &= |2-0| + |1-2| + |3-3| \\ &= 2+1+0 = 3 \end{aligned}$$

Thus it can be concluded that codons *(A, U, G)* and *(C, A, G)* are not connected.

It will be constructed graph $G_i$which is related to amino acids and stop signals where the set of all vertices of G_i is $A \cup \{S\}$and two different vertices are connected by an element side $E(G_i)$if it has a Manhattan distance of one. To make it easier to write the vertices, each amino acid element $A \cup \{S\}$is written in a three-letter symbol and the stopsignal is denoted by a *stop* so that the set of all vertices of $G_i = A \cup \{S\}$ = {Ala, Arg, Asn, Asp, Cys, Glu, Gln, Gly, His, Ile, Leu, Lis, Met, Phe, Pro, Ser, Thr, Try, Tyr, Val, *Stop*}.

Next will be determined the number of $G_i$graphs that can be formed based on multycube *NNN*. Representation of *NNN* that can be formed as many as 24 graphs, with different sorting. Because for every two graphs each obtained from two multicube *NNN* with the same reverse sequencing it can be concluded that the $G_i$graphs that can be formed based on multycube*NNN* are as many as twelve graphs.

Figure 2 and Table 2 it is found that the Ala vertices coded by codons (G,C,C), (G,C,U), (G,C,A),and(G,C,G), respectively conected with the following vertices.

1. $(G,C,C)$is connected to(G,U,C), (A,C,C), and(G,C,U), and has Manhattan distance is one.
   $((G,U,C)$, encodes Val and d $((G,C,C)$, $(G,U,C))$= 1 then the vertex of Ala is connected with Val in graph G_1.
   (A,C,C) encodes Thr and d $((G,C,C)$, (A,C,C)= 1 then the Ala vertex is connected with Thr in the graph G_1.
   Since(G,C,U) and (G,C,C), have Ala code, it can not be deduced from this case.

Table 2. Element of NNN/R_1 With Amino Acids and Stop signal

Class equivalence	Elemen of Class equivalence	Elemen of $A \cup \{S\}$
[(*G*,*C*,*C*)]	(*G*,*C*,*C*), (*G*,*C*,*U*), (*G*,*C*,*A*), and (*G*,*C*,*G*)	Alanin (Ala)
[(*C*,*G*,*C*)]	(*C*,*G*,*C*), (*C*,*G*,*U*), (*C*,*G*,*A*), (*C*,*G*,G), (*A*,*G*,*A*),and(*A*,*G*,*G*)	Arginin (Arg)
[(*A*,*A*,*C*)]	(*A*,*A*,*C*) and (*A*,*A*,*U*)	Asparagin (Asn)
[(*G*,*A*,*C*)]	(*G*,*A*,*C*) and (*G*,*A*,*U*)	Asam aspartat (Asp)
[(*U*,*G*,*C*)]	(*U*,*G*,*C*) and (*U*,*G*,*U*)	Sistein (Cys)
[(*G*,*A*,*A*)]	(*G*,*A*,*A*) and (*G*,*A*,*G*)	Asam glutamat (Glu)
[(*C*,*A*,*A*)]	(*C*,*A*,*A*) and (*C*,*A*,*G*)	Glutamin (Gln)
[(*G*,*G*,*C*)]	(*G*,*G*,*C*), (*G*,*G*,*U*), (*G*,*G*,*A*), and (*G*,*G*,*G*)	Glisin(Gly)
[(*C*,*A*,*C*)]	(*C*,*A*,*C*) and (*C*,*A*,*U*)	Histidin (His)
[(*A*,*U*,*C*)]	(*A*,*U*,*C*), (*A*,*U*,*U*), and (*A*,*U*,*A*)	Isoleusin (Ile)
[(*C*,*U*,*C*)]	(*C*,*U*,*C*),(*C*,*U*,*U*),(*C*,*U*,*A*), (*C*,*U*,*G*), (*U*,*U*,*A*), and (*U*,*U*,*G*)	Leusin (Leu)
[(*A*,*A*,*A*)]	(*A*,*A*,*A*) and (*A*,*A*,*G*)	Lisin (Lis)
[(*A*,*U*,*A*)]	(*A*,*U*,*G*)	Metionin (Met)
[(*U*,*U*,*C*)]	(*U*,*U*,*C*) and (*U*,*U*,*U*)	Phenylalanin (Phe)
[(*C*,*C*,*C*)]	(*C*,*C*,*C*), (*C*,*C*,*U*), (*C*,*C*,*A*), and (*C*,*C*,*G*)	Prolin (Pro)
[(*U*,*C*,*C*)]	(*U*,*C*,*C*),(*U*,*C*,*U*),(*U*,*C*,*A*), (*U*,*C*,*G*), (*A*,*G*,*C*), and (*A*,*G*,*U*)	Serin (Ser)
[(*A*,*C*,*C*)]	(*A*,*C*,*C*), (*A*,*C*,*U*), (*A*,*C*,*A*), and(*A*,*C*,*G*)	Threonin (Thr)
[(*U*,*G*,*G*)]	(*U*,*G*,*G*)	Triptofan (Try)
[(*U*,*A*,*C*)]	(*U*,*A*,*C*) and (*U*,*A*,*U*)	Tirosin (Tyr)
[(*G*,*U*,*C*)]	(*G*,*U*,*C*), (*G*,*U*,*U*), (*G*,*U*,*A*), and(*G*,*U*,*G*)	Valin (Val)
[(*U*,*A*,*A*)]	(*U*,*A*,*A*), (*U*,*A*,*G*), and (*U*,*G*,*A*)	*Stop signal* (*stop*)

2. $(G,C,U)$is connected to (G,U,U), (A,C,U), (G,C,C),and(G,C,A), and has Manhattan distance is one.
   (G,U,U), encodes Val and d $((G,U,U), (G,C,U))$= 1 then the Ala vertex is connected with Val in graph G_1.
   (A,C,U), encodes Thr and d $((G,C,U), (A,C,U))$, = 1 then the Ala vertex is connected to Thr in the graph G_1.
   Because $(G,C,U), (G,C,A)$, and (G,C,C),both encode Ala, it cannot be deduced from this case.
3. (G,U,A), is connected to(G,C,A), (A,C,A), and(G,C,U), and has Manhattan distance is one.

(G, U, A), encodes Val and d $((G, C, A), (G, U, A))$= 1 then the Ala vertex is connected with Val in graph G_1..
(A, C, A), encodes Thr and d $((G, C, A), (A, C, A))$ = 1 then the Ala vertex is connected to Thr in the graph G_1.
Because (G, C, A), and (G, C, U)encode Ala then it can not be deduced from this case.

4. (G, C, G) connected to(G, U, G), (A, C, G), and(G, C, A), and has Manhattan distance is one.
   (G, U, G), encodes Val and d $((G, C, G)(G, U, G))$ = 1 then the vertex Ala is connected with Val in the graph G_1.
   (A, C, G), encodes Thr and d $((G, C, G), (A, C, G))$ = 1 then the Ala vertex is connected to Thr in the graph G_1.
   Because (G, C, G) and (G, C, A), encode Ala, it cannot be deduced from this case.

Based on 1 until 4 it can be concluded that the Ala vertex is only connected with the Val and Thr vertices in the G_1 graph. The same is done for the other twenty vertices to obtain the graph G_1 shown in Figure 3.

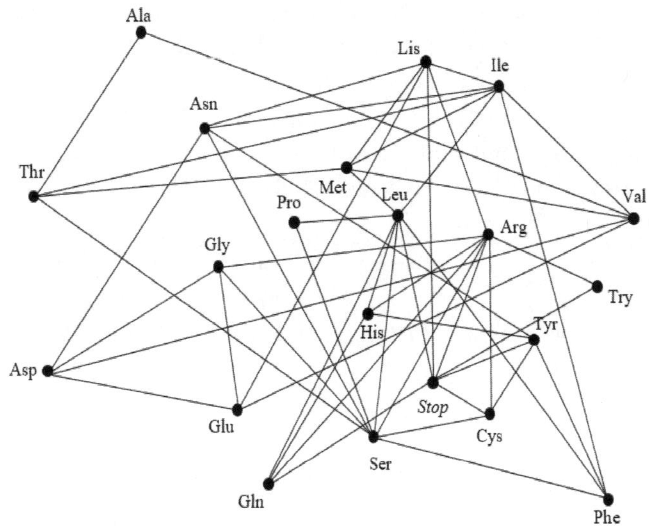

Figure3. Graph G_1 from multicube *NNN*

4. Conclusion

1. The set of DNA has the structure of Algebra as Group, Galois Field and Vector space
2. By using linear transformation and with the help of Geogebra software can be seen representation of mutations geometry in the form of multicube
3. Based on the multicube formed, the chemical compounds of the codon can be represented in graph form so that it can easily see the relationship between the encoded compound

Acknowledgments

The authors thank to the University of Padjadjaran, due to the funding of this work through the scheme of Academic Leadership Grant (ALG) with contract number 855/UN6.3.1/PL/2017.

References

[1] Anton H 2010 *Elementary Linear Algebra Applications Version* (Canada: John Wiley & Sons, inc)

IOP Conf. Series: Materials Science and Engineering **332** (2018) 012026 doi:10.1088/1757-899X/332/1/012026

[2] Akhtar A dan Ali T 2014 Anlysis of Unweigthed Amino Acids Network*International Scholarly Research Notices*
[3] Elrod S L, Stansfield WD 2010*Schaum's Outlines: Genetics*. (New York: Mc Graw Hill)
[4] GalianJ *Contemporary Abstract Algebra*.Sixt Edition 2006 (New York: Houghton)
[5] Jimēnez-Montaño M A, de la Mora-Basañez C R., and Pöschel T 1996 *The hypercube Structure of The Genetic Code Explains Conservative and Non-conservative Amino Acid Substitutions in vivo and in vitro*. *Biosystems***39** pp 117-125
[6] José M V, Morgado E R, dan Govezensky T 2011 Genetic Hotels for The Standard Genetic Code: Evolutionary Analysis Based Upon Novel Three-dimensional Algebraic Models*Bull. Math. Bio*, **73** pp 1443-1476
[7] José M V, Morgado E R, Sánchez R, and Govezensky T 2012 The 24 Possible Algebraic Representations of The Standard Genetic Code in Six and Three Dimensions*Advanced Studies in Biology***4** pp 119-152
[8] José M V, Morgado E R, Guimarães R C, Zamudio G S, Tores S, Bobadilla J R, Sosa D 2014Three-Dimensional Algebraic Models of the tRNA Code and 12 Graph for Representing the Amino Acids*Life*, **4** pp 341-373
[9] Madhulatha T S 2012An Overview on Clustering Methods*IOSR J of Engineering* 2(4)pp 719-725
[10] Marks A D, Smith C, dan Liebermann M 2005*Basic Medical Biochemistry: A Clinical Approach.*Lippinoott Williams and Wilkins
[11] Available from: https://cnx.org/contents/GFy_h8cu@9.87:QEibhJMi@8/The-Genetic-Code[cited 15 June 2017]

Application of improved Vogel's approximation method in minimization of rice distribution costs of Perum BULOG

J Nahar*, E Rusyaman , S D V E Putri

Prodi Matematika, Fakultas MIPA, Universitas Padjadjaran Jl. Raya Sumedang KM 21, Jatinangor, Sumedang 45363, Indonesia

*Email: julitanahar2017@gmail.com

Abstract. This research was conducted at Perum BULOG Sub-Divre Medan which is the implementing institution of Raskin program for several regencies and cities in North Sumatera. Raskin is a program of distributing rice to the poor. In order to minimize rice distribution costs then rice should be allocated optimally. The method used in this study consists of the Improved Vogel Approximation Method (IVAM) to analyse the initial feasible solution, and Modified Distribution (MODI) to test the optimum solution. This study aims to determine whether the IVAM method can provide savings or cost efficiency of rice distribution. From the calculation with IVAM obtained the optimum cost is lower than the company's calculation of Rp945.241.715,5 while the cost of the company's calculation of Rp958.073.750,40. Thus, the use of IVAM can save rice distribution costs of Rp12.832.034,9.

Keywords: Improved Vogel's Approximation Method, Modified Distribution Method, rice distribution cost.

1. Introduction

According to the General Guidelines Rice Subsidies for Poverty Communities In 2016, the rice subsidy program for low-income households or so-called "Beras Miskin" (rice for poor households) is an implementation of Presidential Instruction Number 5 of 2015 about Rice Supply and Distribution Chain Policy by the Government. In order to implement the rice subsidy program, a Coordinating Team was formed which consisted of several state institutions and one of them was Bulog. Perum Bulog is responsible for the supply and distribution of Rice Subsidy for Low Income Society from Bulog warehouse to Distribution Point. Distribution Point is a public facility as a place or location of Rice Subsidy for Low Income Society Subsidy from Perum Bulog to Distribution Officer of Rice Subsidy for Low Income Society in village office / hall, or other location agreed in writing by Regency / city with Bulog.

Perum BULOG Medan Regional Division as the rice program implementer spent considerable fund for rice distribution activity for some areas of Medan. To minimize the cost of this distribution it is necessary to do the planning in the distribution of rice so that the cost is very optimal. One method that can be used to optimize distribution costs is by transportation method.

Transportation problems aim to minimize the cost of transporting certain commodities from a number of sources to a number of destinations (Rao, 2009: 220). All optimal solution algorithms to solve transportation problems require a basic initial feasible solution to produce optimal solutions (Mathirajan, 2003). According to Hamdy A. Taha (2007), in general the Vogel Approach Method produces the best initial solution though not always. In 1990 a method was developed by Kirca and Satir called TOM (Total Opportunity-cost Method) to produce an initial feasible solution on transportation problems. TOM is an effective application of "Best Cell Method" along with some tie-breaking features in the Total Opportunity Cost (TOC) matrix. The reason behind VAM installation with TOC matrix is that VAM usually produces a better starting solution than other baseline feasible basic methods (Meenakshi et al., 2003). This study discusses the Improved VAM (IVAM) VAM which is developed by using TOC Matrix and will be implemented in rice distribution cost optimization for poor families in Bulog Sub Division of Medan.

2. Literature Review

2.1 Transportation Problems

Transportation problems are generally represented by the network in Figure 1.

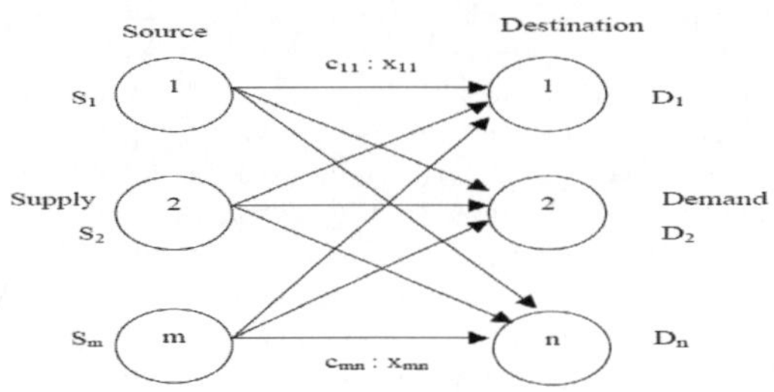

Figure 1. Representation of the Transportation Network Model

The linear programming model that can represent transportation problems in this study is as follows

Minimalization $Z = \sum_{i=1}^{m}\sum_{j=1}^{n} c_{ij}x_{ij}$

Subject to $\sum_{j=1}^{n} x_{ij} \le a_i, \quad i = 1,2,\ldots,m$ (Suply constraints)

$\sum_{i=1}^{m} x_{ij} \ge b_j, \quad j = 1,2,\ldots,n$ (Demand constraints)

$x_{ij} \ge 0.$

If the total supply is equal to the total demand then the problem is said to be balanced for transportation problems.

2.2 Vogel's Approximation Method

To get initial decent starting solution using Vogel's Approximation Method (VAM). VAM is a better method than the method of Least Cost and North West Corner in searching for feasible baseline solutions early. The steps for VAM are:

Step 1 : For each row (column), specify penalty size by subtracting cost the smallest unit on the row (column) of the smallest unit cost thereafter on the line (columns) are the same.

Step 2 : Identify the row or column with the biggest penalty. Decide selection by any. Allocate as many variables as possible with the smallest unit cost in the selected row or column. Adjust the supply and demand, and remove the rows or columns that have been fulfilled. If a row and a column are met simultaneously, only one of two is selected, and if the remaining row (column) is zero supply (demand).

Step 3 : (a) If exactly one row or column with zero supply or demand is left to be selected then iteration stops.

(b) If one row (column) with positive (demand) supply is left for selected, then specify the base variable on the row or column with the least-cost and stop iteration methods.

(c) If all rows and columns that have not been selected have (residual) supply and demand, then specify the zero base variable with the least-cost and stop iteration methods

(d) If not, then go back to step 1.

2.3 Total Opportunity-cost Method
According to Mathirajan and Meenakshi (2003), TOM is an effective application of "best cell methods" along with some tie-breaking features on the total opportunity cost (TOC) matrix. TOC matrix is generated by adding a "cost-matrix row" line (cost opportunity matrix row: for each row, the smallest cost of the row is subtracted with each element on the same row) and "cost opportunity column matrix" (cost opportunity column matrix: for each columns of the original transport cost matrix, the smallest cost of the column is subtracted with each element in the same column). When applying the "best cost method" (least cost method) to TOC matrix.

2.4 Improved Vogel's Approximation Method
According to Serdar and Serkan (2011), VAMs are improved by using the TOC matrix and on alternative allocation costs. The algorithm is to apply the TOC matrix and consider the three highest penalty fees and calculate alternative allocation costs in VAM procedures. Then choose the least of them. Detailed steps are:
Step 1 : Balance the transport problem if total supply> total or total demand supply <total requests.
Step 2 : Create a TOC matrix
Step 3 : Declare penalty fees for each row and column by subtracting the lowest cell cost on the row or column for the lowest cost of the next cell on the same row or column.
Step 4 : Select a row or column with the highest three penalty charges.
Step 5 : Calculate three transportation costs for the three rows or columns selected on step 4 by allocating as much as possible to a viable cell with the lowest transportation cost.
Step 6 : Select the smallest transportation cost of the three allocations in step 5.
Step 7 : Repeat steps 3-6 until all needs are met.
Step 8 : Calculate the total transportation cost for a reasonable allocation by using the original transportation cost matrix that has been balanced.

2.5 Modified Distribution Method
Having determined the initial feasible baseline solution of Improved Vogel's Approximation Method (IVAM) appears likely to be a solution that is not yet optimal, hence another scheme is needed to improve the solution already obtained. The MODI method is considered because the MODI method allows faster index calculation of the improvement for each cell not used without counting all closed paths (Justice K., 2015). Because of this operation, MODI saves time that can be considered from other methods to improve initial feasible basic solutions. MODI works by searching unused routes with the largest path repair index. Once identified, closed paths are tracked out to produce the maximum number of units that can be distributed. In the application, calculate the value for each line example u_1, u_2, u_3if there are three rows and for each column example u_1, u_2, u_3, in the transport table. With notation,
u_i= "value set for line" i
v_j= "value assigned to column" j
c_{ij}= "charge on cell" ij "(shipping cost from supply i to destination j)"
Quoted from Justice K. (2015), the steps of MODI are summarized as follows:
Step 1 : Develop an initial feasible basic solution using IVAM.
Step 2 : Calculate the values u_i and v_j for each row and column by applying the formula $u_i + v_j = c_{ij}$ on each cell that has an allocation.
Step 3 : Calculate the cost change, k_{ij} ,for each blank cell using $c_{ij} - u_i - v_j = k_{ij}$.
Step 4 : Allocate as many empty cells as possible that will result in a decline the best value at the cost in the most negative k_{ij}. Allocate by stepping-stone flow for selected cells.
Step 5 : Repeat steps 2 through 4 until all $k_{ij}$are positive or zero.

3. Discussion and Research Result
The data used in this paper is data on the quantity of rice stocks in Perum BULOG warehouse in July 2013, the amount of rice distributed, and the cost of transport from the warehouse to the distribution point contained in "Application of Transportation Methods in Optimizing Rice Distribution Cost of

IOP Conf. Series: Materials Science and Engineering **332** (2018) 012027 doi:10.1088/1757-899X/332/1/012027

Poor (RASKIN) on Perum Bulog Divre Medan "(Simbolon dkk, 2014). Data can be seen in the following table:

Table 1 Amount of RASKIN Suply

\

No	Warehouse	Location	Total Rice Suply (kg)
1	Mustafa	Medan	3.318.270,00
2	Jemadi	Medan	2.895.437,71
3	Mabar	Medan	3.379.212,29
4	L. Deli	Medan	385.170,00
5	T. Tinggi	T. Tinggi	1.294.575,00
		Total	11.272.665

Table 2 Amount of RASKIN Distribution

No	Warehouse	Point of Distribution	Total Rice (kg)
1	Mustafa	Langkat	524.865,00
		Medan	2.739.405,00
2	Jemadi	D. Serdang	347.777,71
		Sergai	129.390,00
		Langkat	2.418.270,00
3	Mabar	Langkat	218.730,00
		Binjai	166.440,00
4	L. Deli	D. Serdang	2.648.982,29
		Medan	208.410,00
		Binjai	521.820,00
5	T. Tinggi	T. Tinggi	334.575,00
		Sergai	960.000,00
		Total	11.272.665,00

Table 3 Freight Cost RASKIN from Warehouse to Point of Distribution (Rp/kg)

From/To	Medan	Binjai	T. Tinggi
Mustafa	71,22	78,13	100,39
Jemadi	71,62	77,73	99,60
Mabar	73,00	81,69	103,69
L. Deli	76,16	84,45	105,95
T. Tinggi	102,44	111,4	73,50
From/To	**Langkat**	**D. Serdang**	**Sergai**
Mustafa	91,13	87,12	94,53
Jemadi	90,73	86,33	93,74
Mabar	94,69	89,89	97,30
L. Deli	97,45	92,66	100,07
T. Tinggi	119,45	90,98	89,00

From the data in Table 1 and Table 2 will be formed a model that symbolizes each of these variables. A total of five warehouses and six distribution points will be represented by the following notation:

The complete problem of rice price distribution optimization of Perum BULOG can be expressed in equation (1).

$$Min\ Z = \sum_{i=1}^{m}\sum_{j=1}^{n} c_{ij}x_{ij}$$

$$\begin{aligned} Z = {} & 71{,}22x_{11} + 78{,}13\,x_{12} + 100{,}39x_{13} + 91{,}13x_{14} + 87{,}12x_{15} + 94{,}53x_{16} \\ & + 71{,}62x_{21} + 77{,}73x_{22} + 99{,}60x_{23} + 90{,}73x_{24} + 86{,}33x_{25} + 93{,}74x_{26} \\ & + 73{,}00x_{31} + 81{,}69x_{32} + 103{,}69x_{33} + 94{,}69x_{34} + 89{,}89x_{35} \\ & + 97{,}30x_{36} + 76{,}16x_{41} + 84{,}45x_{42} + 105{,}92x_{43} + 97{,}45x_{44} \\ & + 92{,}66x_{45} + 100{,}07x_{46} + 102{,}44x_{51} + 111{,}14x_{52} + 73{,}50x_{53} \\ & + 119{,}45x_{54} + 90{,}98x_{55} + 89{,}00x_{56} \end{aligned}$$

Subject to

$$\begin{aligned} & x_{11} + x_{12} + x_{13} + x_{14} + x_{15} + x_{16} \leq 3.318.270{,}00 \\ & x_{21} + x_{22} + x_{23} + x_{24} + x_{25} + x_{26} \leq 2.895.437{,}71 \\ & x_{31} + x_{32} + x_{33} + x_{34} + x_{35} + x_{36} \leq 3.379.437{,}29 \\ & x_{41} + x_{42} + x_{43} + x_{44} + x_{45} + x_{46} \leq 385.170{,}00 \\ & x_{51} + x_{52} + x_{53} + x_{54} + x_{55} + x_{56} \leq 1.294.575{,}00 \\ & x_{11} + x_{21} + x_{31} + x_{41} + x_{51} + x_{61} \geq 3.001.815 \\ & x_{12} + x_{22} + x_{32} + x_{42} + x_{52} + x_{62} \geq 688.260 \\ & x_{13} + x_{23} + x_{33} + x_{43} + x_{53} + x_{63} \geq 334.575 \\ & x_{14} + x_{24} + x_{34} + x_{44} + x_{54} + x_{64} \geq 3.161.865 \\ & x_{15} + x_{25} + x_{35} + x_{45} + x_{55} + x_{65} \geq 2.996.760 \\ & x_{16} + x_{26} + x_{36} + x_{46} + x_{56} + x_{66} \geq 1.089.390 \\ & x_{ij} \geq 0, \quad i = (1,2,\ldots,5), \quad j = (1,2,\ldots,6) \end{aligned} \tag{1}$$

After getting the optimization model, then look for an initial feasible basic solution with VAM. After nine iterations, obtained:

Table 4 Basic solution is feasible early with VAM

	Dari/Ke	Tujuan						Supply	Penalti 9
		Medan	Binjai	T. Tinggi	Langkat	D. Serdang	Sergai		
Sumber	Mustafa	71.20	78.10	100.39	91.10	87.10	94.50	3,318,270.00	-
		-	688,260.00	-	2,630,010.00	-			
	Jemadi	71.60	77.70	99.60	90.70	86.30	93.70	2,895,437.71	3.00
		-	-	-	531,855.00	2,234,192.71	129,390.00		
	Mabar	73.00	81.60	103.16	94.60	89.80	97.30	3,379,212.29	-
		3,001,815.00	-	-	-	377,397.29	-		
	L. Deli	76.10	84.40	105.92	97.40	92.60	100.07	385,170.00	-
		-	-	-	-	385,170.00	-		
	T. Tinggi	102.40	111.14	73.50	119.45	90.90	89.00	1,294,575.00	-
		-	-	334,575.00	-	-	960,000.00		
Demand		3,001,815.00	688,260.00	-	3,161,865.00	2,996,760.00	1,089,390.00	11,272,665.00	
Penalti 9		-	-	-	90.70	-	93.70		

In the otherhand basic solution is feasible early with IVAM trough three iteration. There is TOC matrix and solution with IVAM.

Table 5 Matrix TOC

From/To	Destination	Supply

IOP Conf. Series: Materials Science and Engineering **332** (2018) 012027 doi:10.1088/1757-899X/332/1/012027

		Medan	Binjai	T. Tinggi	Langkat	D. Serdang	Sergai	
Source	Mustafa	0.00	0.80	29.19	0.80	1.20	7.80	3318270
	Jemadi	0.00	0.00	28.00	0.00	0.00	6.60	2895437.71
	Mabar	0.00	2.50	30.16	2.50	2.10	8.80	3379212.29
	L. Deli	0.00	2.20	29.82	2.20	1.80	8.47	385170
	T. Tinggi	28.90	31.54	0.00	26.85	2.70	0.00	1294575.00
Demand		3001815	688260	334575	3161865	2996760	1089390	11272665

Tabel 6 Third Iteration with IVAM

Dari/Ke		Tujuan						Supply	Penalti 3
		Medan	Binjai	T. Tinggi	Langkat	D. Serdang	Sergai		
Sumber	Mustafa	-	6.90	29.19	19.90	15.90	23.30	3,318,270.00	3.40
		-	688,260.00	-	2,630,010.00	-	-		
	Jemadi	-	6.10	28.00	19.10	14.70	22.10	2,895,437.71	3.00
		-	-	-	531,855.00	2,234,192.71	129,390.00		
	Mabar	-	8.60	30.16	21.60	16.80	24.30	3,379,212.29	-
		3,001,815.00	-	-	-	377,397.29	-		
	L. Deli	-	8.30	29.82	21.30	16.50	23.97	385,170.00	-
		-	-	-	-	385,170.00	-		
	T. Tinggi	28.90	37.64	-	45.95	17.40	15.50	1,294,575.00	-
		-	-	334,575.00	-	-	960,000.00		
Demand		3,001,815.00	688,260.00	-	3,161,865.00	2,996,760.00	1,089,390.00	11,272,665.00	
Penalti 3		0	-	-	0.80	-	23.30		

Use MODI to perform optimality tests. The basic variables $x_{12}, x_{14}, x_{24}, x_{25}, x_{26}, x_{31}, x_{35}, x_{41}, x_{45}, x_{53}, x_{56}$.

The amount of rice allocation obtained by using Improved Vogel's Approximation Method is 688,260 kg from Mustafa to Binjai, 2,630,010 kg from Mustafa to Langkat, 531.855 kg from Jemadi to Langkat, 2,234,192,71 kg from Jemadi to D. Serdang, 129,390 kg from Jemadi to Sergai, 3,001,815 kg from Mabar to Medan, 377,397.29 kg from Mabar to D. Serdang, 385,170 kg from L. Deli to D. Serdang, 334,575 from Tebing Tinggi to Tebing Tinggi and 960,000 kg from Tebing Tinggi to Sergai with total rice distribution cost is Rp 945,241,715,5. Representation in graphical form of optimum solution obtained from optimality test results is shown in Image 4.1

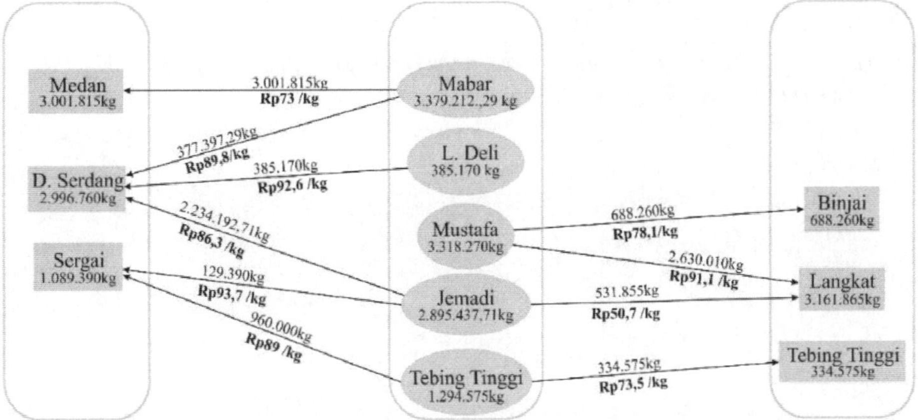

Figure 2 Optimal solution of optimality test results

4. Conclusion

The determination of the optimal solution in the minimization of rice distribution cost on Bulog is done by finding the initial feasible basic solution using Improved Vogel's Approximation Method (IVAM). The steps used in IVAM are the same as the steps on the VAM but with the addition of TOC Matrices. The optimal solution of rice distribution problem in Bulog Divre Medan with 5 (five) warehouses and 6 (six) distribution points in journal Simbolon et al. (2014) The amount of rice allocation obtained by using Improved Vogel's Approximation Method is 688,260 kg from Mustafa to Binjai, 2,630,010 kg from Mustafa to Langkat, 531.855 kg from Jemadi to Langkat, 2,234,192,71 kg from Jemadi to D. Serdang, 129,390 kg from Jemadi to Sergai, 3,001,815 kg from Mabar to Medan, 377,397.29 kg from Mabar to D. Serdang, 385,170 kg from L. Deli to D. Serdang, 334,575 from Tebing Tinggi to Tebing Tinggi and 960,000 kg from Tebing Tinggi to Sergai with total rice distribution cost is Rp 945.241.715,5. The solution obtained from IVAM is the most optimal solution in terms of number of iterations and rice distribution costs.

References

[1] Appati, J. K., Gogovi, G. K., dan Fosn, G. O. 2015. An International Journal of Advanced Computer Technology: *MATLAB Implementation of Vogel's Approximation and the Modified Distribution Methods.*

[2] Chaerani, D. (2011). *Bahan Ajar Matakuliah Pemrograman Linear.* Departemen Matematika FMIPA Unpad.

[3] Hillier, F. S. dan Lieberman, G. J. 2001. *Introduction to Operations Research Seventh Edition.* New York: McGraw-Hill

[4] Hitchcock, F. L. (2006). Journal of Mathematical Physics, Vol.20. *The Distribution of a Product from Several Sources to Numerous Localities*, 224-230.

[5] Korukoglu, S, dan Balli, S. (2011). Mathematical and Computational Application, Vol. 16 No. 2. *An Improved Vogel's Approximation Method for The Transportation Problem, 370-381.*

[6] Mathirajan, M., dan Meenakshi, B. (2004). Asia-Pacific Journal of Operational Research, Vol. 21 No. 4. *Experimental Analysis of Some Variants of Vogel's Approximation Method, 447-462.*

[7] Parlar, M. (2000). *Interactive Operations Research with MAPLE: Methods And Models.* Birkhäuser Boston.

[8] Pedoman Umum Subsidi Beras Bagi Masyarakat Berpendapatan Rendah. 2016. Jakarta: Diperbanyak oleh Kementerian Koordinator Bidang Pembangunan Manusia dan Kebudayaan Republik Indonesia.

[9] Rao, S. S. (1996). *Engineering Optimization Theory and Practice Third Edition.* New Jersey: John Wiley & Sons, Inc.

IOP Conf. Series: Materials Science and Engineering **332** (2018) 012027 doi:10.1088/1757-899X/332/1/012027

[10] Simbolon, L. D., Marihat S., Normalina N. (2014). Saintia Matematika, Vol. 02, No. 03. *Aplikasi Metode Transportasi dalam Optimasi Biaya Distribusi Beras Miskin (Raskin) pada Perum Bulog Sub Divre Medan,* 299-311.

[11] Taha, H. A. (2007). *Operation Research: An Introduction Eight Edition.* Upper Saddle River, New Jersey: Pearson Education, Inc.

Neuro-fuzzy model for estimating race and gender from geometric distances of human face across pose

K Nanaa[1], M N A Rahman[2], M Rizon[3], F S Mohamad[4], M Mamat[5]

[1,2,4,5]Faculty of Informatics and Computing, Universiti Sultan Zainal Abidin, Besut Campus, 22200 Besut, Terengganu, Malaysia
[3]Faculty of Visual Communication Design, National Academy of Arts Culture and Heritage, 50480Kuala Lumpur, Malaysia.

Email: ps0340@putra.unisza.edu.my

Abstract. Classifying human face based on race and gender is a vital process in face recognition. It contributes to an index database and eases 3D synthesis of the human face. Identifying race and gender based on intrinsic factor is problematic, which is more fitting to utilizing nonlinear model for estimating process. In this paper, we aim to estimate race and gender in varied head pose. For this purpose, we collect dataset from PICS and CAS-PEAL databases, detect the landmarks and rotate them to the frontal pose. After geometric distances are calculated, all of distance values will be normalized. Implementation is carried out by using Neural Network Model and Fuzzy Logic Model. These models are combined by using Adaptive Neuro-Fuzzy Model. The experimental results showed that the optimization of address fuzzy membership. Model gives a better assessment rate and found that estimating race contributing to a more accurate gender assessment.

1. Introduction

With advances in face recognition technology, there is a need to further expand the discussion of various sub-areas. These discussions contribute to bridging the gap to reach an advanced recognition system. The main challenges facing recognition systems are the pose, illumination, and occlusion. Furthermore, face recognition among huge database is considered an additional incrementally challenge. Although face recognition seeks to identify the corresponding image, providing a facial classification of subjects facilitates identification process. In another word, the classification of facial images across human groups contributes to the indexing of databases. Thus, the search for the target is narrowed during the identification process. In addition, classification process is useful in the process of constructing 3D face model based on a reference depth map. This is due to the conclusion that depth information does not differ much within the same race [1, 2]. Moreover, gender and race classification can provide a demographic classification system.

To create a system that classifies human groups through images, it first requires detection and representation characteristics of face. This is usually done by linear methods such as Principal Component Analysis (PCA) or Linear Discriminate Analysis (LDA). These methods also contribute to process of identification. They compare characteristics between given image and training samples in order to determine the nearest sample. The facial characteristics are given by shape and appearance. The comparison is usually done using measurement criteria such as Mahalanobis and Support Vector Machines (SVM). However, appearance characteristics play a large role in the process of assessing gender in adults. In contrast, the importance of the shape characteristics is taken into consideration with regard to subjects under the age of 18 years [3].

Combination of linear models was experimented in order to propose a model for classifying the population of Malaysia [4]. The database was created for both genders of Malay, Chinese, and Indians. The experiment was reported to be 90% based on classification rate in face images and it rises to 95.5%in head images. It is also concluded that using linear models in classifying gender provides more accurate results than the classifying race. Generally, non-linear classification methods may be more useful.

Neural networks are utilized to classify a texture data of race of FERET database represented by SVM [5]. The classification rate is reported as 92%, probabilistic Boosting Tree (PBT) is utilized to classify the gender of races: Mongoloid, Caucasian and African [6]. The experimental results show that PBT inclines to classify Mongoloid patterns as females while African patterns as males. This is due to their difference decision boundary of linear model in each race. Furthermore, the classifying gender by PBT on each race separately gave higher accuracy reaches up to 93.71%.

In general, the psychological process of face recognition by the human brain is still somewhat ambiguous. Moreover, humans are less likely to identify individuals across other races [7]. Thus, people rely on facial recognition algorithms according to their race. Therefore, it is useful to index the database in order to get people out of their race. Thus, the identification strategy is enhanced by using algorithms that are more appropriate for each race [8].

In this paper, we utilized non-linier model to classify the gender and race of Asian and Caucasian subjects. The used model is a combination of fuzzy logic and neural network models. Fuzzy logic is among the techniques of artificial intelligence used to indicate truth by inception of Membership Functions MFs. Neural networks are clustering nodes in several layers that are affected by each other Simulates biological neural system. The Adaptive Network based Fuzzy Model (ANFM) is compelled by data, which is known as data driven. It utilizes neural network technique for the arrangement of capacity guess issues. Presumption of the race and gender is achieved with only geometric information. A set of the database is defined as a training set. Then it is clustered into a numeric pattern. Fuzzy rules and memberships are optimally generated to designate the date set. Finally, ANFM approximate proper model for new pattern based on the generated rules.

2. Data Set Preparation

We build data set by combining CAS-PEAL [9] of Asian individuals and PICS [10] of Caucasian individuals. Form each database, we chose 600 images under gender and race equality with different three poses. We split the data set into three sub sets; training set, validation set and test set by the ratio 2:1:1. In ANFM training phrase, training set is repeatedly used to adapt the weights, biases and rules. Validation data is also repeatedly used to estimate the non-training performance error of candidate designs. In addition, it is used to indicate and rebuild structures of ANFM design or stop training according the validation error. Test data is used once and only once on the best design to obtain an unbiased estimate for the predicted error of unseen non-training. All types of data set are considered pairs of input and output.

Input data set is a matrix of victors that reflect geometric distances between landmarks. All landmarks are detected by using Restricted Shape and Abbreviated Appearance Model [11]. It gives more accurate landmark detection because using composite cross-correlation methods [12, 13] in matching term. Based on the landmarks, each victor consists of 12 values; 5 values are vertical distances and 7 values are horizontal distances as shown in Figure 1. To overcome the obstacle of face difference size or distance between the camera and subject during capturing, we have identified a unit of measurement for the normalization of values through it. The unit of measurement is a vertical distance between the left corner of left eye and the left corner of lips γ_0. Consequently, all input victor values reflect the aspect ratio geometric distances to the unit of measurement as shown in Equation (1).

$$\gamma'_{vi} = \frac{\gamma_{vi}}{\gamma_0} \tag{1}$$

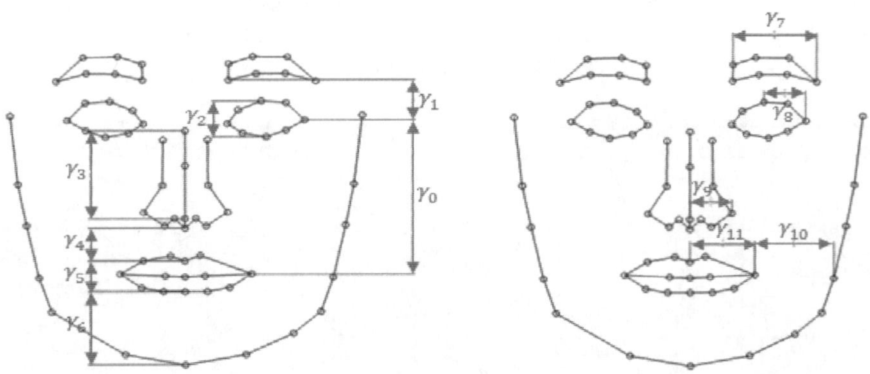

Figure 1. Input data of ANFM from the geometric distances between landmarks.

While horizontal distance between landmarks is changeable through pose, it is useful to mathematically deduce distance value for either frontal or profile pose. Equation (2), Equation (3) and Equation (4) describe the distance through pose θdepending on the value in both frontal and profile pose.

$$\gamma P'_{vi} = \frac{\gamma'_{vi}}{\gamma_0 \sin\theta} - \frac{\gamma F'_{vi}}{\tan\theta} \quad (2)$$

$$\gamma F'_{vi} = \frac{\gamma'_{vi}}{\gamma_0 \cos\theta} - \gamma P'_{vi} \tan\theta \quad (3)$$

$$\gamma'_{vi} = \gamma P'_{vi} \sin\theta + \gamma F'_{vi} \cos\theta \quad (4)$$

3. Forwarding Data through Fuzzy interface system Structure

ANFM is proposed to combine several Fuzzy interface systems (FISs).FIS is structured on several layers. Each layer contains one or more nodes. We utilize FIS for valuating race based on frontal image *FIS_RF*. Other entities of FISs are designed based on the same number of layers but differ in the number of nodes. As shown in Figure 2, there are five layers consist of *FIS_RF*. First layer is input layer with 12 nodes, which is feeding by input data set. Membership function is applied on input data in the second layerinputmf. There are five fuzzy sets for each input node. The rule used to explain membership of an input value ($input_i$) with fuzzy set (A_i) is given by Equation (5) whereas$f(A_i, input_i)$ is a MF.

$$Rule: If\ input_i\ is\ A_i\ \ and \quad then\ inputmf_i\ \ is\ f(A_i, input_i) \quad (5)$$

To simplify, we represented the membership by real values using the premise parameter set as in Equation (6). The parameters set $a_{i,j}$ and $b_{i,j}$ are changeable through training in order to optimize the fuzzy set.

$$inputmf_{i,j} = e^{-\left(\frac{input_i - a_{i,j}}{b_{i,j}}\right)^2} \quad (6)$$

Third layer is considered as rule layer. It has fixed number of nodes correspond to the number of data set. It calculates the strength of the rules for each membership. It is given by Equation (7).

$$Rule_j = \prod_i inputmf_{i,j} \quad (7)$$

The ratio of rules are calculated in the fourth layer outputmf. This layer has the same number of node in previous layer. Each node $outputmf_j$ figures the ratio as shown in Equation (8).

$$outputmf_j = \frac{Rule_i}{\sum_j Rule_j} \quad (8)$$

The last layer has only one node represents the output. This node is connected with the nodes in previous layer by weighted links. The overall output of FIS is given by Equation (9).

$$output = \sum_i w_i \times outputmf_j \tag{9}$$

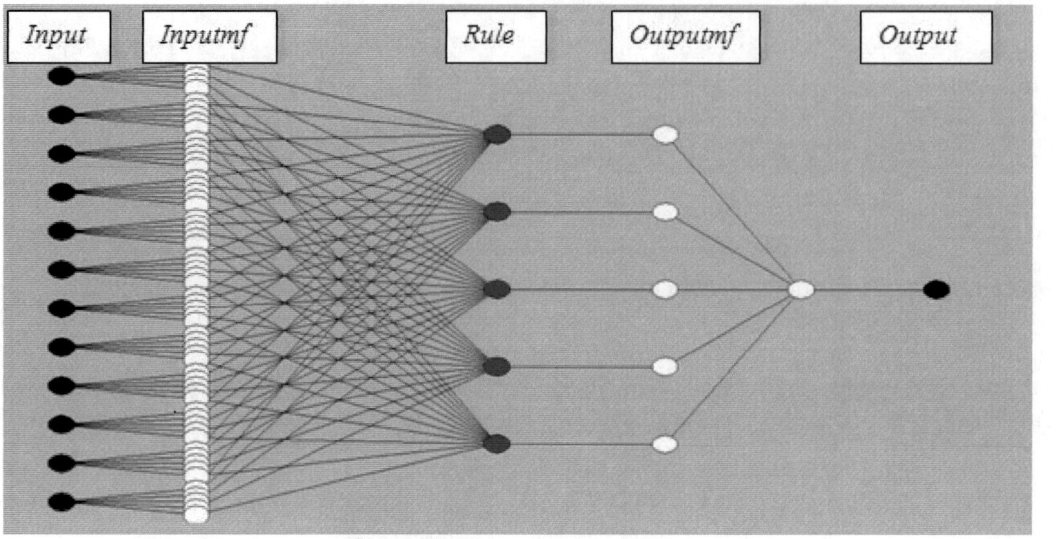

Figure 2. The structure of FIS.

4. Updating Weights and Memberships of FIS

The overall output function of FIS could be represented by a linear combination. In fact, the parameters of the function are the weights and premise parameters; $w_i, a_{i,j}$ and$b_{i,j}$. In training phase, the squared error between the returned output and target is calculated and back propagated. The weights are updated by Equation (10) and Equation (11).

$$\Delta\,(w_i) = w_i \times (output - target)^2 \tag{10}$$

$$new\,(w_i) = old\,(w_i) + \Delta\,(w_i) \tag{11}$$

On other hand, gradient descent method is efficiently used in order to update the premise parameters $a_{i,j}$ and $b_{i,j}$.

5. Performance of FIS

Prudent representing of the target plays an import role to optimizing FIS, in addition to Input data set and training rules. There are four groups of target, covering race and gender. We built FIS_RG_F which are provided by frontal geomantic distances in the input layer. The evaluation of FISs in Figure 3 shows red stars to represent the output and blue dot represents a target sample. The target indicates 1 for female Caucasian, 0.67 for male Caucasian, 0.33 for female Asian and 0 for male Asian. As shown in Figure 3.a, the test set is not accurate especially for Asian gender.

The neuro-fuzzy model FIS_RG_Fcould be divided into two models, FIS_R_F to estimate race and FIS_G_F to estimate gender. Caucasians are indicated by 1 and Asians are indicated by 0 through the target of FIS_R_F. Females are indicated by 1 and Males are indicated by 0 through targets of FIS_G_F. As shown in Figure 3.band Figure 3.c, race is easily estimated while gender still has unsatisfactory results. Therefore, the idea is to rebuild FIS_G_F in addition to providing the input layer with the estimated race target, which is returned by FIS_R_F output. The FIS is indicated by $\overline{FIS_G_F}$and it test results as shown in Figure 3.d.

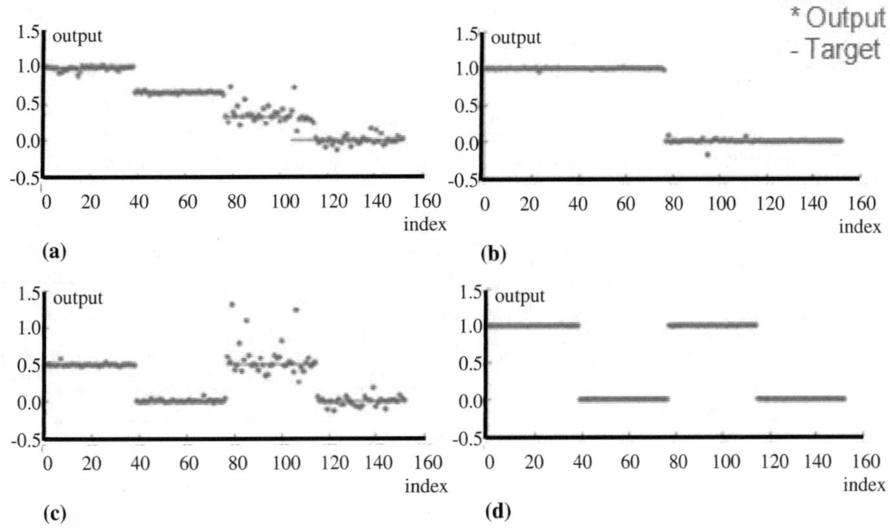

Figure 3. The result of test by training various FISs for estimating race and gender.

Horizontal geometric distances are changeable dramatically between the frontal and profile images of the same subject. Because of that, it is helpful to dedicate FIS to deal with the front pose and the others to deal with the profile pose. Table 1 demonstrates RMS error after training each FIS200 epochs. Results are explaining RMS on training, test, and validation set. We note that the best results are given by the models, which designed to estimate only the race. However, for estimating the gender more accurately, we need a prior result of estimating the race.

.**Table 1.** RMS of various FIS for estimating race and gender after 200 epochs.

FIS	Input data	Output data	RMS		
			Training	Validation	Test
FIS_RG_F	Frontal	Race + Gender	0.005690	0.083657	0.095541
FIS_R_F	Frontal	Race	0.000489	0.009874	0.018080
FIS_G_F	Frontal	Gender	0.019786	0.175846	0.236010
$\overline{FIS\_G_F}$	Frontal+FIS_R_F	Gender	0.000185	0.001047	0.001279
FIS_RG_P	Profile	Race + Gender	0.003857	0.068742	0.086523
FIS_R_P	Profile	Race	0.000353	0.007589	0.014856
FIS_G_P	Profile	Gender	0.013846	0.097854	0.202587
$\overline{FIS\_G_P}$	Profile+FIS_R_P	Gender	0.000154	0.000856	0.001021

Finally, the proposed ANFM structure consists of four sub-FIS model as shown in Figure 4. Two sub FIS are designed to deal with frontal shape to predict the race FIS_R_F and the gender $\overline{FIS\_G_F}$. Other two sub models are designed to deal with profile shape to predict the race FIS_R_P and the gender $\overline{FIS\_G_P}$. The input of FISs responsible for gender prediction is improved by providing the results of race assessment.

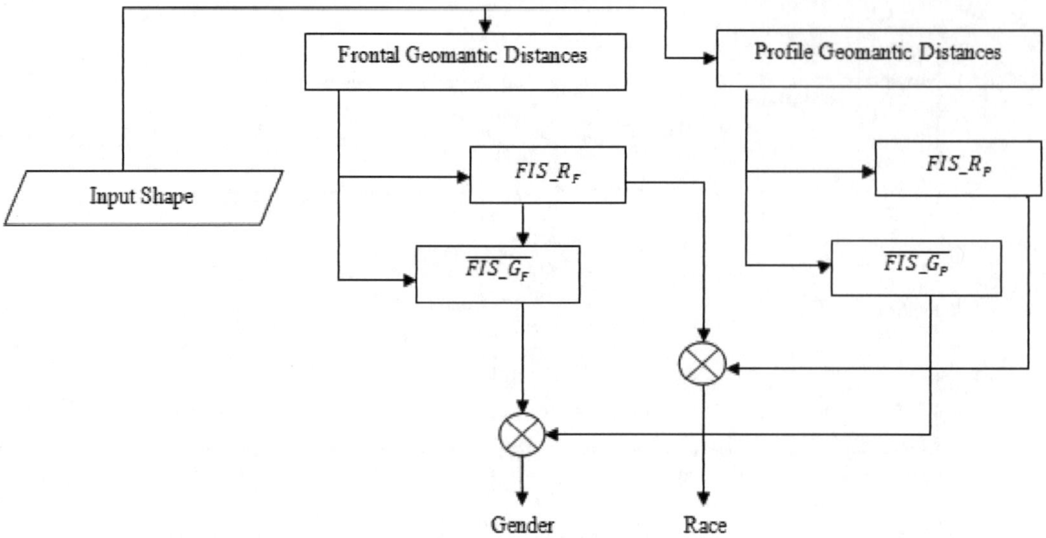

Figure 4. The diagram of the proposed FIS model for estimating race and gender.

6. Discussion and Conclusion

This paper investigates the estimation of race and gender from geometric distances of face across pose face. Based on detecting shape of face images, 12 distances are taken from facial landmarks. Then Frontal and profile distances are calculated in order to provide an input of classification system. The classification process is achieved by ANFM. The used dataset is 600 images taken from CAS and PICS of Asian and Caucasian individuals. After observing several structures of FISs, we concluded that race estimation was easier than gender estimation. For a more accurate gender assessment, we need prior information about the individual's race. The proposed ANFM structure was created based on this conclusion. In this way, we were able to obtain a correct estimate of 100% on the samples of the test and the validation.

However, the previous results were different from some previous study. The present study [4] found that it is easier to classify gender than race on Malaysian dataset. The dataset included samples from three ethnic groups, Malay, Chinese and Indian. In order to compare our results with the study [4], we conducted a study experiment on our database. The rate of classification increased to 97% in the test samples. Furthermore, the race was classified more accurately. The reason for is that gaps between races in our dataset is greater than ethnical groups in Malaysian dataset. Moreover, the use of ANFM is greater in susceptibility than linear models as described in the study [4].

Overall, the results pave way for future work on how to identify races by increasing the size and variety of training samples. Furthermore, it shows the importance of advanced techniques in classify mixed or convergent races.

References

[1] Heo J and Savvides M. 2012 Gender and ethnicity specific generic elastic models from a single 2D image for novel 2D pose face synthesis and recognition. *IEEE transactions on pattern analysis and machine intelligence*, *34* (12), 2341-2350.

[2] Moeini A, Faez K, and Moeini H 2015 Real-world gender classification via local Gabor binary pattern and three-dimensional face reconstruction by generic elastic model *IET Image Processing*, *9*(8), 690-698.

[3] Dantcheva A and Brémond F 2017 Gender estimation based on smile-dynamics. *IEEE Transactions on Information Forensics and Security*, *12*(3), 719-729.

[4] Suit C C 2011 *Gender and ethnic classification of Malaysian face images for face recognition* (Doctoral dissertation, Multimedia University (Malaysia)).

[5] Hsu C W and Lin C J 2002 A comparison of methods for multiclass support vector machines. *IEEE transactions on Neural Networks*, *13*(2), 415-425.

[6] Gao W and Ai H 2009 Face gender classification on consumer images in a multiethnic environment *International Conference on Biometrics* Springer, Berlin, Heidelberg 169-178

[7] Marsh B U, Pezdek K and Ozery D H 2016 The cross-race effect in face recognition memory by bicultural individuals *Acta psychologica*, *169*, 38-44.

[8] Sun G, Zhang G, Yang Y, Bentin S and Zhao L 2014 Mapping the time course of other-race face classification advantage: a cross-race ERP study *Brain topography*, *27*(5), 663-671.

[9] Hancock P 2008 Psychological image collection at stirling (pics). *Web address: http://pics. psych. stir. ac. uk.*

[10] Gao W, Cao B, Shan S, Chen X, Zhou D, Zhang X and Zhao D 2008The CAS-PEAL large-scale Chinese face database and baseline evaluations *IEEE Transactions on Systems, Man, and Cybernetics-Part A: Systems and Humans*, *38*(1), 149-161.

[11] Nanaa K, Rahman M N A, and Rizon, M 2017Restricted Shape and Abbreviated Appearance based Correlation Model for Detecting Facial Landmarks *4th International Conference on Innovations in Engineering, Technology, Computers and Industrial Applications IETCIA-2017.*

[12] Nanaa K, Rizon M, Rahman M N A, Almejrad A, Aziz A Z A and Mohamed S B 2013 Eye detection using composite cross-correlation *American Journal of Applied Sciences*, *10*(11), 1448

[13] Nanaa K, Rizon M and Rahman M N A 2014 Eye detection Using Composite Cross-Correlation form Face Images in Varied Illumination *The International Conference on Artificial Life and Robotics*.

IOP Conf. Series: Materials Science and Engineering **332** (2018) 012029 doi:10.1088/1757-899X/332/1/012029

Implementation of preference ranking organization method for enrichment evaluation (*Promethee*) on selection system of student's achievement

L Karlitasari[1*], D Suhartini[2] and L Nurrosikawati[3]

[1,3]Computer Science Study Program, Pakuan University, Bogor, Indonesia
[2]D3 Program of Management Informatics, Pakuan University, Bogor, Indonesia

*Corresponding author: lita.karlitasari@unpak.ac.id

Abstract. Selection of Student Achievement is conducted every year, starting from the level of Study Program, Faculty, to University, which then rank one will be sent to Kopertis level. The criteria made for the selection are Academic and Rich Scientific, Organizational, Personality, and English. In order for the selection of Student Achievement is Objective, then in addition to the presence of the jury is expected to use methods that support the decision to be more optimal in determining the Student Achievement. One method used is the Promethee Method. Preference Ranking Organization Method for Enrichment Evaluation (Promethee) is a method of ranking in Multi Criteria Decision Making (MCDM). PROMETHEE has the advantage that there is a preference type against the criteria that can take into account alternatives with other alternatives on the same criteria. The conjecture of alternate dominance over a criterion used in PROMETHEE is the use of values in the relationships between alternative ranking values. Based on the calculation result, from 7 applicants between Manual and Promethee Matrices, rank 1, 2, and 3, did not change, only 4 to 7 positions were changed. However, after the sensitivity test, almost all criteria experience a high level of sensitivity. Although it does not affect the students who will be sent to the next level, but can bring psychological impact on prospective student's achievement

1. Introduction

Studying in college, students must be more active in seeking knowledge. Students are people with the ability and opportunity to study in college, so it can be classified as intelligence. Because of the opportunities available, students are expected to be able to act as capable and skilled leaders, either as community leaders or in the workplace. Students are expected to be a dynamic driving force for the process of modernization. The Most outstanding student is a student who successfully achieve high achievement, both academic and non-academic, able to communicate with the Indonesian language and English, be positive, and spirited Pancasila. While students who excel usually have abilities above average in terms of academic, papers, personality, and extracurricular activities and the ability to speak English. There is also the process of selecting outstanding students for this is still subjective. For that field of student affairs and juries who are trusted to test should not make mistakes, this will lead to injustice for participants who are more achievers. It can also be detrimental to the university because with the wrong to send students achievement can reduce the opportunity to score achievement.

2. Methodology

Four criteria used in this study based on maximization rules where the better the topic and the content of scientific work, the more active in the organization, the better the personality and the better the way of communicating each alternative in the English language the better the value will be obtained by the alternative.

Promethee is a ranking stage with the principle of pairwise comparison where each alternative value on the criteria is compared with the value of other alternatives. The PROMETHEE method flow can be seen in Figure 1.

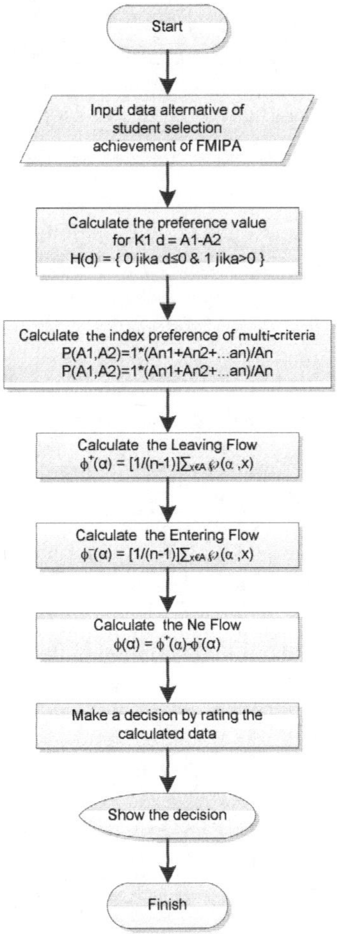

Figure 1. PROMETHEE Method Flow

Criteria and weights used in each criteria for undergraduate students of FMIPA Universita Pakuan degree program are as follows:

Table 1. Criteria and Weights

Criteria	Weights
Scientific Work	30%
Organizational	25%
Personality	20%
English	25%

Alternative data used in this research is the data of student selection of outstanding students of Faculty of Mathematics and Natural Sciences, as follows:

Table 2. Participants of Student Selection of FMIPA

No	Student	Study Program
1	Ikhsan Pratama	Biology
2	Niftrelia SD.	Chemistry
3	Diah Indah P	Mathematics
4	Agvin Pratama	Computer Science
5	Hilda Ayunda	Pharmacy
6	Doni Nugraha	D3 Information System
7	Abdul Basit	D3 Computer Technology

3. Result and Discussion

3.1. Calculate The Preference Value

Calculating the value of preference is done by comparing the value of one alternative with another alternative for each criterion. For example, on the criteria of scientific work we compare the values of A1 with A2, A1 with A3, A1 with A4 to compare A4 with A3. Furthermore, on organizational criteria compare between A1 with A2, A1 with A3 to compare A4 with A3 and so on until the comparison for English ability criteria. There are two provisions to determine the preference value of each alternative comparison: H (d) is 0 if d ≤ 0 and H (d) is 1 if d> 0.

Table 3. Preference Value of Each Criteria

Preference Value of **Scientific Work Criteria**						
(A1,A2) = 0	(A2,A1) = 1	(A3,A1) = 1	(A4,A1) = 1	(A5,A1) = 0	(A6,A1) = 1	(A7,A1) = 1
(A1,A3) = 0	(A2,A3) = 0	(A3,A2) = 0	(A4,A2) = 1	(A5,A1) = 0	(A6,A2) = 0	(A7,A2) = 1
(A1,A4) = 0	(A2,A4) = 0	(A3,A4) = 0	(A4,A3) = 1	(A5,A1) = 0	(A6,A3) = 0	(A7,A3) = 1
(A1,A5) = 1	(A2,A5) = 1	(A3,A5) = 1	(A4,A5) = 1	(A5,A1) = 0	(A6,A4) = 0	(A7,A4) = 0
(A1,A6) = 0	(A2,A6) = 0	(A3,A6) = 0	(A4,A6) = 1	(A5,A1) = 0	(A6,A5) = 1	(A7,A5) = 1
(A1,A7) = 0	(A2,A7) = 0	(A3,A7) = 0	(A4,A7) = 0	(A5,A1) = 0	(A6,A7) = 0	(A7,A6) = 1
Preference Value of **Organizational Criteria**						
(A1,A2) = 0	(A2,A1) = 1	(A3,A1) = 0	(A4,A1) = 0	(A5,A1) = 0	(A6,A1) = 0	(A7,A1) = 0
(A1,A3) = 1	(A2,A3) = 1	(A3,A2) = 0	(A4,A2) = 0	(A5,A1) = 0	(A6,A2) = 0	(A7,A2) = 0
(A1,A4) = 0	(A2,A4) = 1	(A3,A4) = 0	(A4,A3) = 1	(A5,A1) = 1	(A6,A3) = 1	(A7,A3) = 0
(A1,A5) = 1	(A2,A5) = 1	(A3,A5) = 0	(A4,A5) = 1	(A5,A1) = 0	(A6,A4) = 0	(A7,A4) = 0
(A1,A6) = 1	(A2,A6) = 1	(A3,A6) = 0	(A4,A6) = 1	(A5,A1) = 0	(A6,A5) = 0	(A7,A5) = 0
(A1,A7) = 1	(A2,A7) = 1	(A3,A7) = 0	(A4,A7) = 1	(A5,A1) = 1	(A6,A7) = 1	(A7,A6) = 0
Preference Value of **English Criteria**						
(A1,A2) = 0	(A2,A1) = 1	(A3,A1) = 1	(A4,A1) = 0	(A5,A1) = 0	(A6,A1) = 0	(A7,A1) = 0
(A1,A3) = 0	(A2,A3) = 1	(A3,A2) = 0	(A4,A2) = 0	(A5,A1) = 0	(A6,A2) = 0	(A7,A2) = 0
(A1,A4) = 0	(A2,A4) = 1	(A3,A4) = 1	(A4,A3) = 0	(A5,A1) = 0	(A6,A3) = 0	(A7,A3) = 0
(A1,A5) = 1	(A2,A5) = 1	(A3,A5) = 1	(A4,A5) = 1	(A5,A1) = 0	(A6,A4) = 0	(A7,A4) = 0
(A1,A6) = 1	(A2,A6) = 1	(A3,A6) = 1	(A4,A6) = 1	(A5,A1) = 0	(A6,A5) = 0	(A7,A5) = 0
(A1,A7) = 1	(A2,A7) = 1	(A3,A7) = 1	(A4,A7) = 1	(A5,A1) = 1	(A6,A7) = 1	(A7,A6) = 0
Preference Value of **Personality Criteria**						
(A1,A2) = 0	(A2,A1) = 1	(A3,A1) = 0	(A4,A1) = 0	(A5,A1) = 0	(A6,A1) = 0	(A7,A1) = 0
(A1,A3) = 1	(A2,A3) = 1	(A3,A2) = 0	(A4,A2) = 0	(A5,A1) = 0	(A6,A2) = 0	(A7,A2) = 0
(A1,A4) = 0	(A2,A4) = 1	(A3,A4) = 0	(A4,A3) = 1	(A5,A1) = 1	(A6,A3) = 1	(A7,A3) = 0
(A1,A5) = 1	(A2,A5) = 1	(A3,A5) = 0	(A4,A5) = 1	(A5,A1) = 0	(A6,A4) = 0	(A7,A4) = 0
(A1,A6) = 1	(A2,A6) = 1	(A3,A6) = 0	(A4,A6) = 1	(A5,A1) = 0	(A6,A5) = 0	(A7,A5) = 0
(A1,A7) = 1	(A2,A7) = 1	(A3,A7) = 0	(A4,A7) = 1	(A5,A1) = 1	(A6,A7) = 1	(A7,A6) = 0

3.2. Calculate The Preferences Index of Multi-criteria

To calculate the index of criteria preferences can be calculated using the following formula:

$$\wp(a,b) = \sum_{i=1}^{n} \pi \ P_i(a,b) : \forall a,b \in A$$

The result of the calculation of the multicriteria preference index will form a matrix. Here are the results of the calculation of multicriteria preference index based on the calculation of the above preference values:

	A1	**A2**	**A3**	**A4**	**A5**	**A6**	**A7**
A1	-	0	0.5	0	1	0.75	0.75
A2	1	-	0.75	0.75	1	0.75	0.75
A3	0.5	0	-	0.25	0.5	0.25	0.25
A4	0.25	0.25	0.75	-	1	1	0.75
A5	0	0	0.5	0	-	0	0.75
A6	0.25	0	0.5	0	0.25	-	0.75
A7	0.25	0.25	0.25	0	0.25	0.25	-

3.3. Calculate The Leaving Flow

Leaving flow obtained from calculation 1 divided by the number of alternatives minus 1 then multiplied by the number of horizontal preference index (horizontal). To calculate Leaving Flow can be calculated using the formula below:

$$\Phi^{+}(a) = [1/(n-1)] \sum_{x \in A} \wp(a,x)$$

Here are the results of Leaving Flow calculation:

Φ+(A1) = $\frac{1}{7-1}$ $(0 + 0.5 + 0 + 1 + 0.75 + 0.75) = \frac{1}{6}(3) = 0.5$

Φ+(A2) = $\frac{1}{7-1}$ $(1 + 0.75 + 0.75 + 1 + 0.75 + 0.75) = \frac{1}{6}(5) = 0.83$

Φ+(A3) = $\frac{1}{7-1}$ $(0.5 + 0 + 0.25 + 0.5 + 0.25 + 0.25) = \frac{1}{6}(1.75) = 0.29$

Φ+(A4) = $\frac{1}{7-1}$ $(0.25 + 0.25 + 0.75 + 1 + 1 + 0.75) = \frac{1}{6}(4) = 0.67$

Φ+(A5) = $\frac{1}{7-1}$ $(0 + 0 + 0.5 + 0 + 0 + 0.75) = \frac{1}{6}(1.25) = 0.21$

Φ+(A6) = $\frac{1}{7-1}$ $(0.25 + 0 + 0.5 + 0 + 0.25 + 0.75) = \frac{1}{6}(1.75) = 0.29$

Φ+(A7) = $\frac{1}{7-1}$ $(0.25 + 0.25 + 0.25 + 0 + 0.25 + 0.25) = \frac{1}{6}(1.25) = 0.21$

3.4. Calculate The Entering Flow

Entering flow is obtained from calculation 1 divided by the number of alternatives minus 1 then multiplied by the number of multi-criteria preference index vertically (column). To calculate the Entering Flow can be calculated use formula below:

$$\Phi^{-}(a) = [1/(n-1)] \sum_{x \in A} \wp(x,a)$$

Here's the calculation of Entering Flow:

Φ-(A1) = $\frac{1}{7-1}$ $(1 + 0.5 + 0.25 + 0 + 0.25 + 0.25) = \frac{1}{6}(2.25) = 0.375$

Φ-(A2) = $\frac{1}{7-1}$ $(0 + 0 + 0.25 + 0 + 0 + 0.25) = \frac{1}{6}(0.5) = 0.083$

Φ-(A3) = $\frac{1}{7-1}$ $(0.5 + 0.75 + 0.75 + 0.5 + 0.5 + 0.25) = \frac{1}{6}(3.25) = 0.54$

Φ-(A4) = $\frac{1}{7-1}$ $(0 + 0.75 + 0.25 + 0 + 0 + 0) = \frac{1}{6}(1) = 0.17$

IOP Conf. Series: Materials Science and Engineering **332** (2018) 012029 doi:10.1088/1757-899X/332/1/012029

$$\Phi\text{-}(A5) = \frac{1}{7-1}\,(1 + 1 + 0.5 + 1 + 0.25 + 0.25) = \frac{1}{6}(4) = 0.67$$
$$\Phi\text{-}(A6) = \frac{1}{7-1}\,(0.75 + 0.75 + 0.25 + 1 + 0 + 0.25) = \frac{1}{6}(3) = 0.5$$
$$\Phi\text{-}(A7) = \frac{1}{7-1}\,(0.75 + 0.75 + 0.25 + 0.75 + 0.75 + 0.75) = \frac{1}{6}(4) = 0.67$$

3.5. Calculate The Net Flow

Net Flow is the result of the difference of Leaving Flow calculation minus Entering Flow. Here's the calculation:

$$\Phi(A1) = 0.5 - 0.38 = 0.12$$
$$\Phi(A2) = 0.83 - 0.083 = 0.747$$
$$\Phi(A3) = 0.29 - 0.54 = -0.25$$
$$\Phi(A4) = 0.67 - 0.17 = 0.5$$
$$\Phi(A5) = 0.21 - 0.67 = -0.46$$
$$\Phi(A6) = 0.29 - 0.5 = -0.21$$
$$\Phi(A7) = 0.21 - 0.67 = -0.46$$

3.6. Taking Results of Decision

Based on the calculation using PROMETHEE method, the winner of the outstanding students of Faculty of Mathematics and Natural Sciences of Pakuan University is Niftrelia SD from Chemistry Department. The results of student selection of outstanding students in 2014 have 3 appropriate rankings, namely rank 1, 2 and 3, where the level of conformity of the results of calculation of PROMETHEE with manual calculation obtained from the data according to divided by the number of data used and multiplied by 100%. The number of alternative rankings on PROMETHEE according to the manual calculation is 3 then ((3/7) * 100) is 48.2%.

References

[1] Chandra, Andrison Bagus 2002 *Model Pengambilan Keputusan Untuk Mengevaluasi Kinerja Subkontrak Di PT. Astra Honda Motor Melalui Pendekatan AHP Dan PROMETHEE*T (Tesis-Program Pasca Sarjana FT UI, Depok).

[2] J. Lemantara, N.A. Setiawan, and M.N. Aji 2013 *Rancang Bangun Sistem Pendukung Keputusan Pemilihan Mahasiswa Berprestasi Menggunakan Metode AHP dan Promethee* _Jurnal Nasional Teknik Elektro dan Teknik Informatika. Vol.2, No.4, Februari 2013. 20-28. ISSN 2301-4156) PP-20-28.

[3] Nur, Muhammad. 2014 *Studi Komparasi Metode Promethee dan TOPSIS Untuk Memberikan Solusi Terbaik dan Pengambilan Keputusan Menentukan Tingkat Obesitas* (Universitas Pendidikan Indonesia. Skripsi. Jurusan Ilmu Komputer FMIPA UPI, Bandung).

[4] Yulianto, Aan 2014 *Sistem Penunjang Keputusan Pemilihan Mahasiswa Berprestasi Di Fakultas Matematika Dan Ilmu Pengetahuan Alam Universitas Negeri Yogyakarta Dengan Metode AHP Dan TOPSIS* (Skripsi. Jurusan Matematika FMIPA UNY, Yogyakarta).

IOP Conf. Series: Materials Science and Engineering **332** (2018) 012030 doi:10.1088/1757-899X/332/1/012030

Derivative free Davidon-Fletcher-Powell (DFP) for solving symmetric systems of nonlinear equations

M Mamat[1*], M K Dauda[1,3], M A bin Mohamed[1], M Y Waziri[2], F S Mohamad[1], H Abdullah[4]

[1]Faculty of Informatics and Computing, Universiti Sultan Zainal Abidin, Gong Badak, Terengganu, Malaysia.
[2]Department of Mathematical Science, Bayero University, Kano, Nigeria.
[3]Department of Mathematics and Statistics, Kaduna Polytechnic, Kaduna, Nigeria.
[4]Department of Mathematics and Computer Science, Sule Lamido University, Nigeria.

*Corresponding author: musmat567@gmail.com

Abstract: Research from the work of engineers, economist, modelling, industry, computing, and scientist are mostly nonlinear equations in nature. Numerical solution to such systems is widely applied in those areas of mathematics. Over the years, there has been significant theoretical study to develop methods for solving such systems, despite these efforts, unfortunately the methods developed do have deficiency. In a contribution to solve systems of the form $F(x) = 0, x \in R^n$, a derivative free method via the classical Davidon-Fletcher-Powell (DFP) update is presented. This is achieved by simply approximating the inverse Hessian matrix with Q_{k+1}^{-1} to $\theta_k I$. The modified method satisfied the descent condition and possess local superlinear convergence properties. Interestingly, without computing any derivative, the proposed method never fail to converge throughout the numerical experiments. The output is based on number of iterations and CPU time, different initial starting points were used on a solve 40 benchmark test problems. With the aid of the squared norm merit function and derivative-free line search technique, the approach yield a method of solving symmetric systems of nonlinear equations that is capable of significantly reducing the CPU time and number of iteration, as compared to its counterparts. A comparison between the proposed method and classical DFP update were made and found that the proposed methodis the top performer and outperformed the existing method in almost all the cases. In terms of number of iterations, out of the 40 problems solved, the proposed method solved 38 successfully, (95%) while classical DFP solved 2 problems (i.e. 05%). In terms of CPU time, the proposed method solved 29 out of the 40 problems given, (i.e.72.5%) successfully whereas classical DFP solves 11 (27.5%). The method is valid in terms of derivation, reliable in terms of number of iterations and accurate in terms of CPU time. Thus, suitable and achived the objective.

1. Introduction

Consider the system of nonlinear equations

$$F(x) = 0, x \in R^n \tag{1}$$

where the vectors $F : R^n \to R^n$ is a nonlinear mapping assumed to satisfy the following assumptions: (i) There exists an $x^* \in R^n$ such that $F(x^*) = 0$ (ii) F is a continuously differentiable mapping in a neighborhood of x^* of the system and (iii) The Jacobian matrix of F at x given by $J(x) = F'(x)$ is symmetric. Among the many iterative methods for solving (1), Newton's method is the prominent one [1,2,11], it generates a sequence of iterates x_k from a given initial point x_0 via

$$x_{k+1} = x_k - (F'(x_k))^{-1}F(x_k), k = 0,1,2,\cdots \quad (2)$$

where $F'(x_k)$ is the Jacobian matrix of F at x_k.

The Newton method has obvious shortcomings. Computing Jacobian matrix sometimes is difficult, in some cases is even impossible [9,15], in order to alleviate this deficiency, a derivative free method has been devised. This is achieved by modifying the classical Davidon-Fletcher-Powell (DFP). In the proposed method, the H_k of DFP is approximated with $\theta_k I$ where I is the Identity matrix and $\theta_k$the acceleration parameter. Throughout this article, we always assume that the problem (1) is symmetric and can be converted to an equivalent global optimization problem [8,10].

$$\min_{x \in R^n} f(x) \quad (3)$$

with function f defined by

$$f(x) = \frac{1}{2}\|F(x)\|_2^2 \quad (4)$$

Conjugate Gradient methods for solving nonlinear systems of equations generates an iterative points $x_k$from initial given point $x_0$using

$$x_{k+1} = x_k + \alpha_k d_k \quad (5)$$

where $\alpha_k > 0$ is attained via line search and direction $d_k$are obtained using

$$d_k = \begin{cases} -F(x_k) \ \ if\ k = 0 \\ -F(x_{k-1}) + \beta_k d_{k-1} \quad if\ k \geq 1 \end{cases} \quad (6)$$

where $\beta_k$is term as conjugate gradient parameter[514,18]. In the next section, the derivation of the proposed method is presented. In Section 3, the numerical results is established while section 4 presents the conclusion and future work.

2. Derivative free Davidon-Fletcher-Powell (DFDFP) Update

The Davidon-Fletcher-Powell (DFP) Update is an iterative method that generates a sequence of $\{x_n\}_{k\geq 0}$ from a given initial guess $x_0$via the following

$$x_{k+1} = x_k - \alpha_k B_k^{-1}F(x_k), k = 0,1,2,\cdots \quad (7)$$

where $\alpha_k > 0$ is a step length determined by any suitable line search [6,7]. Recall, from DFP formula and using Sherman-Morrison-Woodbury, the inverse hessian approximation $H_k$that corresponds to the DFP update is given by

$$H_{k+1} = H_k - \frac{H_k y_k y_k^T H_k}{y_k^T H_k y_k} + \frac{s_k y_k^T}{y_k^T s_k} \quad (8)$$

$H_k$is updated at each iteration for $k = 0,1,2,\cdots$ .The updated matrix$H_{k+1}$is chosen in such a way that it satisfies the secant equation below

$$s_k = H_{k+1}y_k, \text{ with } s_k = x_{k+1} - x_k \text{and } y_k = F(x_{k+1}) - F(x_k) \quad (9)$$

To improve the efficiency of the classical DFP, the inverse hessian $H_k$is approximated with $\theta_k I$, where $\theta_k = \frac{y_k^T y_k}{y_k^T s_k}$and I is the identity matrix. This makes it derivative free and transforms (8) to

$$H_{k+1} = \theta_k I - \frac{\theta_k I y_k y_k^T \theta_k I}{y_k^T \theta_k I y_k} + \frac{s_k y_k^T}{y_k^T s_k} \quad (10)$$

equation (10) can also be written as

$$H_{k+1} = \theta_k - \frac{\theta_k y_k y_k^T \theta_k}{y_k^T \theta_k y_k} + \frac{s_k y_k^T}{y_k^T s_k} \quad (11)$$

To ensure good approximation, pre-multiply (11) by $g(x_{k+1})$to obtain

$$H_{k+1}g(x_{k+1}) = \theta_k g(x_{k+1}) - \frac{\theta_k y_k y_k^T \theta_k g(x_{k+1})}{y_k^T \theta_k y_k} + \frac{s_k y_k^T g(x_{k+1})}{y_k^T s_k} \quad (12)$$

the direction $H_{k+1}g(x_{k+1})$in (12) can be rewritten as

$$d_{k+1} = -H_{k+1}g(x_{k+1}) \tag{13}$$

Hence from (12) and (13) we have

$$d_{k+1} = -\theta_k g(x_{k+1}) + \frac{\theta_k y_k y_k^T \theta_k g(x_{k+1})}{y_k^T \theta_k y_k} - \frac{s_k y_k^T g(x_{k+1})}{y_k^T s_k} \tag{14}$$

To further ensure derivative free feature of the proposed method, the following equation is used as in [3].

$$g(x_k) \approx \nabla f(x_k) = \frac{F(x_k + \alpha_k F(x_k)) - F(x_k)}{\alpha_k} \tag{15}$$

Now the new direction is obtained via the following

$$d_k = \begin{cases} -F(x_k) \ if \ k = 0 \\ -\theta_k g(x_{k+1}) + \frac{\theta_k y_k y_k^T \theta_k g(x_{k+1})}{y_k^T \theta_k y_k} - \frac{s_k y_k^T g(x_{k+1})}{y_k^T s_k} \quad if \ k \geq 1 \end{cases} \tag{16}$$

where

$$\theta_k = \frac{y_k^T y_k}{y_k^T s_k}, s_k = x_{k+1} - x_k \text{ and } y_k = F(x_{k+1}) - F(x_k)$$

To avoid computing Jacobian Matrix, a non-monotone and derivative free line search from the work of Li and Fukushima [3] is used to compute step size α_k.The line search avoid the necessity of descent directions and guarantee that each iteration is well defined.

The idea is that, the parameter are considered to be $\sigma_1 > 0, \sigma_2 > 0, \sum\eta > 0, r \in \infty(0,1)$ be constants and $\eta_k$be a given positive sequence such that $\eta_k < \infty$, let $\alpha_k = Max\{1, r^k\}$ that satisfy

$$f(x_k + \alpha_k d_k) - f(x_k) \leq -\sigma_1 ||\alpha_k F(x_k)||^2 - \sigma_2 ||\alpha_k d_k||^2 + \eta_k F(x_k) \tag{17}$$

Finally, using the above procedure, the following is the algorithm for the new update.

Algorithm: Derivative free Davidon-Fletcher-Powell (DFDFP) Update.

Step 1: Given $x_0, \alpha > 0, \sigma \epsilon (0,1)$ and $\epsilon > 0$ compute $d_0 = -F(x_0)$, set $k = 0$.

Step 2: Compute $F(x_k)$ and test the stopping criterion, i.e. $||F(x_k)|| \leq \epsilon$, if yes, then stop, otherwise continue with next step.

Step 3: Compute $\alpha_k$by using the line search using (17)

Step 4: Compute$\theta_k = \frac{y_k^T y_k}{y_k^T s_k}$,

Step 5: Compute $x_{k+1} = x_k + \alpha_k d_k$

Step 6: Compute search direction using (16)

Step 7: Set $k = k + 1$ and go to step 2.

3. Numerical Results

This section gives the numerical results of the proposed method. The computational experiment is based on number of iterations and CPU time. The proposed method is denoted as DFDFP, and its performance is compared with the Classical DFP [7,16], this is achieved by solving 40 benchmark problems with their respective initial points using five (5) different dimensions ranging from 10 to 5000. The comparison of the performance between the methods using the benchmarks problems in [12,13], was based on the performance profile presented by Dolan and More [4]. The performance profile $P : R \rightarrow [0,1]$ is defined as follows: Let P and S be the set of problems and set of solvers respectively. For n_s solvers and n_p problems, and for each problem $p \in P$ and for each solver $s \in S$, we define $t_{p,s}$: =(number of iterations required to solve problem p by solver s). The performance ratio is given by

$$r_{p,s} := t_{p,s}/min\{t_{p,s}\}.$$

Then the performance profile is defined by

$$P(\tau) := \frac{1}{n_p} size\{p \in P: r_{p,s} \leq \tau\},$$

for all $\tau \in R$ where $P(\tau)$ is the probability for solver $s \in S$ that a performance ratio $r_{p,s}$is within a factor $\tau \in R$ of the best possible ratio.The code for the proposed method was done using MATLAB 7.1, R2009b programming environment and run on a personal computer 2.4GHz, Intel (R) Core (TM) i7-5500U CPU processor, 4GB RAM memory and on windows XP operator. Both the methods was implemented with the same parameters as $\alpha_1 = 0.01, r = 0.2$, $\sigma_1 = \sigma_2 = 10^{-4}$, and $\eta_k = \frac{1}{\{k+1\}^2}$.The search is stopped if:(i) $\| F(x_k) \| < \epsilon$ with $\epsilon < 10^{-4}$ or (ii) The total number of iteration exceeds 1000. The numerical results of the comparison between the proposed method and the result in [7,16] are presented in table 1 and 2. The meaning of each column in the tables are stated as follows, "P" : stands for Benchmark problem, "ISP" : stands for initial starting points, "n" : stands for dimension of the test problems, "Iter" : the total number of iterations and "CPU" : the CPU time in seconds. In particular problem i, the DFDFP performs better if the number of iteration (iter) or the CPU time in seconds (Time) is less than the number of iteration or the CPU time corresponding to the other methods respectively.

Table 1. Numerical results of problem 1-5 based on iter and CPU

			DFDFP		Classical DFP	
Prob	**ISP**	**Dim**	**Iter**	**CPU**	**Iter**	**CPU**
1	$(0.5,0.5,\cdots,0.5)^T$	10	7	0.001142	8	0.017238
		100	7	0.001654	10	0.001905
		500	7	0.002712	11	0.002654
		1000	7	0.002367	11	0.003932
		10000	8	0.030609	12	0.030811
2	$(0.5,0.5,\cdots,0.5)^T$	10	7	0.002469	10	0.018131
		100	7	0.001869	12	0.001844
		500	8	0.003267	13	0.004150
		1000	8	0.005055	13	0.005513
		10000	8	0.036614	14	0.038679
3	$(0.5,0.5,\cdots,0.5)^T$	10	15	0.035759	15	0.002879
		100	169	0.042314	2	0.000905
		500	14	0.009544	36	0.012241
		1000	14	0.016352	38	0.025325
		10000	22	0.178635	43	0.182390
4	$(0.5,0.5,\cdots,0.5)^T$	10	5	0.014770	9	0.002178
		100	6	0.001694	11	0.001897
		500	6	0.002481	11	0.002557
		1000	6	0.003517	12	0.004300
		10000	6	0.026904	13	0.032316

Table 2. Numerical results of problem 5-8 based on iter and CPU

			DFDFP		Classical DFP	
Prob	**ISP**	**Dim**	**Iter**	**CPU**	**Iter**	**CPU**
5	$(0.5,0.5,\cdots,0.5)^T$	10	10	0.006476	11	0.001915
		100	12	0.002042	13	0.002043
		500	13	0.003809	16	0.002710
		1000	14	0.005711	17	0.004552
		10000	16	0.037686	26	0.052033
6	$(0.5,0.5,\cdots,0.5)^T$	10	8	0.013002	16	0.003420
		100	8	0.001672	18	0.003797
		500	8	0.002510	19	0.004238
		1000	8	0.003307	19	0.008462
		10000	9	0.024195	20	0.059023
7	$(0.5,0.5,\cdots,0.5)^T$	10	3	0.011541	9	0.001636
		100	4	0.001315	10	0.001493
		500	4	0.001569	11	0.002061
		1000	4	0.002279	11	0.002934
		10000	4	0.021024	13	0.021863
8	$(0.5,0.5,\cdots,0.5)^T$	10	7	0.001941	11	0.013392
		100	7	0.001571	12	0.001634
		500	7	0.002544	13	0.001914
		1000	8	0.002835	13	0.003498
		10000	8	0.020552	15	0.021050

From the above tables, the proposed methods solves 38 problems effectively scoring 95% in in terms of number of iterations, while the classical DFP scored 2% only. In terms of CPU times, the proposed method outperformed the classical DFP by scoring 72,5% of the solutions, i.e.29 out of 40 problems was solved effectively by the proposed method, while 27.5% is solved by the classical DFP.

4. Conclusion and Future work

In this article, a derivative free method for solving symmetric nonlinear systems of equations is given. The method achieved its objectives fully, i.e. it is derivative free, effective in terms of number of iterations and faster in terms of CPU time. Thus, DFDFP is a very good alternative for solving symmetric nonlinear systems of equations. For further research work, the DFDFP can be extended to non-smooth nonlinear systems of equations. Also, it can be applied to solve a real life problems.

Acknowledgments

The authors would like to thank the administration of Universiti Sultan ZainalAbidin (UniSZA) for funding this research partially under the fundamental research grant Scheme FRGS/1/2015/ICT03/UniSZA/02/1.

References

[1] Burden, R.L., Faires, J.D 2005 Numerical Analysis *Numerical Solutions of Nonlinear Systems of Equations*, 597-640.

[2] Chong, E.K.P. and Zak, S.H. 2005) An introduction to optimization (2nd ed) *John Wiley and Sons*, New York.

[3] D. H. Li and M. Fukushima 2000 A Derivative-free line search and global convergence of Broyden-like methods for nonlinear equations, *Optimization Methods and Software* 13, 181-201.

[4] E. Dolan and J. More 2002 Bench-marking optimization software with performance profiles, *Math. Program. Ser. A*, 91, 201-213.

[5] Gonglin Yuana and Maojun Zhang 2015 A three-terms Polak-Ribire-Polyak conjugate gradient algorithm for large-scale nonlinear equations, *Journal of Computational and Applied Mathematics* 286, 186-195.

[6] Griva, I., Nash, S.G. and Sofer, A. 2009 Linear and nonlinear optimization (2nd ed). *Society for Industrial and Applied Maths, Philadelphia.*

[7] J. Nocedal and S.J. Wright 2006 Numerical Optimization, 2nd ed., *Springer, New York.*

[8] Jinkui Liu and Shengjie Li 2015 Spectral DY Type Projection Method for Nonlinear Monotone Systems of Equations, *Journal of Computation of Mathematics*, 4, 341-354.

[9] M. Y. Waziri, H. A. Aisha and Mustafa Mamat 2014 A Newton's-Like Method with Extra Updating Strategy for Solving Singular Fuzzy Nonlinear Equations *Applied Mathematical Sciences* 142, 7047-7057.

[10] M. Fatemi 2016 A new efficient conjugate gradient method for unconstrained optimization, *Journal of Computational and Applied Mathematics.*

[11] Mohd Rivaie, Mustafa Mamat and Abdelrahman A. Bashar 2015 A new class of nonlinear conjugate gradient coefficients with exact and inexact line searches, Applied Mathematics and Computation 268, 1152-1163.

[12] Mustafa Mamat, M. K. Dauda, M. Y. Waziri, Fadhilah Ahmad, and Fatma Susilawati Mohamad 2016 Improved Quasi-Newton method via PSB update for solving systems of nonlinear equations, *AIP Conference Proceedings* 1782, 030009; doi:10.1063/1.4966066.

[13] M. K. Dauda, Mustafa Mamat, M. Y. Waziri, Fadhila Ahmad and Fatma Susilawati Mohamad 2016 Inexact CG-Method via SR1 Update for Solving Systems of Nonlinear Equations, *Far East Journal of Mathematical Sciences(FJMS)* 100, Issue 11, 1787-1804.

[14] Neculai Andrei 2013 A simple three-term conjugate gradient algorithm for unconstrained Optimization *Journal of Computational and Applied Mathematics* 241, 19-29.

[15] W. Zhou and D. Shen 2014 An inexact PRP conjugate gradient method for symmetric nonlinear equations, *Numerical Functional Analysis and Optimization*, 35, (3) 370-388.

[16] Y. Ding, E. Lushi and Q. Li, Investigation of Quasi-Newton Methods for Unconstrained Optimization, Simon Fraser University, Canada.

[17] Y.U. Gaohang 2010 Derivative-Free for Method Solving Large-Scale Nonlinear Systems of Equations, *Journal of Management Optimization* 6, 149-160.

[18] Yasushi Narushima and Hiroshi Yabe 2012 Conjugate Gradient Methods Based on Secant Conditions that Generate Descent Search Directions for Unconstrained Optimization, Journal of Computational and Applied Mathematics, 236, 4303-4317.

[19] Zhifeng Dai 2016 Comments on a new class of nonlinear conjugate gradient coefficients with global convergence properties, *Applied Mathematics and Computation* 276, 297-300.

Rabies epidemic model with uncertainty in parameters: crisp and fuzzy approaches

M Z Ndii[1*], Z Amarti[2], E D Wiraningsih[3], A K Supriatna[2]

[1]Department of Mathematics, Faculty of Science and Engineering, University of Nusa Cendana, Kupang-NTT, Indonesia

[2] Department of Mathematics, Faculty of Mathematics and Natural Sciences, University of Padjadjaran, Bandung-West Java, Indonesia

3. Department of Mathematics, Faculty of Mathematics and Natural Sciences, State University of Jakarta, DKI-Jakarta, Indonesia

*Corresponding author: meksianis.ndii@staf.undana.ac.id

Abstract. A deterministic mathematical model is formulated to investigate the transmission dynamics of rabies. In particular, we investigate the effects of vaccination, carrying capacity and the transmission rate on the rabies epidemics and allow for uncertainty in the parameters. We perform crisp and fuzzy approaches. We find that, in the case of crisp parameters, rabies epidemics may be interrupted when the carrying capacity and the transmission rate are not high. Our findings suggest that limiting the growth of dog population and reducing the potential contact between susceptible and infectious dogs may aid in interrupting rabies epidemics. We extend the work by considering a fuzzy carrying capacity and allow for low, medium, and high level of carrying capacity. The result confirms the results obtained by using crisp carrying capacity, that is, when the carrying capacity is not too high, the vaccination could confine the disease effectively.

Keywords: Rabies, mathematical model, crisp and fuzzy approaches, uncertainty

1. Introduction

Rabies is a zoonotic disease that is transmitted to humans via an animal bite in particular dogs. It is estimated that approximately 55,000 cases happens annually, with higher number of cases in Asia and Africa [1]. Rabies is also a problem in several places in Indonesia such as Bali and Flores islands [2, 3]. The fatality rate can reach 100 % once symptoms of disease develop.

A number of control measures such as mass vaccination and culling have been implemented to stop rabies epidemics, but they are not perfectly effective. For example, in Flores Island, Indonesia, although vaccination has been implemented [3], rabies cases are still found. This is likely due to low vaccination uptake level [4]. Therefore, this paper aims to investigate the dynamics of rabies transmission with vaccination strategy.

A mathematical model can be used to investigate the dynamics and the effectiveness of the intervention on reducing disease transmission [5-13]. A number of mathematical models have been developed to understand the rabies epidemics [14-18]. Huo et al. used a mathematical model to investigate rabies epidemics and found that vaccination rate is one of the key components determining rabies transmission [17]. Zinsstag et al. developed a mathematical model to investigate the transmission dynamics. They found that a 70% vaccination coverage level is sufficient to interrupt rabies epidemics [18]. In this paper, we develop a mathematical model to investigate rabies epidemics by including carrying capacity and allowing intrinsic stochasticity in the transmission rate. We explore the effects of carrying capacity and the transmission rate on rabies epidemics. We extend the work by considering fuzzy carrying capacity to investigate the effect of uncertainty on the dynamics of rabies transmission.

This paper is organized as follows. Section 2 presents a formulation of mathematical model. Section 3 presents the parameter exploration, and finally results and discussion are presented.

2. Model Formulation

A deterministic ordinary differential equation model is developed for the transmission of rabies for dogs. The dog's population is divided into four compartments: Susceptible (*S*), Exposed but not yet infectious (*E*), Infectious (*I*) and Vaccinated (*V*). In this model, dogs remain infectious except if they are vaccinated. Therefore, no recovered class is included.

Dogs are born susceptible and their growth is limited by carrying capacity, *C*. Rabies is transmitted to susceptible dogs at rate $\boldsymbol{\beta}$ when there is a contact between infectious and susceptible dogs. The parameter $\boldsymbol{\beta}$ is the transmission rate, which is the contact rate between susceptible and infectious dogs, and the probability that contact successfully transmits disease. The susceptible dogs get vaccinated at a rate *g* and die due to natural death at rate *d*. The exposed dogs progress to infectious class at rate $\boldsymbol{\gamma}$, and die because of natural death rate, $\boldsymbol{d}$, and disease related death, $\boldsymbol{d_r}$. The vaccinated dogs loss immunity at rate *l*, which is then re-susceptible to the virus.

The model is then governed by the following system of differential equations:

$$\frac{dS}{dt} = BN(1 - N/C) - \lambda S - dS - g\epsilon S + lV \qquad (1)$$

$$\frac{dE}{dt} = \lambda S - \gamma E - dE - g\epsilon E \qquad (2)$$

$$\frac{dI}{dt} = \gamma E - (d + d_r)I \qquad (3)$$

$$\frac{dV}{dt} = g\epsilon(S + E) - lV - dV \qquad (4)$$

where $\boldsymbol{\lambda} = \boldsymbol{\beta I}$, and the total dog population is $\boldsymbol{N} = \boldsymbol{S} + \boldsymbol{E} + \boldsymbol{I} + \boldsymbol{V}$.

3. Crisp Approach

Since most parameter values of the model are largely unknown, we explore different parameter values to understand their effects on the dynamics of rabies epidemics. First, we explore different values of carrying capacity and vaccination coverage level. Then, we investigate the effects of the transmission rate on the rabies epidemics by allowing intrinsic stochasticity in the transmission rate. In our exploration, we use carrying capacity 5×10^5, which is around two times the number of dogs in Flores island, Indonesia [3], 7.5×10^5 and 10^6, and vaccination coverage level of 50 % and 75 %. The parameter values are given in Table 1.

In our investigation of stochasticity in the transmission rate, we assigned 1000 different sets of normally distributed transmission rates as

$$\beta = \beta_0 + \sigma_\beta N(0,1)$$

where β_0 is the average transmission rate, σ_β is the variance, and $N(0,1)$ is a Gaussian random variable with a mean of zero and variance equal to one. The values of the parameters β_0 and σ_β are 10^{-5} and 10^{-6}, respectively.

Table 1. *Parameter descriptions,* value and sources. The transmission rate, β, carrying capacity, C, and vaccination coverage level, ϵ, are explored in this paper

Parameters	Description	Value	Unit	Sources
β	Transmission rate	10^{-5}	year^{-1}	Assumed
B	Birth rate	0.013×52	year^{-1}	[18]
C	Carrying capacity	5×10^5	N/A	see text for explanation
d	Natural death rate	0.006638×52	year^{-1}	[18]
d_r	Disease-induced death rate	1	year^{-1}	[19]
g	Vaccine efficacy	0.94	N/A	[18]
ϵ	Vaccination coverage level	0.5	year^{-1}	Explored
l	Vaccination loss rate	0.0081×52	year^{-1}	[18]
γ	Progress from exposed to infectious class	$12/3$	year^{-1}	[1]

In exploring the effects of carrying capacity and vaccination coverage level on rabies epidemics, we found that when carrying capacity is higher, an outbreak occurs although the vaccination coverage level is high (see Figure 1). For example, if carrying capacity is 7.5×10^5 and vaccination coverage level is 75%, an outbreak can still occur, with around 1.98×10^4 cases. The same behaviour is found when carrying capacity is 10^6. A 50% vaccination coverage level is sufficient to stop rabies epidemics if the carrying capacity is low (5×10^5). If the carrying capacity is higher, a higher vaccination coverage level is not sufficient to stop rabies epidemics because of higher number of dogs the population. Carrying capacity affects the growth of dog population. If carrying capacity is high, it is likely that there will be many dogs in the population. Therefore our findings imply that limiting the growth of dogs is necessary in order to interrupt rabies epidemics.

When investigating the effect of the transmission rate with intrinsic stochasticity, we are interested in situations where an outbreak occurs. Therefore, in our exploration, we use carrying capacity of 7.5×10^5 and 10^6. We found that the average outbreak size is $1.9 \times 10^4 \pm 1.96 \times 140.63$ if $K = 7.5 \times 10^5$ and the vaccination coverage level is 50%. If $K = 10^6$ and the vaccination coverage level is 50%, the outbreak size is $3.6341 \times 10^4 \pm 1.96 \times 78.18$. The outbreak sizes are of similar range as when the transmission rate is fixed. This may suggest the role of carrying capacity determining rabies epidemics is a bit stronger than the transmission rate. Table 2 and 3 show the number of infected dogs at time $t = 50$ (Table 2) and $t = 100$ (Table 3) for various values of carrying capacity with the crisp classification (low, medium, and high carrying capacity).

Table 2. The number of infected individuals with different carrying capacity (C) and vaccine scenarios at t = 50. The values for low (C_{low}), medium (C_{medium}) and high (C_{high}) carrying capacity are 5×10^5, 7.5×10^5, and 10^6, respectively.

	Different Vaccine Scenarios		
	No Vaccination	50%	75%
C_{low}	1.7297×10^4	0.0908	1.8×10^{-5}
C_{medium}	3.0203×10^4	1.9826×10^4	1.788×10^3
C_{high}	3.9322×10^4	3.6835×10^4	2.9270×10^4

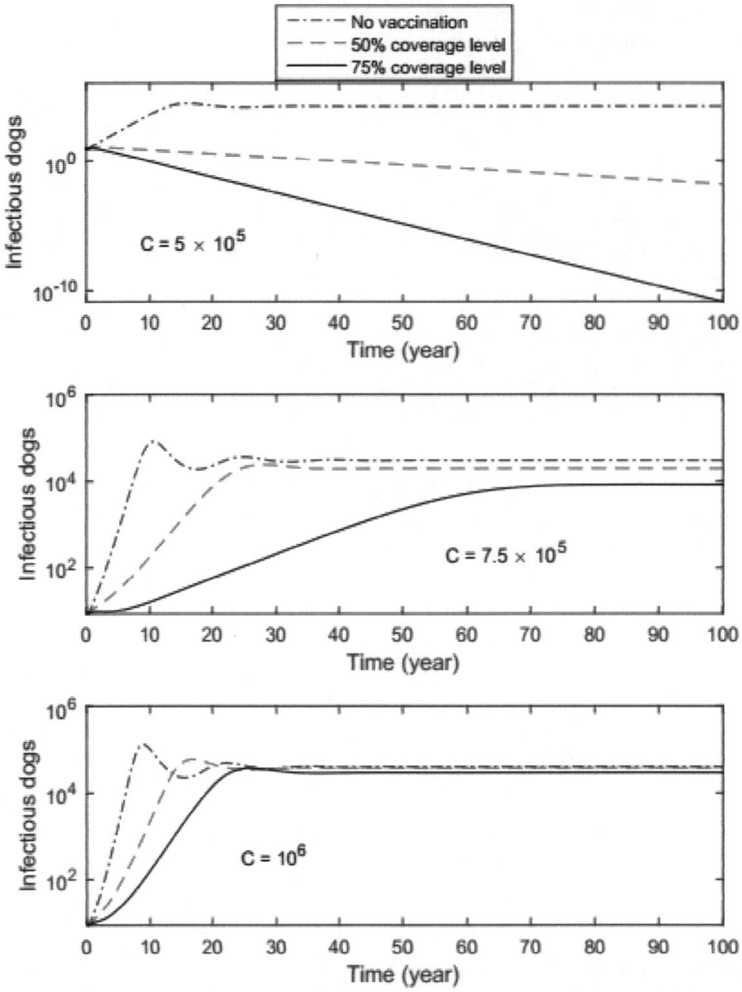

Figure 1. Plot of infectious dogs vs. time with different vaccination coverage levels. The top, middle and bottom plots are for $C = 5 \times 10_5$, $7{:}5 \times 10_5$, and 10_6, respectively.

Table 3. The number of infected individuals with different carrying capacity (C) and vaccine scenarios at t = 100. The values for low (C_{low}), medium (C_{medium}) and high (C_{high}) carrying capacity are 5×10^5, 7.5×10^5, and 10^6, respectively.

	Different Vaccine Scenarios		
	No Vaccination	50%	75%
C_{low}	1.7269×10^4	0.0269	6.3×10^{-11}
C_{medium}	3.0347×10^4	1.979×10^4	8.350×10^3
C_{high}	3.9018×10^4	3.6836×10^4	2.9162×10^4

4. Fuzzy Approach

The number of infected individuals with crisp carrying capacity and initial conditions $I = 50$, $S = 1.5 \times 10^5$ and $V = 0$= 50, are given in Tables 2 and 3. In reality the exact value of carrying capacity is strongly uncertain and hence it is difficult to be determined. In the following discussion we assume that there are three different levels of carrying capacity: low, medium, and high. However, there is no clear way to classify the numerical values of these carrying capacities into the low, medium, and high category. The numerical value of the boundary between the carrying capacity category is blur. To model the levelling of the carrying capacity, we use the fact that the carrying capacity level of 7.5×10^5 is regarded as the medium carrying capacity level. We then use triangular and trapezoidal fuzzy number to make classification. The membership function of the Triangular and Trapezoidal fuzzy number used in the simulation (in 10^5 individuals) are

$$\mu_{low}(C) = \begin{cases} 1, & if\ C \leq 5 \\ \dfrac{C - 7.5}{5 - 7.5}, & if\ 5 < C \leq 7.5 \\ 0, & Otherwise \end{cases}$$

$$\mu_{medium}(C) = \begin{cases} 0, & if\ C \leq 5 \\ \dfrac{C - 5}{7.5 - 5}, & if\ 5 < C \leq 7.5 \\ \dfrac{C - 10}{7.5 - 10}, & if\ 7.5 < C \leq 10 \\ 0, & Otherwise \end{cases}$$

$$\mu_{high}(C) = \begin{cases} 0, & if\ C \leq 7.5 \\ \dfrac{C - 7.5}{10 - 7.5}, & if\ 7.5 < C \leq 10 \\ 1, & Otherwise \end{cases}$$

The graph of these fuzzy numbers is presented in Figure 2.

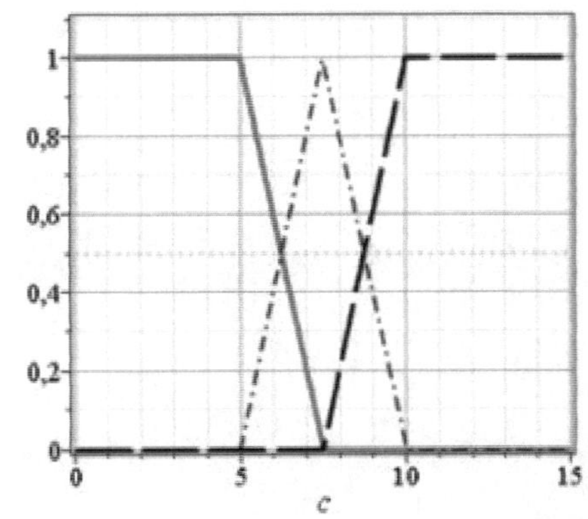

Figure 2. The graph of low-, medium-, and high- level of carrying capacities membership function (in red, blue, and black color, respectively). Horizontal line is in 10^5 individuals.

The number of the infectious dogs when there is no vaccination is presented in Figure 3, for low-, medium-, and high- level of carrying capacities. It is clear that the higher the carrying capacity, the higher the number of infectious dog population. Figures 4 shows the resulting infectious dog populations when there is 50% and 75% of vaccination respectively. The left part

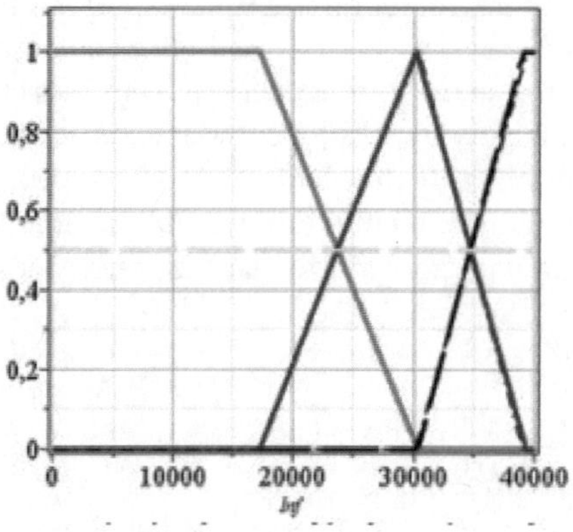

Figure 3. The graph of infectious dog population numbers at $t = 50$ (solid) and $t = 100$ (dashdot) for low-, medium-, and high- level of carrying capacities membership function (in red, blue, and black color, respectively) in the absence of vaccination. The initial values are crisp initial values $I = 50$, $S = 1.5 \times 10^5$ and $V = 0$.

of Figure 4 shows the graph of infectious dog population numbers at $t = 50$ (solid) and $t = 100$ (dash-dot) for low-, medium-, and high- level of carrying capacities membership function (in red, blue, and black color, respectively) for 50% vaccination rate. The graph for $t = 50$ and $t = 100$ coincide, showing that the population is reaching equilibrium. The initial values are the same as in the crisp approach, which are $I = 50$, $S = 1.5 \times 10^5$ and $V = 0$. The right part of Figure 4 shows the same thing as in the left part but for 75% vaccination rate at $t = 50$ (solid) and $t = 100$ (dash-dot). All centroid are lower than that for 50% vaccination indicating the more effective vaccination. The centroid for the low carrying capacity is close to zero.

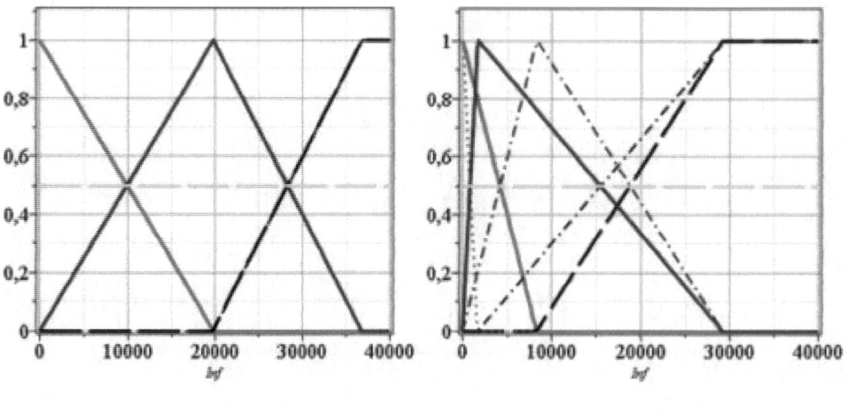

Figure 4. Infected dog population with 50% (left) and 75% (right) vaccination coverage.

5. Conclusions

In this paper, we develop a simple mathematical model and investigate the dynamics of rabies epidemics using crisp and fuzzy approaches to capture uncertainty in the parameters. We found that the results are similar for both approaches. That is, rabies epidemics may be interrupted when the carrying capacity and the transmission rate are not high. Given the complexity of rabies epidemics, a mathematical model for rabies epidemics incorporating human population and other control measures such as culling can be developed to understand the transmission dynamics of rabies. Furthermore, a spatial model can also be formulated to understand the spread of rabies. In general, our findings suggest that if the growth of dog population is limited and the potential contact between dogs is reduced, they may help to stop rabies epidemics.

Acknowledgments
MZN acknowledges funding from Ministry of Research, Technology and Higher Education through Penelitian Pascadoktor (Grant No:222/UN.15.19/LT/2017). AKS acknowledges support from Universitas Padjadjaran, due to funding part of this work through the scheme of Academic Leadership Grant (ALG) with contract number 855/UN6.3.1/PL/2017.

References

[1] WHO 2017 Rabies. (http://who.int/mediacentre/factsheets/fs099/en: World Health Organisation (Accessed on 15 December 2017)
[2] Anak Agung Gde P, Katie H, Janice G, Elly H, Darryn K, Wayan M, Sunny T and Helen S-O 2013 Response to a Rabies Epidemic, Bali, Indonesia, 2008–2011 *Emerging Infectious Disease journal* **19** 648
[3] Wera E, Velthuis A G J, Geong M and Hogeveen H 2013 Costs of Rabies Control: An Economic Calculation Method Applied to Flores Island *PLOS ONE* **8** e83654
[4] Wera E, Mourits M C M and Hogeveen H 2015 Uptake of Rabies Control Measures by Dog Owners in Flores Island, Indonesia *PLOS Neglected Tropical Diseases* **9** e0003589
[5] Ndii M Z, Allingham D, Hickson R I and Glass K 2016 The effect of Wolbachia on dengue outbreaks when dengue is repeatedly introduced *Theoretical Population Biology* **111** 9-15
[6] Ndii M Z, Allingham D, Hickson R I and Glass K 2016 The effect of Wolbachia on dengue dynamics in the presence of two serotypes of dengue: symmetric and asymmetric epidemiological characteristics *Epidemiology and Infection* **144** 2874-82
[7] Ndii M Z, Hickson R I, Allingham D and Mercer G N 2015 Modelling the transmission dynamics of dengue in the presence of Wolbachia *Mathematical Biosciences* **262** 157-66
[8] Aldila D, Nuraini N, Soewono E and Supriatna A K 2014 Mathematical model of temephos resistance in Aedes aegypti mosquito population *AIP Conference Proceedings* **1589** 460-3
[9] Supriatna A K, Nuraini N and Soewono E 2010 *Dengue Virus: Detection, Diagnosis and Control,* ed B Ganim and A Reis: Nova Science)
[10] Ndii M Z and Supriatna A K 2017 Stochastic Mathematical Models in Epidemiology *Information* **20** 6185-96
[11] Ndii M Z, Hickson R I and Mercer G N 2012 Modelling the introduction of *Wolbachia* into Aedes aegypti mosquitoes to reduce dengue transmission *The ANZIAM Journal* **53** 213-27
[12] Tambaru D, Djahi B S and Ndii M Z 2018 The effects of hard water consumption on kidney function: insights from mathematical modelling. In: *Symposium on Biomathematics,* (Bandung, Indonesia: American Institute of Physics)
[13] Ndii M Z, Aggriani N and Supriatna A K 2018 Application of differential tranformation method for solving dengue mathematical model. In: *Symposium on Biomathematics,* (Bandung, Indonesia: American Institute of Physics)
[14] Elmore S A, Chipman R B, Slate D, Huyvaert K P, VerCauteren K C and Gilbert A T 2017 Management and modeling approaches for controlling raccoon rabies: The road to elimination *PLOS Neglected Tropical Diseases* **11** e0005249
[15] Ruan S 2017 Modeling the transmission dynamics and control of rabies in China *Mathematical Biosciences* **286** 65-93
[16] Wiraningsih E D, Agusto F, Aryati L, Lenhart S, Toaha S, Widodo and Govaerts W 2015 Stability analysis of rabies model with vaccination effect and culling in dogs *Applied Mathematical Sciences* **9** 3805-17
[17] Hou Q, Jin Z and Ruan S 2012 Dynamics of rabies epidemics and the impact of control efforts in Guangdong Province, China *Journal of Theoretical Biology* **300** 39-47

IOP Conf. Series: Materials Science and Engineering **332** (2018) 012031 doi:10.1088/1757-899X/332/1/012031

[18] Zinsstag J, Dürr S, Penny M A, Mindekem R, Roth F, Gonzalez S M, Naissengar S and Hattendorf J 2009 Transmission dynamics and economics of rabies control in dogs and humans in an African city *Proceedings of the National Academy of Sciences* **106** 14996-5001

[19] Zhang J, Jin Z, Sun G-Q, Zhou T and Ruan S 2011 Analysis of Rabies in China: Transmission Dynamics and Control *PLOS ONE* **6** e20891

IOP Conf. Series: Materials Science and Engineering **332** (2018) 012032 doi:10.1088/1757-899X/332/1/012032

Feature extraction for face recognition via Active Shape Model (ASM) and Active Appearance Model (AAM)

M Iqtait[*], **F S Mohamad, M Mamat**

Faculty of Informatics and Computing, Universiti Sultan Zainal Abidin, Besut Campus, 22200 Besut, Terengganu, Malaysia

*Email: si1950@putra.unisza.edu.my

Abstract. Biometric is a pattern recognition system which is used for automatic recognition of persons based on characteristics and features of an individual. Face recognition with high recognition rate is still a challenging task and usually accomplished in three phases consisting of face detection, feature extraction, and expression classification. Precise and strong location of trait point is a complicated and difficult issue in face recognition. Cootes proposed a Multi Resolution Active Shape Models (ASM) algorithm, which could extract specified shape accurately and efficiently. Furthermore, as the improvement of ASM, Active Appearance Models algorithm (AAM) is proposed to extracts both shape and texture of specified object simultaneously. In this paper we give more details about the two algorithms and give the results of experiments, testing their performance on one dataset of faces. We found that the ASM is faster and gains more accurate trait point location than the AAM, but the AAM gains a better match to the texture.
Keywords: Face Recognition, Active Shape Model (ASM), Active Appearance Model (AAM).

1. Introduction

With the rapid evolution of information technology, pattern recognition, artificial intelligence and other new technologies, face recognition has a lot of potential computer applications such as social media and multimedia communication, human-computer interaction, human detection, security and access control, which has become one of the important topics of research in recent years. Face recognition mainly includes face detection, feature extraction, texture variance, and classification.

In the face recognition system, locating facial features such as the location points of the eyes, nose, and mouth plays a significant role, because if the facial feature points are situated accurately, the following features extraction and classification stages would be more powerful and efficient [1].

A perfect located method can upgrade the performances of linked research areas, such as a face reconstruction, face recognition and expression recognition [2]. Although humans can easily recognize the accurate location of the facial feature points from a face image, for the computer it is not an easy task, because the computer does not have the complication of the brain of a human. The face has a complicated three-dimensional surface structure, thus, for the formation of a two-dimensional image, the change is very large, particularly for different face pose and facial expression, as well as various lighting conditions, the two-dimensional image distinction is very evident, precise and effective method is a very challenging task. Trait location is very important for the analysis of linked face issues; its accuracy is immediately linked to the reliability of the following application. It is not only to supply a significant geometric information for face image processing and analysis but also plays a significant application for face recognition, facial animation, face synthesis, model-based face image coding, expression analysis of the face pose and so on.

Interpreting images including objects whose appearance can vary is hard, a robust approach has been to employ deformable models, which can perform the variations in shape and/or texture (intensity) of the target objects [3]. Precise and strong location of trait point is a complicated and difficult issue in face recognition. Cootes et al. proposed a Multi-Resolution-Active Shape Models algorithm [4], which could extract specified shape accurately and efficiently. Furthermore, as the improvement of ASM, Active Appearance Models algorithm [5] is proposed to extracts both shape and texture of specified object simultaneously.

To summarize our contribution, we find that a lot of Face recognition algorithms have been proposed. So far, all these algorithms are based on ASM or AAM separately or in cascade [5] [6].However, Previous research has examined the effects of ASM or AAM on recognize the accurate location of the facial feature points from a face image. In contrast, there is a few researches compares the accuracy and efficiency of two algorithms in facial feature detection. In this research we give more details about the two algorithms and give the results of experiments, testing their performance on one dataset of faces. We measure their accuracy by locating landmark points, and their efficiency.

In this paper, we focus on two associated algorithms, Active Shape Model (ASM), which aims to match a collection of model points to an image that was restricted by a statistical model of shape, and the Active Appearance Model (AAM), which looks to match both the position of the model points and a representation of the texture of the object to an image. ASM and AAM use the same underlying statistical model of the shape of purpose objects. This model represents the shape using a group of landmarks, knowing the appropriate ranges of shape variation from a training set of labeled images [7].

1.1. Active Shape Model (ASM)

The shape of an object is represented through landmarks which are one chain of consecutive traits points, each of which is important, point existent in most of the images being considered, for example, the location of the right eye. Enough number of trait points should be provided to cover the comprehensive shape and details. In the proposed model, a total of 68 trait points are defined to explain the shape of a human face, covering the areas of the eyebrows, cheeks, eyes, mouth, and nose. A group of landmarks forms a shape. Meanwhile, the shapes are represented as vectors: all the x-coordinates pursued by the y-coordinates of the points in the form. Align one shape to other with a correspondence transform (allowing rotation, scaling, rotation, and translation) that reduce the Euclidean distance average between shape points. The mean shape is declared the middle of the stratified training shapes [8]. The ASM beginning the search for facial landmarks from the mean shape aligned to the place and size of the face specified by a global face detector.

It then reiterates the following two steps until convergence: (i) propose a temporary shape by determining the positions of shape points through template appropriating from the image texture concerning each point. (ii) Confirms the temporary shape to a universal shape model. The special template suitability is uncertain and the form model gathers the outcomes of the weakened form matches to form a powerful classifier. The complete search is reiterated at each level in an image pyramid, from poor to fine resolution.

1.2. Active Appearance Model (AAM)

AAM model suggested by Cootes et al. [8], AAM is one of the most strong model-based object tracking and detecting algorithms. It is generative, nonlinear, parametric model and can be referred to active contour algorithm ("snakes," [9]), and (ASM) algorithm [7]. Especially, the AAM forms the texture, shape of the object to produce a set of immediate and realistic photos. The AAM has been widely applied in different cases [10-12]. The most persistent implementation of AAM model has been facial tracking and modelling [13].

IOP Conf. Series: Materials Science and Engineering **332** (2018) 012032 doi:10.1088/1757-899X/332/1/012032

ASMs and AAMs vary in the quantity of texture variation they capture. ASMs capture very few texture variations and basically use models of shape, while AAMs include detailed texture (pattern of intensity or color) alteration information are thus completely generative methods that can make photorealistic images. So ASMs match shape models, while AAMs match full models of appearance to an image. In Cootes' report [8] one can detect that Covell explained that the parameters of an Eigen trait model can be applied to drive shape model points to the right place. Cootes' AAM, Applied here, is an extension of this idea. AAMs, consider what are valid shapes (valid forms) and intensity variations from a training set, to generate synthetically examples similar to those in a training-set. The AAM model has a strong modelling and effective ability for fitting the raised complexity due to the high-dimensional texture representation which restricts its application to numerous concepts, such as real time systems.

To prepare the AAM algorithm which is more suitable to actual applications, more potential should be consumed to optimize the calculation of the AAM. Therefore, several enhancements are suggested to attain this aim. Henceforth, various methods are suggested to decrease the distance of the texture, for example, wedge let-based regression tree [15] and Haar wavelet [14]. Moreover, these processes improve the efficiency of the cost of decreasing accuracy or wasting detail information. A frame process is the inverse compositional image alignment (ICIA) [16] algorithm that averts updating texture parameters for each frame and is a fast fitting algorithm for the AAM model. The determination of this model is that it cannot be used to the active shape model that limit the shape and the appearance variance with a lone group of parameters.

2. Methods

The experiment includes training of ASM and AAM algorithms, the experiment flowchart is shown in Figure 3.

2.1. Active shape model(ASM) algorithm

The algorithm of ASM model consists of five steps [4]: (i) labeling the points as shown in figure 3. (ii) Eliciting gray profile to all landmarks. (iii) Aligning training set for ASM. (iv) Calculating statistics on aligned training set on PCA at every resolution. (v) Repeat steps from step 1 to step 4 for all resolution level.

Figure 1. Point's distribution.

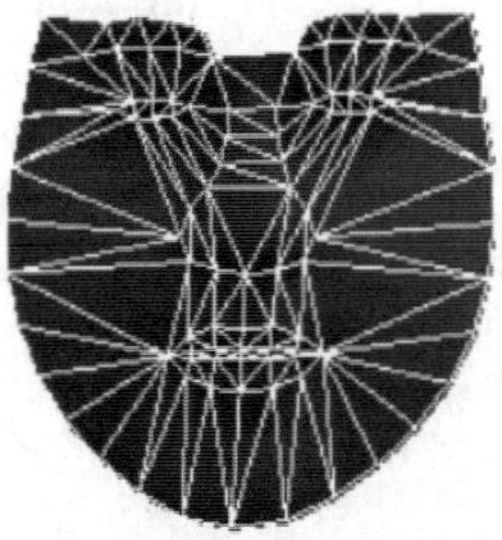

Figure 2. AAM diagram.

After these processes, for each shape X_i can be represented as: $X_i = \overline{X} + P.b$.Where, $\overline{X}$ is the ASM mean of training set, eigenvectors P performs the best significant t methods of variation, b it is a coefficient vector.

2.2. Active Appearance Model(AAM) algorithm

AAM model is originated from ASM model. The full face is splitted into small areas utilizing triangles which vertexes are the ASM landmark points, as shown in Figure 2. The active appearance model connects shape and texture PCA parameters for one group of vectors and a weight vector, then analysis utilizing PCA again.

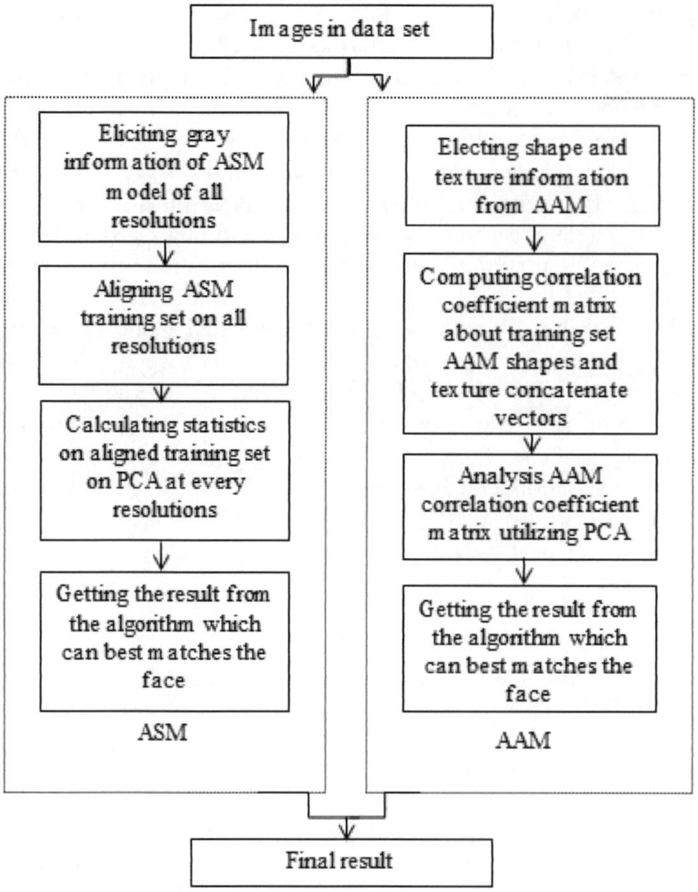

Figure 3. Flowchart of experiment.

The weighing vector is specified by the experiment. It changes with several training set and request many of experiments. Moreover, there is no obvious criterion to estimate the effectiveness of the weight vector. The algorithm of AAM model consists of three steps: (i) connecting shape and texture vectors to each AAM in training set, jointly shape and texture vectors are connected in the same vector: $C_i = (T_i, X_i)^T$, where T_i is a texture vector for m_t pixel, X_i is the position of n landmark point. (ii) Computing the correlation coefficient matrix about connected shape and texture vectors in training set. (iii) Analysis correlation coefficient matrix utilizing PCA. The eigenvectors P of C_{cor} are the hybrid parameters, and they are necessary to monitoring the shape and texture of AAM model. Each pattern in the training group can be proposed as: $C_i = \bar{C} + A.(P.b)$
A represents a diagonal matrix and i element in diametrical contrast correspondent to coefficient of i dimension of C.

2.3. Data sets
We performed the experiments of the ASM and AAM on one data set taken from LFPW database. The data sets contained 100 face images, each marked with 68 points. We did search experiments on the data set, computing the accuracy to relocate the points, how well the texture could be matched and the time consumed to do so.

LFPW database provides us with only the web links and not the real images. Due to broken links, we were able to download only a subset of about 800 out of 1.100 and 224 out of 300 images were obtained for testing and training.

2.4. System requirement
Our experiments were conducted on MATLAB R2016a, PC with processor Intel(R) Core(TM) i5, 4.00 GB RAM memory and Operating System: Windows 7 64-bit.

3. Results

LFPW image database is used to test the efficiency of both algorithms. Total of 100 face images are chosen as training set, each marked with 68 points. The number of searching positions has the most primary influence on the location result. We compared the performance between both algorithms. The experimental results are shown in Figure 4.

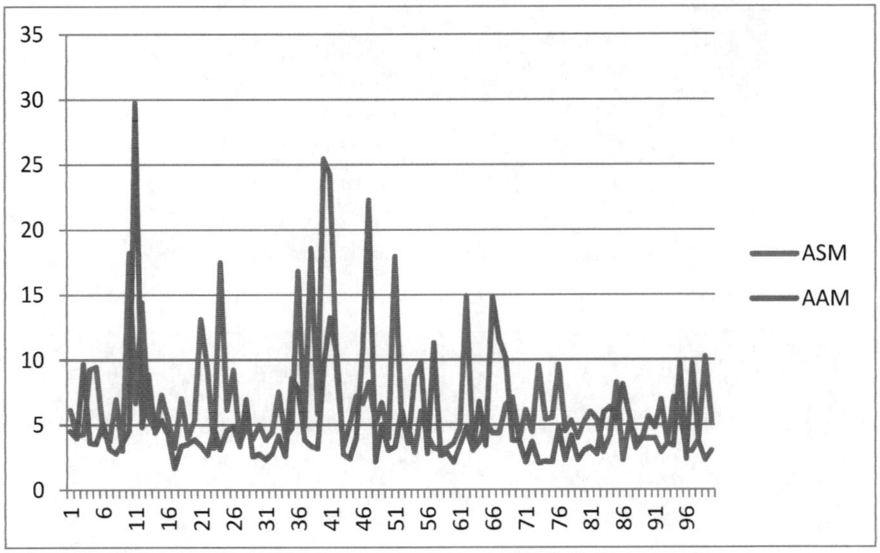

Figure 4. Shows the normalized mean point-to-point error (Euclidean) for AAM and ASM models.

Figure 4 displays resulting point to point errors (the distance between model points and the corresponding points marked in images), the length of point to point error is from 1.6679 to 29.7505. The vertical axis is the point to point error between two algorithms and the Horizontal axis is the image number, the line is the point to point error of all feature points. From above results, it can be seen that, under all conditions, the line of the ASM algorithm is more accurate than AAM algorithm except some cases.

Table 1. summarizes the average of point-to-point error for 100 images using AAM and ASM models.

Model	Point to Point error
ASM	6.502441
AAM	5.238626

IOP Conf. Series: Materials Science and Engineering **332** (2018) 012032 doi:10.1088/1757-899X/332/1/012032

Table 1 compare the average of point to point error for 68 landmarks of training set using both algorithms, the result shows that the difference between ASM and AAM equal 1.263815 so the ASM algorithm locates the points more accurately than the AAM algorithm.

In our experiments, 68 landmark points are labeled on every face image to describe the shape. Figures 5, 6 show the facial feature extraction using ASM, AAM model.

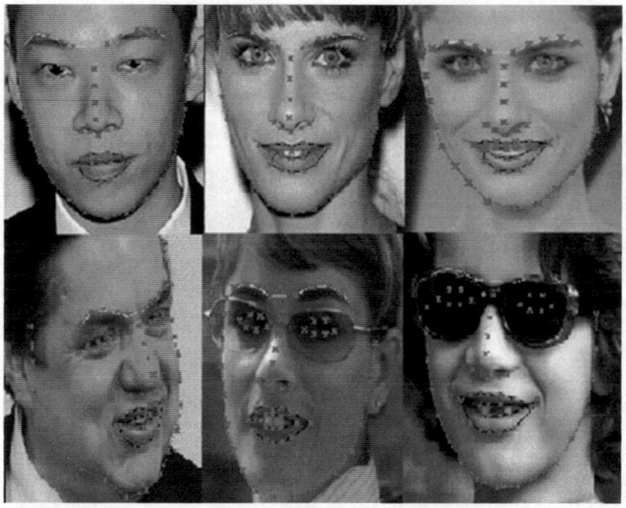

Figure 5. Example of shape composed of landmarks using ASM

Figure 6. Examples of facial feature extraction using AAM model.

3.1. Efficiency

The efficiency was taken regarding time-measurements for running each algorithm on 100 images sequentially. Table 2 shows Time measurements for ASM and AAM algorithms:

Table 2. Comparison of Time measurements for ASM and AAM algorithms.

Model	Time
ASM	133.368243 seconds.
AAM	337.959374 seconds.

Table 2 compare the time measurements for ASM and AAM algorithms, each algorithm performed separately and applied for 100 image sequentially, the experiment conducted on MATLAB R2016a, PC with processor Intel(R) Core(TM) i5, 4.00 GB RAM memory and Operating System: Windows 7 64-bit, The result shows that ASM model takes 133.368243 seconds to perfume full process, AAM model takes 337.959374 seconds to perfume full process, so ASM is faster than AAM.

4. Discussion

As shown in our experiment, see Figure 4, the ASM algorithm is more accurate and faster than AAM algorithm, but AAM algorithm is better in texture matching. ASM algorithm just processes data around the model points and does not take all the information obtainable through an object as the AAM algorithm. So ASM model may be less reliable. The model points proceed to be places of interest (borders or angles) where there is the extreme information. We could train the AAM model to just search using information in areas around strong borders; this would need less image sampling through search so a potentially faster algorithm.

One feature of the AAM is that we can construct a masked model with a comparatively little number of landmarks. Any additional shape variation is extracted in extra methods of the texture model. The ASM requires points about boundaries so as to set appropriate directions for search. Fewer landmarks are required because the significant work only needs to obtain credible image labelling.

5. Conclusion

In this paper, the facial feature extraction using AAM, ASM Algorithms with LFPW face databases. In the experimental results, we found that the ASM model is faster and attains more precise traits point location than the AAM model, and AAM model gives a better match to the image texture. There are three key differences between the ASM and AAM algorithms:

- The AAM attempts to minimise the difference between the target image and the synthesized model image, while the ASM attempts to reduce the distance between model points and the corresponding points found in the image.
- The AAM employs a model of the appearance of the whole of the region, while the ASM employs models of the texture in small regions around every landmark point.
- The AAM model samples the image only under the current position, whereas the ASM model searches around the current position, typically along profiles normal to the boundary.

In the future work this results can be compared to other Models which concerns with face recognition also the researchers can improving and build a combined model using ASM and AAM model that jointly optimizes a precise traits point location and gives a better match to the image texture.

Acknowledgments

This work is supported by the Faculty of Informatics and Computing, Universiti Sultan Zainal Abdin (UniSZA), Terengganu, Malaysia.

References

[1] Wang, N., Gao, X., Tao, D., Yang, H., & Li, X. (2017). Facial feature point detection: A comprehensive survey. Neurocomputing.

[2] Wang, Q., Xie, L., Zhu, B., Yang, T., & Zheng, Y. (2013). Facial Features Extraction based on Active Shape Model. Journal of Multimedia, **8(6),** 747-754.

[3] Cootes, T. F., Edwards, G. J., & Taylor, C. J. (1999, September). Comparing active shape models with active appearance models. In *Bmvc* (Vol. **99**, No. 1, pp. 173-182).

[4] Cootes, T. F., Taylor, C. J., & Lanitis, A. (1994, October). Multi-resolution search with active shape models. In *Pattern Recognition, 1994. Vol. 1-Conference A: Computer Vision & Image Processing., Proceedings of the 12th IAPR International Conference on* (Vol. **1**, pp. 610-612). IEEE.

[5] Cootes, T. F., Edwards, G. J., & Taylor, C. J. (2001). Active appearance models. *IEEE Transactions on pattern analysis and machine intelligence*, **23(6),** 681-685.
[6] Li, S. Z., Zhang, H. J., & Cheng, Q. S. (2002, May). Multi-view face alignment using direct appearance models. In *Automatic Face and Gesture Recognition, 2002. Proceedings. Fifth IEEE International Conference on* (pp. 324-329). IEEE.
[7] Cootes, T. F., Taylor, C. J., Cooper, D. H., & Graham, J. (1995). Active shape models-their training and application. *Computer vision and image understanding*, ***61*(1),** 38-59.
[8] Cootes, T. F., & Taylor, C. J. (2004). Statistical models of appearance for computer vision.
[9] Kass, M., Witkin, A., & Terzopoulos, D. (1988). Snakes: Active contour models. *International journal of computer vision*, ***1*(4)**, 321-331.
[10] Toth, R., & Madabhushi, A. (2012). Multifeature landmark-free active appearance models: application to prostate MRI segmentation. *IEEE Transactions on Medical Imaging*, ***31*(8)**, 1638-1650.
[11] Aykut, M., & Ekinci, M. (2013). AAM-based palm segmentation in unrestricted backgrounds and various postures for palmprint recognition. *Pattern Recognition Letters*, ***34*(9)**, 955-962.
[12] Zhou, M., Liang, L., Sun, J., & Wang, Y. (2010, June). AAM based face tracking with temporal matching and face segmentation. In *Computer Vision and Pattern Recognition (CVPR), 2010 IEEE Conference on* (pp. 701-708). IEEE.
[13] Lee, S. J., Park, K. R., & Kim, J. (2009). A comparative study of facial appearance modeling methods for active appearance models. *Pattern Recognition Letters*, ***30*(14)**, 1335-1346.
[14] Wolstenholme, C. B. H., & Taylor, C. J. (1999, September). Wavelet compression of active appearance models. In *International Conference on Medical Image Computing and Computer-Assisted Intervention* (pp. 544-554). Springer, Berlin, Heidelberg.
[15] Darkner, S., Larsen, R., Stegmann, M. B., & Ersboll, B. K. (2004, March). Wedgelet enhanced appearance models. In *Computer Vision and Pattern Recognition Workshop, 2004. CVPRW'04. Conference on* (pp. 177-177). IEEE.
[16] Baker, S., & Matthews, I. (2004). Lucas-kanade 20 years on: A unifying framework. *International journal of computer vision*, ***56*(3)**, 221-255.
[17] Tzimiropoulos, G., & Pantic, M. (2013). Optimization problems for fast aam fitting in-the-wild. In *Proceedings of the IEEE international conference on computer vision* (pp. 593-600).

IOP Conf. Series: Materials Science and Engineering **332** (2018) 012033 doi:10.1088/1757-899X/332/1/012033

A novel double-convection chaotic attractor, its adaptive control and circuit simulation

M Mamat[1*], S Vaidyanathan[2], A Sambas[3], Mujiarto[3], W S M Sanjaya[4], Subiyanto[5]

[1]Faculty of Informatics and Computing, Universiti Sultan Zainal Abidin, Kuala Terengganu, Malaysia.
[2]Research and Development Centre, Vel Tech University, Avadi, Chennai, India
[3]Department of Mechanical Engineering, Universitas Muhammadiyah Tasikmalaya, Indonesia.
[4]Department of Physics, Universitas Islam Negeri Sunan Gunung Djati, Bandung, Indonesia
[5]Department of Marine Science, Faculty of Fishery and Marine Science, Universitas Padjadjaran, Indonesia.

*Email: must@unisza.edu.my

Abstract. A 3-D novel double-convection chaotic system with three nonlinearities is proposed in this research work. The dynamical properties of the new chaotic system are described in terms of phase portraits, Lyapunov exponents, Kaplan-Yorke dimension, dissipativity, stability analysis of equilibria, etc. Adaptive control and synchronization of the new chaotic system with unknown parameters are achieved via nonlinear controllers and the results are established using Lyapunov stability theory. Furthermore, an electronic circuit realization of the new 3-D novel chaotic system is presented in detail. Finally, the circuit experimental results of the 3-D novel chaotic attractor show agreement with the numerical simulations.

1. Introduction

A chaotic system is commonly defined as a nonlinear dynamical system that is highly sensitive to even small perturbations in its initial conditions [1-4]. In other words, a chaotic system is a nonlinear dynamical system with at least one positive Lyapunov exponent. In the last two decades, many new chaotic systems have been discovered such as Tigan system [5], Li system [6], Jafari system [7], Sundarapandian systems [8-9], Pehlivan system [10], Vaidyanathan systems [11-14], Molaie systems [15], Tacha system [16], Sampath system [17], Volos system [18], Wang system [19], Pham systems [20-22], etc.

Chaos theory has several applications in science and engineering such as oscillators [23-25], chemical reactors [26-28], Tokamak systems [29], voice encryption [30], population biology [31-32], robotics [33-34], neural networks [35-36], memristors [37-38], secure communication system [39-41] etc.

The problem of chaos control of a chaotic system is to find a state feedback control law to stabilize a chaotic system around its unstable equilibrium [1-4]. The problem of synchronization of two chaotic systems deals with synchronizing identical state trajectories of a pair of chaotic systems called master and slave systems asymptotically [1-4]. Some popular methods for chaos control and synchronization of chaotic systems can be listed as an active control, adaptive control, backstepping control, fuzzy control, sliding mode control etc., which are outlined in [1-4].

Rucklidge chaotic system is a popular model in mechanics for nonlinear double convection [42]. When the convection takes place in a fluid layer rotating uniformly about a vertical axis and in the limit of tall thin rolls, convection in an imposed vertical magnetic field and convection in a rotating fluid layer are both modeled by Rucklidge's 3-D system of differential equations, which produces chaotic solutions.

In this work, we modify the dynamics of Rucklidge system [42] and derive a new double-convection chaotic system with an absolute nonlinearity and two quadratic nonlinearities. The phase portraits of the new chaotic system are displayed in Section 2, and the dynamical properties of the new chaotic system such as dissipativity, equilibria analysis, Lyapunov exponents, Kaplan-Yorke dimension, etc. are analyzed in Section 3. Lyapunov exponents of the new chaotic system are obtained as $L_1 = 0.4684$, $L_2 = 0$ and $L_3 = -3.6684$. The Kaplan-Yorke dimension of the new system is found as $D_{KY} = 2.1277$.

Adaptive control and synchronization of the new chaotic system with unknown system parameters are discussed in Sections 4 and 5, respectively. The main adaptive control results derived in this work are established using Lyapunov stability theory [43]. Furthermore, an electronic circuit realization of the new chaotic system is presented in detail in Section 6. The circuit experimental results of the new chaotic attractor show agreement with the numerical simulations. Section 7 contains the conclusions.

2. A new nonlinear double-convection chaotic system

In fluid mechanics modeling, cases of two-dimensional convection in a horizontal layer of Boussinesq fluid with lateral constraints were considered by Rucklidge [42]. When the convection takes place in a fluid layer rotating uniformly about a vertical axis and in the limit of tall thin rolls, convection in an imposed vertical magnetic field and convection in a rotating fluid layer are both modeled by a new third-order set of ordinary differential equations, which produces chaotic solutions.

The Rucklidge chaotic system is described by the 3-D dynamics

$$\begin{cases} \dot{x}_1 = -ax_1 + bx_2 - x_2x_3 \\ \dot{x}_2 = x_1 \\ \dot{x}_3 = -x_3 + x_2^2 \end{cases} \tag{1}$$

where x_1, x_2, x_3 are state variables and a, b are positive constants. In [42], it was established that the system (1) is chaotic for $a = 2.2$ and $b = 6.7$.

In this work, we modify the dynamics of Rucklidge chaotic system (1) and obtain a new dynamics for nonlinear double convection as

$$\begin{cases} \dot{x}_1 = -ax_1 + bx_2 - |x_2| - x_2x_3 \\ \dot{x}_2 = x_1 \\ \dot{x}_3 = -x_3 + x_2^2 \end{cases} \tag{2}$$

where x_1, x_2, x_3 are state variables and a, b are positive constants.

In this paper, we show that the system (2) is *chaotic* for the parameter values

$$a = 2.2, \quad b = 18 \tag{3}$$

For numerical simulations, we take the initial values of the system (2) as

$$x_1(0) = 0.2, \quad x_2(0) = 0.2, \quad x_3(0) = 0.2 \tag{4}$$

IOP Conf. Series: Materials Science and Engineering **332** (2018) 012033 doi:10.1088/1757-899X/332/1/012033

Figure 1 shows the phase portraits strange attractor of the new double-convection chaotic system (2) for the parameter values (3) and initial conditions (4). Figure 1 (a) shows the 3-D phase portrait of the nonlinear double-convection chaotic system (2). Figures 1 (b)-(c) show the projections of the strange attractor of the nonlinear double-convection chaotic system (2) in (x_1, x_2), (x_2, x_3) and (x_1, x_3) coordinate planes, respectively.

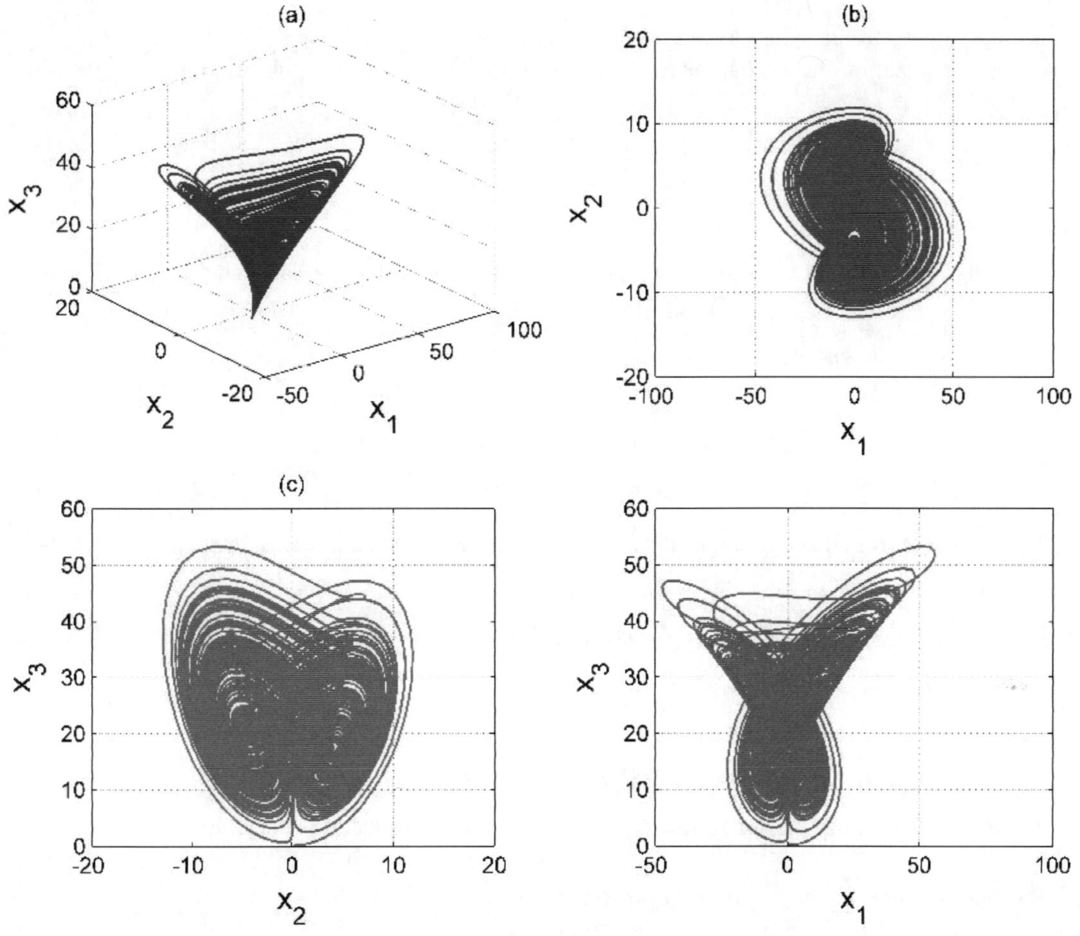

Figure 1. Phase portraits of the new chaotic system (2) for $a = 2.2$, $b = 18$

3. Dynamical properties of the new chaotic system

In this section, we take the parameter values as in the chaotic case, *i.e.* $a = 2.2$ and $b = 18$.

3.1. Dissipativity

If V denotes any volume along the flow of the new chaotic system (2), then

$$\nabla \cdot V = \frac{\partial \dot{x}_1}{\partial x_1} + \frac{\partial \dot{x}_2}{\partial x_2} + \frac{\partial \dot{x}_3}{\partial x_3} = -a - 1 = -\mu < 0 \tag{5}$$

where $\mu = a + 1 > 0$.

This shows that $\dot{V} = -\mu V$. Integrating, we get

$$V(t) = \exp(-\mu t) V(0) \tag{6}$$

Thus, the new chaotic system (2) is dissipative. Hence, the system limit sets are ultimately confined into a specific limit set of zero volume, and the asymptotic motion of the novel chaotic system (2) settles into a strange attractor of the system.

3.2. Equilibrium points

The equilibrium points of the system (2) are obtained by solving the system of equations

$$-ax_1 + bx_2 - x_2x_3 = 0, \quad x_1 = 0, \quad -x_3 + x_2^2 = 0 \tag{7}$$

We take the parameter values as in the chaotic case (3), *i.e.* $a = 2.2$ and $b = 18$.

A simple calculation shows that the new chaotic system (2) has three equilibrium points given by

$$E_0 = \begin{bmatrix} 0 \\ 0 \\ 0 \end{bmatrix}, \quad E_1 = \begin{bmatrix} 0 \\ 4.1231 \\ 17 \end{bmatrix}, \quad E_2 = \begin{bmatrix} 0 \\ -4.3589 \\ 19 \end{bmatrix} \tag{8}$$

To check the stability of the equilibrium points, we calculate the Jacobian of the system (2) as

$$J(x) = \begin{bmatrix} -a & b - \text{sign}(x_2) - x_3 & -x_2 \\ 1 & 0 & 0 \\ 0 & 2x_2 & -1 \end{bmatrix} \tag{9}$$

Let $J_0 = J(E_0)$, $J_1 = J(E_1)$ and $J_2 = J(E_2)$. We find that J_0 has the eigenvalues

$$\lambda_1 = -1, \ \lambda_2 = -5.4829, \ \lambda_3 = 3.2829 \tag{10}$$

This shows that the equilibrium point E_0 is a saddle point. Hence, it is unstable.

Next, we find that J_1 has the eigenvalues

$$\lambda_1 = -4.4335, \ \lambda_{2,3} = 0.6168 \pm 2.6997i \tag{11}$$

This shows that the equilibrium point E_1 is a saddle-focus. Hence, it is unstable.

We also find that J_2 has the eigenvalues

$$\lambda_1 = -4.5512, \ \lambda_{2,3} = 0.6756 \pm 2.8095i \tag{12}$$

This shows that the equilibrium point E_2 is a saddle-focus. Hence, it is unstable.

3.3. Lyapunov exponents and Kaplan-Yorke dimension

The parameters of the new system (2) are taken as $a = 2.2$ and $b = 18$. The initial state of the system (2) is taken as $x_i(0) = 0.2$ for $i = 1, 2, 3$.

The Lyapunov exponents of the system (2) are calculated using Wolf's algorithm [44] (see Figure 2) as

$$L_1 = 0.4684, \ L_2 = 0, \ L_3 = -3.6684. \tag{13}$$

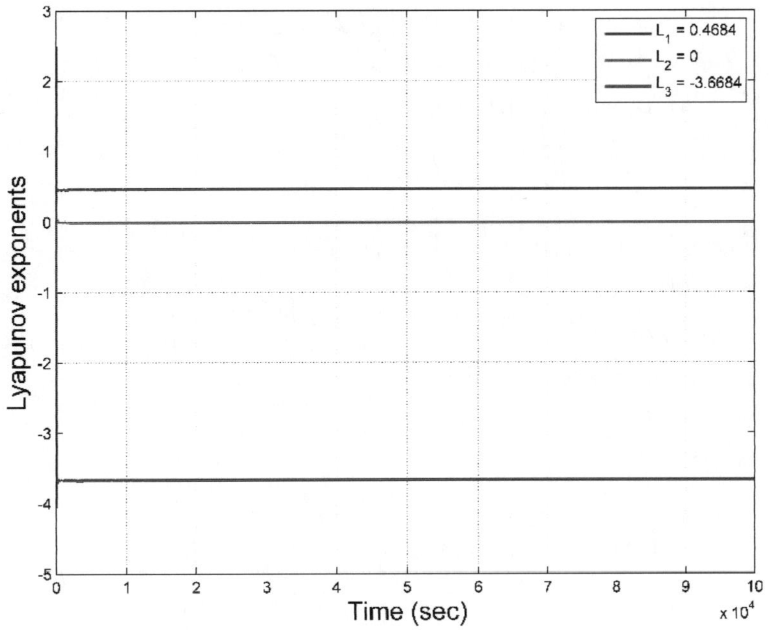

Figure 2. Lyapunov exponents of the new chaotic system (2)

The Kaplan-Yorke dimension of the new chaotic system (2) is calculated as

$$D_{KY} = 2 + \frac{L_1 + L_2}{|L_3|} = 2.1277 \tag{14}$$

3.4. *Symmetry and invariance*

The new chaotic system (2) is invariant under the coordinate transformation

$$(x_1, x_2, x_3) \mapsto (-x_1, -x_2, x_3) \tag{15}$$

This shows that the chaotic system (2) has a rotation symmetry about the x_3-axis. Hence, every non-trivial trajectory of the system (2) must have a twin trajectory.

Also, the x_3-axis is invariant under the flow of the new chaotic system (2). This invariant flow is characterized by the 1-D dynamics $\dot{x}_3 = -x_3$ which is exponentially stable.

4. Adaptive control of the new chaotic system

In this section, we consider the controlled new chaotic system given by

$$\begin{cases} \dot{x}_1 = -ax_1 + bx_2 - |x_2| - x_2 x_3 + u_1 \\ \dot{x}_2 = x_1 + u_2 \\ \dot{x}_3 = -x_3 + x_2^2 + u_3 \end{cases} \tag{16}$$

where x_1, x_2, x_3 are the states and a, b are unknown system parameters.

We consider the adaptive controller defined by

$$\begin{cases} u_1 = \hat{a}(t)x_1 - \hat{b}(t)x_2 + |x_2| + x_2 x_3 - k_1 x_1 \\ u_2 = -x_1 - k_2 x_2 \\ u_3 = x_3 - x_2^2 - k_3 x_3 \end{cases} \tag{17}$$

IOP Conf. Series: Materials Science and Engineering **332** (2018) 012033 doi:10.1088/1757-899X/332/1/012033

where $\hat{a}(t), \hat{b}(t)$ are estimates of a, b respectively and k_1, k_2, k_3 are positive constants.

Substituting (17) into (16), we obtain the closed-loop system

$$\begin{cases} \dot{x}_1 = -[a-\hat{a}(t)]x_1 + [b-\hat{b}(t)]x_2 - k_1 x_1 \\ \dot{x}_2 = -k_2 x_2 \\ \dot{x}_3 = -k_3 x_3 \end{cases} \tag{18}$$

We define parameter estimation errors as follows:

$$\begin{cases} e_a(t) = a - \hat{a}(t) \\ e_b(t) = b - \hat{b}(t) \end{cases} \tag{19}$$

Using (19), the closed loop system (18) reduces to

$$\begin{cases} \dot{x}_1 = -e_a x_1 + e_b x_2 - k_1 x_1 \\ \dot{x}_2 = -k_2 x_2 \\ \dot{x}_3 = -k_3 x_3 \end{cases} \tag{20}$$

Differentiating (19) with respect to t, we get

$$\begin{cases} \dot{e}_a(t) = -\dot{\hat{a}}(t) \\ \dot{e}_b(t) = -\dot{\hat{b}}(t) \end{cases} \tag{21}$$

Next, we consider the Lyapunov function defined by

$$V(x_1, x_2, x_3, e_a, e_b) = \frac{1}{2}\left(x_1^2 + x_2^2 + x_3^2 + e_a^2 + e_b^2\right) \tag{22}$$

which is positive definite on R^5.

Differentiating V along the trajectories of (20) and (21), we obtain

$$\dot{V} = -k_1 x_1^2 - k_2 x_2^2 - k_3 x_3^2 + e_a\left(-x_1^2 - \dot{\hat{a}}\right) + e_b\left(x_1 x_2 - \dot{\hat{b}}\right) \tag{23}$$

In view of (23), we take the parameter update law as

$$\begin{cases} \dot{\hat{a}} = -x_1^2 \\ \dot{\hat{b}} = x_1 x_2 \end{cases} \tag{24}$$

Next, we prove the main theorem of this section.

Theorem 1. The new chaotic system (16) with unknown parameters is globally and asymptotically stabilized by the adaptive control law (17) and the parameter update law (24), where k_1, k_2, k_3 are positive constants.

Proof. The Lyapunov function V defined by (22) is quadratic and positive definite on R^5.

By substituting the parameter update law (24) into (23), we obtain the time-derivative of V as

$$\dot{V} = -k_1 x_1^2 - k_2 x_2^2 - k_3 x_3^2 \tag{25}$$

which is negative semi-definite on R^5.

Thus, by Barbalat's lemma [43], it follows that the closed-loop system (20) is globally asymptotically stable for all initial conditions $x(0) \in R^3$.

Hence, we conclude that the new chaotic system (16) with unknown parameters is globally and asymptotically stabilized by the adaptive control law (17) and the parameter update law (24), where k_1, k_2, k_3 are positive constants. This completes the proof. ■

For numerical simulations, we take the gain constants as $k_i = 10$ for $i = 1, 2, 3$.

We take the parameter values as in the chaotic case (3), i.e. $a = 2.2$ and $b = 18$.

We take the initial conditions of the states of the new chaotic system (16) as $x_1(0) = 18.3$, $x_2(0) = 11.7$ and $x_3(0) = 16.4$.

We take the initial conditions of the parameter estimates as $\hat{a}(0) = 7.5$ and $\hat{b}(0) = 6.8$.

Figure 3 shows the time-history of the states of the new chaotic system (16) after the implementation of the adaptive control law (17) and the parameter update law (24).

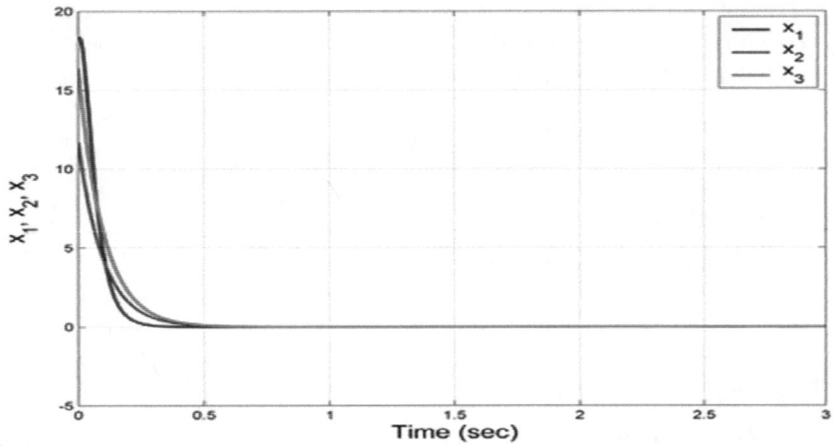

Figure 3. Time-history of the controlled state trajectories of the new chaotic system

5. Adaptive synchronization of the new chaotic system

In this section, we use the adaptive control to synchronize a pair of identical new chaotic systems with unknown state parameters.

As the master system, we consider the new chaotic system given by

$$\begin{cases} \dot{x}_1 = -ax_1 + bx_2 - |x_2| - x_2 x_3 \\ \dot{x}_2 = x_1 \\ \dot{x}_3 = -x_3 + x_2^2 \end{cases} \tag{26}$$

where x_1, x_2, x_3 are the states and a, b are unknown parameters.

As the slave system, we consider the new chaotic system given by

$$\begin{cases} \dot{y}_1 = -ay_1 + by_2 - |y_2| - y_2 y_3 + u_1 \\ \dot{y}_2 = y_1 + u_2 \\ \dot{y}_3 = -y_3 + y_2^2 + u_3 \end{cases} \tag{27}$$

where y_1, y_2, y_3 are the states and u_1, u_2, u_3 are adaptive controls to be designed.

The synchronization error between the systems (26) and (27) is defined as

$$\begin{cases} e_1 = y_1 - x_1 \\ e_2 = y_2 - x_2 \\ e_3 = y_3 - x_3 \end{cases} \tag{28}$$

The error dynamics is obtained as

$$\begin{cases} \dot{e}_1 = -ae_1 + be_2 - |y_2| + |x_2| - y_2y_3 + x_2x_3 + u_1 \\ \dot{e}_2 = e_1 + u_2 \\ \dot{e}_3 = -e_3 + y_2^2 - x_2^2 + u_3 \end{cases} \quad (29)$$

We consider the adaptive control defined by

$$\begin{cases} u_1 = \hat{a}(t)e_1 - \hat{b}(t)e_2 + |y_2| - |x_2| + y_2y_3 - x_2x_3 - k_1e_1 \\ u_2 = -e_1 - k_2e_2 \\ u_3 = e_3 - y_2^2 + x_2^2 - k_3e_3 \end{cases} \quad (30)$$

where k_1, k_2, k_3 are positive gain constants.

Substituting (30) into (29), we obtain the closed-loop system

$$\begin{cases} \dot{e}_1 = -[a - \hat{a}(t)]e_1 + [b - \hat{b}(t)]e_2 - k_1e_1 \\ \dot{e}_2 = -k_2e_2 \\ \dot{e}_3 = -k_3e_3 \end{cases} \quad (31)$$

We define the parameter estimation errors as

$$\begin{cases} e_a(t) = a - \hat{a}(t) \\ e_b(t) = b - \hat{b}(t) \end{cases} \quad (32)$$

Using (32), we can simplify (31) as

$$\begin{cases} \dot{e}_1 = -e_ae_1 + e_be_2 - k_1e_1 \\ \dot{e}_2 = -k_2e_2 \\ \dot{e}_3 = -k_3e_3 \end{cases} \quad (33)$$

Differentiating (32) with respect to t, we obtain

$$\begin{cases} \dot{e}_a(t) = -\dot{\hat{a}}(t) \\ \dot{e}_b(t) = -\dot{\hat{b}}(t) \end{cases} \quad (34)$$

Next, we consider the Lyapunov function defined by

$$V(e_1, e_2, e_3, e_a, e_b) = \frac{1}{2}\left(e_1^2 + e_2^2 + e_3^2 + e_a^2 + e_b^2\right) \quad (35)$$

which is positive definite on $\boldsymbol{R}^5$.

Differentiating V along the trajectories of (33) and (34), we obtain

$$\dot{V} = -k_1e_1^2 - k_2e_2^2 - k_3e_3^2 + e_a\left(-e_1^2 - \dot{\hat{a}}\right) + e_b\left(e_1e_2 - \dot{\hat{b}}\right) \quad (36)$$

In view of (36), we take the parameter update law as

$$\begin{cases} \dot{\hat{a}} = -e_1^2 \\ \dot{\hat{b}} = e_1e_2 \end{cases} \quad (37)$$

Next, we prove the main theorem of this section.

Theorem 2. The new chaotic systems (26) and (27) with unknown parameters are globally and asymptotically stabilized by the adaptive control law (30) and the parameter update law (37), where k_1, k_2, k_3 are positive constants.

Proof. The Lyapunov function V defined by (35) is quadratic and positive definite on R^5.

By substituting the parameter update law (37) into (36), we obtain the time-derivative of V as

$$\dot{V} = -k_1e_1^2 - k_2e_2^2 - k_3e_3^2 \tag{38}$$

which is negative semi-definite on R^5.

Thus, by Barbalat's lemma [43], it follows that the closed-loop system (33) is globally asymptotically stable for all initial conditions $e(0) \in R^3$.

Hence, we conclude that the new chaotic systems (26) and (27) with unknown parameters are globally and asymptotically stabilized by the adaptive control law (30) and the parameter update law (37), where k_1, k_2, k_3 are positive constants. This completes the proof. ■

For numerical simulations, we take the gain constants as $k_i = 10$ for $i = 1, 2, 3$.

We take the parameter values as in the chaotic case (3), i.e. $a = 2.2$ and $b = 18$.

We take the initial conditions of the states of the master system (26) as $x_1(0) = 20.2$, $x_2(0) = 6.4$ and $x_3(0) = 15.1$.

We take the initial conditions of the states of the slave system (27) as $y_1(0) = 18.5$, $y_2(0) = 12.3$ and $y_3(0) = 10.7$.

We take the initial conditions of the parameter estimates as $\hat{a}(0) = 16.3$ and $\hat{b}(0) = 5.8$.

Figure 4 shows the synchronization of the states of the new chaotic systems (26) and (27).

Figure 5 shows the time-history of the synchronization errors e_1, e_2, e_3.

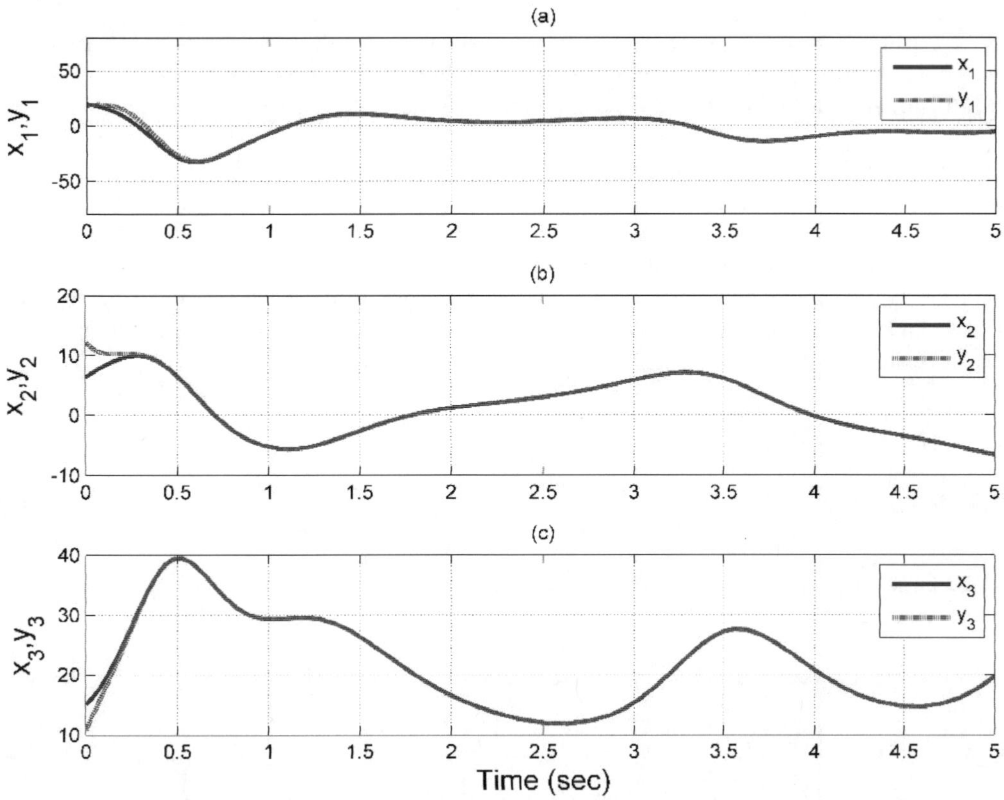

Figure 4. Complete synchronization of the new chaotic systems

IOP Conf. Series: Materials Science and Engineering **332** (2018) 012033 doi:10.1088/1757-899X/332/1/012033

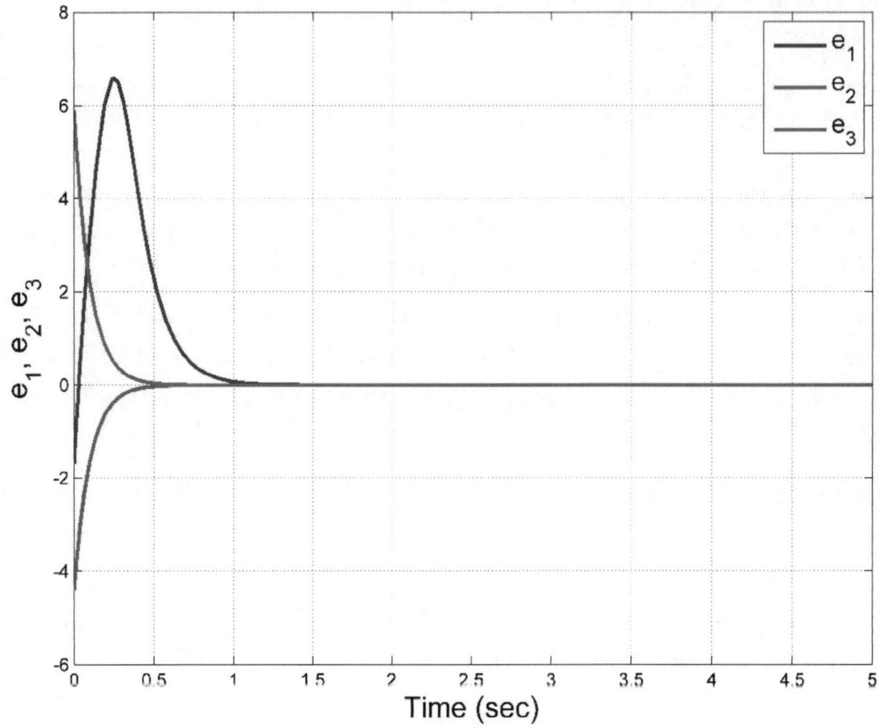

Figure 5. Time-history of the synchronization errors for the new chaotic systems

6. Circuit implementation of the new double-convection chaotic system

Electronic circuit provides an alternative approach to exploring new chaotic system (2). In this section, we design and build an electronic circuit of the system (2) as shown in Figure 6. In more details, there are three integrators (U1A, U3A, U5A), which are created by the operational amplifiers. The circuit consists of simple electronic elements, such as resistors, capacitors, operational amplifiers, analog devices AD633 multipliers and two diodes (1N4148), which provide the signal |X2|. By applying Kirchhoff's laws to the circuit in Figure 6, its circuital equations are derived in the following form:

$$\begin{cases} \dot{x}_1 = -\dfrac{1}{C_1 R_1} x_1 - \dfrac{1}{C_1 R_2} \mid x_2 \mid + \dfrac{1}{C_1 R_3} x_2 - \dfrac{1}{10 C_1 R_4} x_2 x_3 \\ \dot{x}_2 = \dfrac{1}{C_2 R_5} x_1 \\ \dot{x}_3 = -\dfrac{1}{C_3 R_6} z + \dfrac{1}{10 C_3 R_7} x_2^{\ 2} \end{cases} \tag{39}$$

We choose the values of the circuital elements as

$$\begin{cases} R_1 = 4.54\,K\Omega,\, R_3 = 555\,\Omega,\, R_4 = R_7 = 1\,K\Omega \\ R_2 = R_5 = R_6 = R_8 = R_9 = R_{10} = R_{11} = R_{12} = R_{13} = R_{14} = R_{15} = R_{16} = 10\,K\Omega \\ C_1 = C_2 = C_3 = 10\,nF \end{cases} \tag{40}$$

The supplies of all active devices are ± 15 volt. The proposed circuit is implemented in the electronic simulation package MultiSIM. The obtained phase portraits are shown in Figure 1. There is a good agreement between these circuital results and the theoretical ones (see Figures. 7–9).

7. Conclusions

This work proposed a novel three-dimensional double-convection chaotic system with three nonlinearities. The dynamical properties of the new chaotic system were discussed in detail. The qualitative properties included phase portraits, Lyapunov exponents, Kaplan-Yorke dimension, dissipativity, stability analysis of equilibria, etc. Adaptive control and synchronization of the new chaotic system with unknown parameters were achieved via nonlinear controllers and Lyapunov stability theory. Furthermore, an electronic circuit realization of the new 3-D novel chaotic system was proposed and the circuit experimental results of the 3-D novel chaotic attractor showed good agreement with the numerical simulations.

IOP Conf. Series: Materials Science and Engineering **332** (2018) 012033 doi:10.1088/1757-899X/332/1/012033

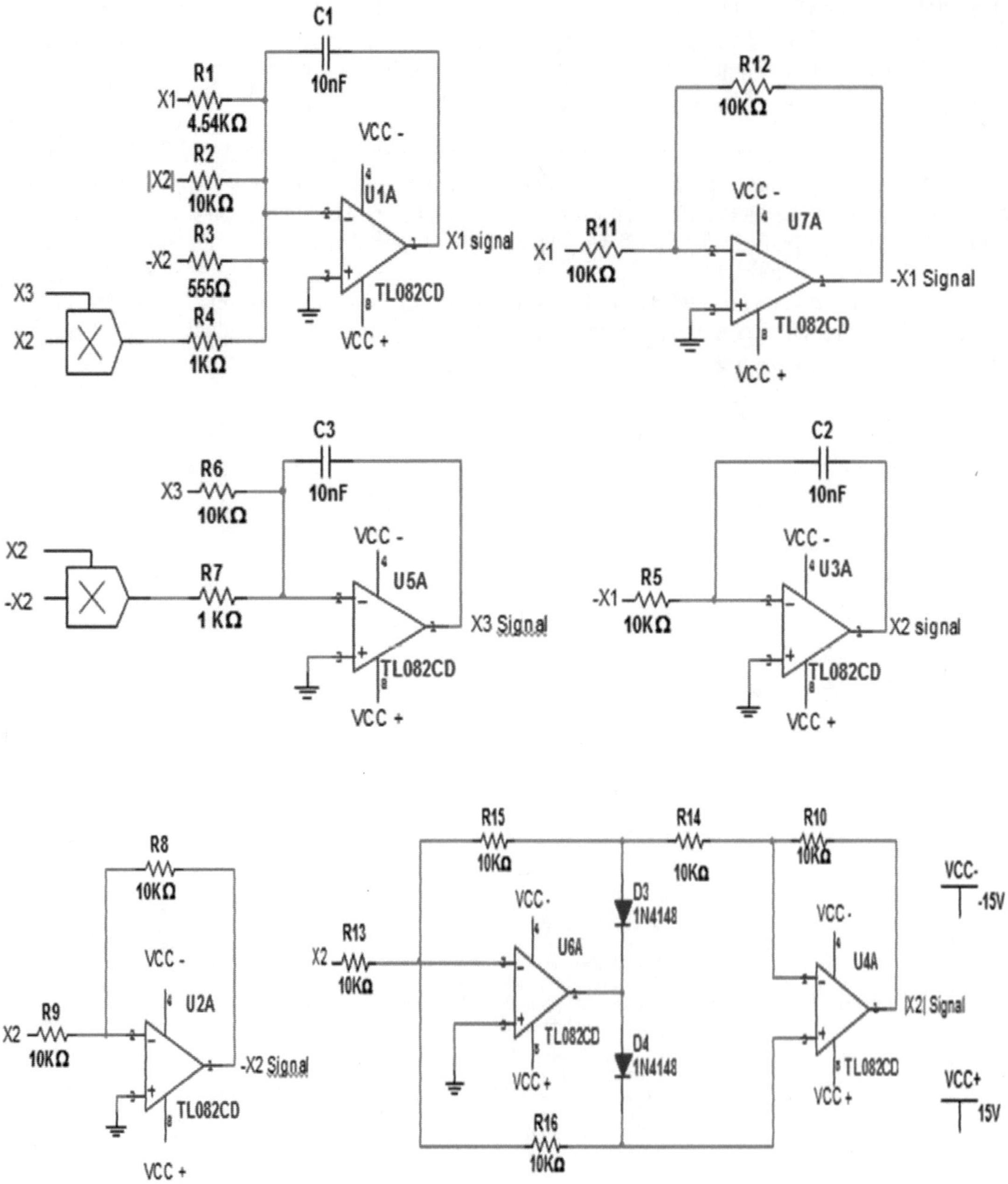

Figure 6 Schematic of the proposed new chaotic system by using MultiSIM

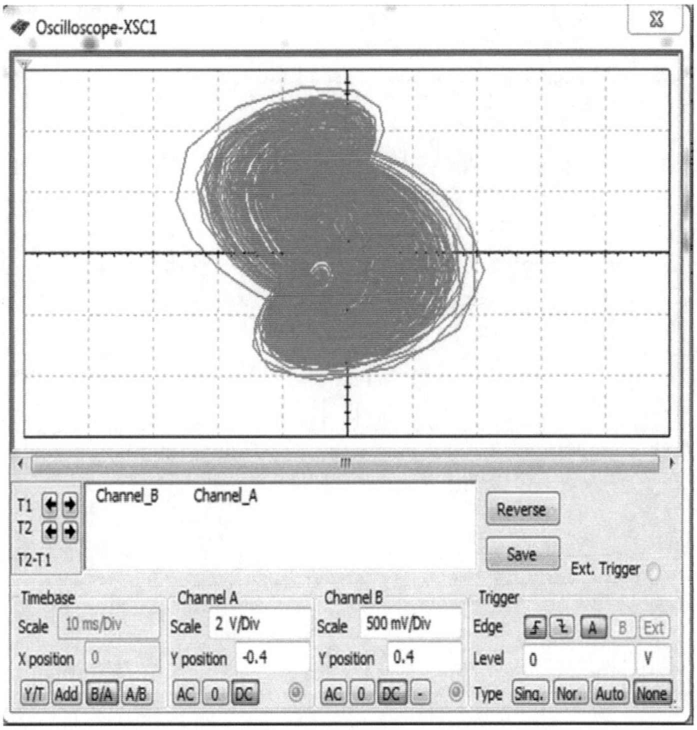

Figure 7 2-D projection of the new chaotic system on the (x_1, x_2) plane

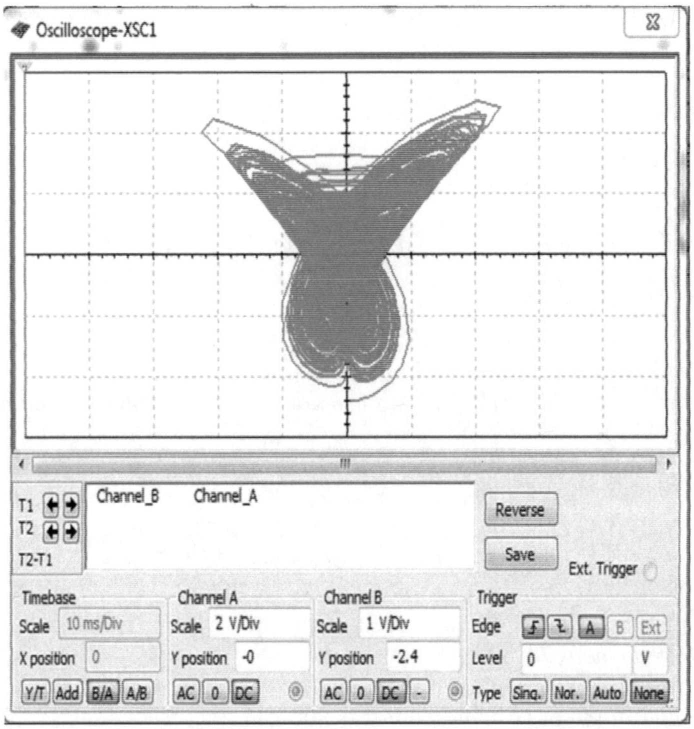

Figure 8 2-D projection of the new chaotic system on the (x_1, x_3) plane

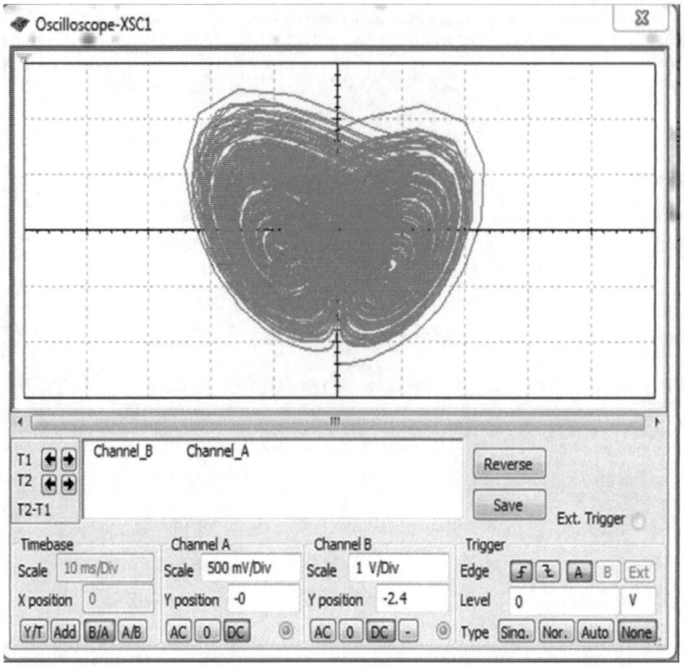

Figure 9 2-D projection of the new chaotic system on the (x_2, x_3) plane

References

[1] Azar A T and Vaidyanathan S 2015 *Chaos Modeling and Control Systems Design* (Berlin: Springer)
[2] Azar A T and Vaidyanathan, S 2016 *Advances in Chaos Theory and Intelligent Control* (Berlin: Springer)
[3] Vaidyanathan S and Volos C 2016 *Advances and Applications in Nonlinear Control Systems* (Berlin: Springer)
[4] Vaidyanathan S and Volos C 2016 *Advances and Applications in Chaotic Systems* (Berlin: Springer)
[5] Tigan G and Opris D 2008 *Chaos, Solitons and Fractals* **36** 1315-1319
[6] Li D 2008 *Physics Letters A* **372** 387-393
[7] Jafari S and Sprott J C 2013 *Chaos, Solitons and Fractals*, **57**, 79-84.
[8] Sundarapandian V and Pehlivan I 2012 *Mathematical and Computer Modelling* **55** 1904-1915
[9] Sundarapandian V 2013 *Journal of Engineering Science and Technology Review* **6** 45-52
[10] Pehlivan I, Moroz I M and Vaidyanathan S 2014 *Journal of Sound and Vibration* **333** 5077-5096
[11] Vaidyanathan S 2014 *International Journal of Modelling Identification and Control* **22** 41-53
[12] Vaidyanathan S 2015 *International Journal of Modelling Identification and Control* **23** 164-172
[13] Vaidyanathan S 2015 *Kyungpook Mathematical Journal* **55** 563-586
[14] Vaidyanathan S 2016 *Archives of Control Sciences* **26** 19-47
[15] Molaie M, Jafari S, Sprott J C and Golpayegani S M R H 2013 *International Journal of Bifurcation and Chaos* **23**, 1350188.
[16] Tacha O I, Volos C K, Kyprianidis I M, Stouboulos I N, Vaidyanathan S and Pham V T 2016 *Applied Mathematics and Computation* **276** 200-217
[17] Sampath S, Vaidyanathan S, Volos C K and Pham V T 2015 *Journal of Engineering Science and*

Technology Review **8** 1-6
[18] Volos C, Maaita J O, Vaidyanathan S, Pham V T, Stouboulos I and Kyprianidis I 2017 *IEEE Transactions on Circuits and Systems II: Express Briefs* **64** 339-343
[19] Wang X and Chen G 2013 *Nonlinear Dynamics*, **71**, 429-436.
[20] Pham V T, Vaidyanathan S, Volos C K, Jafari S, Kuznetsov N V and Hoang T M 2016 *European Physical Journal: Special Topics* **225** 127-136.
[21] Pham V T, Jafari S, Volos C, Giakoumis A, Vaidyanathan S and Kapitaniak T 2016 *IEEE Transactions on Circuits and Systems II: Express Briefs* **63** 878-882
[22] Pham V T, Vaidyanathan S, Volos C, Jafari S and Kingni S T 2016 *Optik* **127** 3259-3265
[23] Kengne J, Chedjou J C, Kenne G and Kyamakya K 2012 *Communications in Nonlinear Science and Numerical Simulation* **17** 2914-2923
[24] Vaidyanathan S 2015 *International Journal of PharmTech Research* **8** 106-116
[25] Itoh M and Chua L 2017 *International Journal of Bifurcation and Chaos*, **27**, 1730005.
[26] Xu C and Wu Y 2015 *Applied Mathematical Modelling* **39** 2295-2310
[27] Bodale I and Oancea V A 2015 *Chaos Solitons and Fractals* **78** 1-9
[28] Vaidyanathan S 2015 *International Journal of ChemTech Research* **8** 73-85
[29] Punjabi A and Boozer A 2014 *Physics Letters A* **378** 2410-2416
[30] Vaidyanathan S, Sambas A, Mamat M and W S Mada Sanjaya 2017 *International Journal of Modelling, Identification and Control* **28** 153-166
[31] Voorsluijs V and Decker Y D 2016 *Physica D: Nonlinear Phenomena* **335** 1-9
[32] Chattopadhyay J, Pal N, Samanta S, Venturino E and Khan Q J A 2015 *Biosystems* **138** 18-24
[33] Iqbal S, Zang X, Zhu Y and Zhao J 2014 *Robotics and Autonomous Systems* **62** 889-909
[34] Sambas A, Vaidyanathan S, Mamat M, Sanjaya W M and Rahayu D S 2016 *Studies in Computational Intelligence* **636** 283-310
[35] Akhmet M and Fen M O 2014 *Neurocomputing* **145** 230-239
[36] Lian S and Chen X 2011 *Applied Soft Computing* **11** 4293-4301
[37] Yang J, Wang L, Wang Y and Guo T 2017 *Neurocomputing* **227** 142-148
[38] Bao B C, Bao H, Wang N, Chen M and Xu Q 2017 *Chaos Solitons and Fractals* **94** 102-111
[39] Sambas A, Sanjaya W S, Mamat M and Prastio R P 2016 *Studies in Fuzziness and Soft Computing* **337** 133-53
[40] Sambas A, Mamat M, W S Mada Sanjaya, Salleh Z and Mohamad F S 2015 *Advanced Studies in Theoretical Physics* **9** 379-394
[41] Sambas A, Sanjaya W S and Halimatussadiyah 2012 *WSEAS Transactions on Systems* **9** 506-515
[42] Rucklidge A M 1992 *Journal of Fluid Mechanics* **237** 209-229
[43] Khalil H K 2002 *Nonlinear Systems* (New York: Prentice Hall)
[44] Wolf A, Swift J B, Swinney H L and Vastano J 1985 *Physica D* **16** 285-317

IOP Conf. Series: Materials Science and Engineering **332** (2018) 012034 doi:10.1088/1757-899X/332/1/012034

The effects of the application of SIMPRO on the completeness and time efficiency of nursing documentation in the outpatient instalation at Dompet Dhuafa Hospital Parung

M Dwisatyadini[1], R T S Hariyati[2], E Afifah[3]

[1]Department of Biology, Faculty of Mathematics and Natural Sciences, Open University of Indonesia.
[2,3]Departement of Nursing, Faculty of Nursing, University of Indonesia.

Corresponding author: mdwisatyadini@gmail.com

Abstract. Nursing documentation is clinical information that has a vital role in nursing services. The nursing process includes assessment, diagnosis, intervention, implementation, and evaluation. The purpose of this study was to determine the effects of the application of SIMPRO on the completeness and the efficiency of nursing documentation in the outpatient installation at Dompet Dhuafa Hospital Parung. This study used quantitative method with pre experimental (pre and posttest without control group) design. The mean of the documentation completeness marks before the application of SIMPRO was 1.87 (SD 0.922), and after SIMPRO was applied increased to 3.61 (0.588). This increase indicated an improvement of the nursing documentation completeness after the implementation of SIMPRO. The mean of time needed by nurses in documenting the nursing care before the application of SIMPRO was 476.13 seconds (SD 78.896). The mean of documenting time decreased more than a half after the application of SIMPRO which was 202.52 seconds (SD 196.723). SIMPRO made a nurse easier to take a decision analysis and decision support system to nursing care plan and documentation.

1. Introduction

Technology reduces the inefficiencies of time to the nurse workflow, reduces the nurse's workload, enhances the client's and families' active participation for health information, and improves nurse time to clients, thereby reducing incidents for unwanted complications such as falling patients, decubitus, and other quality indicators. Technology improves nurse satisfaction, helps overcome the shortage of nursing staff, helps prevent nurse work errors, improves nursing process, and provides health information to nurses to be more confident and improve nurse competence (Bolton, Gassert, Cipriano, 2008). Hariyati (2012) tell's about the effectiveness of model's nursing management information system. If nurse's use model's nursing information systems to documentation of nursing care plan are completeness and relevance. Also effectiveness to decision-making functions and legal aspects. Effectiveness and time efficiency indicated by prior to application system information 25.27% of the total time of one shift (7 hours), after the application system information the time needed 9.60% of the total shift time, in addition from the results of her research also shows control group a values of documentation quality (completeness, relevance, continuity, and decision support system (DSS) function there was not significant difference in first observation and second observation.

IOP Conf. Series: Materials Science and Engineering **332** (2018) 012034 doi:10.1088/1757-899X/332/1/012034

Purwandari (2013) tell's about a completeness of nursing care documentation in the intervention group before and after the application SIMPRO (Management of Nursing Information by Roro) in the adult inpatient room of Dompet Dhuafa Hospital Parung, showed a significant increase of 10.39%, after application of documentation based computer. Also the completeness of documentation and legal aspects after the implementation of SIMPRO (Nursing Management Information System by Roro) showed a difference between the control group of 6.24 (6.75% of the total score of 58.4) of the control group and the legal aspect of deviation was 3.53 (44.12% of total value 8). The benefits of mobile health include better access to health information and health services, improving patient care and patient safety, coordinating care, and empowering patients and families (Snyder, Wu, Miller, Jensen, Bantug & Wolff, 2011).

2. Methods

This study used quantitative method with pre experimental (pre and posttest without control group) design. Implementation of research by assessing 31th documentation, and efficiency time of nursing documentation at outpatient installation of Dompet Dhuafa Hospital before and after SIMPRO implementation. The pre-test data were collected on 28-30 May 2014 and posttest taking was conducted on 16 -18 June 2014. Data analysis was performed with the statistical test of a paired t test, independent *t* test. Simpro using localhost data base.

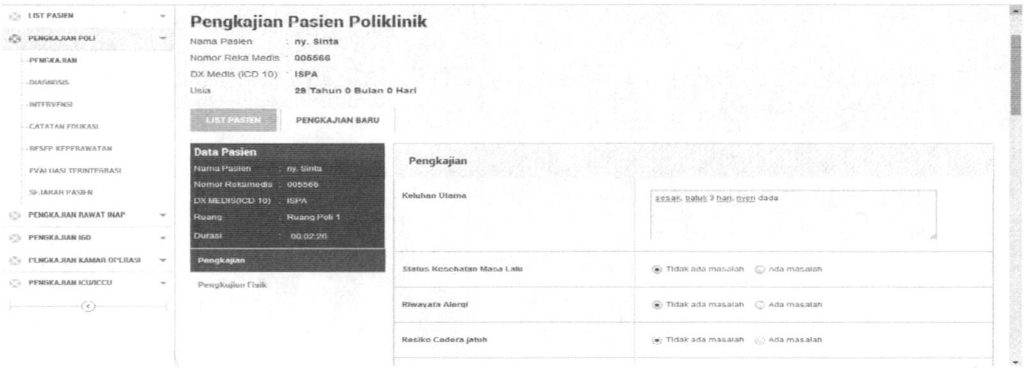

Figure 1 Stages of Nursing Assessment

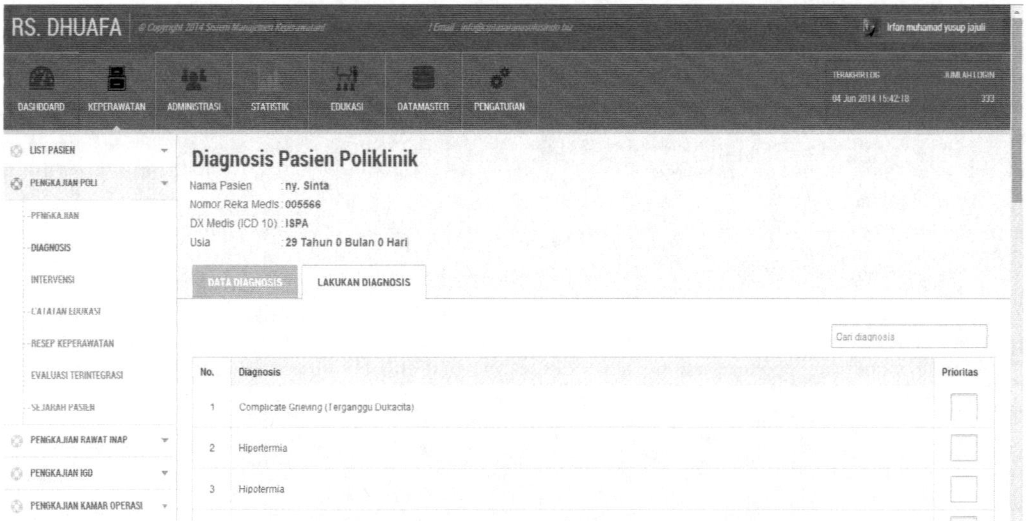

Figure 2 Stages of Nursing Diagnosis

IOP Conf. Series: Materials Science and Engineering **332** (2018) 012034 doi:10.1088/1757-899X/332/1/012034

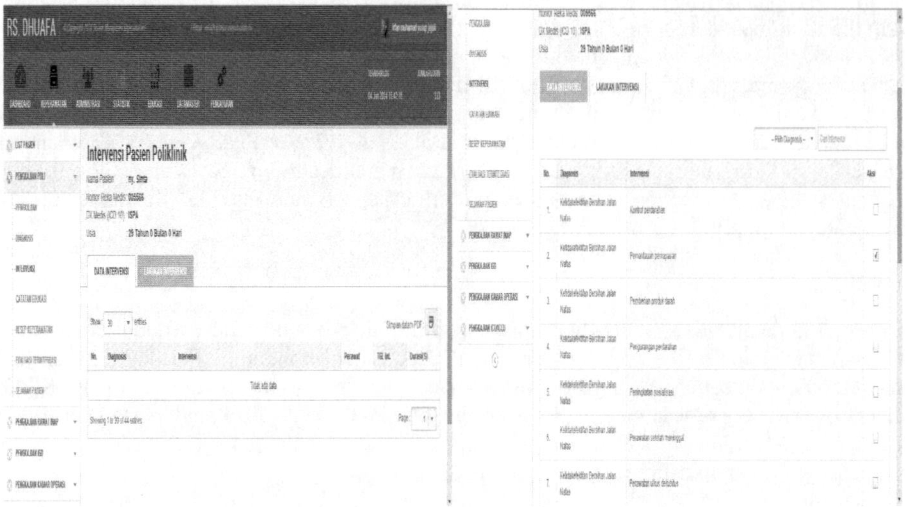

Figure 3 Stages of Nursing Intervention and Implementation

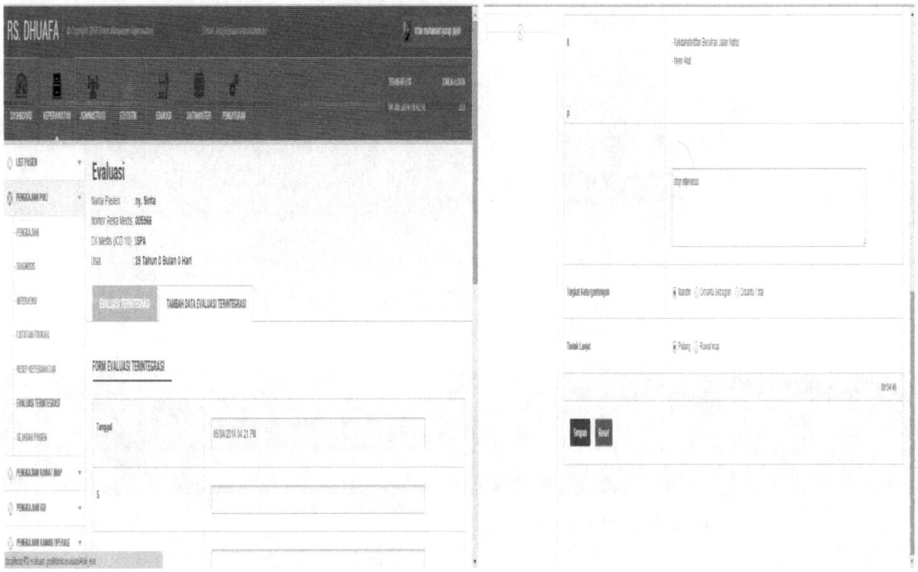

Figure 4 Stages of Nursing Evaluation

3. Results

The univariate result in the first stage of the researcher identifies the completeness and effectiveness of time in nursing documentation on outpatient installation of Dompet Dhuafa Hospital, Parung. Results obtained as follows:

IOP Conf. Series: Materials Science and Engineering **332** (2018) 012034 doi:10.1088/1757-899X/332/1/012034

Table 1 Description of Documentation Completion Before and After Implementation of SIMPRO at Outpatient Dompet Dhuafa Hospital Year 2014 (n = 31)

Completeness Nursing Documentation	Average	SD	Min-Mak	CI 95%	*p value*
Before SIMPRO	1,63	1,188	0-3	0,63-2,62	0,0001*
After SIMPRO	3,13	0,411	2-4	2,59-3,66	

* SD: Standard deviation, * CI: Confidence interval

Based on table 1 shows a mean value of completeness of nursing care documentation before the implementation of SIMPRO is 1.63 (SD 1.188), then after the implementation of SIMPRO increased to 3.13 (SD 0.411).

Table 2 Description of Documentation Time of Nursing Care Before and After Implementation of SIMPRO at Outpatient Dompet Dhuafa Hospital Year 2014 (n = 31)

	SIMPRO Documentation (n=31)				
Effectiveness of time in nursing documentation (seconds)	Average	SD	Min-Mak	CI 95%	*p value*
Before SIMPRO	472,50	93,159	360-600	394,62-550,38	0,0001*
After SIMPRO	189,88	77,832	124,81-254,94	124,81-254,94	

* SD: Standard deviation, * CI: Confidence interval

Based on table 2 shows an average of time nurses needed to documentation a nursing care before the implementation of SIMPRO is 472,50 seconds (SD 93,159). Mean Time Documentation decreased more than half time after SIMPRO implementation to 189.88 seconds (SD 77,832).

Table 3 Completeness and Effectiveness of Time in Documentation Nursing Care Before and After Implementation of SIMPRO at Outpatient Dompet Dhuafa Hospital Year 2014 (n = 31)

Variabel	Before SIMPRO (n=31)				*P value*
	Average	SD	Min-Mak	95%CI	
Completeness Nursing Documentation	1,63	1,188	0-3	0,63-2,62	0,0001*
Effectiveness of time in nursing documentation (seconds)	472,50	93,159	360-600	394,62-550,38	

Variabel	After SIMPRO (n=31)				*p value*
	Average	SD	Min-Mak	95%CI	
Completeness Nursing Documentation	3,13	0,411	2-4	2,59 -3,66	0,0001*
Effectiveness of time in nursing documentation (seconds)	189,88	77,832	82-286	124,81-254,94	

*SD: Standard deviation, * CI: Confidence interval

Based on Table 3 shows increase a mean value of documentation completeness before the implementation of SIMPRO of 1.63 (40.75%) of the total value of 4 to 3.13 (78.25%) of the total value of 4 after the implementation of SIMPRO. Based Effectiveness of Time in Documentation Nursing Care have a decrease time needs before the implementation of SIMPRO is equal to 472,50 second (7,9 min) to 189,88 second (3,2 min) after SIMPRO implementation. The result of statistical test shows a difference between a documentation completeness and efficiency of time in documentation with a significant between before and after the implementation of SIMPRO (p = 0.0001 *). The results of analysis shows the implementation of SIMPRO effective for a nurse because it can improve a completeness of nursing care documentation before and after the implementation of SIMPRO. Its can accelerated the time of nursing care documentation in Outpatient Installation Dompet Dhuafa Hospital before and after the implementation of SIMPRO.

Table 4 Analysis of Average Difference Scores of Completeness and Efficiency of Time for Documentation Nursing Care Before and After The Implementation of SIMPRO at Dompet Dhuafa Hospital Parung Year 2014 (n = 31).

No	Dependen Variabel's	Average	SD	SE	T	*Pvalue*
Completeness Nursing Documentation						
1	Before	1,87	0,922	0,166	-10,431	0,0001*
2	After	3,61	0,588	0,100		
	Deviation	-1,742	0,930	0,167		
Effectiveness of time in nursing documentation (seconds)						
1	Before	476,13	78,896	14,170	7,871	0,0001*
2	After	202,52	196,723	35,333		
	Deviation	273,613	193,547	34,762		

Description *: Meaning on α 0.05

Based on table 4 shows the average value of completeness nursing care documentation before the implementation of SIMPRO 1.87 (46.75%) of the total value of 4. Then after the implementation of SIMPRO 3.61 (90.25%) of the total value 4. The average difference value completeness nursing care documentation -1,742 (43,55%). The average value efficiency of time in documentation before the implemation of SIMPRO was 476.13 (7.9 minutes). Then after the implementation of SIMPRO was 202.52 (3.4 minutes). The average difference value of efficiency of time in documentation 273,613. Based on a results, we made a conclutions, if implementation of SIMPRO have a significant improvement between the completeness nursing care documentation and it has a significant decrease between efficiency of time in documentation before and after the implementation of SIMPRO (p = 0.0001). The results of this analysis shows a differences before and after the implementation of SIMPRO to nursing care documentation. After using a SIMPRO nurse faster for documentation between isnt using it, so its very efficiency of time in documentation. Then it made a completeness nursing care documentation.

4. Discussion

Based on the results of this research is the mean of the documentation completeness marks before the application of SIMPRO was 1.87 (SD 0.922), and after SIMPRO was applied increased to 3.61 (0.588). This increase indicated an improvement of the nursing documentation completeness after the implementation of SIMPRO. The mean of time needed by nurses in documenting the nursing care before the application of SIMPRO was 476.13 seconds (SD 78.896). The mean of documenting time decreased more than a half after the application of SIMPRO which was 202.52 seconds (SD 196.723). The results of this study are corresponding to Iyer and Camp (2005) showing the advantage of documentation using the system can improve the time frame, completeness and quality of nursing documentation, so that nurses can have heaps of time for patient care. Improve the patient's perception of the nurse. Reduced drug delivery errors, accessibility of documentation records easier to access immediately by nurses and other medical staff. Reduce the risk of loss of documentation records and improve the accuracy and quality of documentation records.

Purwandari (2013) was a research in adult inpatient room of Dompet Dhuafa Hospital Parung concerning a completeness of nursing care documentation in the intervention group before and after implementation of SIMPRO (Nursing Management Information System by Roro) showed a significant increase of 10.39%, after implementation of computer - based documentation. However, the research of Wang, Yu, Hailey (2013) in line with this study there is a significant increase of overall completeness value of documentation with paper mean 3.24, SD 0.54 and electronic mean 3.27, SD 0.42. Seen from the results Nurses can take advantage of their clinical skills, knowledge, computer skills, and critical thinking skills to assessment, nursing care plan, intervention, and evaluate nursing care that has been achieved all a patient needs (LeFevre 2009 in Kelley, Brandon & Docherty, 2011).

The nursing process is regarded as a scientific approach in which critical thinking is used to solve problems and this approach was introduced into nursing practice and education (Wang et al., 2011). Nursing documentation offers various options to enable nurses making choices regarding decision making for optimal care (Jefferies et al., 2010). Hariyati (2012) shows control group a values of documentation quality (completeness, relevance, continuity, and decision support system (DSS) function) there was not significant difference in first observation and second observation. In Nursing Practices and Education, after finished nurses performs the assessment a patient, so nurses must think of the right nursing care problem solving, and after implementation of nursing care, nurses need a time for documentation. Implementation of SIMPRO (Nursing Management Information System by Roro) SIMPRO made a nurses completeness their nursing documention, and made a effectiveness of time a nursing documentation. So nurses had a improvement about their critical thinking and their decision analysis to give a optimal nursing care for their patient.

5. Conclusions

Improved Documentation completeness prior to SIMPRO implementation and after SIMPRO implementation. Decrease in time required by nurses in documenting nursing care. Implementation of SIMPRO to the executing nurses has an effect to improve the completeness of nursing care documentation in the Outpatient Installation of Dompet Dhuafa Hospital. Implementation of SIMPRO to the executing nurses has an effect can accelerate the time of documenting the nursing care in the Outpatient Installation of Dompet Dhuafa Hospital. The mean value of differences in completeness of nursing care documentation before and after the implementation of SIMPRO was -1,742 (43.55%). The mean value of time difference between nursing care documentation before and after the implementation of SIMPRO was 273,613 (4.6 minutes).

6. Suggestions

Firstly for Dompet Dhuafa Hospital, this research is expected as input to implement SIMPRO in nursing care documentation in outpatient installation. The management needs to improve the routine monitoring of the implementation of nursing care documentation as a process of documentation evaluation that has been done by the outpatient nurse at least once a month by the outpatient coordinator which later the report will be submitted to the head of the nursing care sub-committee and to the chairman of the nursing committee, nurse performance and quality improvement of nursing service. Improving skills and abilities in documenting nursing care is done through ongoing training per year.

Second for Education, improving training, nursing education and nursing students on SIMPRO in nursing practice, the process of preparing them to use technology as a new strategy of health care.

Third for Researchers, this research as an illustration to conduct further research related to information system development and documentation of nursing care. There is a need for further research on the influence of the application of management information system for the documentation of nursing care (SIMPRO) to the completeness and efficiency of documentation time with longer time with more heterogeneous nurse responders, so it can be seen the effect of SIMPRO on the completeness and efficiency of the time of nursing documentation. In addition, it is suggested to extend the research by examining the effect of the application of management information system to the nursing care documentation (SIMPRO) to the quality in the accuracy of nursing care documentation at Profit and Non Profit Hospital.

References

[1] Bolton, L.B, Gassert, C.A, Cipriano, P.F (2008). Smart Technology, Enduring Solutions, Technology Solutions Can Make Nursing Care Safer and More Efficient. *JHIM*, Volume 22, Number 4.
[2] Hariyati .R.T. S (2012). *The effectiveness of a computer-based nursing management information system.* Dissertation.FIK- UI.
[3] Iyer, P.W & Camp. N.H (2005). *Nursing Documentation a nursing process approach. Ed. 3.* Jakarta: EGC.
[4] Jefferies, D., Johnson, M., Griffiths, R. (2010). A meta-study of the essentials of quality nursing documentation. International Journal of Nursing Practice, 16(2), 112-124
[5] Purwandari R. (2012). *Influence of "SIMPRO" to the Completeness and Legal Aspects of Nursing Care Documentation in Inpatient Room Adult Dompet Dhuafa Hospital, Parung Bogor.* Tesis. FIK UI.
[6] Snyder C.F, Wu A.W, Miller R.S, Jensen R.E, Bantug E.T, & Wolff A.C. (2011). The Role of Informatics in Promoting Patient-Centered Care. *The Cancer Journal*, Volume 17, Number 4.
[7] Wang. N, Yu. P, Hailey. D (2013). Description and comparison of documentation of nursing assessment between paper-based and electronic systems in Australian aged care homes. http://dx.doi.org/10.1016/j.ijmedinf.2013.05.002.
[8] Wang. N, Yu. P, Hailey. D (2011). Quality of nursing documentation and approaches to its evaluation: a mixed-method systematic review. Journal of Advanced Nursing 00(0), 000–000. doi: 10.1111/j.1365-2648.2011.05634.

IOP Conf. Series: Materials Science and Engineering **332** (2018) 012035 doi:10.1088/1757-899X/332/1/012035

A fuzzy mathematical model of West Java population with logistic growth model

N S Nurkholipah[1*], Z Amarti[2], N Anggriani[3], A K Supriatna[4]

[1,2]Master Student of Mathematics Study Program, Faculty of Mathematics and Natural Sciences, Universitas Padjadjaran, Indonesia.
[3,4]Departement of Mathematics, Faculty of Mathematics and Natural Sciences, Universitas Padjadjaran, Indonesia.

*Corresponding author: nendensitinurkholipah@yahoo.co.id

Abstract. In this paper we develop a mathematics model of population growth in the West Java Province Indonesia. The model takes the form as a logistic differential equation. We parameterize the model using several triples of data, and choose the best triple which has the smallest Mean Absolute Percentage Error (MAPE). The resulting model is able to predict the historical data with a high accuracy and it also able to predict the future of population number. Predicting the future population is among the important factors that affect the consideration is preparing a good management for the population. Several experiment are done to look at the effect of impreciseness in the data. This is done by considering a fuzzy initial value to the crisp model assuming that the model propagates the fuzziness of the independent variable to the dependent variable. We assume here a triangle fuzzy number representing the impreciseness in the data. We found that the fuzziness may disappear in the long-term. Other scenarios also investigated, such as the effect of fuzzy parameters to the crisp initial value of the population. The solution of the model is obtained numerically using the fourth-order Runge-Kutta scheme.

1. Introduction

Population growth is an important issue that a government in any country should pay attention. It is needed as a bases for population management action by the government, e.g. to make decision on how many new jobs should be provided to reduce the number of poor population. As an example, countries in West Africa which is a category of poor countries have an average annual population growth rate of 2,6% whereas ideally population growth rate is bellow1%[1]. This information can be used to either in providing new jobs or in reducing the growth rate of the populations.

In West Java, Indonesia, based on the result of the population census every ten years from 1980-2010 [2], the population is 23.434.003, 29.415.723, 35.723.473 and 43.053.732, respectively. It can be seen that the population of West Java Province every ten years has increased. The increasing number of population in a region causes many problems in various aspects, such as in social aspect, health aspects and environmental aspect. To overcome these problems, the government must prepare appropriate action for the future. One of them by knowing the projected population. The projection of the population of West Java Province can be done through mathematical modelling. Mathematical modelling is used here to estimate the population in the future, to know the population growth rate, and to know the maximum population limitation (the population carrying capacity).

IOP Conf. Series: Materials Science and Engineering **332** (2018) 012035 doi:10.1088/1757-899X/332/1/012035

The term of population projection has seen discussed in several studies, e.g. by the authors in [1], [3], [4], [5], [6]. In this regard, we developed a mathematical model of population growth in West Java Province of Indonesia. The model used is the form of logistic differential equations. Sampling method in logistic model is done by varying sampling interval. The smallest MAPE interval becomes a consideration for projected population in West Java Province.

Furthermore, because the population census data may be containing uncertainty, we use fuzzy set theory with triangle membership function. This is done by considering the fuzzy initial value of the obtaining data, instead of the crispinitial value of the data. Hence the resulting mathematical model contains uncertainty in some sense.In this paper we investigate the effect of this fuzzy initial value to the population projection. We use numerical solution using the fourth-order Runge-Kutta method to develop the projection. Other scenario is also investigate, i.e. the effect of a fuzzy parameter on a crisp initial value of the population. In the following section we provide a brief introductory to some mathematical concept used in the subsequent discussion.

2. Preliminaries

This section briefly describes the main theoretical tools that will be used in developing the mathematical model for a population growth.

2.1. Logistic Growth Model

This model is called Verhulst model, a refinement of the exponential model and was first introduced by Pierre Verhulst in 1838 [5]. Verhulst points out that population growth depends not only on the size of the population but the extent of this size from its upper limit as carrying capacity. The logistic equation can be written as follow:

$$\frac{dP}{dt} = rP\left(1 - \frac{P}{K}\right), \tag{1}$$

Where r represents the intrinsic growth rate, P represents the total population and K represents the carrying capacity. From equation (1) the following solution can be obtained:

$$P(t) = \frac{K}{1+\left(\frac{K}{P_0}-1\right)e^{-rt}}. \tag{2}$$

The values of parameters r and Kcan be estimated from the population $P(t)$when three data for three consecutive years with the same time-space are known. Let us assume that P_0be the population at $t = 0$, $P_T$be the population at $t = T$ and P_{2T} be the population at $t = 2T$, then from equation (2) it is obtained:

For $t = T$:

$$\frac{\beta}{\alpha}(1 - e^{-\alpha T}) = \frac{1}{P_T} - \frac{e^{-\alpha T}}{P_0}. \tag{3}$$

For $t = 2T$:

$$\frac{\beta}{\alpha}(1 - e^{-2\alpha T}) = \frac{1}{P_{2T}} - \frac{e^{-2\alpha T}}{P_0}. \tag{4}$$

By dividing equation (4) by (3) to eliminate K, it is obtained:

$$e^{-\alpha T} = \frac{P_0(P_{2T}-P_T)}{P_{2T}(P_T-P_0)}, \tag{5}$$

so the population growth rate(r) and carrying capacity (K)can be written expicitly as:

$$r = \frac{1}{T}\ln\frac{P_{2T}(P_0-P_T)}{P_0(P_T-P_{2T})}, \tag{6}$$

$$K = \frac{P_T(P_0P_T-2P_0P_{2T}+P_TP_{2T})}{{P_T}^2-P_0P_{2T}}. \tag{7}$$

These formulas is the main ingridient in the analysis of the data in the paper.

2.2 Mean Absolute Percentage Error (MAPE)

Mean Absolute Percentage Error (MAPE) is a statistical evaluation used to assess the suitability of a model in population projection. MAPE is expressed as a percentage. The concept of MAPE is very simple but becomes very important in choosing the best model from other models. A model with a smaller MAPE is a model chosen from other models. The mathematical from of MAPE is as follow:

$$MAPE = \frac{1}{N}\sum_{t=1}^{N}\left|\frac{x_t-\hat{x}_t}{x_t}\right| \times 100\% \qquad (8)$$

where $x_t$represent the actual population at time t, $\hat{x}_t$ represents the population projection at time t and N is the number of observations of each population. The smallest MAPE value is the best because of the smaller percentage of error generated from a model. For the interpretation of MAPE value that if a MAPE value of less than 10% is very accurate projection, 10% to 20% is a very good projection, 21% to 50% is reasonable projection and over 51% is forecasting not accurate [7].

Table 1. Membership Function of Fuzzy Set Theory

Function	Membership Function	Function	Membership Function
Linear Representations (a1) 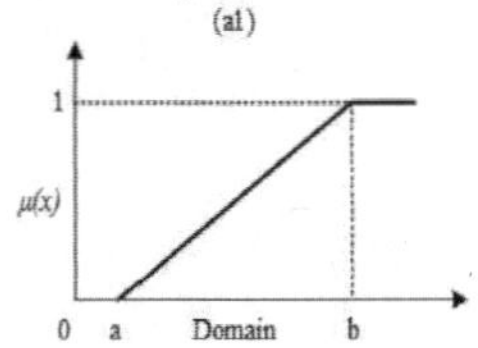 (a2) 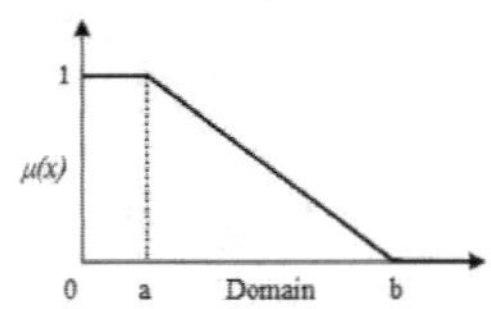 	$\mu(x) = \begin{cases} 0, x \le a \\ \frac{x-a}{b-a}, a \le x \le b \\ 1, x \ge b \end{cases}$ $\mu(x) = \begin{cases} 1, x \le a \\ \frac{b-x}{b-a}, a \le x \le b \\ 0, x \ge b \end{cases}$	**Triangular Curve Representation** (b) 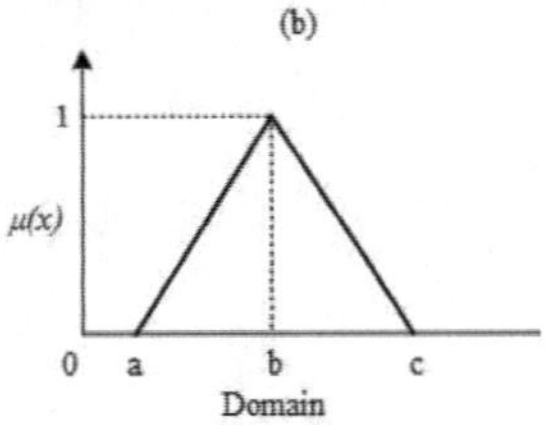 	$\mu(x) = \begin{cases} 0, x \le a \text{ atau } x \ge c \\ \frac{x-a}{b-a}, a \le x \le b \\ \frac{c-x}{c-b}, x \ge b \end{cases}$
Trapezoidal curve Representation (c) 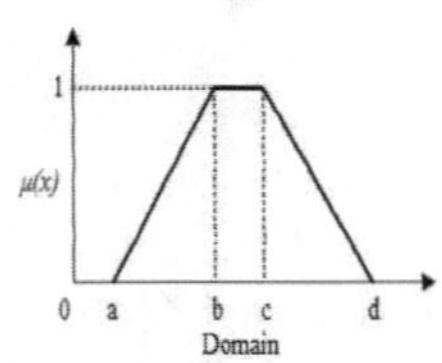 	$\mu(x) = \begin{cases} 0, x \le a \text{ atau } x \ge c \\ \frac{x-a}{b-a}, a \le x \le b \\ 1, b \le x \le c \\ \frac{d-x}{d-c}, c \le x \le d \end{cases}$	**Gaussian Curve Representation** (e) 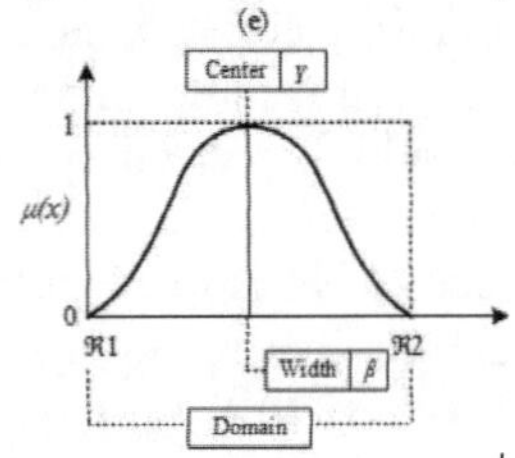 	$G(x;\beta,\gamma) = e^{-\frac{(x-\gamma)^2}{2\beta^2}}$
Combination Curve Representation (d) 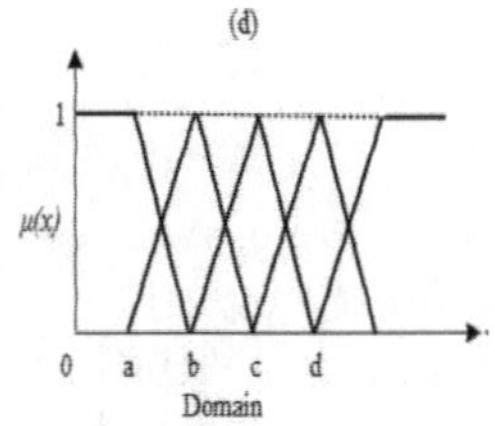 	Combination of linear membership functions (up and down) and triangle membership functions.		

2.3 Fuzzy Set Theory

The fuzzy set theory is designed to represent the uncertainty, inaccuracies and vagueness of the data/information available. The fuzzy set is a generalization of the ordinary set concept. For the universe of discourse U, the fuzzy set is determined by the membership function that maps the member u to the membership range in the interval [0,1]. For the set of membership fuctional members of discrete value 0 and 1.Membership function is a curve showing the mapping of data input points into their membership values (often called membership degrees) that have intervals between 0 and 1. One way to get membership value is through a functional approach. Some forms of membership function are linear representations (top and bottom), triangular curve representation, triangle curve, trapezoidal curve and Gaussian curve representations, Combination Curve [9].The graph of membership function for each curve can be seen in the following Table 1. In this paper we only concentrate in the triangular fuzzy number, since other forms can be use analogously.

2.4 The Fourth-Order Runge-Kutta Method

The most commonly used Runge-Kutta method is the fourth-order classical Runge-Kutta method, along with the equation of the classical fourth-order Runge-Kutta method [8].

$$y_{i+1} = y_i + \left[\frac{1}{6}(k_1 + 2k_2 + 2k_3 + k_4)\right] h, \tag{9}$$

with

$$k_1 = f(x_i, y_i),$$
$$k_2 = f\left(x_i + \frac{1}{2}h,\ y_i + \frac{1}{2}hk_1\right),$$
$$k_3 = f\left(x_i + \frac{1}{2}h,\ y_i + \frac{1}{2}hk_2\right),$$
$$k_4 = f(x_i + h,\ y_i + hk_3),$$

where k is recursively computed, i.e. k_1 is needed in k_2 calculation, k_2 is needed in k_3 calculation and k_3 is needed in k_4 calculation.

3. Results and Discussion

To select the most appropriate projection in the Logistic model, the population projection of 12 different samples of three points are taken with the same spaces between the points , i.e. the sampling interval 1, 2, 3 up to interval 12. The resulting projection is then compared with the actual data. The best projection is the projection that produces the smallest MAPE. For each interval on sampling, the population data taken as much as three pieces. If P_0 is the population at $t = 0$, P_T is the population at $t = T$ and P_{2T} is the population at $t = 2T$, so for interval 1, data was taken in 1990 as P_0, 1991 as P_T and 1992 as P_{2T}. At interval 2, data was taken in 1990 as P_0, 1992 as P_T and 1994 as P_{2T}. Likewise until at interval 12.In the logistic model, interval 12 is chosen as the best interval because it has a projection close to the actual data. The following is the resulting growth rate and carrying capacity of interval 12 to represent the best growth rate and carrying capacity of the model, namely

$$r^* = 0.040718403, \quad K^* = 75692359.51, \quad \text{with } MAPE = 1.410076124. \tag{10}$$

Using the values of parameters in (10), the projection by the logistic equation is presented in Figure 1. The logistic solution is compared to the actual data. The figure shows the projection up to time t = 100 and the actual data is up to time t = 25. The projection from the logistic model is pretty close. This is clarified by the resulting MAPE from the projection solution to the actual data which is 1.410076124. In the next section we look for the numerical solution of the best model by using the fourth-order Runge-Kutta method for various scenario of the fuzziness of the initial values and the model parameters. We assume that initial value of the population is a triangular fuzzy number. Other scenario is we assume that carrying capacity (K) and growth rate (r) are also fuzzy numbers with triangular membership functions.

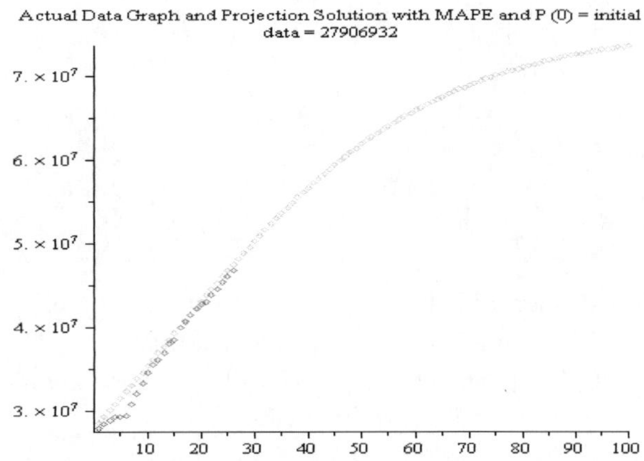

Figure 1. Graph of projection solution and actual data

3.1 Projection with the Fuzzy Initial Value

Suppose the initial value of the population (P_0) is a fuzzy number in the form of triangular membership $P_0 = (P_{0_1}, P_{0_2}, P_{0_3})$ with $P_{0_1} = 26906932, P_{0_2} = 27906932$ dan $P_{0_3} = 28906932$. Using the fourth-order Runge-Kutta method it is obtained the projection as in Figures 2, 3, 4 and 5.

(2a) (2b)

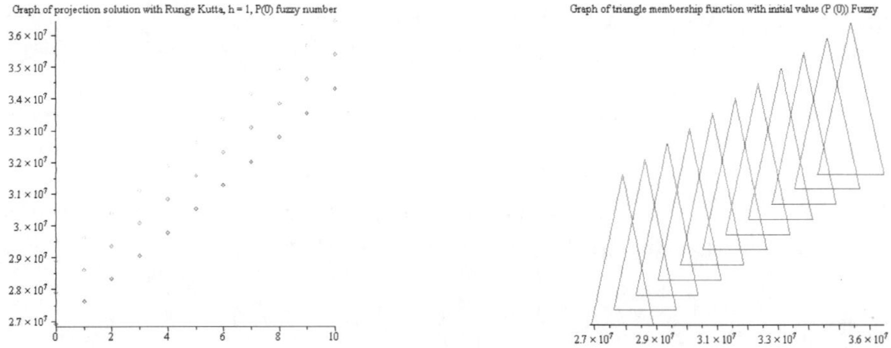

Figure 2. (2a) Numerical solution with the triangular fuzzy number with $P_{0_1} = 26906932, P_{0_2} = 27906932$ and$P_{0_3} = 28906932$ for $T = 10$; (2b) Graph of triangle membership function for the model solution.

In Figures 2 to 5, specifically figure (2a), (3a), (4a) and (5a) it is illustrated the numerical solution with the triangular fuzzy number P_0 with $P_{0_1} = 26906932, P_{0_2} = 27906932$ and$P_{0_3} = 28906932$ for projection up to 10, 100, 510 and 1000 iterations of time. The graph is increasing and grows asymptotically to the carrying capacity. Further, in Figures 2 to 5, specifically figure (2b), (3b), (4b) and (5b) it is illustrated the graph of triangle membership function for the model solution for projection up to 10, 100, 510 and 1000 iterations of time.

IOP Conf. Series: Materials Science and Engineering **332** (2018) 012035 doi:10.1088/1757-899X/332/1/012035

(3a) (3b) (3c)

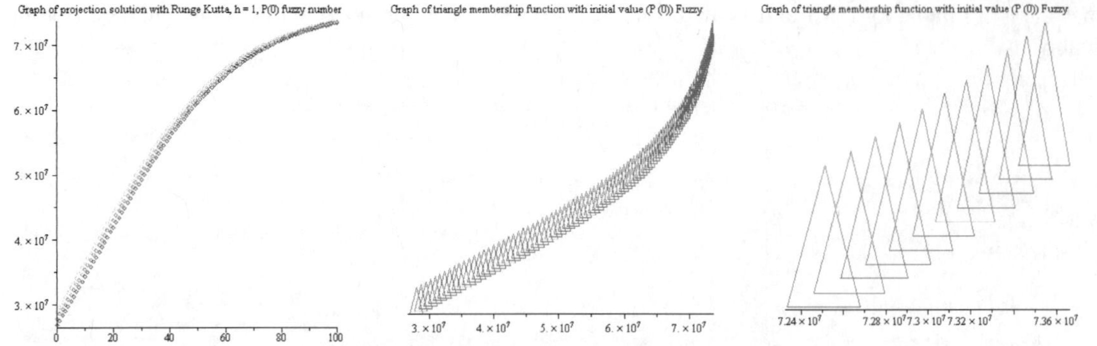

Figure 3. (3a) Numerical solution with the triangular fuzzy number with $P_{0_1} = 26906932, P_{0_2} = 27906932$ and $P_{0_3} = 28906932$ for $T = 100$; (3b) Graph of triangle membership function for the model solution with $T = 100$; (3c) Graph of triangle membership function for the model solution with $T = 90 \cdots 100$.

(4a) (4b) (4c)

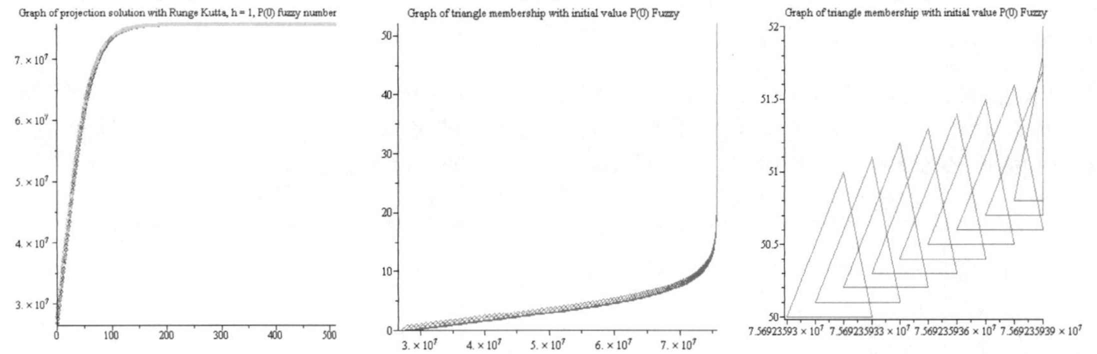

Figure 4. (4a) Numerical solution with the triangular fuzzy number with$P_{0_1} = 26906932, P_{0_2} = 27906932$ and $P_{0_3} = 28906932$ for $T = 510$; (4b) Graph of triangle membership function for the model solution with $T = 510$; (4c) Graph of triangle membership function for the model solution with $T = 500 \cdots 510$.

(5a) (5b) (5c)

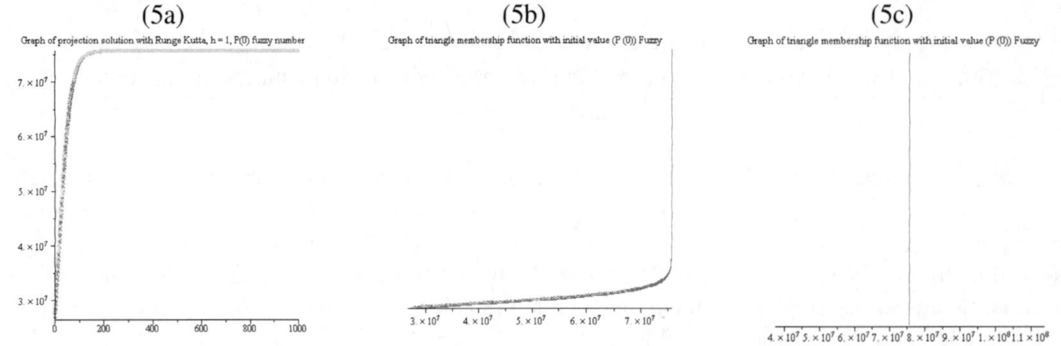

Figure 5. (5a) Numerical solution with the triangular fuzzy number with $P_{0_1} = 26906932, P_{0_2} = 27906932$ and $P_{0_3} = 28906932$ for $T = 1000$; (5b) Graph of triangle membership function for the model solution with $T = 1000$; (5c) Graph of triangle membership function for the model solution with $T = 990 \cdots 1000$.

We note that the support of the resulting triangular fuzzy number is getting smaller and smaller, until finally disappear and collapse to form a crisp number as indicated in Figure 4. This means that even though at first there is uncertainty in the initial value but in the long-term this uncertainty is dissapearing. To see that the resulting Runge-Kutta solution is indeed a fuzzy triangular number, we plot the solution at iteration 90 to 100 (Figure 3c). Figure (4c) is a graph of triangular membership function which in its 510 uncertainty is already disappearing, and at iteration 990 to 1000 the uncertainty is completely dissapear.

3.2 Projection with the Fuzzy Growth Rate (r)

In this section we assume that the parameters in the model is fuzzy. Suppose the growth rate of the population (r) is a fuzzy number in the form of triangular membership $r = (r_1, r_2, r_2)$ with $r_1 = -1.040718403, r_2 = 0.040718403$ and $r_3 = 1.040718403$. Using the fourth-order Runge-Kutta method it is obtained the projection as in Figure 6, 7 and 8. Figures 6 to 8, specifically figure (6a), (7a) and (8a) illustrate the numerical solution with the three initial values of the population given equally $P_{0_1} = 27906932$, $P_{0_2} = 27906932$ and $P_{0_3} = 27906932$ but r subjectively based on fuzzy set theory, i.e. $r_1 = -1.040718403, r_2 = 0.040718403$ and $r_3 = 1.040718403$ for projection up to 10, 100, and 500 iterations of time. The population increases when $r_2 = 0.040718403$ and $r_3 = 1.040718403$ grows asymptotically towards carrying capacity. Whereas when $r_1 = -1.040718403$ the population will go down and eventually extinct.

(6a) (6b) (6c)

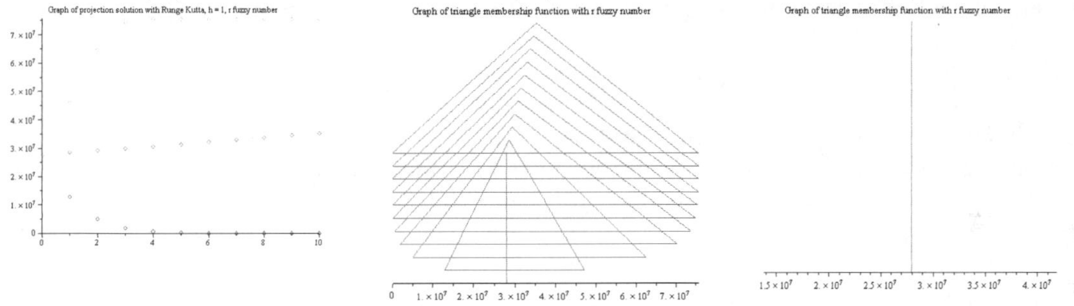

Figure 6. (6a) Numerical solution with the triangular fuzzy number with $r_1 = -1.040718403, r_2 = 0.040718403$ and $r_3 = 1.040718403$ for $T = 10$; (6b) Graph of triangle membership function for the model solution with $T = 10$; (6c) Graph of triangle membership function for the model solution with $T = 0$

(7a) (7b) (7c)

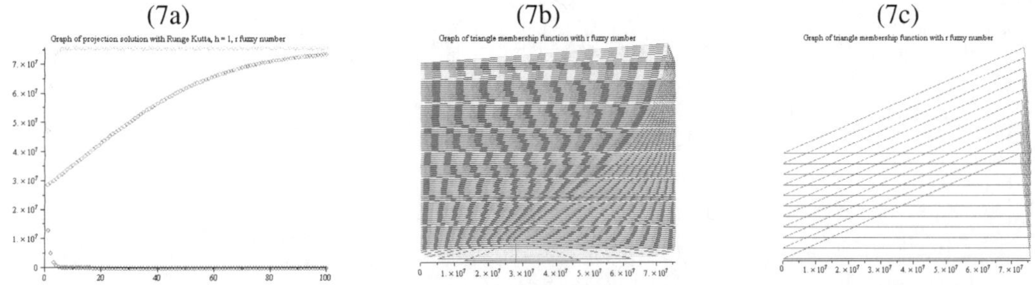

Figure 7. (7a) Numerical solution with the triangular fuzzy number with $r_1 = -1.040718403, r_2 = 0.040718403$ and $r_3 = 1.040718403$ $T = 100$; (7b) Graph of triangle membership function for the model solution $T = 100$; (7c) Graph of triangle membership function for the model solution $T = 90 - 100$

IOP Conf. Series: Materials Science and Engineering **332** (2018) 012035 doi:10.1088/1757-899X/332/1/012035

(8a) (8b) (8c)

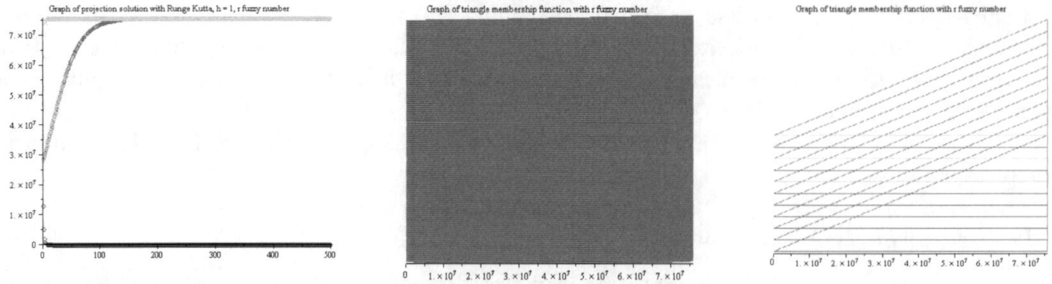

Figure 8. (8a) Numerical solution with the triangular fuzzy number with $r_1 = -1.040718403, r_2 = 0.040718403$ and $r_3 = 1.040718403$ for $T = 500$; (8b) Graph of triangle membership function for the model solution $T = 500$; (8c) Graph of triangle membership function for the model solution $T = 490 - 500$;

In Figures 6 to 8, specifically Figure (6b), (7b), and (8b) it is illustrated the graph of triangle membership function for the model solution for projection up to 10, 100 and 500 iterations of time. This figure shows the larger triangular membership function. In Figure (6c) it is shown a graph of triangle membership function at iteration 0, in which it is shown that is does not contain uncertainty on data. Figure (7b) is a graph of triangle membership function at iteration 0 to 100 and Figure (8b) is a graph of triangle membership function at iteration 0 to 500. In Figure (8b) is almost identical to Figure (7b), but because of the higher iteration so that the membership function in Figure (8b) coincides and looks like a blocked red. Figure (7c) is a graph of triangle membership function at iteration 90 to 100 and Figure (8c) is a graph of triangle membership function at iteration 490 to 500.

3.3 Projection with the Fuzzy Carrying Capacity (K)

Suppose the carrying capacity of the population (K) is a fuzzy number in the form of triangular membership $K = (K_1, K_2, K_2)$ with $K_1 = 74692359.51$, $K_2 = 75692359.51$ and $K_3 = 76692359.51$. Using the fourth-order Runge-Kutta method it is obtained the projection as in Figures 9, 10 and 11. In Figures 9 to 11, specifically figures (9a), (10a) and (11a), it is illustrated the numerical solution with the three initial values of the population given $P_{0_1} = 27906932, P_{0_2} = 27906932$ and $P_{0_3} = 27906932$ but K subjectively based on fuzzy set teory, i.e. $K_1 = 74692359.51$, $K_2 = 75692359.51$ and $K_3 = 76692359.51$ on iterations 10,100 and 500. Graphs 9, 10, 11 show that the population always rises and grows asymptotically to 3 different carrying capacities.

(9a) (9b) (9c)

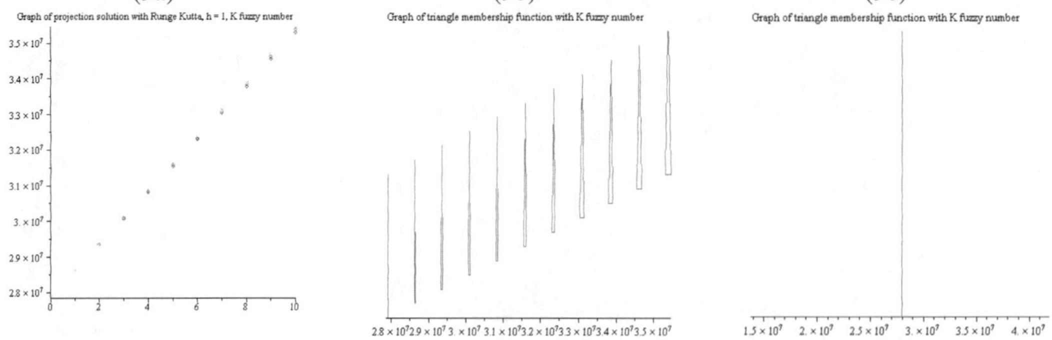

Figure 9. (9a) Numerical solution with the triangular fuzzy number with $K_1 = 74692359.51$, $K_2 = 75692359.51$ and $K_3 = 76692359.51$ for $T = 10$; (9b) Graph of triangle membership function for the model solution with $T = 10$; (9c) Graph of triangle membership function for the model solution with $T = 0$

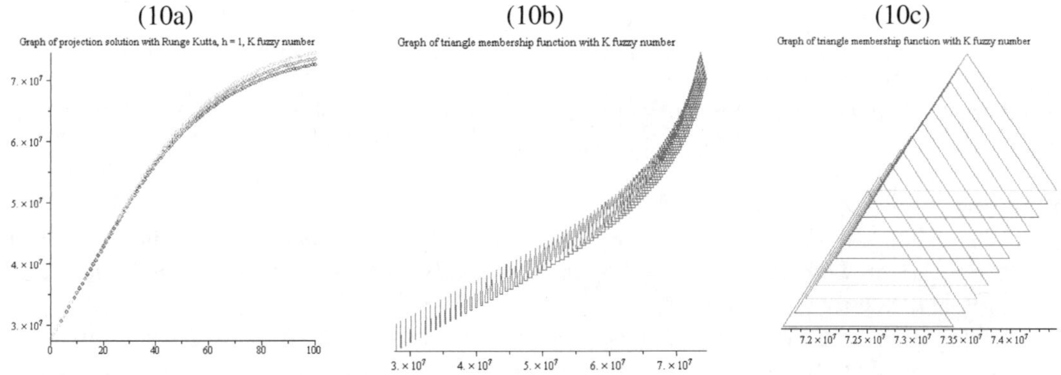

Figure 10. (10a) Numerical solution with the triangular fuzzy number with $K_1 = 74692359.51$, $K_2 = 75692359.51$ and $K_3 = 76692359.51$; (10b) Graph of triangle membership function for the model solution with $T = 100$; (10c) Graph of triangle membership function for the model solution with $T = 90 \cdots 100$

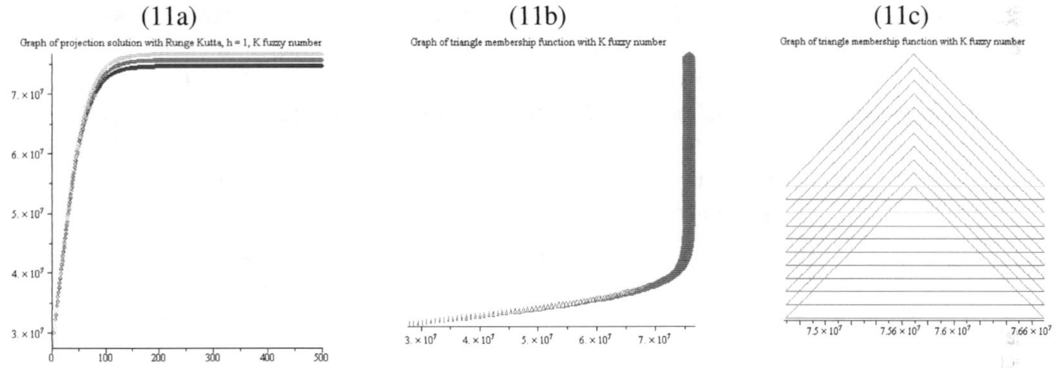

Figure 11. (11a) Numerical solution with the triangular fuzzy number with $K_1 = 74692359.51$, $K_2 = 75692359.51$ and $K_3 = 76692359.51$; (11b) Graph of triangle membership function for the model solution with $T = 500$; (11c) Graph of triangle membership function for the model solution with $T = 490 \cdots 500$

While in Figures (9b), (10b), and (11b) it is illustrated the graph of triangle membership function for the model solution for projection up to 10, 100 and 500 iterations of time. In Figure (9c) it is shown the graph of triangle membership function at iteration 0. It shows that it does not contain uncertainty on data. Figure (10c) is a graph of triangle membership function at iteration 90 to 100 and figure (11c) is a graph of triangle membership function at iteration 490 to 500 where uncertainty becomes apparent.

4. Conclusion

The population growth model used in this paper is the Logistic model. The value of carrying capacity (K), and growth rate (r) is obtained from the best sampling interval projection on the growth model of population logistic in West Java Province. We then look for the numeric solution of the model by using fourth-order Runge-Kutta method. To overcome uncertainty on the data, the fuzzy set theory approach is used. In this case, the initial population number (P_0), growth rate (r) and carrying capacity (K) are approximated by the triangular membership function. From the result of the discussion, it can be concluded that as far as mathematical model concern, fuzzy number initial values are affected the fuzziness of the projection, but in the case of logistic model, this fuzziness is dissapearing in the long-term. We also noted that fuzzy theory approach is considered appropriate to overcome uncertainty on the data in solving numerical solution of the model.

Acknowledgments

The authors thank to the University of Padjadjaran, due to the funding of this work through the scheme of Academic Leadership Grant (ALG) with contract number 855/UN6.3.1/PL/2017.

References

[1] Zulkarnain D. 2014. "*Proyeksi Populasi Penduduk Kota Bandung Menggunakan Model Pertumbuhan Populasi Verhulst dengan Memvariasikan Interval Pengambilan Sampel*". UIN SGD Bandung, Vol VIII No. 1

[2] Katalog BPS. 2016. *Jawa Barat Dalam Angka 2016.* Penerbit BPS Provinsi Jawa Barat.

[3] Zheng, Songfeng, *Methods of Evaluating Estimator*. Lecture notes Missouri State University.

[4] T R Malthus. 1978. *An Essay on The Pinciple of Population, Introduction by Philip Appleman, and assorted commentary on Malthus edited by Appleman*, Norton Critical Edition, ISBN 0-393-09202-X.

[5] Bacaer, Nicolas. 2011. *A Short History of Mathematical Population Dynamics*. Spinger-Verlag London Limited.

[6] Wali, Augustus, et al. 2011. Mathematical Modeling of Rwanda's Population Growth. *Applied Mathematical Science*, Vol 5, No. 53, 2617-2628.

[7] Lewis C D. 1982. *International and business forecasting method; A practical guide to eksponential smooting and curve fitting*. Butterworth Scientific, London.

[8] Chapra S C and R P Canale. 1991. Terjemahan, Metode Numerik untuk Teknik. UI-Press.

[9] Kusumadewi S and H Purnomo, 2010. Aplikasi Logika Fuzzy untuk Pendukung Keputusan, 2 ed., Yogyakarta: Graha Ilmu.

IOP Conf. Series: Materials Science and Engineering **332** (2018) 012036 doi:10.1088/1757-899X/332/1/012036

Optimal control issues in plant disease with host demographic factor and botanical fungicides

N Anggriani[1*], M Mardiyah[1], N Istifadah[2], A K Supriatna[1]
[1]Department of Mathematics, Faculty of Mathematics and Natural Sciences, Universitas Padjadjaran, Jl. Raya Bandung-Sumedang km 21 Jatinangor 45363, Indonesia
[2]Department of Plant Pathology, Faculty of Agriculture, Padjadjaran University, Indonesia

*Corresponding author: nursanti.anggriani@gmail.com

Abstract. In this paper, we discuss a mathematical model of plant disease with the effect of fungicide. We assume that the fungicide is given as a preventive treatment to infectious plants. The model is constructed based on the development of the disease in which the monomolecular is monocyclic. We show the value of the Basic Reproduction Number (BRN) $\mathcal{R}_0$ of the plant disease transmission. The BRN is computed from the largest eigenvalue of the next generation matrix of the model. The result shows that in the region where $\mathcal{R}_0$ greater than one there is a single stable endemic equilibrium. However, in the region where $\mathcal{R}_0$ less than one this endemic equilibrium becomes unstable. The dynamics of the model is highly sensitive to changes in contact rate and infectious period. We also discuss the optimal control of the infected plant host by considering a preventive treatment aimed at reducing the infected host plant. The obtaining optimal control shows that it can reduce the number of infected hosts compared to that without control. Some numerical simulations are also given to illustrate our analytical results.

1. Introduction

Epidemiology of plant diseases studies the development of plant disease populations in space and time. This area implies that the epidemiology of plant diseases is a change of disease, not a disease state at a given time and space. The Plant disease control has been shown the process of change in relation to disease cycles and disease populations. Both are interrelated because the cycle of disease can change the population of the disease [1]. The disease cycle in plants generally consists of infection, sporulation and dissemination. Plant diseases caused by fungi in the infection process consist of germination, penetration, and colonization sub-processes; the sporulation process consists of sporophore production sub-processes, spore production, spore maturation; as well as the dissemination process comprising sub-process of spore release, spore transport, and spore deposition. In the first cycle of disease, the infection process begins with the spores that act as the primary inoculum.

The host plant not only passes the infection caused by the primary inoculum alone. Where the consequence of the time required to complete one cycle of disease is whether the inoculum can still or not re-infect by the primary inoculum. If there are a fragile host plant and an appropriate environmental condition, the first cycles of the disease will produce a secondary inoculum that will re-start the disease cycle. Secondary inoculum which has the chance to re-infect is said to be the secondary or secondary infection.

As a primary pathogen, the fungus can infect the host tissue before there is another pathogenic fungal attack and may cause symptoms. As a secondary pathogen it appears when the fungus infects the host plant after another pathogen attack, so the attack rate becomes so severe [2]. The epidemic which affected by host's demography is very short compared with the pathogen. The dynamics like that are common for many plant diseases where comprise the natural population unit among which a pathogen spreads [8]. Based on the problem, it is required for farmers to control the disease by applying synthesis exterminator that is fungicide. Control of plant diseases with the use of synthetic fungicides continuously result in environmental pollution and poisoning in humans. In various studies, chemicals used have negative effects on humans and animals, such as various dangerous diseases such as cancer and disability [4]. Therefore, better control and preventive strategies are required to prevent plant disease transmission [5].

Mathematical models provide a powerful tool used to understand the dynamics of disease spread through a population and in decision making in regards to disease prediction and disease control [6]. A number of control and preventive strategies have been implemented such as the use of host plants resistant and chemical control [7]. Kermack and McKendrick in [8] developed a classical model for micro parasite host interactions in mammals which forms the basis of plant epidemiological models. The first models of temporal development of epidemic plant diseases were developed by Van der Plank in [9] and have formed the basis for plant disease modeling. Subsequently, various models have been developed. A mathematical model is commonly developed to examine the plant disease dynamics including the effects of control and preventive strategies on plant disease dynamics [6-23]. A number of researches studied the epidemiological models of the plant involving fungicides with curative treatment [13] and protection factor [12]. Furthermore, a discussion of the theory and plant epidemiology can be found in [14].

Mathematical models have been developed to investigate the dynamics of plant disease with and without roguing mechanism [16] and found that applying roguing mechanism can prevent plant disease transmission. Earlier theories on vector- borne diseases in [17, 18, 19, 20] highlight the importance of the direct interactions between the vector's dynamics and disease prevalence in the host. A model with roguing and replanting has been developed [16] and solved using backward Euler method. Furthermore, mathematical models have been developed to study the effects of plant disease transmission dynamics with roguing, replanting and additional treatment such as curative [21] and preventive treatments of plants [22].

A mathematical model discussed in [10] investigate the combination of non-dimensionalization to simplify an SIRX model with reducing the numbers of free parameters to assist biological interpretation. By using $\mathcal{R}_0$ the author in [10] also analyze disease invasion and endemic stability to characterize disease persistence [11]. In this paper we investigate the effectiveness of botanical fungicides by looking at the dynamics between epidemiology and host demographics with the density of protected host based on [11]. A protected host population is defined as a population that have been given botanical fungicides. Host demographics are shown from the monomolecular host growth function.

2. Dynamical Model

To develop the mathematical model, we consider the following assumptions:

1. There is a dependence between the host growth rate with the maximum host population
2. The infected host population is divided to two compartments: the primary infection caused by inoculum primary and the secondary infection caused by inoculum secondary
3. The susceptible host population do not affect the infected primer host population
4. The susceptible plant is given botanical fungicide
5. Primary infection population is a monocyclic process with primary inoculum as a source of infection and is characterized by symptoms that occur first in a season.
6. The population consists of healthy hosts (vulnerable) who have been given preventative botanical fungicides.

Based on the assumption above, the interaction between all sub-population can be depicted as in Figure 1.

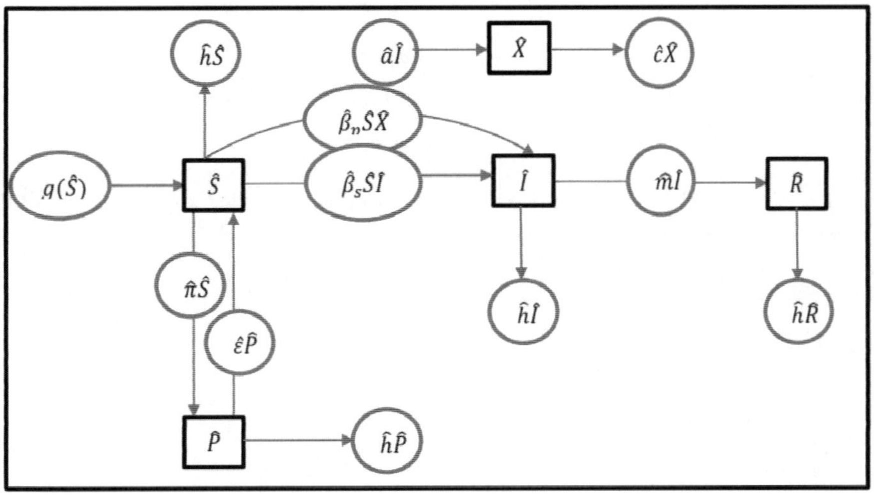

Figure 1. Transmission Diagram

Table 1. Definition of Variables and Parameters

Parameter/ Variable	Definition	Dimension Analog	Value ($\mathcal{R}_0 < 1$)	Value ($\mathcal{R}_0 > 1$)
$g(\hat{S})$	production of susceptible hosts			
$f(S)$	The dynamic of vulnerable hosts without infection, $f(\hat{S}) = g(\hat{S}) - \hat{h}\hat{S}$, $f(\hat{S}) = \frac{f(S)}{\hat{b}}$			
S	Density of susceptible host			
I	Density of Infected host			
R	Density of Removed Host			
X	Density of primary infection Host			
P	Density of Protection Host			
β_p	Rate of Primary imfection	$\beta_p = \frac{\widehat{\beta_p}\hat{a}\hat{K}}{\hat{b}^2}$	0.05	0.5
β_s	Rate of secondary infection	$\beta_s = \frac{\widehat{\beta_s}\hat{K}}{\hat{b}}$	0.075	0.75
m	rate of disease-induced mortality for infected hosts			
h	rate of removal (all classes of host)	$h = \frac{\hat{h}}{\hat{b}}$	0.25	0.25
b	Birth rate		0.5	0.5
k	carrying capacity		0.3	0.3
a	rate of production of inoculum by infected hosts		0.25	0.25
c	rate of decay of inoculum	$c = \frac{\hat{c}}{\hat{b}}$	0.8	0.8
$\hat{\pi}$	the effectiveness of botanical fungicides	$\pi = \frac{\hat{\pi}}{\hat{b}}$	0.1	0.1
$\hat{\varepsilon}$	The rate of damage to the botanical fungicide given to the plant host	$\varepsilon = \frac{\hat{\varepsilon}}{\hat{b}}$	0.01	0.01

Therefore, their dynamics can be described by the following autonomous system:

$$\frac{d\hat{S}}{d\hat{t}} = g(\hat{S}) + \hat{\varepsilon}\hat{P} - (\hat{\beta}_p\hat{X} + \hat{\beta}_s\hat{I})\hat{S} - \hat{\pi}\hat{S} - \hat{h}\hat{S}$$

$$\frac{d\hat{I}}{d\hat{t}} = (\hat{\beta}_p\hat{X} + \hat{\beta}_s\hat{I})\hat{S} - \hat{m}\hat{I} - \hat{h}\hat{I}$$

$$\frac{d\hat{R}}{d\hat{t}} = \hat{m}\hat{I} - \hat{h}\hat{R}$$
$$\frac{d\hat{X}}{d\hat{t}} = \hat{a}\hat{I} - \hat{c}\hat{X} \tag{1}$$
$$\frac{d\hat{P}}{d\hat{t}} = \hat{\pi}\hat{S} - \hat{\varepsilon}\hat{P} - \hat{h}\hat{P}$$

where $\hat{S} + \hat{I} + \hat{R} + \hat{X} + \hat{P} = \hat{\kappa}$.

The description of all variables and parameters are given in Table 1 with the values of all parameters are non-negative.

2.1. Non-dimensionalization

To proceed further, we consider the following dimensionless variables:

$S = \frac{\hat{S}}{\hat{\kappa}}$, $= \frac{\hat{I}}{\hat{\kappa}}$, $X = \frac{\hat{b}\hat{X}}{\hat{a}\hat{\kappa}}$, $R = \frac{\hat{b}\hat{R}}{\hat{m}\hat{\kappa}}$, $P = \frac{\hat{P}}{\hat{\kappa}}$ and $t = \hat{b}\hat{t}$

with the normalized parameter:

$\beta_p = \frac{\widehat{\beta_p}\hat{a}\hat{\kappa}}{\hat{b}^2}$, $\beta_s = \frac{\widehat{\beta_s}\hat{\kappa}}{\hat{b}}$, $\mu = \frac{\hat{\mu}}{\hat{b}}$, $c = \frac{\hat{c}}{\hat{b}}$, $h = \frac{\hat{h}}{\hat{b}}$, $\varepsilon = \frac{\hat{\varepsilon}}{\hat{b}}$, $\pi = \frac{\hat{\pi}}{\hat{b}}$ where $\hat{\mu} = \hat{m} + \hat{h}$

to obtain a reduced and normalized model of (1) in the form:

$$\frac{dS}{dt} = f(S) - (\beta_p X + \beta_s I)S - \pi S + \varepsilon P$$
$$\frac{dI}{dt} = (\beta_p X + \beta_s I)S - \mu I$$
$$\frac{dR}{dt} = I - hR \tag{2}$$
$$\frac{dX}{dt} = I - cX$$
$$\frac{dP}{dt} = \pi S - \varepsilon P - hP$$

Note that here we have:

$$\hat{S} + \hat{I} + \hat{R} + \hat{X} + \hat{P} = \frac{\hat{S}}{\hat{\kappa}} + \frac{\hat{I}}{\hat{\kappa}} + \frac{\hat{R}}{\hat{\kappa}} + \frac{\hat{X}}{\hat{\kappa}} + \frac{\hat{P}}{\hat{\kappa}} = \frac{\hat{\kappa}}{\hat{\kappa}} = 1.$$

2.2. Monomoleculer Model

The monomolecular model is the SIRXP model with the monomolecular function given by:

$$f(\hat{S}) = \hat{b}(\hat{\kappa} - q(\hat{S})) \tag{3}$$

which after normalization becomes

$$f(S) = 1 - S \tag{4}$$

Substitute (4) into (2) to obtain:

$$\frac{dS}{dt} = 1 - S - (\beta_p X + \beta_s I)S - \pi S + \varepsilon P$$
$$\frac{dI}{dt} = (\beta_p X + \beta_s I)S - \mu I$$
$$\frac{dR}{dt} = I - hR \tag{5}$$
$$\frac{dX}{dt} = I - cX$$

$$\frac{dP}{dt} = \pi S - \varepsilon P - hP$$

3. Mathematical Analysis

3.1. Equilibrium Point

There are two different equilibrium points for model (1). First is the disease free equilibrium point where all infected compartment are equal to zero, given by

$$E_{M0} = (S_0, I_0, R_0, X_0, P_0) = \left(\frac{\varepsilon+h}{\varepsilon+h+\pi h}, 0, 0, 0, \frac{\pi}{\varepsilon+h+\pi h}\right) \qquad (6)$$

Second is the non-trivial endemic equilibrium point which is given by

$$E_{M1} = (S_*, I_*, R_*, X_*, P_*) = \begin{pmatrix} \dfrac{\mu c}{\beta_p+\beta_s c}, \\ \dfrac{\beta_p\varepsilon+\beta_p h+\beta_s c\,\varepsilon+\beta_s ch-\mu c\varepsilon-\mu ch-\mu c\pi h}{\mu\left(\beta_p\varepsilon+\beta_p h+\beta_s c\varepsilon+\beta_s ch\right)}, \\ \dfrac{\beta_p\varepsilon+\beta_p h+\beta_s c\lambda+\beta_s ch-\mu c\varepsilon-\mu ch-\mu c\pi h}{\mu h\left(\beta_p\varepsilon+\beta_p h+\beta_s c\,\varepsilon+\beta_s ch\right)}, \\ \dfrac{\beta_p\varepsilon+\beta_p h+\beta_s c\,\varepsilon+\beta_s ch-\mu c\varepsilon-\mu ch-\mu c\pi h}{\mu c\left(\beta_p\varepsilon+\beta_p h+\beta_s c\varepsilon+\beta_s ch\right)}, \\ \dfrac{\mu c\pi}{\beta_p\varepsilon+\beta_p h+\beta_s c\,\varepsilon+\beta_s ch} \end{pmatrix} = \begin{pmatrix} \dfrac{\mu c}{\beta_p+\beta_s c}, \\ \dfrac{1}{\mu} - \dfrac{c}{R_0}, \\ \dfrac{1}{\mu h} - \dfrac{c}{hR_0}, \\ \dfrac{1}{\mu c} - \dfrac{1}{R_0}, \\ \dfrac{\mu c\pi}{\left(\beta_p+\beta_s c\right)(\varepsilon+h)} \end{pmatrix} \qquad (7)$$

3.2. Basic Reproduction Ratio

The Basic Reproduction Ratio represents the expected numbers of secondary cases produced by a typical infected individual during its entire period of infectiousness in completely susceptible population (see the detail in [23,24]). The Basic Reproduction Ratio, denoted by $\mathcal{R}_0$, is taken from spectral radius of the next generation matrix [24,25,26] of system (5),

$$G = \begin{bmatrix} \beta_s\left(\dfrac{\varepsilon+h}{\varepsilon+h+\pi h}\right)c + \beta_p\left(\dfrac{\varepsilon+h}{\varepsilon+h+\pi h}\right) & \beta_p\left(\dfrac{\varepsilon+h}{\varepsilon+h+\pi h}\right)\mu \\ 0 & 0 \end{bmatrix}$$

Taking the spectral radius of matrix G will bring us to the form of the Basic Reproduction Ratio

$$\mathcal{R}_0 = \left(\beta_s c + \beta_p\right)\left(\frac{\varepsilon+h}{\varepsilon+h+\pi h}\right)$$

3.3. Local Stability Analysis

Theorem 1

The disease-free equilibrium point of the system (1)-(5) is locally asymptotically stable if $\mathcal{R}_0 < 1$.

Proof :

By following [23] and from (6), the Jacobian matrix for non-endemic equilibrium point is

$$JT = \begin{bmatrix} -1-\pi & \dfrac{-\beta_s(\varepsilon+h)}{\varepsilon+h+\pi h} & 0 & \dfrac{-\beta_p(\varepsilon+h)}{\varepsilon+h+\pi h} & \varepsilon \\ 0 & & 0 & & 0 \\ 0 & \beta_s S-(\varepsilon+h)-\mu & -h & \beta_p S & 0 \\ 0 & 1 & 0 & 0 & 0 \\ \pi & 1 & 0 & -c & -\varepsilon-h \\ & 0 & & 0 & \end{bmatrix}$$

The polynomial characteristic of JT is $P(\lambda) = (\lambda+h)P_1(\lambda)P_2(\lambda))$ with:

$$P_1(\lambda_{2,3}) = \frac{1}{\varepsilon+h+\pi h}\Big((\varepsilon + h + \pi h)\lambda^2 + (c\varepsilon + hc + \pi hc - \beta_s\varepsilon - \beta_s h + \mu\varepsilon + \mu h + \pi\mu h)\lambda - \beta_s c\varepsilon - \beta_s ch + \mu c\varepsilon + \mu ch + \mu c\pi h - \beta_p\varepsilon - \beta_p h = 0\Big)$$

$$P_2(\lambda_{4,5}) = \lambda^2 + (\varepsilon + h + \pi + 1)\lambda + \varepsilon + h + \pi h$$

The eigenvalues of P_T are $-\alpha$ and the roots of the polynomial $p_1(\lambda)$ and $p_2(\lambda)$. The coefficient $q_1 > 0$ if $\mathcal{R}_0 < 1$, hence all the roots of polynomial P_T have negative real part when $\mathcal{R}_0 < 1$. This means that the non-endemic equilibrium point $E_{M0} = (S_0, I_0, R_0, X_0, P_0)$ is locally asymptotically stable when $\mathcal{R}_0 < 1$ [23,27]. This proves the theorem.

Theorem 2

The endemic equilibrium point of the system (1)-(5) is locally asymptotically stable if $\mathcal{R}_0 > 1$.

Proof:

The Jacobian matrix of the endemic equilibrium (7) gives the characteristic polynomial:

$$Pe(\lambda) = (\lambda + h)P_3(\lambda)$$

$$P_3(\lambda) = a_0\lambda^4 + a_1\lambda^3 + a_2\lambda^2 + a_3\lambda + a_4$$

The eigenvalues of P_e are $-h$ and the roots of the polynomial $P_3(\lambda)$. From Routh-Hortwitz criterian we have $a_0, a_1, a_2, a_3, a_4 > 0$, $a_2a_3 - a_1a_4 > 0$, and $a_1a_2a_3 - {a_1}^2a_4 - a_0{a_3}^2 > 0$, hence all the roots of polynomial P_{T_*} have negative real part when $\mathcal{R}_0 > 1$ [20]. This means that the non-endemic equilibrium point $E_{M1} = (S_*, I_*, R_*, X_*, P_*)$ is locally asymptotically stable when $\mathcal{R}_0 > 1$ [23, 27]. This proves the theorem.

4. Optimal Control Problem

To obtain a decrease in the number of infected populations, the dynamic model of the spread of plant diseases is given a preventive treatment of botanical fungicides. Therefore, optimal control is required on the treatment given to the cost in order to keep the infected plant population at a minimum. In completion of this optimal control, the Pontryagin Maximum Principle is used with the following objective functions [28]:

$$J = \int_{t_0}^{t_f} [A_0 I(t) + A_1 u^2(t)]\, dt$$

The variable $u(t)$ is the control variable of the preventive treatment given (i.e botanical fungicide) and A_0 is the weight of infected population balance and A_1 is the weight of the preventive cost equilibrium cost. The state variable for the model $x(t) = \begin{bmatrix} S(t) \\ I(t) \\ R(t) \\ X(t) \\ P(t) \end{bmatrix}$ and the constraint:

$$\frac{dS}{dt} = 1 - S - (\beta_p X + \beta_s I)S - u\pi S + \varepsilon P$$

$$\frac{dI}{dt} = (\beta_p X + \beta_s I)S - \mu I$$

$$\frac{dR}{dt} = I - hR$$

$$\frac{dX}{dt} = I - cX$$

$$\frac{dP}{dt} = u\pi S - \varepsilon P - hP$$

The system should satisfy the condition:

$$0 \le u(t) \le 1, 0 \le t \le t_f, S(0) = S_0 > 0, I(0) = I_0 \ge 0, R(0) = R_0 \ge 0, X(0) = X_0 \ge 0, P(0) = P_0 \ge 0$$

We define Hamiltonian function as $H = f(x,u,t) + \lambda' g(x,u,t)$, which equivalent to :

$$H = A_0 I(t) + A_1 u^2(t) + \lambda_1\left(1 - S - (\beta_p X + \beta_s I)S - u\pi S + \varepsilon P\right) + \lambda_2\left((\beta_p X + \beta_s I)S - \mu I\right) + \lambda_3\,(I - hR) + \lambda_4(I - cX) + \lambda_5(u\pi S - \varepsilon P - hP)$$

where $\lambda_1(t), \lambda_2(t), \lambda_3(t), \lambda_4(t)$, and $\lambda_5(t)$ are the *co-state* variable or the Lagrange multiplier of the optimization problem. The necessary conditions that an optimal control is achieved, it must satisfy the following Pontryagin's Maximum Principle:

- State equation

$$\dot{S} = \frac{\partial H}{\partial \lambda_1} = 1 - S - (\beta_p X + \beta_s I)S - u\pi S + \varepsilon P$$
$$\dot{I} = \frac{\partial H}{\partial \lambda_2} = (\beta_p X + \beta_s I)S - \mu I$$
$$\dot{R} = \frac{\partial H}{\partial \lambda_3} = I - hR$$
$$\dot{X} = \frac{\partial H}{\partial \lambda_4} = I - cX$$
$$\dot{P} = \frac{\partial H}{\partial \lambda_5} = u\pi S - \varepsilon P - hP$$

- Co-state equation

$$\dot{\lambda}_1 = -\frac{\partial H}{\partial S} = \lambda_1(1 + \beta_p X + \beta_s I + u\pi) - \lambda_2(\beta_p X + \beta_s I) - u\pi\lambda_5$$
$$\dot{\lambda}_2 = -\frac{\partial H}{\partial I} = -A_0 + \beta_s S\lambda_1 + \mu\lambda_2 - \lambda_3 - \lambda_4$$
$$\dot{\lambda}_3 = -\frac{\partial H}{\partial R} = h\lambda_3$$
$$\dot{\lambda}_4 = -\frac{\partial H}{\partial X} = \beta_p S(\lambda_1 - \lambda_2) + c\lambda_4$$
$$\dot{\lambda}_5 = -\frac{\partial H}{\partial P} = h\lambda_5 + \varepsilon\lambda_5 - \varepsilon\lambda_1$$

- Stationer condition

$$\frac{\partial H}{\partial u} = 0 \rightarrow u = \frac{(\lambda_1 - \lambda_5)\pi S}{2A_1}$$

Since $0 \le u \le 1$, then we get : $u = \min\left\{\max\left(0, \frac{(\lambda_1 - \lambda_5)\pi S}{2A_1}\right), 1\right\}$ as the optimal control of the system.

5. Numerical Result

In order to illustrate the dynamics of each compartment, we give numerical example with control and without control. We use the values of the parameters shown in Table 1 and the initial condition for each compartment is respectively, $S(0) = 1$, $I(0) = 0.8$, $R(0) = 0$, $X_0(0) = 0.85$, $P_0(0) = 0$.

From figure 2 and figure 3, the primary and the secondary infected susceptible rates increase until day 25 for $\mathcal{R}_0 > 1$, so the host infected population or the host removed plant will be increase at the same time. When $\mathcal{R}_0 < 1$ the host of primary infected plant and the secondary infected plants decreases or does not occur outbreak. The host protected plant by the fungicide will be decreases when $\mathcal{R}_0 > 1$ and increases when $\mathcal{R}_0 < 1$.

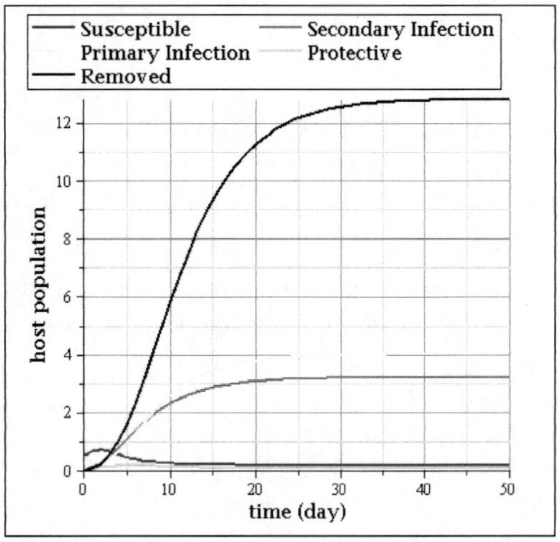

Figure 2. The dynamical of population plant when $\mathcal{R}_0 > 1$

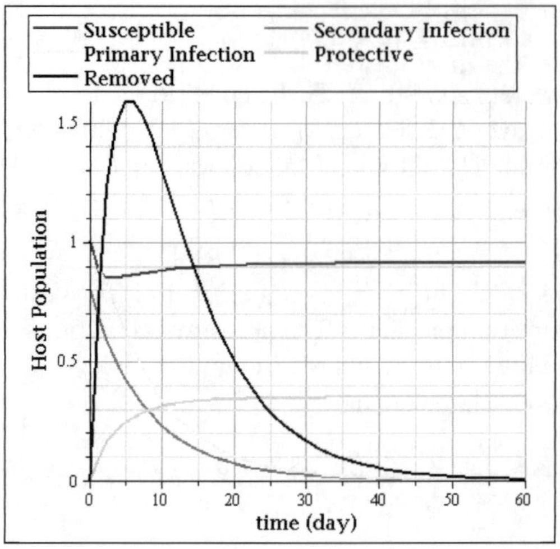

Figure 3. The dynamical of population plant when $\mathcal{R}_0 < 1$.

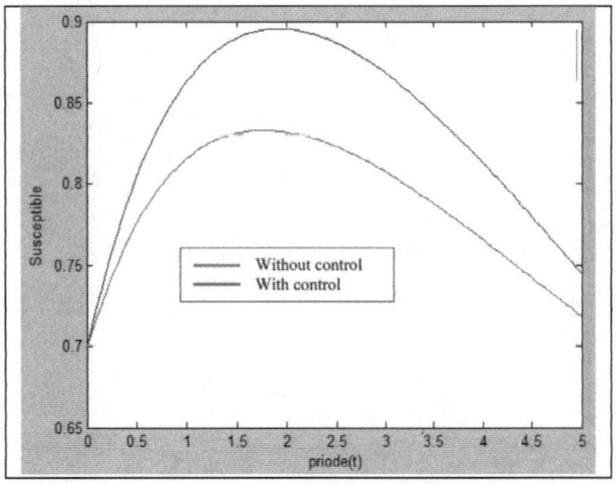

Figure 4. Dynamic of Susceptible Plant with control and without control.

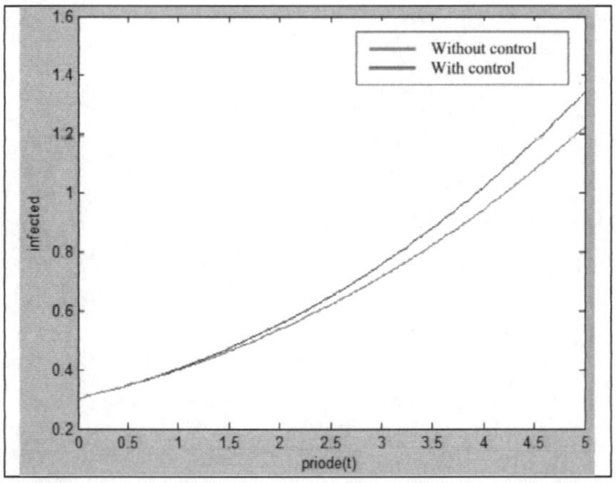

Figure 5. Dynamic of Infected Plant with control and without control.

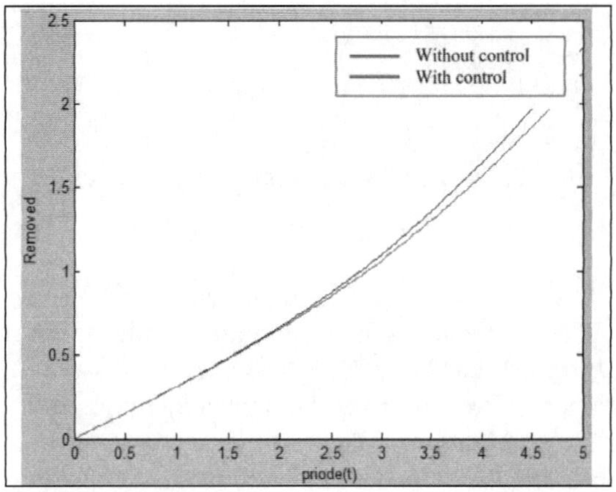

Figure 6. Dynamic of Removed Plant with control and without control.

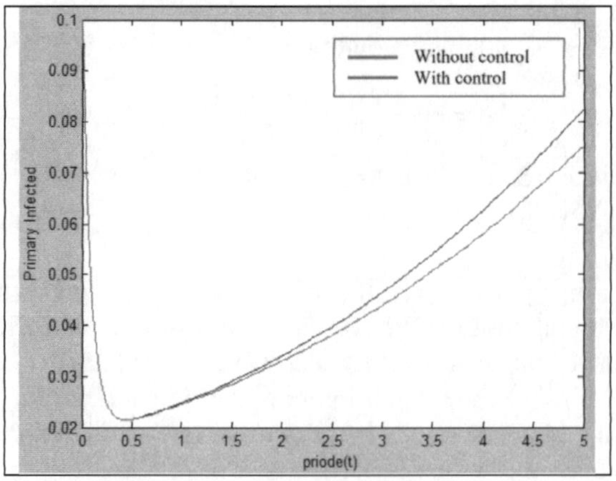

Figure 7. Dynamic of Primary Infected Plant with control and without control.

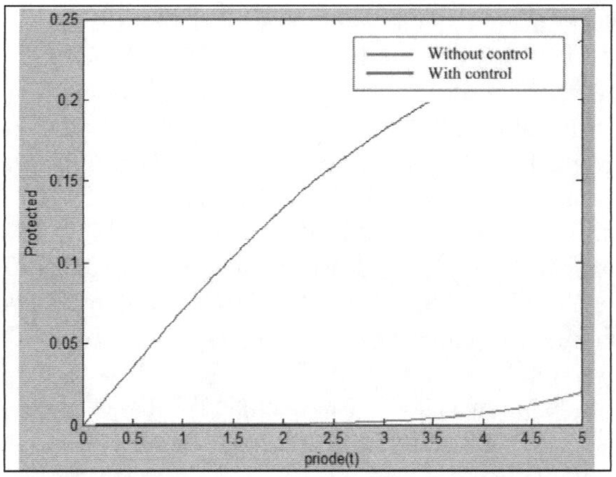

Figure 8. Dynamic of Protected Plant with control and without control.

Figure 4 show that with control, the dynamics of susceptible host plant will increase at 2.5 months then decrease until harvesting or retraction. In figure 7 show that the dynamics of primary infected host plant will decrease until at a half of month and then will be increasing this is caused the infected secondary plant increasing. The dynamics of removed host plant and the protective host plant with the fungicide will be increasing after control treatment. Figure 4 until figure 8 show that the host plant will decrease after control treatment.

6. Conclusion

In this paper, we develop a mathematical model of plant disease with the effect of fungicide. It has been shown numerically, that the optimal control strategy exists. The numerical example shows that the dynamics of primary infected host plant will decrease until at a half of month and then it will be increasing. This eventually causes the infected secondary plant increasing. The dynamics of removed host plant and the protective host plant with the fungicide will be increasing after control treatment.

Acknowledgments

The primary work is carried out with funding from PUPT RISTEKDIKTI 2017, with contract number 718/UN6.3.1/PL/2017.

References

[1] Jeger M 2009 *Epidemiology of Plant Disease* Wiley Online Library
[2] Miller J et al 2014 *Late Blight Management Action Plan for Potatoes* University of Idaho https://www.cals.uidaho.edu/edcomm/pdf/CIS/CIS1132.pdf
[3] Yung J et al 2017 *Plant Fungal Pathogenesis* BioMed Research International
[4] Shim Y K Mlynarek SP and Wijngaarden E 2009 *Parental Exposure to Pesticides and Childhood Brain Cancer* U.S. Atlantic Coast Childhood Brain Cancer Study Environmental Health Perspectives 117(6).
[5] Agrios G N 2005 *Plant Pathology* San Diego California: Academic Pres
[6] Zhou F and Yao H 2014 Global Dynamics of a Host-Vector-Predator mathematical model *J. of App Mathematics.* 59.
[7] Rida S Z et al 2016 Mathematical Model of Vector-Borne Plant Disease with Memory on the Host and The vector *Progr.Fract. Differ. Appl.*2 (4) p 277-285.
[8] Kermack WO and Mckendrick A G 1927 A combination to the mathematical theory of epidemics, *Proceeding of the Royal Society of London* (115) p700-721.
[9] Van der Plank J E 1963 *Plant disease: Epidemics and control* New York USA: Academics Pres

[10] Gilligan C A 2002 An epidemiological framework for disease management *In Advances in botanical research* vol 38 pp 1–64 London UK Academic Press (doi:10.1016/S0065-2296(02)38027-3)
[11] Cunniffe N J and Gilligan C A 2009 Invasion, Persistence, and Control in Epidemic Models for Plant Pathogens : The Effect of Host Demography *J. R. Soc. Interface* (*7*), p *439-451*
[12] Anggriani N Istifadah N Hanifah M and Supriatna A K 2016 A Mathematical Model of Protectant and Curative Fungicide Application and its Stability Analysis. *IOP Conf. Series: Earth and Environmental Science*, 31 (2016) 012014 doi:10.1088/1755-1315/31/1/012014.
[13] Anggriani N Nurul Putri L and Supriatna AK 2015 Stability analysis and optimal control plant fungal: An explicit model with curative factor *AIP Conference Proceedings* 1651 (40); doi: 10.1063/1.4914430
[14] Zang X S Holt J and Colvin J 2000 Mathematical Models of Host Plant Infection by Helper-Dependent Virus Complexes: Why Are Helper Viruses Always Avirulent? *Analytical and Theoretical Plant Pathology* 90(1) p 85-93
[15] Madden L V Jeger M J and Van den Bosch F 2000 A theoretical assessment of the effects of vector-virus transmission mechanism on plant virus disease epidemics *Phytopathology* 90 p 576-594. URL : https://doi.org/10.1094/PHYTO.2000.90.6.576
[16] Chan M S and Jeger M J 1994 An analytical model of plant virus disease dynamics with roguing and replanting. *Journal of Applied Ecology* (31) p 413–427
[17] Gourley S A Liu R and Wu J 2007 Eradicating Vector-borne disease via age-structured curling *J. Math Biol* (54) p 309-335.
[18] Holt J et al 1997 An epidemiological model incorporating vector population dynamics applied to African cassava mosaic virus disease *J. Appl. Ecol.* (34)793806
[19] Murwayi A L M Onyango T and Owour B 2017 Mathematical Analysis of Plant Disease Dispersion Model that Incorporates wind Strength and Insect Vector at Equilibrium *British Journal of Mathematics and Computer Science* 22(5) p 1-17
[20] Anggriani N Yusuf M Supriatna A 2017 The Effect of Insecticide on the Vector of Rice Tungro Disease: Insight from a Mathematical Model, *Information: an International Interdisciplinary Journal* 20 (9A)
[21] Arumi D Anggriani N Hertini N 2017 *Proceedings of 2th Nasional Conference* UPH
[22] Anggriani N et al 2017 Dynamical Analysis of Plant Disease Model with Roguing, Replanting and Preventive Treatment, *Proceedings of 4th International Conferenve on Research, Implementation, and Education of Mathematics and Science*
[23] Brauer F and Castillo-Chavez C 2001 *Mathematicals Models in Population Biology and Epidemology* Springer-Verlag
[24] Diekmann O and Heesterbeek J A P 2000 *Mathematical Epidemiology of Infectious Diseases: Model building, analysis and interpretation* Wiley: New York
[25] Diekmann O Heesterbeek J A P and Metz J A J 1990 On the definition and computation of the basic reproduction ratio R0 in models for infectious diseases in heterogeneous populations. *Journal of Mathematical Biology*, 28 (4) p 356-382
[26] Castillo-Chavez C Feng Z and Huang W 2006 On the Computation of R_0 and its Role on Global Stability. Mathematical Approaches for Emerging and Reemerging Infectious Disease: An Introduction *IMA Springer-Verlag* 125 p 229-250
[27] S Wiggins 2000 *Introduction to Applied Nonlinear Dynamical Systems and Chaos* (*Texts in Applied Mathematics*) *2nd Edition*, Springer-Verlag : Heidelberg
[28] Leinhart S and Workman J T 2007 *Optimal Control Applied to Biological Models* CRC Press, taylor & Francis Group

IOP Conf. Series: Materials Science and Engineering **332** (2018) 012037 doi:10.1088/1757-899X/332/1/012037

Linear regression based on Minimum Covariance Determinant (MCD) and TELBS methods on the productivity of phytoplankton

N Gusriani*, Firdaniza

Department of Mathematics, Faculty of Mathematics and Science, Padjadjaran University, Jalan Raya Bandung-Sumedang Km 21 Jatinangor, Indonesia

*Corresponding author: nurul.gusriani@unpad.ac.id

Abstract. The existence of outliers on multiple linear regression analysis causes the Gaussian assumption to be unfulfilled. If the Least Square method is forcedly used on these data, it will produce a model that cannot represent most data. For that, we need a robust regression method against outliers. This paper will compare the Minimum Covariance Determinant (MCD) method and the TELBS method on secondary data on the productivity of phytoplankton, which contains outliers. Based on the robust determinant coefficient value, MCD method produces a better model compared to TELBS method.

1. Introduction

Outlier is one of the problems in linear regression analysis that gives a very big influence on the model if it is used the Least Squares method (LS) [8]. The existence of the outliers in the data will cause the regression line to shift toward the outliers. This will cause the resulting error to be large so that the Gaussian assumption of normality may not be met. In LS method, outliers are usually set aside from the data set. Setting aside this data is not a wise procedure because sometimes the data provides information that other data cannot provide. To deal with this, the robust method is used as an alternative of LS method. The robust procedure is intended to accommodate the data weirdness, as well as eliminate the identification of data outliers and also automatic in tackling outliers [2].

Some of the robust methods that have been introduced include M estimator proposed by Huber in 1973, Least Median Square (LMS) and Least Trimmed Square (LTS) proposed by Rousseeuw in 1984 [6]. Birch introduced the robust method that was a generalization of the M estimator [1] called the Bounded-Influence estimator (B-I estimator). Furthermore, the regression of Minimum Covariance Determinant (MCD) proposed by Rousseeuw in 2004 and further deepened by Hubert [6]. The latest method is the TELBS method [11]. Similar to the MCD method, the TELBS method can deal with the problem of calling either in space **X** or space **Y**.

Based on the similarity in its ability to handle the problem of outliers, this paper examines TELBS and MCD methods applied to secondary data on the productivity of phytoplankton. The model obtained from both of these methods will be compared based on the robust coefficient of determination to determine the best model. The data centralize measurement used in robust determinant coefficients is median in accordance with the nature of data containing the outliers.

2. Linear regression analysis

Regression analysis is a statistical application that can build the relationship model between the independent variable (X) and the dependent variable (Y). In the linear model, the commonly used method for estimating the parameter is the Least Squares method (LS). The LS method is the easiest method to perform with certain assumptions called the Gaussian assumption [4]. If all Gaussian assumptions are met then $\boldsymbol{\beta}$ is a linear, non-biased and has minimum variance estimator.

Model of linear regression sample in matrix form, expressed by:

$$\mathbf{y} = \boldsymbol{X}\boldsymbol{\beta} + \mathbf{e} \tag{1}$$

where $\mathbf{y}$ is the dependent variable observation vector (n x 1), $\boldsymbol{X}$ is the independent variable observation matrix (n x (k + 1)), $\boldsymbol{\beta}$ is the parameter to be estimated ((k + 1) x 1) and $\mathbf{e}$ is a vector residual (n x 1). Using the LS method, the $\boldsymbol{\beta}$ component estimated by the principle is to minimize the sum of the residual squares to produce the following estimates:

$$\hat{\boldsymbol{\beta}} = \left(\boldsymbol{X}^t\boldsymbol{X}\right)^{-1}\boldsymbol{X}^t\mathbf{y} \tag{2}$$

The use of the LS method in estimating parameter parameters is a very well known method because it is easy to understand and to calculate. However, the use of this method is usually influenced by the presence of outliers. Outlier is the furthest observation of data centre that may have a major effect on regression coefficients [10]. The condition will arise if the sample data presented is varied and extreme. Every time there is an outlier, the model will usually change, so if used as a prediction model will get poor results and give a wrong interpretation. The existence of outliers can cause errors and variance of data to be large. Consequently, the parameter interval becomes large.

Outlier is rejected if after searching it results in errors in measurement or analysis, inaccuracy of data recording, and damage to the measuring tool. If it is not the result of such mistakes, careful investigation must be made [3]. An investigation or diagnosis in regression analysis is one way to monitor problems arising in relation to data or models. One way of diagnosing data related to outlier is to use a hat matrix [7] defined as follows:

$$\boldsymbol{H} = \boldsymbol{X}\left(\boldsymbol{X}^t\boldsymbol{X}\right)^{-1}\boldsymbol{X}^t \tag{3}$$

The matrix hat essentially transforms the observed value $\mathbf{y}$ vector to the estimated value $\mathbf{y}$ vector. The main diagonal of the hat matrix is that h_{ii} will show a value greater than $2p / n$, if the data is the outlier where $p = k + 1$.

2.1. Minimum Covariance Determinant (MCD) method

The Minimum Covariance Determinant (MCD) estimator is one of the first affine equivariant and highly robust estimators of multivariate location and scatters [6]. MCD method begins by determining the subset of matrix $\boldsymbol{X}$ as much as h observations are:

$$h = \frac{(n+k+1)}{2} \tag{4}$$

So there will be a combination of observations as much as $a = C_h^n$. The subsets of the matrix $\boldsymbol{X}$ are denoted by:

$$\boldsymbol{H}_b = \begin{bmatrix} x_{11} & x_{12} & \cdots & x_{1k} \\ x_{21} & x_{22} & \cdots & x_{2k} \\ \vdots & \vdots & \ddots & \vdots \\ x_{h1} & x_{h2} & \cdots & x_{hk} \end{bmatrix}, b = 1,2,\ldots,a. \quad (5)$$

From each $\boldsymbol{H}_b$, call as $\boldsymbol{H}_{bm}$, the average vector and the covariance matrix are determined by:

$$\mathbf{t}_{bm} = \frac{1}{h}(\boldsymbol{H}_{bm})^t \mathbf{v} \quad (6)$$

$$\boldsymbol{C}_{bm} = \frac{1}{h}\left[\boldsymbol{H}_{bm} - \mathbf{v}\mathbf{t}_{bm}^t\right]^t \left[\boldsymbol{H}_{bm} - \mathbf{v}\mathbf{t}_{bm}^t\right] \quad (7)$$

where $\mathbf{v}$ is the vector column is worth 1 (h x 1). Next for each b is calculated the value of det($\boldsymbol{C}_{bm}$). For m = 1, if det($\boldsymbol{C}_{bm}$) $\neq$ 0, then the distance of the Mahalanobis, is the element in the diagonal of the matrix, is determined by:

$$d_{MD}^2 = \left(\boldsymbol{X}^* - \mathbf{v}^*\overline{\boldsymbol{X}^*}\right)\boldsymbol{C}^{-1}\left(\boldsymbol{X}^* - \mathbf{v}^*\overline{\boldsymbol{X}^*}\right)^t \quad (8)$$

where $\boldsymbol{X}^*$ is the matrix of the observed independent variables (n x k), $\mathbf{v}^*$ is column vector is worth 1 (n x 1). $\boldsymbol{C}^{-1}$ is the inverse of the covariance matrix (k x k), and $\overline{\boldsymbol{X}^*}$ is the average matrix of each row (1 x k). The distance of the mahalanobis is sorted from the smallest value and taken as h, then the selected number of observations is identified. After that, a new $\boldsymbol{X}^*$ matrix is formed as $\boldsymbol{H}_{bm}$ ($m = m + 1$) and the new $\mathbf{t}_{bm}$ and $\boldsymbol{C}_{bm}$ values are determined. The next step is to compare the value of det($\boldsymbol{C}_{bm}$) and det($\boldsymbol{C}_{bm}$-1). If:

1. det($\boldsymbol{C}_{bm}$) $\neq$ det($\boldsymbol{C}_{bm}$-1), using equation (8) recovered the value of the distance of the new mahalanobis, then new $\boldsymbol{X}^*$ is formed to obtain new $\boldsymbol{H}_{bm}$ matrix;
2. det($\boldsymbol{C}_{bm}$) = det($\boldsymbol{C}_{bm}$-1), data processing is done from by taking another subset.

This process is done until all of the $\boldsymbol{H}_b$ subsets are out. Further, the smallest determinant of covariance of all determinants of the subsets of $\boldsymbol{H}_b$ is selected. The subset with the smallest covariance determinant is called $\boldsymbol{H}_{MCD}$, so further can be also determined $\mathbf{t}_{MCD}$ and $\boldsymbol{C}_{MCD}$. The $\boldsymbol{H}_{MCD}$ matrix element is the independent variable of the selected sample unit based on the smallest order of the mahalanobis distance so that it can be determined also the pair of the dependent variable.
The variance-covariance matrix of the variables X and Y variables whose chosen probes can be obtained by the following formula:

$$\widehat{\Sigma} = \begin{pmatrix} \widehat{\Sigma}_{XX} & \widehat{\Sigma}_{XY} \\ \widehat{\Sigma}_{YX} & \widehat{\Sigma}_{YY} \end{pmatrix} \quad (9)$$

Then we can determine covariance of error from:

$$\widehat{\Sigma}_{\varepsilon} = \widehat{\Sigma}_{YY} - \widehat{\theta}^t_{MCD}\widehat{\Sigma}_{XX}\widehat{\theta}_{MCD} \quad (10)$$

with $\widehat{\theta}_{MCD} = \widehat{\Sigma}_{YY}\widehat{\Sigma}_{XY}$.

The main principle in the MCD method is to calculate the weights $w_{ii} = 1$ if $d(e_{MCDi}) \leq \sqrt{\chi^2_{k,(1-\alpha)}}$ with $d(e_{MCD}) = \sqrt{\mathbf{e}_{MCD}(\widehat{\Sigma_\varepsilon})^{-1}(\mathbf{e}_{MCD})^t}$ so $\boldsymbol{W}_{MCD}$ matrix (n x n) is

$$\boldsymbol{W}_{MCD} = \begin{bmatrix} w_{11} & w_{12} & \cdots & w_{1n} \\ w_{21} & w_{22} & \cdots & w_{2n} \\ \vdots & \vdots & \ddots & \vdots \\ w_{n1} & w_{n2} & \cdots & w_{nn} \end{bmatrix} \tag{11}$$

with the matrix entries $w_{ij} = 0$ for $i \neq j$. This matrix is the robust weight matrix and is used to obtain sensitivity parameter estimates with the following equations:

$$\hat{\boldsymbol{\beta}}_{\text{MCD}} = \left(\boldsymbol{X}^t\boldsymbol{W}_{MCD}\boldsymbol{X}\right)^{-1}\boldsymbol{X}^t\boldsymbol{W}_{MCD}\mathbf{y} \tag{12}$$

2.2. TELBS method

One of the regression methods to estimate regression parameters when there are outliers is the TELBS method. TELBS estimator is better than the least squares method, M and MM estimator and its performed by minimizing objective function [11]:

$$\min_\beta \sum_{i=1}^{n} \frac{\rho(t_i)}{L_i} = \min_\beta \sum_{i=1}^{n} \frac{1-\text{sech}\left(\omega.t_i\right)}{L_i} \tag{13}$$

with

$$t_i = \frac{(y_i - x_i^t\hat{\boldsymbol{\beta}})(1-h_{ii})}{\sigma} \tag{14}$$

$$L_i = \sum_{j=1}^{k} \max\left(M_j, |x_{ij}|\right) \tag{15}$$

$$M_j = \text{median}\ \left\{|x_{1j}|, |x_{2j}|, \ldots, |x_{nj}|\right\} \tag{16}$$

ω are positive real numbers called constants of constancy (value 0.628). Estimator value σ can be obtained from the equation as follows:

$$\hat{\sigma} = 1.1926\, \underset{i;1\leq i\leq n}{\text{median}}\left(\underset{j;1\leq j\leq n}{\text{median}}(r_i - r_j)\right) \tag{17}$$

The selection of the constant 1.1926 makes σ a near-unbiased approximation of the limited sample [9]. To minimize equation (13) it must be derived to the parameter $\boldsymbol{\beta}$ and equal to zero, so that equation (13) can be expressed as:

$$\sum_{i=1}^{n} \frac{\psi(t_i)}{L_i} \frac{(1-h_{ii})}{\sigma} \frac{\partial(y_i - \beta_0 - \beta_1 x_{i1} - \beta_2 x_{i2} - \ldots - \beta_k x_{ik})}{\partial \beta_j} = 0 \tag{18}$$

with $\psi(x) = \frac{\partial \rho(x)}{\partial x} = \omega Sech(\omega x) Tanh(\omega x)$.
The weighted function is defined (11):

$$w_{ii}^* = \frac{\psi(t_i)(1-h_{ii})}{\sigma e_i L_i} \tag{19}$$

then equation (18) can be written to be:

$$\sum_{i=1}^{n} w_{ii}^* e_i \frac{\partial(y_i - \beta_0 - \beta_1 x_{i1} - \beta_2 x_{i2} - \ldots - \beta_k x_{ik})}{\partial \beta_j} = 0 \tag{20}$$

Equation (20) if it is derived to β_0 and β_j will produce the matrix form $(\mathbf{X^t W^* X})\boldsymbol{\beta} = \mathbf{X^t W^* y}$ so we obtain an estimate of TELBS methods are:

$$\hat{\boldsymbol{\beta}}_{TELBS} = (\boldsymbol{X^t W^* X})^{-1} \boldsymbol{X^t W^*} \mathbf{y} \tag{21}$$

where $\boldsymbol{W}^*$ is the square matrix with the diagonal element is w_{ii}^* in equation (19) and the matrix entry $w_{ij}^* = 0, \text{for } i \neq j$. In the TELBS method the estimated value obtained by the iteration process until the convergence is achieved.
The robust determination coefficient is then determined to see the suitability of the model as follows [11]:

$$R^2 = 1 - \left(\frac{\underset{i:1 \le i \le n}{Median} |e_i|}{\underset{i:1 \le i \le n}{Median} \left| y_i - \underset{i:1 \le i \le n}{Median}\, y_i \right|} \right)^2 \tag{22}$$

3. Data

The secondary data used is the result of research on the primary productivity of phytoplankton with the physicochemical factors of aquatic in aqua culture with floating net [5]. From the result of the research, there are twenty data pair, with primary productivity (gr.L/m^2) as an independent variable (Y) and three independent variables namely light intensity (kkal/m^2) as X_1, PH (X_2), and phytoplankton density (ind/mL) as X_3.

4. Results and discussion

The hat matrix is used to detect the presence of the outlier. The main diagonal of the hat matrix is shown in Figure 1. It appears that the 17th, 18th, and 19th data are the outliers represented by a value greater than $2p/n = 0.3$. If forced by using LS, the value of determination coefficient is 0.57762, meaning that only 57.762% variability of primary production of phytoplankton can be explained by the factors that influence it.

IOP Conf. Series: Materials Science and Engineering **332** (2018) 012037 doi:10.1088/1757-899X/332/1/012037

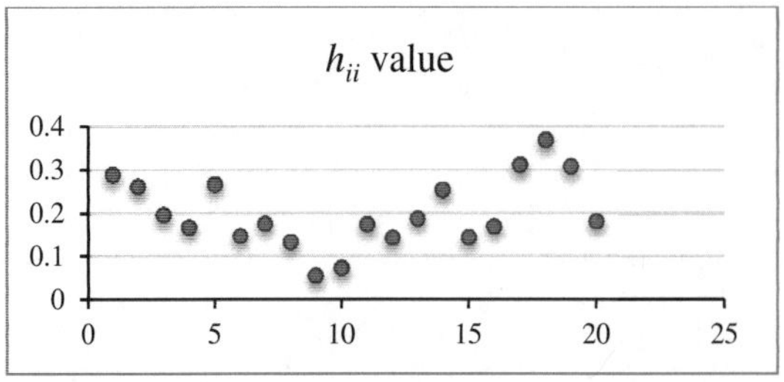

Figure 1. h_{ii} value of phytoplankton productivity

The small coefficient of determination in LS method becomes the reason for modeling using MCD method and TELBS method. Using Maple 15, the results are presented in the following table:

Table 1. The result with MCD and TELBS methods

Component	MCD method	TELBS method
β_0	0.5855	-0.4259
β_1	10^{-4}	$1.75.10^{-4}$
β_2	0.0826	0.0584
β_3	3.10^{-4}	$2.232.10^{-4}$
R^2	0.8994	0.8781

The results of Table 1 above show that based on the coefficient of determination, MCD produces a better model than TELBS method. This is reinforced by the residues values illustrated in figure below which shows that the residues generated by MCD tend to be less than the TELBS model.

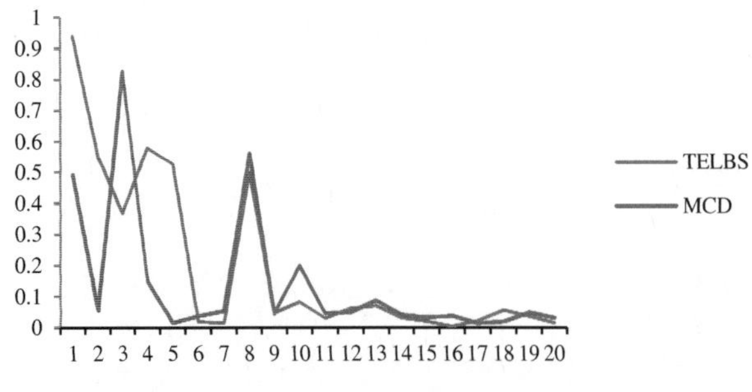

Figure 2. Residues based on MCD and TLBS methods

Thus the best model for the phytoplankton productivity is as follows:

$$\hat{y}_i = 0.5855 + 10^{-4} x_{i1} + 0.0826 x_{i2} + 3.10^{-4} x_{i3} \tag{23}$$

The coefficient of determination with MCD method gives a value of 0.8994. This means that 89.94% of the primary production variability of phytoplankton can be explained by the factors of light intensity, pH, and phytoplankton density. The remaining 10.06% is determined by other factors, which are not included in the study.

5. Conclusion

In this paper, we develop a linear model of the productivity of phytoplankton. It has been shown, that based on robust coefficient determinant, the MCD model is better than TELBS when data is contaminated by outlier.

Acknowledgments
The authors thank to the Padjadjaran University, due to the funding of this work through the scheme of Riset Fundamental Unpad (RFU) with contract number 872/UN6.3.1/LT/2017.

References
[1] Birch J B 1992 Estimation and Inference in Multiple Regression Using Robust Weight: A Unified Approach *Technical Report* 92-2, Departemen of Statistics Virginia Polytechnic Institute and State University, Blackburg Virginia.
[2] Cahyawati D 2009 Efektifitas Metode Regresi Robust Penduga Welsch dalam Mengatasi Pencilan pada Pemodelan Regresi Linear Berganda, *Jurnal Penelitian Sains*, Vol.12, no.1(A), Unsri, Sumatera Selatan.
[3] Chandraningtyas S, dkk. 2013 Regresi Robust MM-*Estimator* untuk Penanganan Pencilan pada Regresi Linier Berganda. *Jurnal Gaussian*, Vol. 2, no.4, 395-404, Undip, Semarang.
[4] Gujarati D N 1998 *Basic Econometrica*, 2nd Edition, New York, McGraw-Hill Inc.
[5] Gusriani, Firdaniza, Ardelina D 2011 Kajian Penaksir *Bounded-Influence* dan Metode *Minimum Covariance Determinant* pada Analisis Regresi Linier Berganda untuk Kasus Pencilan, Laporan Penelitian, Jurusan Matematika FMIPA Unpad, Bandung.
[6] Hubert Mia et al. 2008 High-Breakdown Robust Multivariate Methods *Statistical Science* Vol. 23 No.1 (online: http://arxiv.org/pdf/0808.0657).
[7] Myers R H 1990 *Classical and Modern Regression With Applications* 2nd edition Boston PSW-KENT Publishing Company.
[8] Rousseeuw P J and Leroy A M 1986 *Robust Regression and Outlier Detection* England: John John Wiley & Sons Ltd.
[9] Rousseeuw P J and Croux 1993 Alternative to the Median Absolute Deviation *American Statistical Association*. 1993, Vol. 88, No. 424.
[10] Sembiring R K 2003 *Analisis Regresi*, Edisi 2. Bandung: ITB.
[11] Tabatabai M A et al. 2012 TELBS Robust Linear Regression Method Open Access Medical Statistics, USA, Dove Medical Press.

IOP Conf. Series: Materials Science and Engineering **332** (2018) 012038 doi:10.1088/1757-899X/332/1/012038

A hybrid Dantzig-Wolfe, Benders decomposition and column generation procedure for multiple diet production planning under uncertainties

A Udomsungworagul[1*], P Charnsethikul[2]

[1]Network Development Division, Strategic Retail Marketing Department, PTT Public Company Limited, Bangkok, Thailand
[2]Department of Industrial Engineering, Faculty of Engineering, Kasetsart University, Bangkok, Thailand

*Corresponding author: atit.u@pttplc.com

Abstract. This article introduces methodology to solve large scale two-phase linear programming with a case of multiple time period animal diet problems under both nutrients in raw materials and finished product demand uncertainties. Assumption of allowing to manufacture multiple product formulas in the same time period and assumption of allowing to hold raw materials and finished products inventory have been added. Dantzig-Wolfe decompositions, Benders decomposition and Column generations technique has been combined and applied to solve the problem. The proposed procedure was programmed using VBA and Solver tool in Microsoft Excel. A case study was used and tested in term of efficiency and effectiveness trade-offs.

1. Introduction

In the animal feed commerce or industry has a majority costs in feed mix. Feed producers need to consider finding the optimal feed mix with the minimum cost fulfilling nutrients requirements. The classical diet model was first studied by (Stigler. 1945). Since then, the feed mix production planning has also been an important decision for avoiding stock out or over stock manufactured that effect its related cost.

This paper propose methodology for solving stochastic linear programming in form of two-stage linear programming (Wagner. 1975) combining two type of problem, diet problem under uncertainty of nutrient in feed mix by (Thammanivit and Charnsethikul. 2013) and inventory problem under uncertainty of mixed products demand in finite multiple time period by (Pisit *et al.* 2008). The assumption of allowing to produce many product formulas in same time period is added. Dantzig-Wolfe decomposition, column generation and Benders decomposition algorithms have been used to solve this proposed model. Dantzig-Wolfe decomposition (Dantzig and Wolfe. 1960) has been used to solve large scale block angular problem of production planning part in the model. Column generation by Charnsethikul (2011) was proposed on case of relaxing the infinite number decision variables problem controlled from product tolerances allowed to manufacture possibility. Benders decomposition (Infanger. 1994) can also be applied to the case of large scale diet problem.

2. Mathematical model

The objective of model in this article (Atit and Peerayuth 2017) is to seek for the production plan of manufacturing and to store feed mix products in each period against uncertain demands. The production planning part dealt with raw materials purchasing cost, raw materials holding cost, finished products holding cost and finished product back ordering cost with inventory balance constraint as equations (1)-(4) are considered. Set up cost between successive periods are not considered in this model. The corresponding diet model part is related with corrective action cost of rework for out of specification nutrients manufactured with the constraint of nutrition mixed from uncertain nutrients of raw materials and the constraint of raw materials proportion boundary. The hypothesis of allowing to produce many product formulas in the same time period is assumed and presented in equations (5)-(12).

$$\text{Minimize z} = \sum_{t=1}^{T}\left(\sum_{j=1}^{\infty} c_{jt}x_{jt} + g_{R_{i,t}}I_{R_{i,t}} + h_{R_{i,t}}Q_{i,t} + g_{D_t}I^{+}_{D_{k,t}} + h_{D_t}I^{-}_{D_{k,t}}\right) \tag{1}$$

Subject to

$$\sum_{j=1}^{\infty} a_{ijt}x_{jt} + I_{R_{i,t}} - Q_{i,t} - I_{R_{i,t-1}} = 0 \; ; t = 1,...,T \; , i = 1,2,...,m \; , \; I_{R_{i,0}} \text{ is a constant} \tag{2}$$

$$I^{+}_{D_{k,t-1}} - I^{-}_{D_{k,t-1}} + \sum_{j=1}^{\infty} x_{jt} - I^{+}_{D_{i,k,t}} + I^{-}_{D_{k,t}} = d_{k,t} \; ;k = 1,..., K \; , t = 2,...,T \; ,I^{+}_{D_{k,0}} \; I^{-}_{D_{k,o}} \text{ are constant} \tag{3}$$

$$x_{jt}, I_{R_{i,t}}, Q_{i,t}, I^{+}_{D_{k,t}}, I^{-}_{D_{k,t}} \geq 0 \;\; \forall j,k,t \tag{4}$$

where

T : Number of time period considered,
K : Number of possible demand scenario considered,
x_{jt} : Decision variable of producing product formula j quantity at time period t,
a_{ijt} : Raw material i percentage in product formula j at time period t,
$I_{R_{i,t}}$: Decision variable quantity of holding raw material i at time period t,
$Q_{i,t}$: Decision variable quantity of ordering raw material i at time period t,
$g_{R_{i,t}}$: Raw material i holding cost per unit at time period t,
$h_{R_{i,t}}$: Raw material i ordering cost per unit at time period t,
$I^{+}_{D_{k,t}}$: Decision variable quantity of holding finished product at scenario k time period t,
$I^{-}_{D_{k,t}}$: Decision variable quantity of back ordering finished product at scenario k time period t,
g_{D_t} : Finished product holding cost per unit at time period t,
h_{D_t} : Finished product back ordering cost per unit at time period t,
$d_{k,t}$: Finished product demand at scenario k time period t and
c_{jt} : Corrective action cost of manufacturing product formula j at time period t coming from reworking in the case of under and over nutrients requirement as equations (6)-(7).

$$\text{Let } a_{ijt} = x_i^{*(jt)} \tag{5}$$

$$c_{jt} = c^{(jt)} \tag{6}$$

$$c^{(jt)} = \sum_{s=1}^{N}\sum_{l=1}^{m}(g_{ls1}u_{ls1}^{(jt)} + h_{ls1}v_{ls1}^{(jt)}) + \sum_{s=1}^{N}\sum_{l=1}^{m}(g_{ls2}u_{ls2}^{(jt)} + h_{ls2}v_{ls2}^{(jt)}) \tag{7}$$

$x_i^{*(jt)}$ can be found from following conditions :

$$\sum_{i=1}^{m} a_{ils}^{*} x_i^{*(jt)} + u_{ls1}^{(jt)} - v_{ls1}^{(jt)} = b_{l1}^{*}\ ; l = 1,2,...,p\ \ ; s = 1,2,...,N \tag{8}$$

$$\sum_{i=1}^{m} a_{ils}^{*} x_i^{*(jt)} + u_{ls2}^{(jt)} - v_{ls2}^{(jt)} = b_{l2}^{*}\ ; l = 1,2,...,p\ \ ; s = 1,2,...,N \tag{9}$$

$$l_i^{*} \le x_i^{*(jt)} \le u_i^{*} \tag{10}$$

$$\sum_{i=1}^{m} x_i^{*(jt)} = 1\ \ \forall j,t \tag{11}$$

$$x_i^{*(j)}, u_{kl1}^{(j)}, v_{kl1}^{(j)}, u_{kl2}^{(j)}, v_{kl2}^{(j)} \ge 0\ \ \forall_{i,j,k,l} \tag{12}$$

where

N : A number of possible scenarios considered,

$u_{ls1}^{(jt)}$: Slack of the amount of minimum nutrient in nutrient l scenario s product formula j at time period t,

$v_{ls1}^{(jt)}$: Surplus of the amount of minimum nutrient in nutrient l scenario s product formula j at time period t,

$u_{ls2}^{(jt)}$: Slack of the amount of maximum nutrient in nutrient l scenario s product formula j at time period t,

$v_{ls2}^{(jt)}$: Surplus of the amount of maximum nutrient in nutrient l scenario s product formula j at time period t,

g_{ls1} : Expected cost per unit of $u_{ls1}^{(jt)}$,

h_{ls1} : Expected cost per unit of $v_{ls1}^{(jt)}$,

g_{ls2} : Expected cost per unit of $u_{ls2}^{(jt)}$,

h_{ls2} : Expected cost per unit of $v_{ls2}^{(jt)}$,

a_{ils}^{*} : Amount per unit of nutrient l in raw material i scenario s,

$\tilde{b}_{l1}^{*}$: Minimum acceptable quantity of nutrient l,

$\tilde{b}_{l2}^{*}$: Maximum acceptable quantity of nutrient l,

l_i^{*} : Smallest allowable fraction of raw material i and

u_i^{*} : Largest allowable fraction of raw material i.

3. Methodology

Three major algorithms were used to solve the model described in previous section. Dantzig-Wolfe decomposition is applied for solving large scale two-stage linear programming for inventory problem (equations (1)-(4)). The column generation technique is used for the problem of infinite number of manufacturing quantity decision variables controlled from possibility in continuous fraction formula equations (6)-(12) while Benders decomposition can be utilized for solving large scale diet problem equations (7)-(12).

3.1. Dantzig-Wolfe Decomposition

Consider equations (1)-(4). This can be solved by using algorithm for two-stage linear programming such as Benders decomposition or Dantzig-Wolfe decomposition. Dantzig-Wolfe decomposition has been chosen in this step because it is easier to find directly both primal and dual solutions together. Equations (1)-(4) need to be transformed to the corresponding dual problem as equations (13)-(19).

$$\text{Maximize } w=\sum_{i=1}^{m} I_{R_{i,0}} y_{R_{i,1}} + \sum_{k=1}^{K}\sum_{t=1}^{T} d_{k,t} y_{D_{k,t}} \tag{13}$$

Subject to

$$\sum_{i=1}^{m} a_{ijt} y_{R_{i,t}} + \sum_{k=1}^{K} y_{D_{k,t}} \le c_{jt} \quad ; j=1,...,J \ ; t=1,...,T \tag{14}$$

$$y_{R_{i,t}} - y_{R_{i,t+1}} \le g_{R_{i,t}} \quad ; t=1,...,T-1 \ ; \forall i \tag{15}$$

$$y_{R_{i,T}} \le g_{R_{i,T}} \quad ; \forall i \tag{16}$$

$$y_{R_{i,t}} \ge -h_{i,t} \quad ; \forall i,t \tag{17}$$

$$-h_{D_t} \le y_{D_{k,t}} - y_{D_{k,t+1}} \le g_{D_t} \quad ; t=1,...,T-1 \ ; \forall k \tag{18}$$

$$-h_{D_T} \le y_{D_{k,T}} \le g_{D_T} \quad ; \forall k \tag{19}$$

Equations (13)-(19) are combined from two independent constraints set, Equations (14)-(17) and Equations (18)-(19), and linking with coupling constraint equations (14) and J is a finite number of feed mix formula initially used and sampled from infinite number of alternatives. This pattern can be transformed to convex combination model with initial solutions set $\theta = 1,...,\eta$ as equations (20)-(23).

$$\text{Maximize } w=\sum_{\theta=1}^{\eta}\Big(\sum_{i=1}^{m} I_{R_{i,0}} \vec{y}^{\alpha}_{R_{i,1}} + \sum_{k=1}^{K}\sum_{t=1}^{T} d_{k,t} \vec{y}^{\alpha}_{D_{k,t}}\Big)\hat{\lambda}_{\theta} \tag{20}$$

$$\sum_{\theta=1}^{\eta}\Big(\sum_{i=1}^{m} a_{ijt} \vec{y}^{\theta}_{R_{i,t}} + \sum_{k=1}^{K} \vec{y}^{\theta}_{D_{k,t}}\Big)\hat{\lambda}_{\theta} \le c_{jt} \quad ; j=1,...,n \ ; k=1,...,K \tag{21}$$

$$\sum_{\theta=1}^{\eta} \hat{\lambda}_{\theta} = 1 \tag{22}$$

$$\hat{\lambda}_{\theta} \ge 0 \quad \forall \theta \tag{23}$$

where
η: Number of solutions set considered,
$\vec{y}^{\theta}_{R_{i,t}}$: Optimal solutions of $\dot{y}_{D_{k,t}}$ equations (15)-(17) solutions set θ and
$\vec{y}^{\theta}_{D_{k,t}}$: Optimal solutions of $\dot{y}_{D_{k,t}}$ equations (18)-(19) solutions set θ.

According to equations (20)-(23), Initial solutions set are needed then searching for new potential solutions set $\eta+1$ iteratively. Searching algorithm is still using equations (13)-(19) concept by relaxing the coupling constraint equation (14) as Lagrange multiplier terms in the objective function as equation (23).

$$\text{Maximize } w=\sum_{i=1}^{m} I_{R_{i,0}} \vec{y}^{\theta}_{R_{i,1}} + \sum_{k=1}^{K}\sum_{t=1}^{T} \dot{d}_{k,t} \vec{y}^{\theta}_{D_{k,t}} - \sum_{t=1}^{T}\sum_{j=1}^{n} \lambda'_{jt}\Big(\sum_{i=1}^{m} \dot{a}_{ijt} \vec{y}^{\theta}_{R_{i,t}} + \sum_{k=1}^{K} \vec{y}^{\theta}_{D_{k,t}} - c_{jt}\Big) \tag{23}$$

Equation (23) can be reformed as equation (24).

$$\text{Maximize } w=\sum_{i=1}^{m}\Big(I_{R_{i,0}} - \sum_{t=1}^{T}\sum_{j=1}^{n} \lambda'_{jt} \dot{a}_{ijt}\Big) y_{R_{i,t}} + \sum_{k=1}^{K}\sum_{t=1}^{T}\Big(\dot{d}_{k,t} - \sum_{j=1}^{n} \lambda'_{jt}\Big) y_{D_{k,t}} + \sum_{t=1}^{T}\sum_{j=1}^{n} c_{jt} \lambda'_{jt} \tag{24}$$

Subject to

$$\vec{y}^{\theta}_{R_{i,t}} - \vec{y}^{\theta}_{R_{i,t+1}} \le g_{R_{i,t}} \quad ; t=1,...,T-1 \ ; \forall i \tag{25}$$

$$y_{R_{i,t}}^{\theta} \leq g_{R_{i,T}} \quad ;\forall i \tag{26}$$

$$y_{R_{i,t}}^{\theta} \geq -h_{i,t} \quad ;\forall i,t \tag{27}$$

$$-h_{D_t} \leq y_{D_{k,t}} - y_{D_{k,t+1}} \leq g_{D_t} \quad ; t=1,\ldots,T-1 \quad ;\forall k \tag{28}$$

$$-h_{D_T} \leq y_{D_{k,T}} \leq g_{D_T} \quad ;\forall k \tag{29}$$

where λ'_{jt} is a dual decision variable of equation (14) formula j at time period t.

Solving equations (20)-(23) can only find the optimal solutions λ'_{jt} for the solutions set $\theta = 1,\ldots,\eta$. Adding the optimal solutions of equations (24)-(29) into solutions set $\eta+1$ equations (20)-(22) can improve the objective function value. Repeating these steps until the objective function value equations (20) converges will make the solutions close to optimal.

Equations (24)-(29) can be separated into two independent problems, constraints (25)-(27) and (28)-(29), equation with constraints (25)-(27) can be solved by using the direct simplex method but equation constraints (28)-(29) may not be proper because of the larger size involved. This pattern can written as equations (30)-(33) and solved by using Special purpose method (Apisak *et al.* 2016) introducing the new variable u'_{kt} as follows.

$$\text{Maximize } w = \sum_{k-1}^{K}\sum_{e-1}^{T-1}\left(\sum_{t-1}^{e}(d_{k,t} - \sum_{j-1}^{n}\lambda'_{jt})u'_{kt}\right) + \sum_{k-1}^{K}\left(\left(\sum_{t-1}^{T}(d_{k,t} - \sum_{j-1}^{n}\lambda'_{jt})\right)\dot{y}_{D_{k,T}}\right) \tag{30}$$

Subject to

$$y_{D_{k,t}} = y_{D_{k,t+1}} + u'_{k,t} \quad ;t=1,\ldots,T-1 \tag{31}$$

$$-h_{D_T} \leq y_{D_{k,T}} \leq g_{D_T} \tag{32}$$

$$-h_{D_t} \leq u'_{k,t} \leq g_{D_t} \quad ;t=1,\ldots,T-1 \tag{33}$$

According to the new structure as described by equations (30)-(33), All of decision variables $\dot{y}_{D_{k,T}}, u'_{k,t}$ are independent. Thus, the decisions are depended upon on the objective function coefficients. If the coefficients are positive the variables are equal to $g_{D_t}, \forall t$. In contrast, if the coefficients are negative the variables are equal to $-h_{D_t}, \forall t$.

Finally, an optimal solution of equations (13)-(19) are equal to the product of the solutions set and their optimal weight as shown by equations (34)-(35).

$$y_{R_{i,t}} = \sum_{\theta=1}^{\eta}\hat{\lambda}_{\theta}\vec{y}_{R_{i,t}}^{\theta} \tag{34}$$

$$y_{D_{k,t}} = \sum_{\theta=1}^{\eta}\hat{\lambda}_{\theta}\vec{y}_{D_{i,t}}^{\theta} \tag{35}$$

In conclusion, the optimal solution for both primal problem equations (1)-(4) and dual solutions λ'_{jt} from equations (21) can be simultaneously found using the above procedure.

After the optimal solutions x_{jt} from equations (1)-(4) are obtained. $I_{R_{i,t}}, Q_{i,t}, I^{+}_{D_{k,t}}$ and $I^{-}_{D_{k,t}}$ can be calculated in their relative equations. This method can be used with only finite decision variables problem ($j = 1,...,n$). Therefore, the column generation approach is needed to expand the subproblem size to approximate closely to the master problem.

3.2. Column generation

In other issues, equations (1)-(4) still have the problem of infinite number of decision variables of manufactured quantity consequence from continuous coefficient equations (7)-(12). This can be solved by using the column generations technique (Charnsethikul. 2011); create the restricted master problem and solve for minimum reduced cost of non-considering decision variables then add an improved one to the restricted master problem iteratively until all reduced cost of non-considering decision variables are nonnegative, To solve the minimum reduced cost search equations for expanding the size of restricted master problem. Benders decomposition is also needed as a sub-procedure when the number of scenarios becomes too large (Artit and Peerayuth, 2016).

Reduced cost of any x_{jt} from equations (1)-(4) can be written as the right hand side of equation (36). By maximizing the objective function, if there exist no positive R , the current best solution is already the optimal solution. Otherwise, the solution of equations (36) with most positive R will lead solution of equations (1)-(4) to a better optimal solution (Charnsethikul, 2011).

$$\text{Maximize } R = (\sum_{i=1}^{m} y_{R_{i,t}} x_i^{*(jt)} + \sum_{k=1}^{K} y_{D_{k,t}}) - \sum_{s=1}^{N}\sum_{l=1}^{m}(g_{ls1}u_{ls1}^{(jt)} + h_{ls1}v_{ls1}^{(jt)}) \qquad (36)$$

Objective function equation (36) can be written as equation (37) to standardize two-stage linear programming form (Wagner. 1975) as follows.

$$\text{Minimize } R = -\sum_{i=1}^{m} y_{R_{i,t}} x_i^{*(jt)} + \sum_{s=1}^{N}\sum_{l=1}^{m}(g_{ls1}u_{ls1}^{(jt)} + h_{ls1}v_{ls1}^{(jt)}) - \sum_{k=1}^{K} y_{D_{k,t}} \qquad (37)$$

Solving equation (37) is still constrained by raw material fraction usage condition equation (8)-(12). Adding this solution to restricted master problem of equations (1)-(4) and resolve equations (37) until no negative objective value detected will lead to the solution optimality.

3.3. Benders decomposition

In order to solve equation (37) constrained by equations (8)-(12), the direct Simplex method may not be the proper choice because the problem can be too large. This depends on number of nutrient scenarios considered. Benders decomposition can be applied to this case. Given the first stage decision variables be any feasible constant solution $x_i^{*(jt)} = x_i^{**(jt)}$. Reduced cost search equations are formed into equations (38)-(41) as follows.

$$\text{Minimize } R = \sum_{s=1}^{N}\sum_{l=1}^{m}(g_{ls1}u_{ls1}^{(jt)} + h_{ls1}v_{ls1}^{(jt)}) - \sum_{i=1}^{m} y_{R_{i,t}} x_i^{**(jt)} - \sum_{k=1}^{K} y_{D_{k,t}} \qquad (38)$$

Subject to

$$u_{ls1}^{(jt)} - v_{ls1}^{(jt)} = b_{l1}^{*} - \sum_{i=1}^{m} a_{ils}^{*} x_i^{**(jt)} \quad ; l = 1,2,...,p \quad ; s = 1,2,...,N \qquad (39)$$

$$u_{ls2}^{(jk)} - v_{ls2}^{(jk)} = b_{l2}^{*(k)} - \sum_{i=1}^{m} a_{ils}^{*} x_i^{**(jk)} \quad ; l = 1,2,...,p \quad ; s = 1,2,...,N \qquad (40)$$

$$u_{kl1}^{(j)}, v_{kl1}^{(j)}, u_{kl2}^{(j)}, v_{kl2}^{(j)} \geq 0 \quad \forall_{i,j,k,l} \tag{41}$$

The dual problem of equations (38)-(41) can be formulated as following equations (42)-(44).

$$\text{Maximize } \dot{w} = \sum_{l=1}^{p}\sum_{s=1}^{N}(b_{ls1} - \sum_{i=1}^{m} a_{ils}^{*} x_i^{**(jt)}) y_{lsq1}^{*(jt)} + \sum_{l=1}^{p}\sum_{s=1}^{N}(b_{ls2} - \sum_{i=1}^{m} a_{ils}^{*} x_i^{**(jt)}) y_{lsq2}^{*(jt)} - \sum_{i=1}^{m} y_{R_{i,t}} x_i^{**(jt)} - \sum_{k=1}^{K} y_{D_{k,t}} \tag{42}$$

Subject to

$$-h_{ls1} \leq y_{lsq1}^{*(jt)} \leq g_{ls1} \tag{43}$$

$$-h_{ls2} \leq y_{lsq2}^{*(jt)} \leq g_{ls2} \tag{44}$$

where
$\dot{w}$: Dual objective value of equations (38)-(41),
$y_{lsq1}^{*(jt)}$: Dual decision variable of nutrient lower bound constraint for nutrient l scenario s computing in round q of formula j at time period t from equation (39).
$y_{lsq2}^{*(jt)}$: Dual decision variable of nutrient upper bound constraint for nutrient l scenario s computing in round q formula j at time period t from equation (40).

The optimal solutions of equations (42)-(44) are trivial. Independent constraint causes the solutions depending upon the corresponding objective function coefficients. If the coefficient is positive the solution is equal to the related upper bound. Otherwise, the solution is equal to the related lower bound.

After solving equations (42)-(44), the next step is to solve the cuts equations for deciding the new first stage decision variables $x_i^{*(jt)} = x_i^{**(jt)}$ which can be shown in the following equations (45)-(46).

$$\text{Minimize } \dot{T} \tag{45}$$

Subject to

$$T + \sum_{l=1}^{p}\sum_{i=1}^{m}\sum_{s=1}^{N}((y_{lsq1}^{*(jt)} + y_{lsq2}^{*(jt)}) a_{ils}^{*} + y_{R_{i,t}}) x_i^{*(jt)}$$
$$\geq \sum_{l=1}^{p}\sum_{s=1}^{N}(b_{ls1}^{*} \dot{y}_{lsq1}^{*(jt)} + b_{ls2}^{*} y_{lsq2}^{*(jt)}) - \sum_{i=1}^{m} \dot{y}_{R_{i,t}} x_i^{**(jt)} - \sum_{k=1}^{K} y_{D_{k,t}} \quad \forall q \tag{46}$$

where
$\dot{T}$: The smallest upper bound of the reduced cost search problem.

Replacing new $x_i^{*(jt)}$ by $x_i^{**(jt)}$ in the equation (38)-(41) then resolve with the same algorithm until $T \to R$ will lead to the converged optimal solutions of equations (37) subjecting to constraint equations (8)-(12). In summary, figure 1 is illustrated the proposed method flowchart.

4. Computational test

To test the method capability, Visual basic for application in Microsoft Excel has been developed. Working parallel with automatic computing in spread sheets cells and Solver tool. In this study, the computer processor is Intel(R) Core(TM) i5-3450S CPU @ 2.80GHz. The experimental case consists of 8 types of raw materials. Nutrients percentage in raw materials, Nutrients requirement, raw materials fraction boundary, corrective action cost and corrective action cost per unit data are based on the pig diet problem by (Wanida *et al.* 2008) shown in appendix (Table: 1-Table: 4).

As (Wanida *et al.* 2008), assume the cost of raw materials equal to 42, 31, 34, 65, 22, 87, 120 and 15 baht per unit respectively. The assumption of no corrective action cost in the case of producing nutrient at its lower bound and under the upper bound has been used. Testing with 6 time periods, inventory holding cost are 5, 15, 20, 5, 5, 10, 15, 5 baht per unit per time period respectively, Inventory level at initial time period are 200, 300, 150, 150, 500, 150, 700, 650 units respectively. Finished products holding cost in each time period are 10, 15, 10, 12, 10, 20 baht per unit. Back ordering cost in each time period are 30, 24, 20, 16, 25, 22 baht per unit. Finished product demand in each time periods are 1,000±20, 1,200±20, 1,600±20, 1,800±10, 900±10, 1,300±40 units. Uncertainties of nutrients in raw material and demand are both simulated as discrete uniform distribution with 10,000 scenarios simulation size.

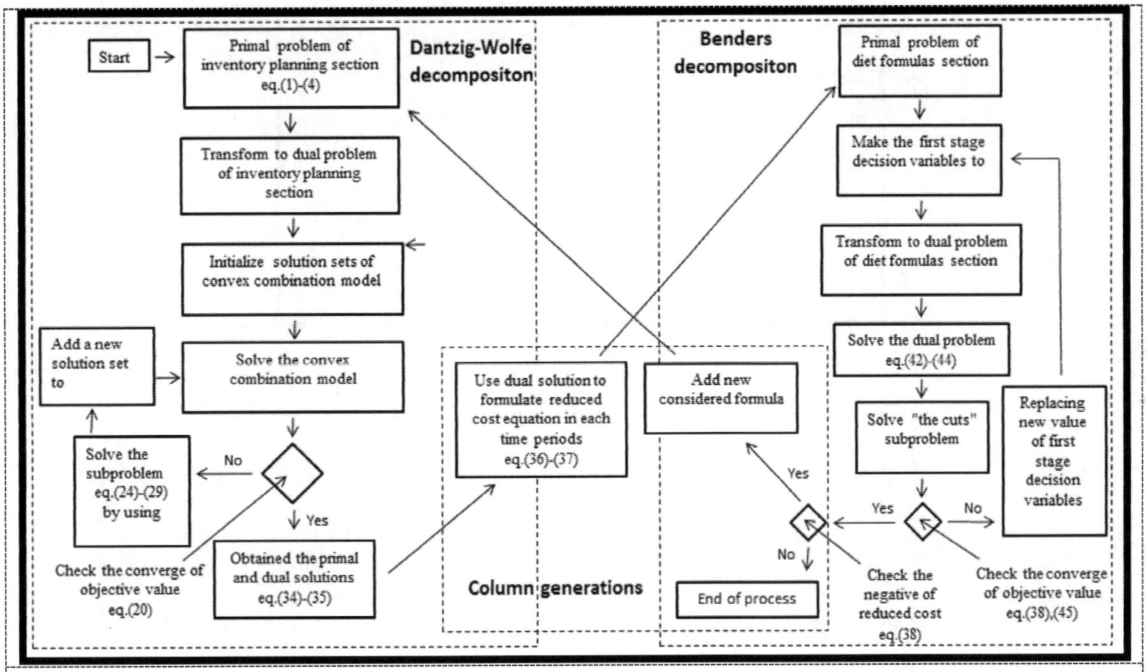

Figure 1. Proposed method flow chart.

5. Result and discussion

The solution obtained is to produce only the first 3 out of 6 periods shown that holding cost impact is less than back ordering cost. The optimal overall expected value is 524,373 baht with 2 iterations of column generations being used while 15, 14 and 11 iterations of Dantzig-Wolfe decomposition has been used in each iteration of the column generation method. Figure 2 shows the objective value updating result.

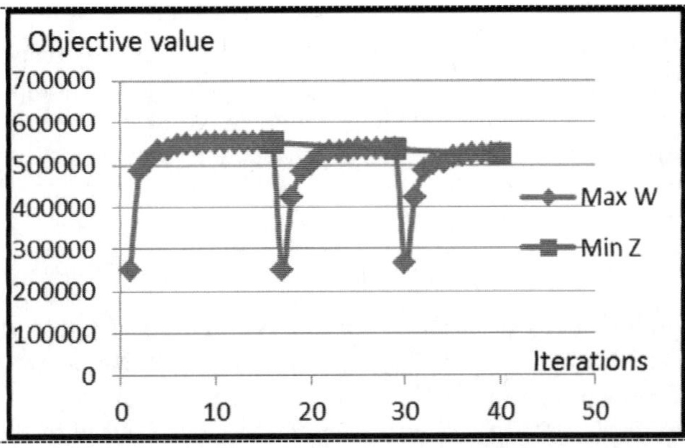

Figure 2. Objective value updating thru iterations.

The total runtime is 3525.39 second including from 3 iterations of column generation working with Benders decomposition and 3 iterations of Dantzig-Wolfe decomposition. Runtime per iteration and cumulative runtime are shown in figures 3 and 4 indicating that Dantzig-Wolfe decomposition consume more computing time because of the lack of any special propose technique to replace using simplex method in equations (24)-(27).

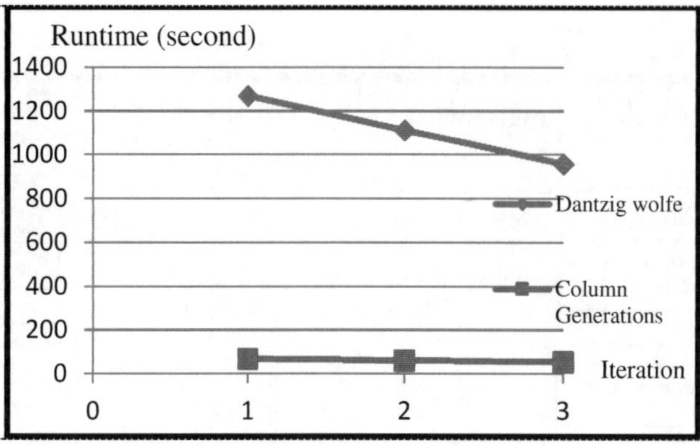

Figure 3. Runtime per iteration.

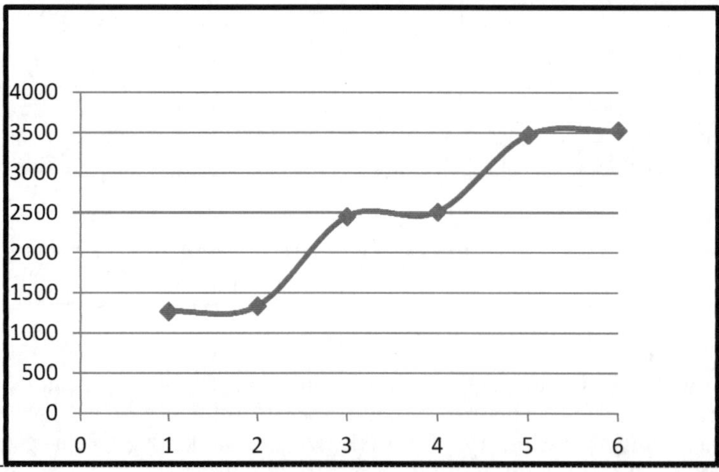

Figure 4. Cumulative runtime.

6. Conclusion

Two stage linear programming can be applied with the simultaneous inventory and diet planning problem. Dantzig-Wolfe decomposition and Benders decomposition are compatible in this case. The column generation technique is helpful for relaxing infinite number of related decision variables. These three algorithms can also be working as a hybrid approach. The preliminary on both computing time consumed and the solution quality when solving the case problem are reasonable and can be further improved.

References

[1] Apisak W, Peerayuth C and Pichet P 2016 Benders Decomposition Combining with a Special Purpose Method Solving Related Sub Problems for Production/Inventory Planning Having Uncertain Demand within a Finite Horizon. *Operations research network conference* 23-31.

[2] Artit U and Peerayuth C 2016 A Column Generation Combined with Benders' Decomposition for Multi-Product-Mix Stochastic Linear Programming with a Case Study of the Diet Problem.

Operations research network conference 70-76.
[3] Atit U and Peerayuth C 2017 Dantzig-Wolfe decomposition Benders decomposition and Column generation for Multi-Diet Production Planning under Uncertainty *SIMMOD 2017* 39-46.
[4] Charnsethikul P 2011 A column generations approach for the products mix based semi-infinite linear programming model *International Journal of Management Science and Engineering Management* 6(2), 106-109.
[5] Dantzig G B and Wolfe P 1960 Decomposition Principle for Linear Programs. *Operations Research 8,* 101–111.
[6] Infanger G 1994 Planning Under Uncertainty. The Scientific Press Series. Boyd and Fraser.
[7] Pisit S, Pitsanu T, Somsak C and Peerayuth C 2008 Aggregate Planning under Uncertain Demands by Stochastic programming technique. *Operations research network conference*, 25-32.
[8] Stiger G J 1945 The cost of Subsistence. *Journal of Farm Economics* 27(2), 303-314.
[9] Thammaniwit S and Charnsethikul P 2013 Application of Benders Decomposition Solving a Feedmix Problem among Supply and Demand Uncertainties *International Transaction Journal of Engineering, Management & Applied Sciences & Technologies* 4(2), 111-128.
[10] Wagner H M 1975 Principle of Operations Research, Prentice-Hall, Inc.
[11] Wanida L, Supatchaya C, Silprapa M and Peerayuth C 2008 Solving Stochastic Programming Problem by Alternative Aggregation Method with a Case Study of the Diet Problem. *Operations research network conference* 1-7.

Appendix

Table 1. Nutrient percentage in raw materials.

Materials	Corn	Fat(%)	Calcium(%)	Phosphorus(%)	Energy (kcal)
Corn	40±1	22±1	11±1	7±2	1,800±50
Millet	5±0.25	4±1	8±1	5±0.5	2,200±100
Cassava	1±0.01	5±2	12±2	14±1	2,000±25
Fishmeal	60±2	10±1	15±2	10±1	1,750±50
Rice bran	-	13±2	6±1	5±1	1,800±50
Concentrated feed	-	-	17±2	15±3	500±20
Vitamin	-	-	25±1	24±2	200±10
Salt	7±1	5±1	30±2	21±1	1,200±100

Table 2. Nutrients requirement.

Nutrients	Requirement
Protein (%)	16-22
Fat (%)	≤ 14
Calcium (%)	≥ 7
Phosphorus (%)	≥ 5
Energy (kcal)	1700-2200

IOP Conf. Series: Materials Science and Engineering **332** (2018) 012038 doi:10.1088/1757-899X/332/1/012038

Table 3. Raw material fraction boundary.

Materials	Boundary (%)
Corn	15-25
Millet	20-30
Cassava	5-15
Fishmeal	10-15
Rice bran	15-20
Concentrated feed	2-3
Vitamin	1-2
Salt	4-8

Table 4. Corrective action cost.

Nutrients	Corrective action cost (Baht/unit)
Protein (%)	3
Fat (%)	5
Calcum (%)	4
Phosphorus (%)	6
Energy (kcal)	0.25

IOP Conf. Series: Materials Science and Engineering **332** (2018) 012039 doi:10.1088/1757-899X/332/1/012039

Scheduling nurses' shifts at PGI Cikini Hospital

J C T Nainggolan, R D Kusumastuti*

Department of Management, Faculty of Economics and Business, Universitas Indonesia, Kampus UI Depok 16424, Indonesia

*Corresponding author: ratih.dyah@ui.ac.id

Abstract. Hospitals play an essential role in the community by providing medical services to the public. In order to provide high quality medical services, hospitals must manage their resources (including nurses) effectively and efficiently. Scheduling of nurses' work shifts, in particular, is crucial, and must be conducted carefully to ensure availability and fairness. This research discusses the job scheduling system for nurses in PGI Cikini Hospital, Jakarta with Goal Programming approach. The research objectives are to identify nurse scheduling criteria and find the best schedule that can meet the criteria. The model has hospital regulations (including government regulations) as hard constraints, and nurses' preferences as soft constraints. We gather primary data (hospital regulations and nurses' preferences) through interviews with three Head Nurses and distributing questionnaires to fifty nurses. The results show that on the best schedule, all hard constraints can be satisfied. However, only two out of four soft constraints are satisfied. Compared to current scheduling practice, the resulting schedule ensures the availability of nurses as it satisfies all hospital's regulations and it has a higher level of fairness as it can accommodate some of the nurses' preferences.

1. Introduction

The healthcare industry is growing significantly in Indonesia. According to data from the Ministry of Health (MOH), the number of hospitals in Indonesia has increased annually in the average of 5.3% from 2013 to 2016 [1]. In the Jakarta province (the capital city of Indonesia), in particular, the number of hospitals is about 190 in 2016, consisting of general and specialized hospitals [1]. Concerning the number of patients, 39 state-owned hospitals in Indonesia serve about 6.8 million patients in 2016 [2]. As the number of patients is predicted to increase significantly, the availability of human resources, especially nurses, becomes crucial to ensure high-quality health services. The main challenge faced by healthcare service providers is how to allocate and manage resources efficiently to anticipate demand fluctuation [3].

Scheduling nurses' shifts are still considered as a complex problem [4]. According to Kumar et al., scheduling has not been able to provide an optimal solution to the hospital, that can minimize labor costs and satisfy nurses' preferences, as well as distribute well-balanced work shift among nurses, and meet workplace constraints [5].

Good scheduling will increase the quality of healthcare services provided to the customers, as well as increasing nurses' job satisfaction. Generally, shift scheduling is conducted by complying with hospitals' regulations related to workload and by considering nurses' preferences. Work shifts may influence employee's performance, personally, socially and psychologically [6]. Good scheduling allows nurses to have sufficient time for themselves and their families, and provide fairness as each nurse has a similar number of shifts, and thus, allows them to work optimally.

This research is concerned with scheduling nurses' shifts at PGI Cikini Hospital, a private hospital in Jakarta. In this hospital, the nurses' shifts scheduling is performed manually by the Head Nurse in each unit, at the end of every month. This task is still considered as tedious and challenging for the Head Nurse, and the resulting schedule has not been able to provide fairness to nurses. Furthermore, it has not been able to meet all hospital's regulations and accommodate nurses' preferences. Thus, there is a need for a scheduling approach that can provide a better shift schedule for the nurses in the hospital. As different hospitals may have different regulations and nurse preferences, existing nurse scheduling models cannot be directly applied to the problem.

Therefore, the objectives of this research are to identify factors that affect scheduling (hospital regulations and nurses' preferences), and then apply goal programming approach to schedule nurses' work shifts, and compare the resulting schedule with current schedule implemented by the hospital. Goal programming approach is used as it attempts to optimize multiple objective functions simultaneously [7], thus, it can be used to develop nurses' scheduling that must take into accounts hospital regulations and nurses' preferences.

The remainder of the paper is presented as follows. Section 2 discusses the methodology, while Section 3 shows the results and discussions, and Section 4 presents conclusions and recommendations.

2. Methodology

According to Azaiez and Al Sharif, human resources scheduling decisions in health institutions, such as hospitals, can be classified into four interrelated decisions, namely staffing, scheduling, allocation, and assignment decisions [7]. Nurse scheduling, in particular, is the process of determining the schedule for nurses so that the health institution can meet the customer demand [8]. Nurse scheduling is a complex problem that affects hospital personnel on a daily basis, and thus, it is essential to develop it efficiently, and to balance the workload among nurses, and to attempt to satisfy personal preferences [9].

Azaiez and Al Sharif [7] develop a 0-1 goal programming model for nurse scheduling. The model takes into account hospital objectives and nurses' preferences (which are obtained via a survey), and applied to a hospital in Saudi Arabia. The model assumes that a working day consists of two twelve-hour shifts, and the schedule is developed for every planning period. The model is applied to a case with 28 days scheduling period and 15 nurses. Jenal et al. [10] later modify model in [7], and develop a cyclical shift scheduling approach. A cyclic schedule consists of a set of work patterns that are rotated among a group of workers over a set of scheduling horizon [10]. The model by Jenal et al. assumes that one working day consists of three eight-hour shifts, and creates schedule patterns that will be used by Head Nurses to allocate shifts for nurses for the whole year. The model is applied to a case with 21 days scheduling period and 18 nurses.

In this paper, we develop the nurses' work shifts schedule for PGI Cikini Hospital using goal programming approach. The model has hospital regulations (including government regulations) as hard constraints, and nurses' preferences as soft constraints. We modify the model from [7] and [10] as hospital regulations and nurses' preferences in the PGI Cikini Hospital are slightly different from those discussed in the aforementioned models. The scheduling period is 28 days, and the model can be easily modified to a cyclic schedule as in [10]. We gather primary data (hospital regulations and nurses' preferences) through interviews with three Head Nurses and distributing questionnaires to fifty nurses.

2.1. Problem definition

In the current scheduling practice at PGI Cikini Hospital, the Head Nurse in each unit is responsible to create, implement, revise, and documenting the nurse schedule. As previously mentioned, the schedule is created manually at the end of each month, and the process is considered as challenging by the Head Nurse and takes a lot of time to complete. The created schedule often has work shifts distributed unevenly among nurses.

Thus, the objective of this research is to develop the work shift schedule that ensures availability and provides more fairness to the nurses. The schedule is created for 16 nurses in one unit. The main assumptions of the model are as follows: (1) The number of nurses is constant throughout the scheduling period; (2) There are three shifts in a day, morning shift (07.00-14.00), evening shift (14.00-21.00), and night shift (21.00-07.00); (3) The number of minimum nurses in each shift is constant, namely 4 nurses in morning shift, 5 nurses in evening shift, and 3 nurses in night shift; (4) The number of shifts on weekdays (Monday-Saturday) is similar to weekend (Sunday); (5) The planning period is for 28 days (4 weeks), and the first day is Monday; (6) There is no public holiday in the planning period; (7) The schedule does not take into account nurses' skill level or experience.

2.2. Model Formulations

2.2.1. Hard constraints. The hard constraints of the model represent the hospital rules (obtained through interviews with three Head Nurses at the hospital), namely: (1) The number of nurses in each shift satisfies the minimum requirement; (2) A nurse can only work in one shift in a day; (3) If a nurse work two night-shifts in two consecutive days, he/she will get two days-off afterwards; (4) A nurse works 5 to 6 days in a week.

2.2.2. Soft constraints. The soft constraints represent the nurses' preferences (obtained from questionnaires). They are: (1) All nurses have similar workdays in the scheduling period; (2) Each nurse at least gets one day-off on weekend in the scheduling period; (3) The total number of morning and evening shits is more than the number of night shifts; (4) There is no working day in between days-off.

2.2.3. The model. The notations for indices, parameters, and decision variables are presented in Table 1, followed by the model formulations.

Table 1. Model notations

Symbols	Descriptions
n	Number of days in a scheduling period
m	Number of nurses available in the respective division/unit
i	Days, $i = 1, 2, 3, ..., n$
k	Nurses, $k = 1, 2, 3, ..., m$
P_i	Number of nurses required for morning shift on day i, $i = 1, ..., n$
T_i	Number of nurses required for evening shift on day i, $i = 1, ..., n$
M_i	Number of nurses required for night shift on day i, $i = 1, ..., n$
$X_{i,k}$	= 1, if nurse k gets morning shift on day i = 0, otherwise
$Y_{i,k}$	= 1, if nurse k gets evening shift on day i = 0, otherwise
$Z_{i,k}$	= 1, if nurse k gets night shift on day i = 0, otherwise
$C_{i,k}$	= 1, if nurse k gets day-off on day i = 0, otherwise

Based on the interview results, below are hospital regulations (hard constraints) that must be considered by the schedule:

1. The number of nurses on morning, evening and night shifts must meet the minimum requirement.

$$\sum_{k=1}^{m} X_{i,k} \geq P_i, i = 1,2, \dots, n \quad (1)$$

$$\sum_{k=1}^{m} Y_{i,k} \geq T_i, i = 1,2, \dots, n \quad (2)$$

$$\sum_{k=1}^{m} Z_{i,k} \geq M_i, i = 1,2, \dots, n \quad (3)$$

2. Each nurse may only work one shift in day:

$$X_{i,k} + Y_{i,k} + Z_{i,k} + C_{i,k} = 1, i = 1,2, \dots, n \text{ and } k = 1,2, \dots, m \quad (4)$$

3. Each nurse that gets two night-shifts in a row, will get two consecutive days-off afterwards:

$Z_{1,k} + Z_{2,k} + C_{3,k} + C_{4,k} + Z_{11,k} + Z_{12,k} + C_{13,k} + C_{14,k} + Z_{21,k} + Z_{22,k} + C_{23,k} + C_{24,k} = 12, k = 1$ (5)

$Z_{1,k} + Z_{2,k} + C_{3,k} + C_{4,k} + Z_{11,k} + Z_{12,k} + C_{13,k} + C_{14,k} + Z_{23,k} + Z_{24,k} + C_{25,k} + C_{26,k} = 12, k = 2$ (6)

$Z_{1,k} + Z_{2,k} + C_{3,k} + C_{4,k} + Z_{13,k} + Z_{14,k} + C_{15,k} + C_{16,k} + Z_{23,k} + Z_{24,k} + C_{25,k} + C_{26,k} = 12, k = 3$ (7)

$Z_{3,k} + Z_{4,k} + C_{5,k} + C_{6,k} + Z_{13,k} + Z_{14,k} + C_{15,k} + C_{16,k} + Z_{23,k} + Z_{24,k} + C_{25,k} + C_{26,k} = 12, k = 4$ (8)

$Z_{3,k} + Z_{4,k} + C_{5,k} + C_{6,k} + Z_{13,k} + Z_{14,k} + C_{15,k} + C_{16,k} + Z_{25,k} + Z_{26,k} + C_{27,k} + C_{28,k} = 12, k = 5$ (9)

$Z_{3,k} + Z_{4,k} + C_{5,k} + C_{6,k} + Z_{15,k} + Z_{16,k} + C_{17,k} + C_{18,k} + Z_{25,k} + Z_{26,k} + C_{27,k} + C_{28,k} = 12, k = 6$ (10)

$Z_{5,k} + Z_{6,k} + C_{7,k} + C_{8,k} + Z_{15,k} + Z_{16,k} + C_{17,k} + C_{18,k} + Z_{25,k} + Z_{26,k} = 10, k = 7$ (11)

$Z_{5,k} + Z_{6,k} + C_{7,k} + C_{8,k} + Z_{15,k} + Z_{16,k} + C_{17,k} + C_{18,k} + Z_{27,k} + Z_{28,k} = 10, k = 8$ (12)

$Z_{5,k} + Z_{6,k} + C_{7,k} + C_{8,k} + Z_{17,k} + Z_{18,k} + C_{19,k} + C_{20,k} + Z_{27,k} + Z_{28,k} = 10, k = 9$ (13)

$Z_{7,k} + Z_{8,k} + C_{9,k} + C_{10,k} + Z_{17,k} + Z_{18,k} + C_{19,k} + C_{20,k} + Z_{27,k} + Z_{28,k} = 10, k = 10$ (14)

$Z_{7,k} + Z_{8,k} + C_{9,k} + C_{10,k} + Z_{17,k} + Z_{18,k} + C_{19,k} + C_{20,k} = 8, k = 11$ (15)

$Z_{7,k} + Z_{8,k} + C_{9,k} + C_{10,k} + Z_{19,k} + Z_{20,k} + C_{21,k} + C_{22,k} = 8, k = 12$ (16)

$Z_{9,k} + Z_{10,k} + C_{11,k} + C_{12,k} + Z_{19,k} + Z_{20,k} + C_{21,k} + C_{22,k} = 8, k = 13, 14$ (17)

$Z_{9,k} + Z_{10,k} + C_{11,k} + C_{12,k} + Z_{21,k} + Z_{22,k} + C_{23,k} + C_{24,k} = 8, k = 15$ (18)

$Z_{11,k} + Z_{12,k} + C_{13,k} + C_{14,k} + Z_{21,k} + Z_{22,k} + C_{23,k} + C_{24,k} = 8, k = 16$ (19)

4. Each nurse works 5 to 6 days in a week:

$\sum_{i=1}^{n}(X_{i,k} + Y_{i,k} + Z_{i,k}) \geq 5, k = 1,2, \dots, m$ (20)

$\sum_{i=1}^{n}(X_{i,k} + Y_{i,k} + Z_{i,k}) \leq 6, k = 1,2, \dots, m$ (21)

Based on the questionnaire results, the following are nurses' preferences (soft constraints):

1. All nurses have similar number of workdays in every scheduling period:

$\sum_{i=1}^{n}(X_{i,k} + Y_{i,k} + Z_{i,k}) = 22, k = 1,2, \dots, m$ (22)

2. Each nurse at least has one day-off on weekend in every scheduling period:

$C_{7,k} + C_{14,k} + C_{21,k} + C_{28,k} \geq 1, k = 1,2, \dots, m$ (23)

3. The number of morning and evening shifts must be larger than the number of night shifts for each nurse in every scheduling period:

$\sum_{i=1}^{n}(X_{i,k} + Y_{i,k}) - \sum_{i=1}^{n} Z_{i,k} \geq 1, k = 1,2, \dots, m$ (24)

4. Avoid workday in between days-off:

$C_{i,k} + X_{i+1,k} + Y_{i+1,k} + Z_{i+1,k} + C_{i+2,k} \leq 2, i = 1,2, \dots, n-2$ and $k = 1,2, \dots, m$ (25)

According to the Head Nurse, all nurses' preferences have the same priority. Thus, the goals in the model are as follows:

- Goal 1: all nurses have similar number of workdays in each scheduling period. Thus, the goal is to minimize the positive and negative deviations from the Goal 1. The negative deviation for nurse k is $\eta 1_k$, while $\rho 1_k$ is its positive deviation.

  $\sum_{i=1}^{n}(X_{i,k} + Y_{i,k} + Z_{i,k}) + \eta 1_k - \rho 1_k = 22, k = 1,2, \dots, m$ (26)

- Goal 2: Each nurse has at least one day-off on weekend. Thus, the goal is to minimize negative deviation from Goal 2. The negative deviation for nurse k is $\eta 2_k$, while $\rho 2_k$ is its positive deviation.

  $C_{7,k} + C_{14,k} + C_{21,k} + C_{28,k} + \eta 2_k - \rho 2_k \geq 1, k = 1,2, \dots, m$ (27)

- Goal 3: The number of morning and evening shifts must be larger than the number of night shifts for each nurse in every scheduling period. Thus, the goal is to minimize negative deviation from the Goal 3. The negative deviation for nurse k is $\eta 3_k$, while $\rho 3_k$ is its positive deviation.

  $\sum_{i=1}^{n}(X_{i,k} + Y_{i,k}) - \sum_{i=1}^{n} Z_{i,k} + \eta 3_k - \rho 3_k \geq 1, k = 1,2, \dots, m$ (28)

- Goal 4: Avoid workday in between days-off. Thus, the goal is to minimize positive deviation from the Goal 4. The negative deviation for nurse k is $\eta 4_k$, while $\rho 4_k$ is its positive deviation.
$C_{i,k} + X_{i+1,k} + Y_{i+1,k} + Z_{i+1,k} + C_{i+2,k} + \eta 4_k - \rho 4_k \leq 2, i = 1,2,\dots,n-2$ and $k = 1,2,\dots,m$ (29)

Hence, the goal programming formulations are like the following:

$$\text{Minimize } \left(\sum_{k=1}^{m}(\eta 1_k + \rho 1_k), \sum_{k=1}^{m}\eta 2_k, \sum_{k=1}^{m}\eta 3_k, \sum_{k=1}^{m}\rho 4_k\right) \quad (30)$$

Subject to:

Formulations (1) - (21);
Formulations (26) - (29);
$X_{i,k}$, $Y_{i,k}$, $Z_{i,k}$, $C_{i,k}$ are 0 or 1, for all i and k (31)
$\eta 1_k, \rho 1_k, \eta 2_k, \rho 2_k, \eta 3_k, \rho 3_k, \eta 4_k, \rho 4_k \geq 0$, for all k (32)

2.3. Solution Method

The model is formulated using MS. Excel, and solved with Open Solver. The goal programming model is solved to satisfy the hard and soft constraints, and the objective is to minimize the deviations from the nurses' preferences. All preferences have similar priority. So, the model is solved by alternately making each preference as the main priority.

3. Results and discussions

The results show that the best schedule is obtained when the objective function is minimizing positive deviation from Goal 4 subject to hard and soft constraints. Figure 1 presents the best schedule. The sixteen columns in the middle of the figure present the 28-day schedule for 16 nurses, while the three columns on the far right explain the number of nurses in morning, evening, and night shifts in each day. Lastly, the five rows on the bottom of the figure give the total number of morning, evening, and night shifts, along with the total number of workdays and days-off for each nurse. The best schedule indicates that:

	Nurse																Number of Nurses in Each Shift		
Day	1	2	3	4	5	6	7	8	9	10	11	12	13	14	15	16	Morning (M)	Evening (E)	Night (N)
1	N	N	N	E	M	M	M	E	E	M			E	E		M	5	5	3
2	N	N	N	E	M	M	M	M	M	E	E	E	M	E	E	E	6	7	3
3				N	N	N	E	E	M	M	M	M		E	E	E	4	5	3
4				N	N	N	E	M	M		E	M	M	E	E	E	4	5	3
5	E	M	M				N	N	N	M	M	E	E	E	E	M	5	5	3
6	M	E	E				N	N	N	E	E	E	M	M	M		4	5	3
7	E	E	E	M	E	M				N	N	N	E		M	M	4	5	3
8	M	M		E	E	M				N	N	N	E	E	E	M	4	5	3
9	M	E	E	E	E	M	M	E	E				N	N	N	M	4	6	3
10	M	E	E		E	M	M	M	E				N	N	N	E	4	5	3
11	N	N	E		E	M	M	E	M	M	E	E				N	4	5	3
12	N	N	M	E		E	E	M	M	E	E	M				N	4	5	3
13			N	N	N	E	E	M		E	E	E	M	M	M		4	5	3
14			N	N	N		M	M	M	E	M	E	E	E	E		4	5	3
15	E	E				N	N	N	E	E	E	M	M	M		M	4	5	3
16	E	E				N	N	N	E	E	E	M	M	M	M	M	5	5	3
17		M	E	E	E				N	N	N	E	M	M	M	E	4	5	3
18	E		E	M	E				N	N	N	E	M	M	M	E	4	5	3
19	M	E	E	M	E	M	E	E				N	N	N	M		4	5	3
20	E	E	E	E	M	M	M	E				N	N	N	M	M	5	5	3
21	N	E	M	E	E	E	E	M	M	M	E				N	N	4	6	3
22	N	M	E	E	E	M	E		E	M	M				N	N	4	5	3
23		N	N	N	E	E	E	E	M		M	M	M	E			4	5	3
24		N	N	N	M	E	E	M	M	M	E	E	E	E			4	6	3
25	M				N	N	N	E		M	E	E	M	M	E	E	4	5	3
26	E				N	N	N	E	M	M	E	E	M	M	M	E	5	5	3
27	E	E	E	E				N	N	N		E	M	M	M	M	4	5	3
28	E	E	E	M				N	N	N	M	M		E	E	M	4	5	3
Total Morning Shifts	6	4	3	4	4	10	7	8	10	9	6	7	12	9	10	10			
TotalEvening Shifts	9	11	12	10	11	5	9	9	6	7	12	12	6	10	8	8			
Total Night Shifts	6	6	6	6	6	6	6	6	6	6	4	4	4	4	4	4			
Total Workdays	21	21	21	20	21	21	22	23	22	22	22	23	22	23	22	22			
Total Days-off	7	7	7	8	7	7	6	5	6	6	6	5	6	5	6	6			

Figure 1. The best schedule

- Goal 1 is not achieved: 9 out of 16 nurses do not have similar number workdays, which is 22 days. Their number of workdays are in the range of 20-23 days.
- Goal 2 is not achieved: The negative deviation is 5, which means there will be 5 nurses that will not have a day-off on weekends.
- Goal 3 is achieved: All nurses will have larger number of morning and evening shifts than night shifts.
- Goal 4 is achieved: There is no workday in between days-off for all nurses.

These mean that in the best schedule, all hard constraints are satisfied, however, only soft constraints number 3 (the number of morning and evening shifts must be larger than the number of night shifts for each nurse) and number 4 (avoid workday in between days-off) are satisfied.

Compared to current scheduling practice that the Head Nurse does (by creating the schedule manually at the end of each month), the resulting schedule ensures the availability of nurses because it satisfies all hospital's regulations and it has higher level of fairness to all nurses as it can accommodate some of the nurses' preferences (see figure 2).

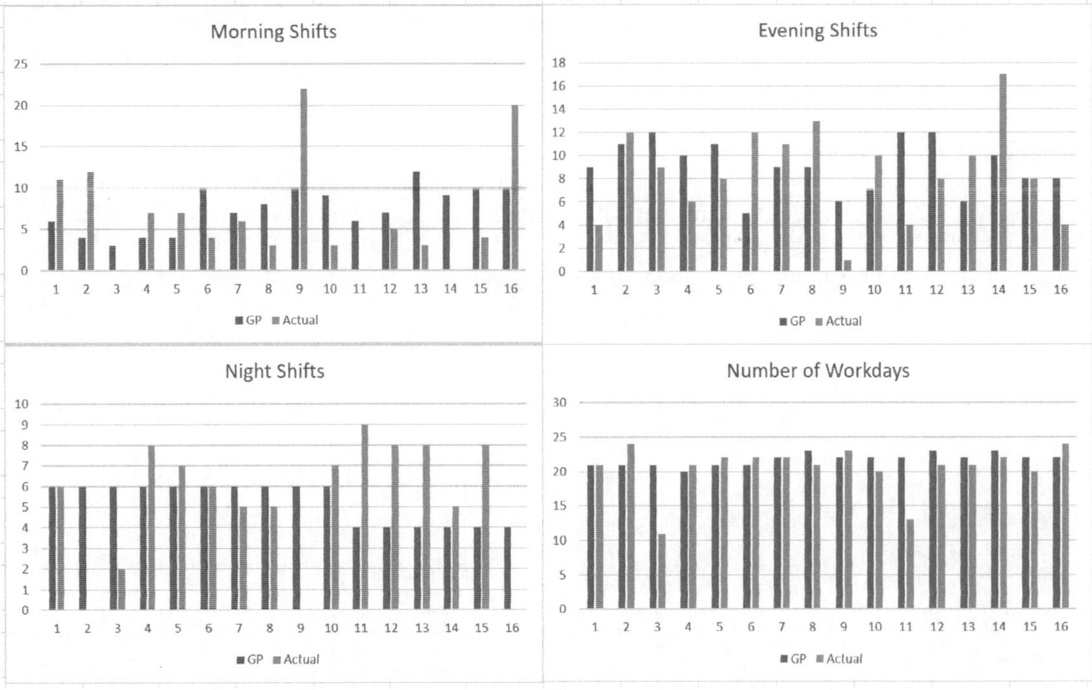

Figure 2. Comparisons of actual and GP schedules

As can be seen in figure 2, the actual schedule (Actual) shows little fairness to the nurses as the numbers of workdays across nurses are in the range of 13 to 24 days. The distribution of shifts also shows little fairness as there are nurses that do not get morning or night shifts at all. These conditions may affect nurses' quality of work and job satisfaction. The resulting schedule (GP) on the other hand, is relatively consistent and has a high level of fairness to nurses. Even though nurses do not get similar number of workdays, the total number of workdays is in the range of 20-23 days with more evenly distributed shifts.

4. Conclusions

Scheduling nurses is a very important task in the hospital and must be done repetitively. Current scheduling practice at PGI Cikini Hospital indicates unfairness and uneven distribution of shifts as the schedule is created manually. Thus, it is very difficult to create a balanced schedule that has the same number of shifts or workdays for all nurses.

The paper presents nurses scheduling at the hospital using goal programming approach. The scheduling is conducted by considering the hospital's regulations/requirements to ensure that the number of minimum nurses available in each shift is sufficient, each nurse gets one shift in a day, nurses that are assigned two night-shifts in a row will get two consecutive days-off, and each nurse works for five to six days in a week. The proposed scheduling approach also considers four nurses' preferences which are obtained from the questionnaires.

The best schedule satisfies all the hospital's regulations, however, only two out of four nurses' preferences can be met. Compared to the manually-created schedule, the proposed schedule offers more fairness to all nurses as the shifts are more evenly distributed, and the differences in the number of workdays between nurses are smaller.

The proposed scheduling approach has some limitations. Firstly, in this approach, nurses are assumed to have similar level of skill. They are not classified based on their educational background and experience. In reality, in each shift, nurses are assigned based on their experiences. Secondly, we also assume that the nurses' preferences have similar priority that may not be the actual condition. The future research may address these limitations by taking into account the levels of skill of nurses, as a set of different nurse skills may be required in each shift. Furthermore, the rank of nurses' preferences can be obtained through a survey, so that the priority of each preference can be set, and the resulting schedule will better represent the nurses' preferences priority.

References

[1] MOH 2017a *Profil Kesehatan Indonesia Tahun 2016 (Health Profile of Indonesia Year 2016)* (Jakarta: MOH RI)

[2] MOH 2017b *Laporan Akuntabilitas Kinerja Instansi Pemerintah: Direktorat Jendral Pelayanan Masyarakat Tahun 2016 (Accountability Report of Government Institution Performance: Directorate General of Community Service Year 2016)* www.yankes.kemkes.go.id

[3] Zurn, P, Dolea, C & Stilwell, B 2005 *Nurse retention and recruitment: Developing a motivated workforce* http://www.icn.ch/images/stories/documents/publications/GNRI/Issue4_Retention.pdf

[4] Ismail, W R, Jenal, R & Hamdan, N A 2012 Goal programming based master plan for cyclical nurse scheduling *J Theor Appl Inf. Technol* **46** 499-504

[5] Kumar, B S, Nagalakshmi, G & Kumaraguru, S 2014 A shift sequence for nurse scheduling using linear programming problem *IOSR-JNHS* **3** 24-8

[6] Tayyari, F & Smith, J L 1997 *Occupational ergonomics: Principles and applications* (London: Chapman Hall)

[7] Azaiez, M N & Al-Sharif, S S 2005 A 0-1 goal programming model for nurse scheduling *Comput Oper Res* **32** 491-507

[8] Ernst, A T, Jiang, H & Krishnamoorthy, M 2004 Staff scheduling and rostering: A review of application models *Eur J Oper Res* **153** 3-27

[9] Burke, E K, De Causmaecker, P, Vanden Berghe, G & Van Landeghem, H 2004 The state of the art of nurse rostering *J Scheduling* **7** 441-9

[10] Jenal, R, Ismail, W R, Liong, C Y & Oughalime, A 2011 A cyclical nurse schedule using goal programming *ITB J Sci* **43A** 151-161

Pricing of premiums for equity-linked life insurance based on joint mortality models

Riaman*, **K Parmikanti, I Irianingsih, S Supian**

Departemen Matematika, FMIPA Unpad, Bandung 45363, INDONESIA

*E-mail: riaman@unpad.ac.id

Abstract. Life insurance equity - linked is a financial product that not only offers protection, but also investment. The calculation of equity-linked life insurance premiums generally uses mortality tables. Because of advances in medical technology and reduced birth rates, it appears that the use of mortality tables is less relevant in the calculation of premiums. To overcome this problem, we use a combination mortality model which in this study is determined based on Indonesian Mortality table 2011 to determine the chances of death and survival. In this research, we use the Combined Mortality Model of the Weibull, Inverse-Weibull, and Gompertz Mortality Model. After determining the Combined Mortality Model, simulators calculate the value of the claim to be given and the premium price numerically. By calculating equity-linked life insurance premiums well, it is expected that no party will be disadvantaged due to the inaccuracy of the calculation result

1. Introduction

Equity-linked life insurance is one of the development products of traditional life insurance. Traditional life insurance only benefits when the insured person (the insured) is at risk. Surely it is related to the benefits of protection offered by life insurance. But as financial developments, life insurance allows not only the benefits of protection but also investment. This is driven by the increasing public awareness of the importance of investment. Insurance that provides protection and investment benefits is then called equity-linked life insurance. In equity-linked life insurance, there are two risks to be studied, namely payoff uncertainty and time uncertainty [11].

To overcome payoff uncertainty, commonly used stock option pricing models such as model [4] or [10]. However, to cover uncertainty of time, always use the death table to determine the life insurance premium associated with equity. It appears that the use of mortality tables is less relevant as medical advances in the medical field and decline in birth rates. There are two main areas of weakness from the use of mortality tables. First, the mortality table shows only historical data, so it can`t get Forecasting mortality rates to evaluate insurance. Secondly, mortality tables do not allow for highly accurate mortality information between the two ages because mortality tables are presented discretely, whereas the death process occurs continuously. Based on these reasons, it is assumed that the premium price offered is lower than it should be. This in the long term can lead to even bankruptcy losses for life insurance companies [11].

To overcome this, a combined mortality table was used to replace the role of mortality tables in the calculation of death and survival chances. Based on the journals written by [1] and [6], it is classified that the *Weibull* type mortality function for children, the *Inverse-Weibull* type for teenagers, and

IOP Conf. Series: Materials Science and Engineering **332** (2018) 012040 doi:10.1088/1757-899X/332/1/012040

Gompertz *types*. This is evident from the analysis of mortality charts of the United States and Taiwan. In this study, Indonesia mortality chart 2011 was used by AAJI (Association of Indonesian Life Insurance) to determine the Combined Mortality Model suitable for use in Indonesia and its effect on the calculation model of equity-linked life insurance premium. Then a simulation of the value of claims and equity-linked life insurance premium price with combined Mortality Model was obtained.

2. Mortality Models

The combined mortality model is a model formed on the basis of a combination of mortality. This combined mortality model uses the mortality function of the Weibull, Inverse-Weibull, and Gompertz distributions. Based on [11] as well as a combined mortality study by [6] which states that the mortality rate of children follows the Weibull distribution, adolescents follow the Inverse-Weibull distribution, and Adult following the Gompertz distribution. The combined mortality model has the following formula:

$$s(x) = \sum_{k=1}^{m} \psi_k s_k(x) \qquad 1$$

Where $s(x)$ is a combined of m survival function, and ψ_k indicates the probability of death by cause k, that satisfy $\psi_1 + \psi_2 + \cdots + \psi_k = 1$, and $s_k(x)$ is an opportunity of survival until the age of x, cause of death k. In this case $k = 1,2,3$, Where k = 1 indicates mortality due to children factor, k = 2 indicates death due to adolescent factor, and k = 3 shows death due to adult factor. $s_k(x)$ is the probability of survival of factors k in age x. For children, weibull distribution is used, while adolescents use Inverse-Weibull distribution, and adults use Gompertz distribution, then the following mortality model is obtained [5]:

$$s_1(x) = \exp\left(-\left(\frac{x}{m_1}\right)^{\frac{m_1}{\sigma_1}}\right) \qquad 2$$

$$s_2(x) = 1 - \exp\left(-\left(\frac{x}{m_2}\right)^{-\frac{m_2}{\sigma_2}}\right) \qquad 3$$

$$s_3(x) = \exp\left(e^{\frac{-m_3}{\sigma_3}} - e^{\frac{(x-m_3)}{\sigma_3}}\right) \qquad 4$$

Of course to determine the model, overall need to know the value of the parameters contained in it, that is $\psi_1, \psi_2, \psi_3, m_1, m_2, m_3, \sigma_1, \sigma_2, \sigma_3$. To determine the value of each of these parameters, a method such as [6] is used, that is, by minimizing the loss function as follows:

$$\sum_{x=0}^{111} (1 - (\hat{q}_x / q_x))^2 \qquad 5$$

Where q_x is the probability of death taken from Indonesian Mortality Table 2011. And $\hat{q}_x$ represents the probability of death to be estimated and given the symbol $\hat{q}_x = 1-(\hat{s}(x+1)/\hat{s}(x))$. For $\hat{s}(x)$, a combined mortality model is used, i.e. [3]

$$s(x) = \psi_1 \exp\left(-\left(\frac{x}{m_1}\right)^{\frac{m_1}{\sigma_1}}\right) + \psi_2\left(1-\exp\left(-\left(\frac{x}{m_2}\right)^{-\frac{m_2}{\sigma_2}}\right)\right) + \psi_3 \exp\left(e^{\frac{-m_3}{\sigma_3}} - e^{\frac{(x-m_3)}{\sigma_3}}\right) \qquad 6$$

The result that the minimum value of the loss function is obtained by estimating the parameters are as follows:

Table 1 Parameter Value of Combined Mortality Model

No.	Parameters	Values
1	ψ_1	0,498128988633006
2	ψ_2	0,501871002123818
3	ψ_3	$7,87648632878604 \cdot 10^{(-9)}$
4	m_1	0,00221415081475328
5	m_2	0,0166257778526924
6	m_3	0,00644796904262000
7	σ_1	0,839543372660222
8	σ_2	5.39644740308355
9	σ_3	$2,60032623812453 \cdot 10^{(-8)}$

Substitute the values of these parameters to $s(x)$ so that the combined mortality model based on the Indonesian mortality table is as follows:

$$s(x) = 0.498128988633006 \cdot \exp\left(-\left(\frac{x}{0.00221415081475328}\right)^{2.637327489 \cdot 10^{-3}}\right) \qquad 7$$

$$+0.501871002123818\left(1-\exp\left(-\left(\frac{x}{0.0166257778526924}\right)^{-3.080874622 \cdot 10^{-3}}\right)\right)+$$

$$7.87648632878604 \cdot 10^{(-9)} \cdot \exp\left(e^{-247967.6953} - e^{\frac{(x-0.00644796904262000)}{2.60032623812453 \cdot 10^{(-8)}}} \right)$$

The model is a combined mortality model that will be used to determine the premium value of equity-linked life insurance. The model obtained is certainly capable of measuring the specific chance of death and survival, which is of course useful to determine the possibility of a claim occurring and useful in the calculation of equity-linked life insurance premiums. This can be seen from the use of the Combined Mortality Model in the model of equity-linked life insurance premium pricing on equations, [7], [9].

$$D_t = R_{pv}(t,T)s(T \mid t) + \int_t^T R_{pv}(t,u) f(s \mid u) ds$$

Here are assumptions used in numerical simulations:

1. The interest used is constant, i.e., the average of the BI rate data obtained, that is equal to 0.067 or 6.7%.
2. Purchase of shares is only 1 lot (100 pieces).
3. Allocated value for risk free investment of IDR500,000.00.
4. Age of insurance entry is 20 years, 25 years, 35 years old with 55 years of retirement age.
5. The correlation between portfolio and interest rate is 0.

Simulation of Equity-Linked Life Insurance Premium Price Determination by using Numerically Combined Mortality Model is as follows:

First, we will determine the value of the expected present value of the claim. Calculations performed directly for three ages of pre-determined insurance. Thus, with [1], we will find the value of $R_{pv}(0,35), R_{pv}(0,30), R_{pv}(0,25)$. For all three have the same portfolio, namely:

$$y(o) = 100 \cdot S(0) + 500000$$

Based on the stock data obtained, used the value until the last close of the data, which is $S(0) = 11675$. So obtained value of the $y(o) = 1667500$.

Next will be calculated the value of the variance between the portfolio and the interest rate, namely:

$$\sigma_x^{\ 2} = \sigma_y^{\ 2} + \sigma_r^{\ 2}(s) - 2\rho\sigma_y\sigma_r(s)$$

Since $p = 0$ and r are constant, then we get:

$$\sigma_x^{\ 2} = \sigma_y^{\ 2} + \sigma_r^{\ 2}$$

Based on the analysis of BI rates data, it is found that $\sigma_r^{\ 2} = 0{,}0000082892$. Based on the analysis of stock data, it is found that the variance of stock is $0{,}022089325$. Then calculated the variance of the portfolio, i.e. by[2]:

$$n^2 \operatorname{var}(S(t)) = 100^2 \cdot 0{,}022089325 = 220{,}89325$$

So obtained:

$$\sigma_x^{\ 2} = \sigma_y^{\ 2} + \sigma_r^{\ 2} = 0{,}0000082892 + 220{,}89325 = 220{,}8932582892$$

Calculated the value of the standard deviation, namely:

$$\sigma_x = \sqrt{220{,}8932582892} = 14{,}8624782$$

Next will be determined zero coupon bond to be used [8]. It was previously assumed that the interest rate was constant so obtained:

$$B(t,u)=E\left(\frac{1}{e^{\int_t^u r(w)dw}}\right)=E\left(e^{r(u-t)}\right)=e^{r(u-t)}$$

In this numerical simulation, M=500000 is used, so that
$\theta = M / y(t) = 500000/1667500 = 0,299850075$ is obtained

The amount of claim can be specified for each entry age as follows:

$$R_{pv}(0,30)=1667500\cdot 1-0,1873081795\cdot 500000\cdot 0+0,1873081795\cdot 500000=1761154,09$$

Furthermore, based on the value of expected present value, will be determined the premium value of each age category entered. By using a computer program, the following results are obtained:

Table 2 Numerical Calculation Results

No.	Entry Age	Value of Claim	Price of Premium
1	20 years	IDR 1715423,601	IDR 844883,5799
2	25 years	IDR 1734494,337	IDR 854557,5769
3	30 years	IDR 1761154,09	IDR 868002,1250

3. Conclusion

Based on the analysis of the Indonesian mortality rate in 2011, we obtained a composite mortality model in accordance with the Indonesian Mortality Model with model parameters that determine the amount of claims emerging and the calculation of the premium price. With the Combined Mortality Model, the model of claim calculation and premium is performed numerical simulation with various assumptions and obtained results as shown in Table 2. These results are only the initial stage, not yet until the analysis phase.

4. Acknowledgments

The Authors would like to thank to DRPMI, directorate of research and community service, Padjadjaran University, who gave funding for this research and dissemination of this paper via RFU.

References

[1] Baxter, M., and Rennie, A. 1996. Financial Calculus an Introduction to Derivative Pricing. Cambridge: Cambridge University Press.
[2] Black, F. and Scholes, M. 1973. The Pricing of Options and Corporate Liabilities. The Journal of Political Economy, Vol. 81, No. 3, 637-654. The University of Chicago Press.
[3] Bowers, N., Gerber, H., Hickman, J., Jones, D., dan Nesbitt, C. 1997. Actuarial Mathematics. Schaumburg, IL: Society of Actuaries.
[4] Brennan, M. J., and Schwartz, E. S. 1976. The pricing of equity-linked life insurance policy with an asset value guarantee. Journal of Financial Economics, 3, 195-213.
[5] Capinski, M., and Zastawniak, T. 2003. Mathematics for Finance an Introduction to Financial Engineering. London: Springer.
[6] Carriere, J. F. 1992. Parametric models for life tables. Transactions of the Society of Actuaries, 44, 77-99.
[7] Chriss, N. A. 1997. Black-Scholes and Beyond, Option Pricing Models. United States of America: McGraw-Hill.
[8] Dickson, David C.M. 2005. Insurence Risk and Ruin. Cambridge : Cambridge University Press.

IOP Conf. Series: Materials Science and Engineering **332** (2018) 012040 doi:10.1088/1757-899X/332/1/012040

[9] Sudjana. 2005. Metoda Statistika. Bandung: Penerbit TARSITO Bandung.Pramesti, Getut. 2011. Distribusi Rayleigh Untuk Klaim Agregasi. Media Statistika Vol. 4, No. 2. Desember 2011 : 105-112.

[10] Savelli, N., Clemente, G. P. 2010. A Collective risk model for outstanding claims reserve distribution (online). (http://www.ica2010.com/docs/176_PPT_ Savelli. pdf, diakses 24 Juli 2012)

[11] Tsai, M. S., Lee, S. C., Chen, J. L., and Wu, S. L. 2014. A New Method to Evaluate Equity-Linked Life Insurance. Contemporary Management Research Pages 23-32, Vol. 10, No.1, March 2014.

IOP Conf. Series: Materials Science and Engineering **332** (2018) 012041 doi:10.1088/1757-899X/332/1/012041

Time prediction of failure a type of lamps by using general composite hazard rate model

Riaman*, E Lesmana, B Subartini, S Supian

Departemen Matematika, FMIPA Unpad, Bandung 45363, Indonesia

*Corresponding author: riaman@unpad.ac.id

Abstract. This paper discusses the basic survival model estimates to obtain the average predictive value of lamp failure time. This estimate is for the parametric model, General Composite Hazard Level Model. The random time variable model used is the exponential distribution model, as the basis, which has a constant hazard function. In this case, we discuss an example of survival model estimation for a composite hazard function, using an exponential model as its basis. To estimate this model is done by estimating model parameters, through the construction of survival function and empirical cumulative function. The model obtained, will then be used to predict the average failure time of the model, for the type of lamp. By grouping the data into several intervals and the average value of failure at each interval, then calculate the average failure time of a model based on each interval, the p value obtained from the test result is 0.3296.

1. Introduction

This model is the application of the theory for the estimation of the basic survival model of the data, as a result of development [6]. In the development grouped into two groups, namely parametric model and non parametric model [1]. Which each group has its own roles and functions. Models with exponential distributions are usually often selected for a random variable model of failure time [2]. [3],[4]. In the case here presented model examples which are simple models of composite hazard rate model and composite hazard rate model [5], [6]. The main purpose of this study was to obtain an estimate value for survival function, cumulative hazard function, along with the following derivatives. This appraisal is done for S and λ and then the result, through the regression model using the least squares method, is used to obtain estimates for λ, β, and *y*. The problem discussed in this paper involves data on the timing of failure of a material type. In which case for model simulation used data with sample size of 49 pieces.

2. Related Concepts and Results

The simplest model of the composite hazard rate model is the exponential piecewise model. The exponential piecewise model is defined by the Hazard Rate Function of function as follows [5], [7].

$$\lambda(t) = \begin{cases} \lambda_1 & : 0 \le t < y \\ \lambda_2 & : \quad t \ge y \end{cases} \qquad 1$$

The survival function is:

$$\lambda(t) = \begin{cases} e^{-t\lambda 1} & 0 \le t < y \\ e^{-y\lambda t} e^{-(t-y)\lambda 2} & t \ge y \end{cases} \qquad 2$$

And its density function is

$$f(t) = \begin{cases} \lambda_1 e^{-t\lambda 1} & 0 \le t < y \\ \lambda_2 e^{-y\lambda t} e^{-(t-y)\lambda 2} & t \ge y \end{cases} \qquad 3$$

Each segment of the composite hazard rate model has a constant hazard ratio. The simple case can be seen in the following figure, in which case the hazard increases according to the duration of time/age.

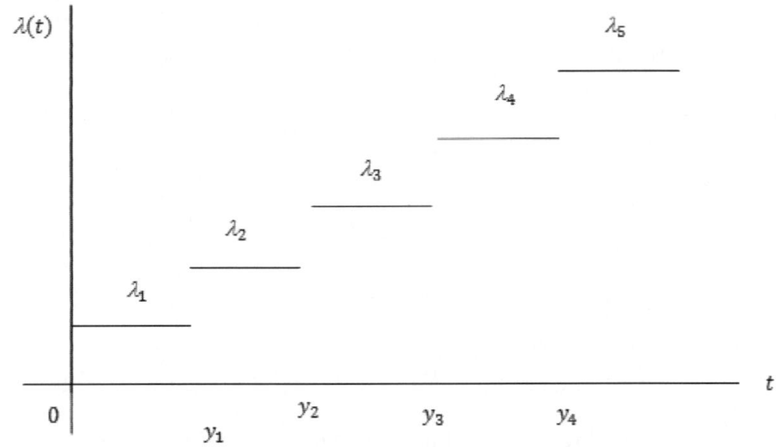

Figure 1. Hazard Rate function

In the discussion it will be presented the data in advance, in which case the model parameters will be estimated, the data in the form of long lifetime of energy-saving lamps. A sample of size n = 49, which is the lifespan of energy-saving lamps. The failure time, in hours, is presented in the following table by observation.

Table1. Failure time (in hours)

No	Time	No	Time	No	Time	No	Time	No	Time	No	Time	No	Time
1	1051	8	4006	15	5905	22	8108	29	10205	36	11608	43	14110
2	1337	9	4012	16	5956	23	8546	30	10396	37	11745	44	12296
3	1389	10	4063	17	6068	24	8666	31	10861	38	11762	45	15395
4	1921	11	4921	18	6121	25	8831	32	11026	39	11895	46	16179
5	1942	12	5445	19	6473	26	9106	33	11214	40	12044	47	17092
6	2322	13	5620	20	7501	27	9711	34	11362	41	13520	48	17568
7	3629	14	5817	21	7886	28	9806	35	11604	42	13670	49	17568

To estimate the parameters of the model, the first is to establish an empirical survival function based on the observed data, $S^0(t)$ and empirical CHF, i.e. $\Lambda^0(t) = -\ln S^0(t)$, the results are presented in Table 2.

IOP Conf. Series: Materials Science and Engineering **332** (2018) 012041 doi:10.1088/1757-899X/332/1/012041

Table 2. Data on Estimated Results

t	$S^0(t)$	$\Lambda^0(t)$
1000	1.000	0.000
2000	0.898	0.108
3000	0.876	0.131
4000	0.857	0.154
5000	0.776	0.254
6000	0.673	0.395
7000	0.612	0.491
8000	0.571	0.560
9000	0.490	0.714
10000	0.429	0.847
11000	0.367	1.001
12000	0.204	1.589
13000	0.184	1.695
14000	0.143	1.946
15000	0.102	2.282
16000	0.082	2.506
17000	0.061	2.973
18000	0.000	∞

Using the minimum area criterion, the estimated value of the optimum for y is $\hat{y} = 3400$. Based on the value of these estimates, the values of λ and β are estimated by the least squares method of the regression model

$$c = \lambda x_1 + \beta x_2 \qquad 4$$

where

$$c = \Lambda^0(t),\ x_1 = t,\ \text{and}\ x_2 = \frac{1}{3}(t - y)^3.I(t)$$

For this data, $\hat{\lambda} = 0.0490$ and β = 0.0038, while the data matching test can be performed using Chi-square test, by grouping the data into several intervals and the mean value of failure time at each interval, for then calculate the average failure time of the model based on each interval. The results are presented in Table 3.

Table 3. Data test results match test

Failure time (in thousands)	Number of observations of failure time	Average time of failure
0.0 – 1.5	3	3.53
1.5 – 3.5	3	4.31
3.5 – 5.5	6	4.34
5.5 – 7.5	7	5.94
7.5 – 9.5	7	8.04
9.5 – 11.5	8	8.83
11.5 – 13.5	6	7.30
13.5 – 15.5	5	4.34
15.5 -	4	2.37

3. Conclusion

Based on the above discussion results obtained $\hat{\lambda} = 0.0490$ and $\beta = 0.0038$, while the data matching test was performed using Chi-square test, by grouping the data into several intervals and the mean value of failure time at each interval, for then calculate the average failure time of the model based on each interval, with the p value obtained from the test result is 0.3296.

Acknowledgments
The Authors would like to thank to DRPMI, directorate of research and community service, Padjadjaran University, who gave funding for this research and dissemination of this paper, via Riset Fundamental Unpad (RFU).

References

[1] Riaman et all, 2017, Prediksi Waktu Kegagalan Suatu Jenis Lampu dengan General Composite Hazard Rate Model, Prociding of Mathematics and Education Mathematics, Kuningan University, West Java.
[2] Batten, R.W, 1978, Mortality Table Construction. Englewood Cliffs : Prentice Hall, Inc.
[3] Benjamin, B. And J.H. Pollard, 1980, *The Analysis of Mortality and Other Statistics*. London : William Heinemann Ltd.
[4] Cox, D.R. and Oakes, D. 1984, *Analysis of Survival Data*. New York :Chapman and Hall
[5] Elandt-Johnson, R.C., and N.L. Johnson, 1980, *Survival Models and Data Analysis*. New York: John Willey and Son.
[6] Klein, John, P., and Moeschberger, Melvin, L., 1998, *Survival Analysis: Techniques for Censored and Truncated Data*. New York: Springer Verlag Inc.
[7] London, D., 1997, *Survival Model and Their Estimation*, 3rd Edition. Winsted: Actex Publication Inc.

Optimization of hybrid model on hajj travel

R Cahyandari[1*], R L Ariany[2], Sukono[3]

[1]Department of Mathematics, Faculty of Science and Technology, UIN Sunan Gunung Djati, Bandung, Indonesia
[2]Department of Mathematics Education, Faculty of Tarbiyah and Teacher Training, UIN Sunan Gunung Djati Bandung, Indonesia
[3]Department of Mathematics, Faculty of Mathematics and Natural Science, Universitas Padjadjaran, Indonesia

[*]Email: rini_cahyandari@uinsgd.ac.id

Abstract. Hajj travel insurance is an insurance product offered by the insurance company in preparing funds to perform the pilgrimage. This insurance product helps would-be pilgrims to set aside a fund of saving hajj with regularly, but also provides funds of profit sharing (*mudharabah*) and insurance protection. Scheme of insurance product fund management is largely using the *hybrid* model, which is the fund from would-be pilgrims will be divided into three account management, that is personal account, *tabarru'*, and *ujrah*. Scheme of *hybrid* model on hajj travel insurance was already discussed at the earlier paper with titled "*The Hybrid Model Algorithm on Sharia Insurance*", taking the example case of Mitra Mabrur Plus product from Bumiputera company. On these advanced paper, will be made the previous optimization model design, with partition of benefit the *tabarru'* account. Benefits such as compensation for 40 critical illness which initially only for participants of insurance only, on optimization is intended for participants of the insurance and his heir, also to benefit the hospital bills. Meanwhile, the benefits of death benefit is given if the participant is fixed die.

Keywords: *insurance of hajj travel, hybrid model, tabarru', rider*

1. Background

Islamic insurance industry in Indonesia was born in 1994, marked by the establishment of family *takaful* insurance and *takaful* insurance general [4]. From birth until now, it was growth and development, namely the increasing number of increasingly sharia insurance industry, the increasing number of asset investment, developments that are experiencing positive trends, and variations diverse insurance products adjust the needs of each person.

Increasing the growth and development of sharia insurance industry, automatically have an impact on increasing business competition, so this industry is vying interests of insurance as many participants. One of the strategies that must be considered in the interest of the insurance participants is an insurance product innovation that offers many benefits, and packaged in an attractive marketing and right on target.

Example of product innovation, including the use of sharia insurance mutual policy, this means that the premium payment to an insurance participant policy can be used by other family members so that the payment becomes lighter [2, 3]. The next innovation was the benefits offered insurance products, for example on hajj travel insurance. The benefits that are offered not only compensation for the death, but compensation 40 critical illnesses, hospital bills and cash value.

Hajj travel insurance is a product that is intended for participants who have plans to travel on pilgrimage to the Holy Land of Mecca. These insurance products not only help set aside funds on a regular basis, but also offers for results (*mudharabah*) investments and benefits in the form of insurance protection [4]. However, generally the benefits provided are only valid for one person only, that is insurance participants who also prospective worshipers.

Study of mathematical model about Sharia insurance, first time researchers convey on IORA International Conference on Operation Research 2016, then discussed on the first paper with the title: *A Design of Mathematical Modelling for The Mudharabah Scheme in Shariah Insurance* and it has been publicated [1]. Next, the second paper written with the title: *The Hybrid Model Algorithm on Sharia Insurance* and it has been presented on International Conference on the Science and Engineering Computation in 2017, in the process of review for publication. In this second paper the researchers discuss on algorithm *hybrid* model as a mathematical model schemes of insurance fund management from hajj travel of Mitra Mabrur Plus Bumiputera company, where the funds are paid in premiums by insurance participants will be divided into three accounts by the company, namely personal account, *tabbaru'* account and *ujrah*. The granting of the benefits derived from *tabbaru'* account already invested, and the benefits provided not only compensation for the death, there's also a compensation for 40 critical illnesses and compensation for the hospital bills.

Based on the discussion, it can be concluded that the *hybrid* model is used by hajj travel insurance of Mitra Mabrur Plus Bumiputera have already good enough in terms of fund management as seen from the proportion of personal accounts that increases and decreases of u*jrah* year to year. But there is still a bit short in the management if it is to continue to compete in the future, that is some of the benefits provided by the company are only valid for attendees only. Therefore it is interesting to do further research, by designing optimization model of *hybrid* travel insurance of Mitra Mabrur Plus through the partition of benefits from *tabbaru'* accounts so few benefits do not apply only to participants, but applies also to his heir.

2. Methodology

Stages of research conducted, including the following:

- Visit and interview with actuary of Bumiputera insurance company about the data and the scheme of *hybrid* model from Mitra Mabrur Plus insurance.
- Translate *hybrid* model scheme into a mathematical equation by entering some preliminary values that have been set by the company.
- Present a *hybrid* model with substitution the value of the specified company's premiums to the mathematical equations on stage 2.
- Perform the optimization with the benefits of partitioned *tabbaru'* account so that some benefits are more optimally, that apply to participants and his heir.

There are below is the data of hajj travel insurance of Mitra Mabrur Plus company with policy limited for 5 years and some initial values are set, among them:

Age of the Policyholder : 38 Years
Death Coverage : 6.000.000
Hospitall Bills : 600.000
40 Critical illness : 3.000.000
Monthly profit in first year : 201

Table 1. Mitra Mabrur Plus's Insurance Data

		Year				
	Year of Policy	1	2	3	4	5
1	Contributions per year	1,200,000.00				
	Contributions per month (P)	100,000				
2	% Basis of *Tabarru'*	7.47	7.47	7.47	7.47	7.47
3	*% Ujrah*	19.35	11.84	9.34	9.34	9.34
4	*% Personal Account*	73.18	80.69	83.19	83.19	83.19

Table 2. Mathematical equations for *hybrid* model scheme of Mitra Mabrur Plus Insurance

1	2	3	4	5	6	7	8	9	10
Month	**age**	**contribution**	**Tabarru' account**	**Ujrah**	**Personal Account**	**accumu lation**	**monthly profit**	**Yearly Profit**	**Surrender value**
1	38	P	0.0747*P	0.1935 *P	0.7318*P	73,180	201		73,381
2-11	38	P	0.0747*P	0.1935 *P	0.7318*P	X_1	Y_1		X_1+Y_1
12	38	P	0.0747*P	0.1935 *P	0.7318*P	X_1	Y_1	0.05*(0.7318*P)	(X_1+Y_1)+(0.05*(0.7318*P))
13	39	P	0.0747*P	0.1184 *P	0.8069*P	(0.7318 *P) + (0.8069 *P)	Y_2		((0.7318*P) + (0.8069*P)) + Y_2
14-23	39	P	0.0747*P	0.1184 *P	0.8069*P	X_2	Y_2		X_2+Y_2
24	39	P	0.0747*P	0.1184 *P	0.8069*P	X_2	Y_2	0.05*(0.8069*P)	(X_2+Y_2)+ (0.05*(0.8069*P))
25	40	P	0.0747*P	0.0934 *P	0.8319*P	(0.8069 *P) + (0.8319 *P)	Y_3		((0.8069*P) + (0.8319*P))+ Y_3
26-35	40	P	0.0747*P	0.0934 *P	0.8319*P	X_3	Y_3		X_3+Y_3
36	40	P	0.0747*P	0.0934 *P	0.8319*P	X_3	Y_3	0.05*(0.8319*P)	(X_3+Y_3)+ (0.05*(0.8319*P))
37-47	41	P	0.0747*P	0.0934 *P	0.8319*P	X_4	Y_4		X_4+Y_4
48	41	P	0.0747*P	0.0934 *P	0.8319*P	X_4	Y_4	0.05*(0.8319*P)	(X_4+Y_4)+ (0.05*(0.8319*P))
49-59	42	P	0.0747*P	0.0934 *P	0.8319*P	X_5	Y_5		X_5+Y_5
60	42	P	0.0747*P	0.0934 *P	0.8319*P	X_5	Y_5	0.05*(0.8319*P)	(X_5+Y_5)+(0.05*(0.8319*P))

Table 3. Mathematical equations for *hybrid* model scheme of Mitra Mabrur Plus Insurance (continued)

1	2	11	12	13	14
Month	**age**	**40 critical Illness**	**death coverage**	**Hospital Bills**	**Death Benefit**
1	38	0.5*((P*12)*5)	(P*12)*5	0.1*((P*12)*5)	((P*12)*5) + 73,381
2-11	38	0.5*((P*12)*5)	(P*12)*5	0.1*((P*12)*5)	((P*12)*5) + (X_1+Y_1)
12	38	0.5*((P*12)*5)	(P*12)*5	0.1*((P*12)*5)	((P*12)*5) + ((X_1+Y_1)+(0.05*(0.7318*P)))
13	39	0.5*((P*12)*5)	(P*12)*5	0.1*((P*12)*5)	((P*12)*5) + (((0.7318*P)+(0.8069*P)) + Y_2)
14-23	39	0.5*((P*12)*5)	(P*12)*5	0.1*((P*12)*5)	((P*12)*5) + (X_2+Y_2)

1	2	11	12	13	14
Month	**age**	**40 critical Illness**	**death coverage**	**Hospital Bills**	**Death Benefit**
24	39	0.5*((P*12)*5)	(P*12)*5	0.1*((P*12)*5)	$((P*12)*5) + ((X_2+Y_2)+ (0.05*(0.8069*P)))$
25	40	0.5*((P*12)*5)	(P*12)*5	0.1*((P*12)*5)	$((P*12)*5) + (((0.8069*P)+(0.8319*P))+ Y_3)$
26-35	40	0.5*((P*12)*5)	(P*12)*5	0.1*((P*12)*5)	$((P*12)*5) + (X_3+Y_3)$
36	40	0.5*((P*12)*5)	(P*12)*5	0.1*((P*12)*5)	$((P*12)*5) + ((X_3+Y_3)+ (0.05*(0.8319*P)))$
37-47	41	0.5*((P*12)*5)	(P*12)*5	0.1*((P*12)*5)	$((P*12)*5) + (X_4+Y_4)$
48	41	0.5*((P*12)*5)	(P*12)*5	0.1*((P*12)*5)	$((P*12)*5) + ((X_4+Y_4)+ (0.05*(0.8319*P)))$
49-59	42	0.5*((P*12)*5)	(P*12)*5	0.1*((P*12)*5)	$((P*12)*5) + (X_5+Y_5)$
60	42	0.5*((P*12)*5)	(P*12)*5	0.1*((P*12)*5)	$((P*12)*5) + ((X_5+Y_5)+(0.05*(0.8319*P)))$

There are several terms that are often used in insurance, including:

1. Policy : The insurance agreement between the insurer and the policyholder.
2. Premium : An amount of money that is listed in the policy that approved by the policyholder, for paid to insurance companies according the agreement so that the policy remains active.
3. Personal Account : Collection of funds that belong to policyholders and payable if the agreement ends, the participant resigns, or dies.
4. *Tabarru'* Account : The collection of funds that intended by the policyholder as alms with the purpose to help each other and will be paid in the event of claims.
5. Surrender Value : The amount of money that have to be paid to policyholders, if the insurance agreement is terminated before the expiry of the agreement.
6. Rider : Conditions that attached to the insurance policy which gives additional benefits or restrictions.
7. *Ujrah* : Wages or rewards.
8. Hospital Bills : Rider (benefits) in the form of reimbursement of the cost of hospital.
9. Critical Illness : Rider (benefits) in the form of sum assured in case of critical illness.
10. Death Coverage : An amount of money that are prepared by the insurance company as compensation if the policyholder dies.
11. Death Benefit : An amount of money that will be paid if the policyholder dies before and or after the period of the insurance agreement terminates.

3. Result

Using the data of Mitra Mabrur Plus Insurance, the value of premium is Rp. 100.000 was substituted into the Table 3, then retrieved the table of *hybrid* model bellows:

Tabel 4. *Hybrid* Model Table of Mitra Mabrur Plus Insurance

1	2	3	4	5	6	7	8	9	10	11	12	13	14
Month	**age**	**layout payment**	**Tabarru' account**	***Ujrah***	**personal account**	**accumulation**	**monthly profit**	**yearly profit**	**surrender value**	**40 critical Illness**	**death coverage**	**hospital bills**	**Death Benefit**
1	38	100,000	7,470	19,350	73,180	73,180	201		73,381	3,000,000	6,000,000	600,000	6,073,381
2	38	100,000	7,470	19,350	73,180	146,360	201		146,561	3,000,000	6,000,000	600,000	6,146,561
3	38	100,000	7,470	19,350	73,180	146,360	201		146,561	3,000,000	6,000,000	600,000	6,146,561
4	38	100,000	7,470	19,350	73,180	146,360	201		146,561	3,000,000	6,000,000	600,000	6,146,561
5	38	100,000	7,470	19,350	73,180	146,360	201		146,561	3,000,000	6,000,000	600,000	6,146,561

1	2	3	4	5	6	7	8	9	10	11	12	13	14
Month	**age**	**layout payment**	**Tabarru' account**	***Ujrah***	**personal account**	**accumulation**	**monthly profit**	**yearly profit**	**surrender value**	**40 critical Illness**	**death coverage**	**hospital bills**	**Death Benefit**
6	38	100,000	7,470	19,350	73,180	146,360	201		146,561	3,000,000	6,000,000	600,000	6,146,561
7	38	100,000	7,470	19,350	73,180	146,360	201		146,561	3,000,000	6,000,000	600,000	6,146,561
8	38	100,000	7,470	19,350	73,180	146,360	201		146,561	3,000,000	6,000,000	600,000	6,146,561
9	38	100,000	7,470	19,350	73,180	146,360	201		146,561	3,000,000	6,000,000	600,000	6,146,561
10	38	100,000	7,470	19,350	73,180	146,360	201		146,561	3,000,000	6,000,000	600,000	6,146,561
11	38	100,000	7,470	19,350	73,180	146,360	201		146,561	3,000,000	6,000,000	600,000	6,146,561
12	38	100,000	7,470	19,350	73,180	146,360	201	3,659	150,220	3,000,000	6,000,000	600,000	6,150,220
13	39	100,000	7,470	11,840	80,690	153,870	38,372		192,242	3,000,000	6,000,000	600,000	6,192,242
14	39	100,000	7,470	11,840	80,690	161,380	38,372		199,752	3,000,000	6,000,000	600,000	6,199,752
15	39	100,000	7,470	11,840	80,690	161,380	38,372		199,752	3,000,000	6,000,000	600,000	6,199,752
16	39	100,000	7,470	11,840	80,690	161,380	38,372		199,752	3,000,000	6,000,000	600,000	6,199,752
17	39	100,000	7,470	11,840	80,690	161,380	38,372		199,752	3,000,000	6,000,000	600,000	6,199,752
18	39	100,000	7,470	11,840	80,690	161,380	38,372		199,752	3,000,000	6,000,000	600,000	6,199,752
19	39	100,000	7,470	11,840	80,690	161,380	38,372		199,752	3,000,000	6,000,000	600,000	6,199,752
20	39	100,000	7,470	11,840	80,690	161,380	38,372		199,752	3,000,000	6,000,000	600,000	6,199,752
21	39	100,000	7,470	11,840	80,690	161,380	38,372		199,752	3,000,000	6,000,000	600,000	6,199,752
22	39	100,000	7,470	11,840	80,690	161,380	38,372		199,752	3,000,000	6,000,000	600,000	6,199,752
23	39	100,000	7,470	11,840	80,690	161,380	38,372		199,752	3,000,000	6,000,000	600,000	6,199,752
24	39	100,000	7,470	11,840	80,690	161,380	38,372	4,035	203,787	3,000,000	6,000,000	600,000	6,203,787
25	40	100,000	7,470	9,340	83,190	163,880	51,983		215,863	3,000,000	6,000,000	600,000	6,215,863
26	40	100,000	7,470	9,340	83,190	166,380	51,983		218,363	3,000,000	6,000,000	600,000	6,218,363
27	40	100,000	7,470	9,340	83,190	166,380	51,983		218,363	3,000,000	6,000,000	600,000	6,218,363
28	40	100,000	7,470	9,340	83,190	166,380	51,983		218,363	3,000,000	6,000,000	600,000	6,218,363
29	40	100,000	7,470	9,340	83,190	166,380	51,983		218,363	3,000,000	6,000,000	600,000	6,218,363
30	40	100,000	7,470	9,340	83,190	166,380	51,983		218,363	3,000,000	6,000,000	600,000	6,218,363
31	40	100,000	7,470	9,340	83,190	166,380	51,983		218,363	3,000,000	6,000,000	600,000	6,218,363
32	40	100,000	7,470	9,340	83,190	166,380	51,983		218,363	3,000,000	6,000,000	600,000	6,218,363
33	40	100,000	7,470	9,340	83,190	166,380	51,983		218,363	3,000,000	6,000,000	600,000	6,218,363
34	40	100,000	7,470	9,340	83,190	166,380	51,983		218,363	3,000,000	6,000,000	600,000	6,218,363
35	40	100,000	7,470	9,340	83,190	166,380	51,983		218,363	3,000,000	6,000,000	600,000	6,218,363
36	40	100,000	7,470	9,340	83,190	166,380	51,983	4,160	222,523	3,000,000	6,000,000	600,000	6,222,523
37	41	100,000	7,470	9,340	83,190	166,380	56,744		223,124	3,000,000	6,000,000	600,000	6,223,124
38	41	100,000	7,470	9,340	83,190	166,380	56,744		223,124	3,000,000	6,000,000	600,000	6,223,124

1	2	3	4	5	6	7	8	9	10	11	12	13	14
Month	**age**	**layout payment**	**Tabarru' account**	***Ujrah***	**personal account**	**accumulation**	**monthly profit**	**yearly profit**	**surrender value**	**40 critical Illness**	**death coverage**	**hospital bills**	**Death Benefit**
39	41	100,000	7,470	9,340	83,190	166,380	56,744		223,124	3,000,000	6,000,000	600,000	6,223,124
40	41	100,000	7,470	9,340	83,190	166,380	56,744		223,124	3,000,000	6,000,000	600,000	6,223,124
41	41	100,000	7,470	9,340	83,190	166,380	56,744		223,124	3,000,000	6,000,000	600,000	6,223,124
42	41	100,000	7,470	9,340	83,190	166,380	56,744		223,124	3,000,000	6,000,000	600,000	6,223,124
43	41	100,000	7,470	9,340	83,190	166,380	56,744		223,124	3,000,000	6,000,000	600,000	6,223,124
44	41	100,000	7,470	9,340	83,190	166,380	56,744		223,124	3,000,000	6,000,000	600,000	6,223,124
45	41	100,000	7,470	9,340	83,190	166,380	56,744		223,124	3,000,000	6,000,000	600,000	6,223,124
46	41	100,000	7,470	9,340	83,190	166,380	56,744		223,124	3,000,000	6,000,000	600,000	6,223,124
47	41	100,000	7,470	9,340	83,190	166,380	56,744		223,124	3,000,000	6,000,000	600,000	6,223,124
48	41	100,000	7,470	9,340	83,190	166,380	56,744	4,160	227,284	3,000,000	6,000,000	600,000	6,227,284
49	42	100,000	7,470	9,340	83,190	166,380	57,953		224,333	3,000,000	6,000,000	600,000	6,224,333
50	42	100,000	7,470	9,340	83,190	166,380	57,953		224,333	3,000,000	6,000,000	600,000	6,224,333
51	42	100,000	7,470	9,340	83,190	166,380	57,963		224,333	3,000,000	6,000,000	600,000	6,224,333
52	42	100,000	7,470	9,340	83,190	166,190	57,963		224,333	3,000,000	6,000,000	600,000	6,224,333
53	42	100,000	7,470	9,340	83,190	166,380	57,953		224,333	3,000,000	6,000,000	600,000	6,224,333
54	42	100,000	7,470	9,340	83,190	166,380	57,953		224,333	3,000,000	6,000,000	600,000	6,224,333
55	42	100,000	7,470	9,340	83,190	166,380	57,953		224,333	3,000,000	6,000,000	600,000	6,224,333
56	42	100,000	7,470	9,340	83,190	166,380	57,953		224,333	3,000,000	6,000,000	600,000	6,224,333
57	42	100,000	7,470	9,340	83,190	166,380	57,953		224,333	3,000,000	6,000,000	600,000	6,224,333
58	42	100,000	7,470	9,340	83,190	166,380	57,953		224,333	3,000,000	6,000,000	600,000	6,224,333
59	42	100,000	7,470	9,340	83,190	166,380	57,953		224,333	3,000,000	6,000,000	600,000	6,224,333
60	42	100,000	7,470	9,340	83,190	166,380	57,953	4,160	228,493	3,000,000	6,000,000	600,000	6,228,493

In the Table 4, it is showed the details of the management of the fund contribution to policyholders during a five-year contract. Through this table could be seen nominal amount each account, made per month or per year, and the benefits that become accrues of policyholders in case of claims during the year of policy runs or if the contract is stopped in the middle for some reason. The benefits of 40 critical illness, hospital bills, and the death benefit is offered in table 3 are only intended for insurance participants, so with a quantity *tabbaru'* account that fixed will be do an optimization of *hybrid* model of through partition its benefits.

Quantity of *tabbaru'* account is Rp. 7.470 (fixed) and will be divided for the provision of compensation for 40 critical illness as much as two people (participants and his heir), compensation for hospital bills by as much as two people (participants and his heir), and a death benefit as much as one person (if the participant died).

Table 5. *Hybrid* Model's Benefit Partition of Mitra Mabrur Plus Insurance

Benefit	*40 Critical Illness* (5X)	*Death benefit* (20X)	*Hospital Bills* (1X)
Benefits of partition ratio	$5X$ (2 people) = $10X$	$20X$ (1 people) = $20X$	$1X$ (2 people) = $2X$
Total compensation for each of the benefits	233.44 (10) = 2,334.4	233.44 (20) = 4,668.8	233.44 (2) = 466.88

Based on Table 5 above, a total of partition benefits equal to 32 with the value X = Rp. 300,000 (per unit), and if the fund's total compensation for each of the aggregated benefit then retrieved: Rp, 2, 334.4 + Rp.4, 668.8 + Rp 466.88 = 7, 470.08 ≅ 7.470 (*tabbaru'* account). This means that with a quantity *tabbaru'* account were fixed, the travel hajj insurance of Mitra Mabrur Plus company can offer the benefits of 40 critical illness and hospital bills for participants and his heir. While the benefits of a death benefit only to heirs if the participant died.

4. Summary

Optimization of the *hybrid* model of hajj travel insurance of Mitra Mabrur Plus can be a design innovation of sharia insurance product, because this optimization can be done on other sharia insurance product, which becomes more value in order of competition the business interests of insurance participant. The *hybrid* model of hajj travel insurance of Mitra Mabrur Plus after optimization, the mechanisms to manage them is getting better due to the amount of increasing personal account and a decreasing *ujrah* quantities each year. These conditions will be profitable for the insurance participants. Meanwhile the amount of *tabarru'* accounts that dixed during the contract year of policy is profitable to the insurance company due to minimize loss that caused by fluctuations in investment performance and profitable also for participants because of two insurance benefits that applies to participants and his heir.

5. Acknowledgment

We thank LP2M Universitas Sunan Gunung Djati Bandung, which has been providing facilities for research and publication. Thanks to Mr. Windhu Wijatmoko, Drs, MM, AAIJ, FSAI as actuary of the AJB Bumiputera 1912 company that willing take the time to discuss about Mitra Mabrur Plus insurance, and Annisa Azizah, S.Si has helped directly in the writing of this paper. Thanks were also given to all who has helped directly or indirectly so that these paper can be resolved properly.

References

[1] Cahyandari, R, et.al. "*A design of mathematical modelling for the mudharabah scheme in shariah insurance*". IOP Conference Series: Materials Science and Engineering, Volume 166.

[2] Ghazali, P . L, et.al. "*Integration Model In Premium Life Table Of Family Takaful*". Journal of Applied Science, 8(7): 3763-3776 (2012).

[3] Ghazali, P . L, et.al. "*Implementation of Integration Model for All*". Journal of Applied Science Research, 8(3): 1802-1812 (2012).

[4] Sula, M. S. "*Asuransi Syariah (Life and General) Konsep dan Sistem Operasional*". Gema Insani. Jakarta (2004).

IOP Conf. Series: Materials Science and Engineering **332** (2018) 012043 doi:10.1088/1757-899X/332/1/012043

Improving the quality of learning in science through optimization of lesson study for learning community

S Setyaningsih
Faculty of Mathematics and Natural Sciences, Pakuan University, Bogor, Indonesia

Email: sri_setya@yahoo.com

Abstract. Lesson Study for Learning Community is one of lecturer profession building system through collaborative and continuous learning study based on the principles of openness, collegiality, and mutual learning to build learning community in order to form professional learning community. To achieve the above, we need a strategy and learning method with specific subscription technique.. This paper provides a description of how the quality of learning in the field of science can be improved by implementing strategies and methods accordingly, namely by applying lesson study for learning community optimally. Initially this research was focused on the study of instructional techniques. Learning method used is learning model Contextual teaching and Learning (CTL) and model of Problem Based Learning (PBL). The results showed that there was a significant increase in competence, attitudes, and psychomotor in the four study programs that were modelled. Therefore, it can be concluded that the implementation of learning strategies in Lesson study for Learning Community is needed to be used to improve the competence, attitude and psychomotor of science students.

1. Background and Purpose of the Research

Learning design is a way to plan a system of learning environment after previously first set learning strategies. Learning strategy is a strategy or technique that must be owned by the lecturers. It is very necessary because it will greatly determine the quality or properness of learning done by a lecturer, because the learning process requires knowledge, art, and expertise in delivering materials to students according to goals, efficient, and effectiveness.

Lecturers as professional educators must have idealism and high fighting spirit, as well as professional performance, especially in designing the teaching program and implementation of the learning process. Miriam Zukas emphasizes pedagogic meaning for lecturers very important, because lecturers can be viewed from three angles: teacher as charismatic subject, teacher as component craftperson, and teacher as reflective practitioner [1].

Perrenoud in Bipoupout Jean Calvin states that pedagogical competence involves the mastery of educators, in learning planning and evaluation, presentation skills (knowledge of learning materials), technical skills, managerial skills, and communication skills for evaluation [2]. During this time, in the implementation of learning for FMIPA students there are still many obstacles. This is evident from the results of surveys and monitoring conducted by the Institute for Development and Increased Instructional Activities (LP2AI) on the learning process. In the pedagogic aspect, the frequency of negative perception is still high with the grading of the ability to enliven the classroom (12%), the use of media and learning technology 10.3%, and clarity in the material and answer to the questions in class (7.5%). In the aspect of student involvement competence in research and/or study/ engineering/design (11%), and clarity and use of communication technology explain the subject (16,8%).

To achieve the quality of learning as well as mentioned above, a strategy that includes plan, method, or series of activities is designed to accomplish a particular education goal. Lecturer as a communicator must have a high psychomotor so that students get optimal results in learning and can encourage students to be able to find their own ideas. Improving the empowerment of lecturers through lesson study according to Zaki Malik and Athman Bouguettaya is one of the strategies to create more innovative lecturers for high quality to gain competitive advantage [3].

Therefore, based on the description as mentioned above, it is necessary to establish a strategy that can overcome the optimal learning and one of them is through the implementation of Lesson study for community in FMIPA.

2. Research Method

The research method used in this research is action research method through improvement or improvement of quality, either micro or macro. Collection is done by observation, test, field record, and documentation. Data analysis techniques performed with data reduction, data presentation, and conclusion. The process of data selection which is data reduction is obtained from the subject matter and field notes, while the collection of information as the presentation of data and then done the conclusion of data based on the results of reduction and presentation of data.

The research was conducted for two semesters, namely: even semester 2016/2017 and odd semester 2017/2018 on Biology, Chemistry, Computer Science and Pharmacy. The research stage consists of three cycles and the observation sheet consists of the stages of planning (Plan), implementation (Do), and reflection (See). As the object of research is the student. Parameters observed were student learning activities, collaboration and communication among students during the learning process took place. The reflection was performed based on the use of lesson study model based on CTL and PBL in the classroom at the end of the cycle and conducted through three cycles.

3. Result and Discussion

Lawrence Ingvarson quotes Fenstermacher and Richardson in Linda E. Martin et al that there are two distinctions between good teachers and successful teaching [4], as follows:

> *"By good teaching we mean that the content taught accords with disciplinary standards of adequacy and completeness, and that the methods employed are age appropriate, morally defensible, and undertaken with the intention of enhancing the leaner's competence with respect to the content studied. By successful teaching we mean that the learner actually acquires, to some reasonable and acceptable level of proficiency, what the teacher is engaged in teaching".*

The implementation of collaborative, communicative, and innovative learning models is the theme of this research. Preparation of chapter design and lesson designed is done for open class preparation and data retrieval is done in 3 cycles. Implementation of Cycle I Action is designed in accordance with the concept and strategy that has been planned from the beginning. The relationship diagram of approaches, strategies, methods, and instructional techniques are presented in figure 1 below.

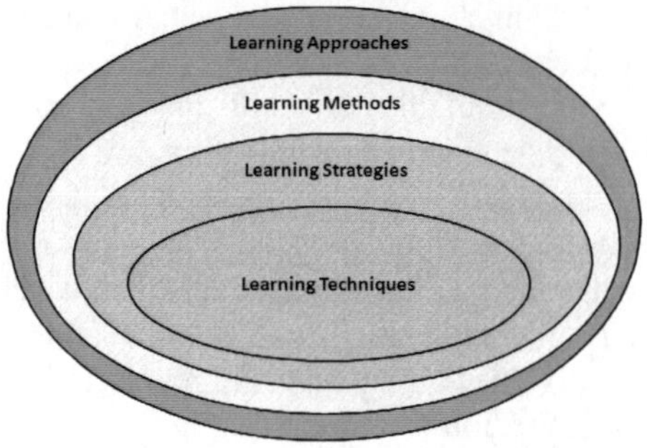

Figure 1. Relationship between approaches, strategies, methods, and learning techniques

The stages plan in cycle 2 and cycle 3 are implemented based on the recommendations of the reflection of cycle 1. The learning outcomes are observed through the five different capacities/capabilities achieved, which include intellectual skills, verbal information, cognitive models, attitudes, and motor skills. Election of course is done by choosing the relevant subject has previous learner prerequisite so that will facilitate us in observation of competency, affective, and psychomotor aspect of student. Observations made by observers at cycles 1, 2, and 3 and through two even and odd semesters obtained the results of student enthusiasm on task and vice versa off task as shown in Figure 2.

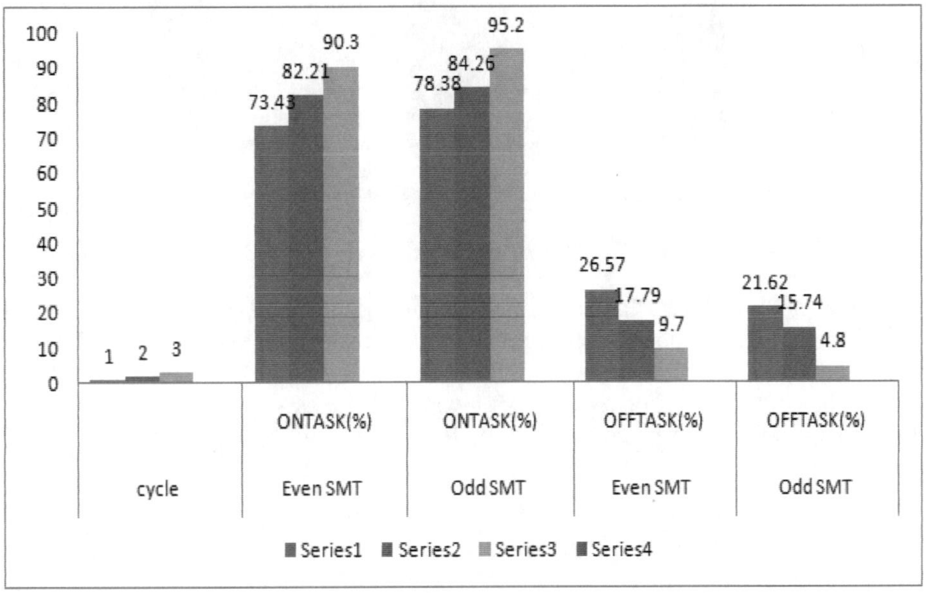

Figure 2. Students' enthusiasm in cycles 1, 2, and 3 even and odd semesters

Based on the table and graph of learning enthusiasm data during the learning process on cycles 1, 2 and 3 of the even semester increased on-task from 73.43% to 82.21% and on the third cycle 90.30% and in the odd semester on -task seen with the first cycle has begun an increase compared to the even semester that is in the first cycle 78.38%, the second cycle 84.26% and the third cycle also seen there is a significant change that is equal to 95.20%. The data indicate that there is an increase in learning enthusiasm in even and odd semester averaged for two semesters in example for one-on-task cycle of 7.33% and on the second and third cycles of 9.55%. This means that from cycle one to cycle 3 has a significant increase of 16.85%, while the off-task of two semesters on average decreased from cycle one to cycle two of 7.23%, then from cycle 2 to cycle 3 9.56%. It shows that from cycle one to cycle three has decreased significantly, that is equal to 16.85%. For example, from Computer Introduction study, based on cognitive test at the end of cycles 1, 2, and 3 there are 7 students or about 21%, 6 students or 18%, and in cycle 3 there are 3 students or about 10% did not reach the Minimum Exhaustiveness Criteria score (KKM75), while 29 students or about 91% of students had reached KKM.

Steps in the application of CTL and PBL contained in it initiation stage and acquired, students are divided into groups. Furthermore, in the elaboration stage students are given the task to be done in groups and discussions. It is realized in the form of concept explanation. Affective learning is one of the variables that must be improved in the implementation of Lesson Study for Learning Community in FMIPA. Students who have interest in learning and a number of attitudes will feel happy even challenged to study certain courses so that will gain higher success.

IOP Conf. Series: Materials Science and Engineering **332** (2018) 012043 doi:10.1088/1757-899X/332/1/012043

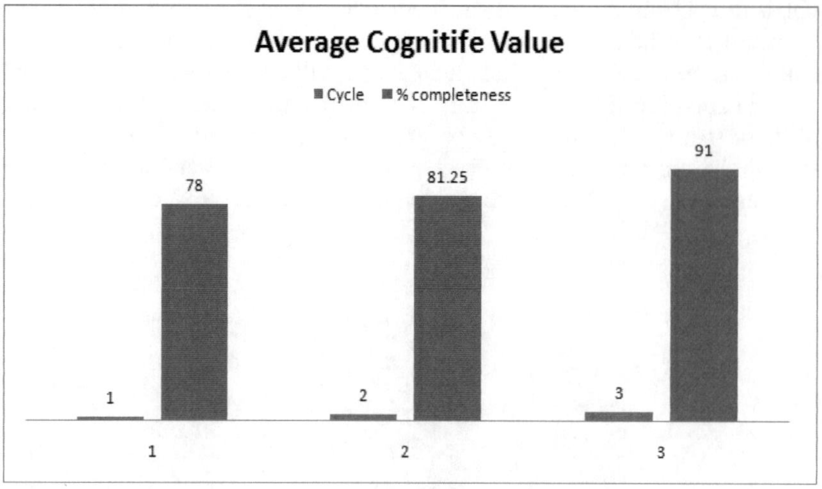

Figure 3. Completion of Student Cognitive Value from Cycles 1, 2, and 3

David R. Krathwohl stated that "In the original Taxonomy, the Knowledge category embodied both noun and verb aspects [5]. The noun or subject matter aspect was specified in Knowledge's extensive subcategories. The verb aspect was included in the definition given to Knowledge in that the student was expected to be able to recall or recognize knowledge". Results of affective learning or student attitudes toward the course of Software Engineering from cycle 1 to cycle 2 and then to cycle 3 can be seen in Figure 4.

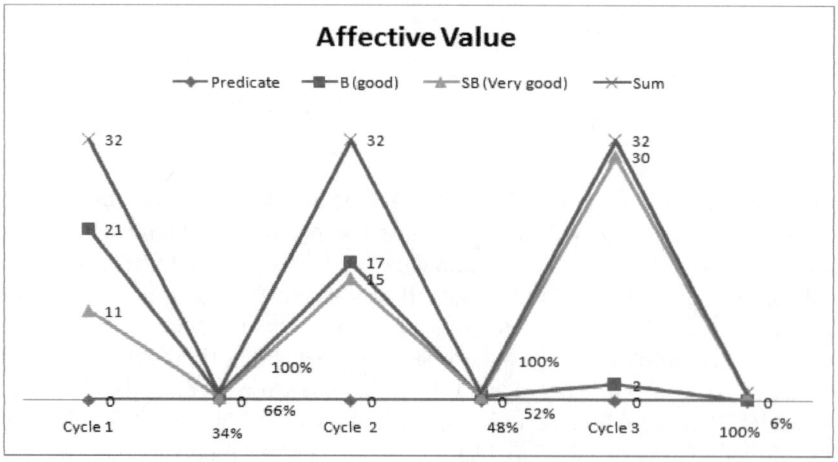

Figure 4. Affective values of students

Application of Lesson Study method based on Contextual Teaching and Learning and Problem Based Learning can improve students' skill in discussion. The visualization can be seen in Figure 5. The average value of students per psychomotor indicator of cycles 1, 2, and 3 in two semesters in Software Engineering courses can be seen in Figure 5.

4. Conclusion

The results showed that there was a significant increase in competence, attitudes and psychomotor from research conducted on four study programs. The success criteria in each step on the benchmark of KKM achievement with the value of more than 75 and the classical reach 80% of students has reached the value of more than 75.

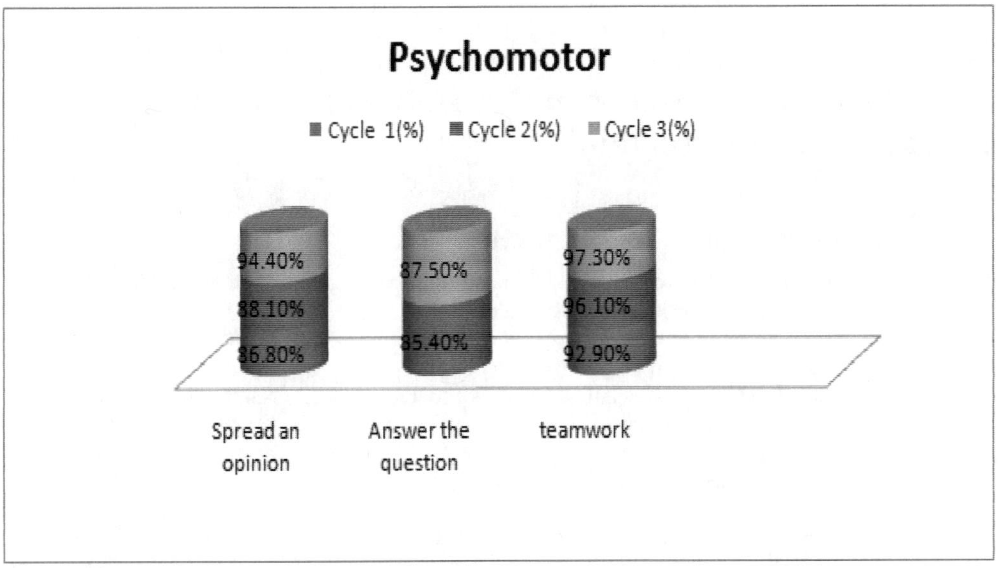

Figure 5. Average student scores per psychomotor indicators cycles 1, 2, and 3

Guidelines for classical assessment of attitudes of the average score of 80 and psychomotor complete up to 19-25 (very good). In the four average programs per aspect of attitude assessment in the cycle that includes the aspects of receiving, responding, valuing, organization and characterization by value set indicate that there is a significant increase so that the strategy of CTL and PBL is the right strategy for the sphere of attitude.

Enthusiastic increase in the third cycle on-task 92.75%, while off-task 7.25%. It can be concluded that the implementation of learning strategy in Lesson study for Learning Community can be used to improve the competence, affectivity, and psychomotor of students in the field of science.

The quality of learning innovation, the development of learning communities between the lecturers and between students showed an increase in student enthusiasm in the third cycle on-task of 92.75%, while off-ask decreased by 7.25%. Therefore, it can be concluded that the learning model of Lesson Study based on CTL and PBL is good enough as an innovative learning model that is suitable with the condition and situation in FMIPA. Thus it can be concluded that the method of Lesson Study for Learning Community is an appropriate and effective learning strategy to improve the quality of learning in the field of Science, especially in Pakuan University FMIPA.

References

[1] Zukas Miriam 2006, "Pedagogic Learning in the Pedagogic Workplace: Educators Lifelong Learning and Learning future, International Journal of Pedagogies & Learning 2.3, p.71-80.

[2] Calvin Bipout Jean & Nuefo Evelyn Chumba 2011, "Teacher's pedagogic competence and pupil's academic performance in English in Francophone schools", Educational Research (ISSN: 2141-5161).

[3] Malik, Zaki and Bouguettaya Athman 2009, "Trust Management for Service - Oriented Environment", London: Business Media.

[4] Ingvarson Lawrence, "Standards - Based Professional Learning and Certification: by The Profession for the Profession", Bauserman (editor), the Guliford Press, 2014, h.392.

[5] David R. Krathwohl 2002, "A Revision of Bloom's Taxonomy: An Overview THEORY INTO PRACTICE", Volume 41, Number 4, Autumn College of Education, The Ohio State University.

CO-CD base management model of Universitas Terbuka community service program

S W Kridasakti[1*], Sudirah[1], H Siregar[2]

[1]Department of Sociology and Humaniora, Faculty of Law, Politics and Social Sciences, the Open University, Indonesia.
[1]Department of Sociology and Humaniora, Faculty of Law, Politics and Social Sciences, the Open University, Indonesia.
[2]Department of Mathematics, Faculty of Mathematics and Natural Science, the Open University, Indonesia.

*Corresponding author: sriwahyu@ecampus.ut.ac.id

Abstract. These studies were to respond whether the UT social-aid management had been executed under CO-CD principles (Ife J. 1995) and what CO-CD base community service management model can be built. The goals of these evaluational studies were UT social-aid managerial performance profile (2011-2013) and CO-CD management model development. The methods used were *Survey and FGD*. For data collection were involving the UT officers, the counterparts, and the documents. The analysis used combination between the Performance Analysis (Irawan P., 2003) and the CIPP (*Stuffelbeam, D, L., & Shinkfield, A, J.*, 1985). The findings showed that the quantitative targeting in program completion was credible in achievement (85%). However, the "*qulitative targeting*" of the management goals was indicating far from a good-stage (≤5.0_*Interval-Force:* 1-10 Scale). The "Gap" was due to the absent of socialization_needs-analysis_maintenance_release factors on the UT social-service grand-policy. The trial of CO-CD Base Management Model had been imposed to the community that turned out to be very effective to self-help, and the ensuing SOP had been successfully defined. Conclusion, 'CO-CD Principles' were not designed in UT community service programs management. However, if efficiency and effectivity likely to be achieved, the SOP of 'CO-CD Base Management Model has to be adopted.

1. Introduction

The implementation of community engagement program is mandatory for all universities. Through the involvement their lecturers, it is expected to be run effectively. The objectives of UT's community engagement program so called 'Abdimas-Bansos' are to develop the capacity of the marginal community and to revitalize the environmentally critical areas.

The management of Abdimas program, accountabelly has to be subjected to Permenkeu No. 13 Year 2003, and substantially based on the community empowerment the principles known as CO-CD (Community Organization - Community Development). The monitoring and evaluation activities of the UT Abdimas program have been limited to the general performance issues of administration and financial accountability. Similarly, the substantive aspects of intervention in community empowerment and improvement of critical areas are also still weak in design. The weakness of the conceptual basis of CO-CD on the community engagement program management have been still low in its effectivity and inefficiency pursuing its objectives of the Abdimas program.

The strength of the UT strategic plan policy on the Abdimas program is an appropriate momentum for the development of the CO-CD base management model for UT Abdimas program.

2. Problem Statement

What look a like is the profile of the UT Abdimas-Bansos 2011-2013 management performance? And what management model is more effective that needs to be built to fit the principles of CO-CD on UT community engagement program?

3. Scope

The scope of this research and development included review of UT Abdimas-Bansos program management performance through performance factors mapping and development of CO-CD base management model.

4. Goal and Benefit

The purpose of this study was to obtain a better level of effectiveness on the management of UT Abdimas-Bansos program. While the benefits would be a built CO-CD Base Management Model for effective UT Abdimas-Bansos program implementation.

5. Theoretical Framework

The high population growth, malnutrition, and high illiteracy rate still illustrate that the nation is still poor, which indicated by the low value of HDI in Indonesia (Nugroho, Y. 2007). The people of Indonesia to this day still do not enjoy the development results evenly (Hadiyanti, P. 2006 and EFA, 2006 in Samhadi, Kompas 18 Maret 2008). The economic development alone that address poverty will not add any value without being followed by social welfare improvement (Midgley, 1995: 23). The argument according to Elliot (Isbandi Rukminto Adi, 2002: 23) is that social development is essentially proactive and preventive in order to empower the various potential laxities that exist in the community, as well as implementing a multisystem-wide intervention strategy. In the view of the NGOs, the success of development is when members of society with their social institutions can carry out their improvement from within the society itself (Tesoriero F, Samuel M and Annadurai P, 2006). Community development model is a community development activity that is considered able to increase community access to achieve social, economic, and cultural conditions (Rudito B., et al., 2003: 40). Thus, according to Ife (1995: 182), empowerment is to prepare the community with various resources, opportunities or opportunities, knowledge, and expertise to increase people's capacity in determining their future.

Empowerment is a collaborative process (Wibowo A. 2009), benefactors and the communities must work together as partners. Some important things in the design of the management of the empowerment program to be undertaken must consider some determinant factors inside the CO-CD to its success (Ife, 1995). Principles of implementation of community empowerment programs should be undertaken on the basis of CO-CD principles through implementation stages starting from the socialization stage, organizing activities, identifying needs, applying training and non-training, maintanance and release (Kusuma, E., 2006). The design of community empowerment interventions at least includes three important factors, which must always be targeting, empowering, and community (Rothman J, Erlich J & Tropman J, 2007).

Various government and non-government institutions including universities such as UT are expected to facilitate marginalized communities that allow the potential of vulnerable groups to grow. The UT Social Aid Program is a manifestation of UT's role in the success of the Government's program in alleviating marginalized communities so that they can afford to help themselves.

Conceptually, this study had been done through the use of a combination of concepts between Performance Analysis model (Irawan P, 1998, 2003), with the Context-Input-Process-Product (CIPP)-Evaluation model (Stuffelbeam, D, L., & Shinkfield, A, J., 1985), and the use of CO-CD principles released by Kusuma E. (2006) and Ife (1995). In the basic concept of performance analysis described there were seven 'cells' that must be done through research, namely (1) identifying performance standards, (2) formulating actual performance, (3) identifying performance gaps, (4) identifying problems, (5) identify the evidence of the problem, (6) identify the cause of the problem, (7) formulate the solution alternative. The Milles and Hubermann Interactive Model of Analysis (Biglen & Bogdan, 1998) was also employed to facilitate the data analysis process through out the study.

6. Research and Development Model

In diagram the research and development model looks like this:

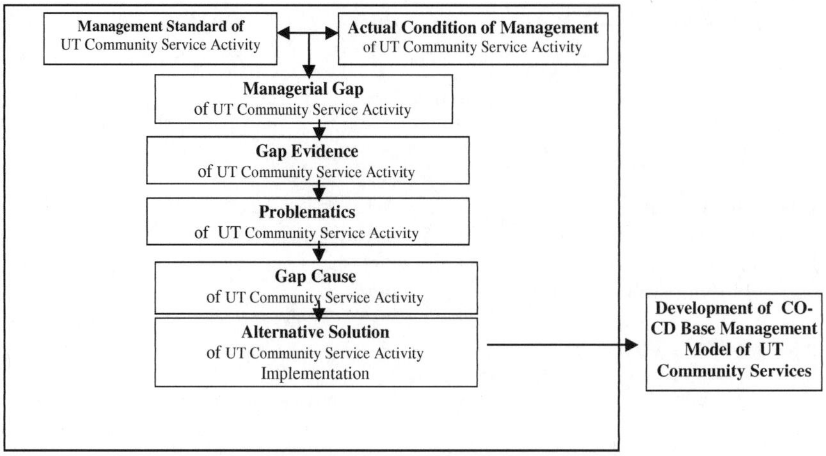

Figure 1. Research and Development Model

7. Basic Assumption

By optimizing the integrative community ang enabler resource management, between the needs base and the appropriate 'CO-CD' intervention capability will certainly increase the positive impact on the community development outcomes.

8. Research Locus

This study covered the identification of various managerial problems of UT Abdimas-Bansos program 2011-2012-2013, which samples were drawn from the total population of 46 program activities that had been selected in 15 communities spread over Jabodetabek and 3 greening-partners in 4 reforestation areas on the island of Sulawesi and Java.

9. Methodology

The main methods used were a survey for performance analysis and FGD for the CO-CD Management Base development. The performance analysis of UT Abdimas-Bansos program management was basically descriptive qualitative, and the data collection techniques used were Depth-Interview, Observation, and Document Review.

10. Population and Sample

The population in this study were all managers within the PPM-LPPM-UT and partners of the UT Abdimas-Bansos program organizer. The description of the relationship among samples and data collection techniques to be implemented in this study is shown in the following diagram.

11. Data Analysis

The explanation of the use of research tools had focused on answering the two research questions, such as the following:

The principles of data analysis were used Performance Analysis (Irawan P., 2003) to analyze and illustrate the value of UT Abdimas-Bansos management performance through which responded by the interview guidline instrument, that covered the standard issues of UT Abdimas-Bansos management, the actual management performance, the implementation gap, the implementation problematics, the cause of problematis, and alternative solution of the UT Abdimas-Bansos management. Then in parallel CIPP analysis instruments (Stuffelbeam, D, L., & Shinkfield, A, J., 1985) had been employed to identify stage by stage of managerial performance processes. Miles-Huberman Interactive Model Model Data Analysis had also been used in the process of checking the validity of data through the check and recheck process of key informant responses, through phasing and clustering of data collected (Cluster-1 and Cluster-2) before each of them is conclusively drawn.

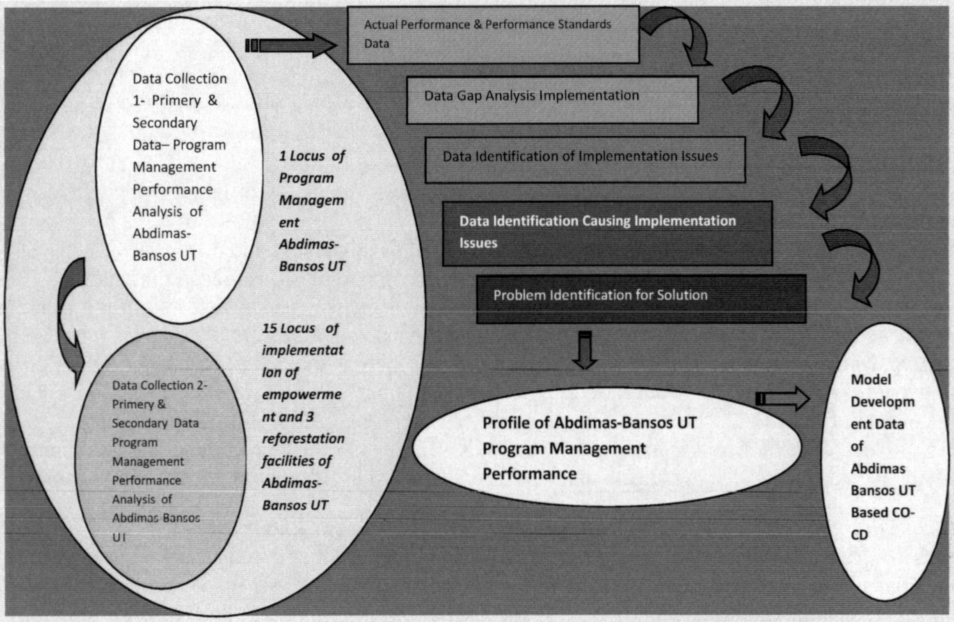

Figure 2. Data Collection Process According to Research Model

12. Performance Analysis Results

12.1 General Results of UT Community Service Programs Perfomance Management 2011-2013

The realization of the program in general is the absence of socialization stage by the managers of PPM-UT to the partners causing the emergence of the disefficiency and the effectiveness of program implementation. The results of the training intervention on training product and the fulfillment of institutional needs directly can be realized, but this performance is not incapable of creating Sustainably Self-help condition. This is because the coordinating partners of NGO/Community Group Leaders only follow the TOR (Terms of Reference) as the work reference in the implementation of the program of Bimus-Bansos Th 2011-2013. While TOR is not based on CO-CD. There is no 'Maintenance' or 'Release' phase in any unit of intervention program implementation to ensure the achievement of self-help of beneficiary communities. Therefore, it is necessary to draw up a TOR script that explicitly mentions the indicative targets of the 'Maintenance' and 'Release' stages.

12.2 Results of Interview with the Working Partner on Greening Program 2011-2013.

The interview results taken from the NGO Working Partners to do with their management performance are presented in Table 1[1] in the Appendix.

[1] The data contained in Table 1 has similar profiles (Data Sharing) to the data profiles compiled by Nurcholis H, Kridasakti SW, Siregar H 2014 Impact Analysis of the UT Community Service Program 2013 on the Benefeciary Community and the Target Areas, Fundamental Research Reports PP- LPPM UT and Dikti. The difference was laid on the data collection technic, one used management-performance base the other one used program impact base. See also the research design in Kridasakti S W 2014 Analisis Kinerja Pengelolaan Program Abdimas-Bansos Universitas Terbuka Tahun 2012 dan Dampaknya Terhadap Komunitas Masyarakat Penerima Manfaat (*Performance Analysis of The Management of Community Services through Social-Aid Program of Universitas Terbuka 2012 to the Benefiiary Community*) Thesis Graduate Study University of Muhammadiyah Jakarta

12.3 Document Review Results of UT Abdimas-Bansos 2011-2013 Management Performance
The result of the 2011-2013 document review taken from the Abdimas reports indicated that quantitatively had succeeded 85% accountable. The remaining 15% was a failure related to the failed performance and commitment of the Working-Partners in executing the program. However, qualitatively the management performance results measured by CO-CD principles were still showing substantial weakness. The overall, the management of the UT Abdimas-Bansos 2011-2013 programs were not executed according to CO-CD principles. As with no socialization phase, weak needs-assesment and organizing practices such as poor scheduling and loose quality outcomes were the major crucial issue of the performance. Of the 11 (eleven) Abdimas-Bansos UT Th 2011-2013 programs, not all can be executed thoroughly, and only 7 (seven) program packages can be completed.

From the perspective of Needs Analysis component, the weakness was inconsistent application of variables and indicators of needs-assessment parameters to the candidate-communities. The perspective of Program-Implementation component, the emergence weaknesses was in the technical implementation of the intervention, which case was of the difficulty creating consistency between the planning and the program realization. The next, in the perspective of Maintenance and Release components of CO-CD, UT did not adopt these components. The Maintenance and Release components of the CO-CD intervention program were not recognized in UT Abdimas-Bansos 2011-2013. So that, of the six crucial components in the CO-CD management procedure, UT was limited to only three components e.g.: organizational, needs analysis, and program implementation component.

12.4 Results of the Observation of UT Abdimas-Bansos Management Performance
The UT 'Greening' projects, which were especially taken place at Situ-Gintung Tangerang Selatan and Pandeglang (Banjarnegara and Kadu Hejo), after 2 years of running had shown no impact at all or 'Fail'. The other problem was the low rate of success of the reforestation programs, since the average intervention programs on the critical areas weres not based on the community-based reforestation. On top of it of, the impact of the greening program was very difficult to verify due to the lack of clear greening design, whether it involves verifying technic to the number of trees planted in such land-countour or the supervising techniques by PPM-UT. Below is the result of aggregate observation of the executed programs 2011-2012-2013[2].

Table 1. Observation Results of the UT Community-Service Management Performance 2011-2013

Number of Community Members	73 people	
Number of Communities	18 (eighteen) _Benefeciary Communities	
	3 (three Working-Partners) _Greening of Reforestation	
Observation Object	18 (eighteen) Benefeciary Communities with 73 members.	
Observation Time	April-June 2013.	
No	**Effectivity of the UT Community Service Program Management**	**Scores of Performance Quality Calculated with Forced Interval** 1-2-3-4-5-6-7-8-9-10
1	Program policy and planning socialization of UT community service programs related to CO-CD principles	Organized **0** Disorganized
2	Development of policy analysis and resource management of UT community service programs related to CO-CD principles	Organized **5** Disorganized

[2] The data contained in Table 2 shares similar profiles (Data Sharing) to the data profiles gathered by Nurcholis H, Kridasakti SW, Siregar H 2014 Impact Analysis of the UT Community Service Program 2013 on the Benefeciary Community and the Target Areas, Fundamental Research Reports PP- LPPM UT and Dikti. See also for replication of the design in Kridasakti S W 2014 Analisis Kinerja Pengelolaan Program Abdimas-Bansos Universitas Terbuka Tahun 2012 dan Dampaknya Terhadap Komunitas Masyarakat Penerima Manfaat (*Performance Analysis of The Management of Community Services through Social-Aid Program of Universitas Terbuka 2012 to the Benefiiary Community*) Thesis Graduate Study University of Muhammadiyah Jakarta

3	Organizing activity of UT community development programs over coordination, communication and information accessability of the programs related to CO-CD principles	Organized **5** Disorganized
4	Needs analysis implementation in regard the SOP to the actual condition of the potential beneficiaries communities related to CO-CD principles	Consistent **5** Inconsistent
5	Community service programs implementation through intervention (training-nontraining) in regard consistency between programs planning and realization related to CO-CD principles (self-help)	Consistent **5** Inconsistent
6	Performance in maintenance of the implemented community service program to beneficiary communities (self-help)	Consistent **0** Inconsistent
7	Assurance of beneficiary communities' opportunity for having self-help sustainability	Consistent **0** Inconsistent

Comment:

- The domination of UT community-service programs 2011-2013 almost all areas of community development were focused on welfare sector, and none of programs adopted the maintenance and release phase.
- UT community-service program on Greening and Forestation sector, in average after 2 years of running had shown ineffective at all. The low rate of success of the reforestation programs was due to the poor needs analysis factor and the results very difficult to be objectively verified by PPM-UT.

12.5 Result of Self Observation on the Program Implementation

The self-observation results showed that in the Socialization component, there was no socialization activity prior to its implementation to the beneficiary partners. So the Working-Partners found it difficult to understand the background and purpose of program. The various organizing activities were troublesome, especially the difficulty of understanding the procedures of financial accountability under the scheme of 'Pure Bansos or Non-pure Bansos'). The implementation of needs analysis activities became difficult to understand, because the stage of socialization was not adopted. There was unsuitable needs assement parameter set by PPM-UT applied to the candidate-communities, whose needs did not meet criteria of assement parameter. So it must undergo the process of adjusting the proposal to the criteria determined by UT as the donor.

The overall implementation of activities carried out in the allocation is very limited in time wise, so that the results becoming less maximal. There were no 'Maintenance' and 'Disengagement' activities to be implemented, after being given further aid of training-nontraining assistance.

12.6 Trial Results of the CO-CD Base Management Model

The trial implementation of the CO-CD Base Management Model of UT Abdimas program was conducted from March to November 2015. The goals, besides testing the implementability of the CO-CD management model were also developing a learning studio for the poor children (30 members) in Cilengok village, Padarincang sub-district, Serang regency, to access learning resources. The capacity development program was called "Science and Technology for Society: Learning Center for Unfortunate Children". The implementation of the trial CO-CD Base Management program was consistently following the CO-CD principles designed by the Enabler Team (lecturers) from UT. This intervention design-pattern was arranged into 6 phases of activities, namely: Socialization Phase, Organizing Phase, Needs Analysis Phase, Implementation Phase, Maintenance Phase, and Release Phase. The CO-CD intervention design was a pilot-project in synergy with the research program of the CO-CD Base Management Model development program granted by DIKTI 2015. The

IOP Conf. Series: Materials Science and Engineering **332** (2018) 012044 doi:10.1088/1757-899X/332/1/012044

colaborative synergy between the UT Abdimas program and the DIKTI research program was to improve the efficiency and effectivity of both given project-programs. On the side of the CO-CD Base Management Model program development, a full trial that based on the six phases of CO-CD intervention principles was simulated to the beneficiary community.

The trial result of the CO-CD Base Management Model through the "Science and Technology for Society: Learning Center for Unfortunate Children" project in Cilengo Village, Padarincang Sub-district, Serang Regency was been reportedly successful. All six phases of CO-CD principles had been successfully executed to the community up to the self-help level (Reports can be seen on Webbsite: Simpen UT).

12.7 Manuscript Dovelopment Results of CO-CD Base Management Model

This CO-CD Base Model Management manuscript had been developed for three types of UT community engagement for enablers, e.g.: Lecturers as individual (Abdipelmas-Mandiri), Lecturers at regional office, and Lecturers at headquater. These three types of community engagement have different characteristics, so the management model is also slightly done differently but remains in the same based on CO-CD principles. This CO-CD Base Management Model is aims to provide a stronger leverage to the impact of the results of the implementation of the UT community engagement program. The manuscript of UT Abdimas CO-CD Base Management Model provides technical guidance to all UT stakeholders such as the managers of the UT Abdimas working unit (PPM-LPPM-UT), the lecturers in community engagement implementation that start with the proposal submission, reporting activities, procedures for obtaining certificates, and the potential working-partners of the beneficiary communities. This manuscript can be accessed on UT Webbsite: Simpen UT, which systematically is as follows:

Table 2. The Systematic of UT CO-CD Base Community-Engagement Management Model

		Page
FOREWORD		i
TABLE OF CONTENTS		ii
CHAPTER I	FOREFRONT	3
	A. Background	3
	B. The Purpose of CO-CD Base Management Model	3
	C. The Scope of the UT Community-Engagement Program CO-CD Base Management	4
	D. Implementation Procedure of UT- CO-CD Base Management	5
	E. The Scheme of UT CO-CD Base Management	8
CHAPTER II.	THEORETICAL FRAMEWORK	9
	A. Platform of Theory and Framework of the CO-CD	9
	B. Program Design and the Framework for the Development of the CO-CD Base Management Model	10
	C. Basic Assumptions	11
CHAPTER III.	DEVELOPMENT METHOD	11
	A. Locus of Trial and Development	11
	B. Method of Trial and Development	12
	C. Results of Trial and Development	13
CHAPTER IV.	UT COMMUNITY-ENGAGEMENT CO-CD BASE MANAGEMENT MODEL	20
	A. Phase-1: Socialization of Empowerment	20
	B. Phase-2: Organization of Empowerment	32
	C. Phase-3: Needs Analysis of Empowerment	76
	D. Phase-4: Implementation of Empowerment	79
	E. Phase-5: Maintenance of Empowerment	81
	F. Phase-6: Release of Empowerment	95
CHAPTER V.	CONCLUSIONS AND RECOMMENDATIONS	98
ATTACHMENT		99

The more detail of the CO-CD principles in developing the model of the proposal would look like this below in Tabel 3[3].

Table 3. The Design of Community Development Proposal Based on CO-CD Pinciples

Component	Description
Title	The title writing is formulated and reflects the substance and locus of activities undertaken in the territory of the community who obtained the social-aid program. For example: Development of a Reading Garden for Community Students of Community Programs of Illiterate Eradication in groups ... Villages / Sub-districts / Sub-districts ... or Abdimas Improving Domestic Waste Processing Skills into Compost for Community Groups .... Villages / Sub-districts ... ...
Sitution Analysis	1) Describe briefly, clearly and as fully as possible the current partner condition 2) Display data both qualitatively and quantitatively in support of partner condition information 3) Explain the meaning of partner's existence to the environment (impact of its existence)
Community Problematics	1) Partners or communities are people who will get abdimas or community programs that are willing to be involved in the abdimas program. 2) Identify problems faced by partners. θ Write down clearly your justification with your partner in determining priority issues to address. 3) Keep the problem specific, concrete and really a partner issue.
Strategies and Solutions offered	1) Describe the strategies and approaches offered to support the implementation of the Abdimas program, shall be based on CO-CD procedures and principles, including: Socialization, Organizing, Critical Needs Analysis of partners / communities, Implementation / Training-nontraining, Maintenance of Intervention Results, and Release - Self-help sustainability. 2) Write down each stage of activity in accordance with CO-CD procedures / principles that demonstrate concrete steps to the solution of mutually agreed issues to the disposal stage, 3) Describe how partner participation in program implementation is at least 8 CD-CD determinant factors.
Targeting Outputs	1) Write the type of output that will be produced in accordance with the activity plan either in the form through Training and Nontrainig; 2) If the output is in the form of product or goods or certificate and the like, state also its specifications based on the principles of Critical Needs of partners / communities, Training-nontraining interventions Strategy, and Self-help Sustainability according to 8 factors dterminan CO-CD.
Feasibility of Implementing Community Service at the Region	1) Describe the Implementation Team Qualification, Team Skills Relevance, Team Synergy and Community Experience, especially for the Team Leader; 2) Schedule of Activities, covering all stages of CO-CD-based implementation procedures.
Budgeting	1) Create a Table showing the relationship between activity and cost 2) Feasibility of Proposed Fees specified from item 1) in the transport and reporting classification (maximum 20%), consumables, equipment, other expenses; 3) Program costs include PPn (10% Added-Value Tax), PPh (Income Tax) Article 21.

13. Discussion

The overall data taken from the interviews and observations that are presented in Table 1 and Table 4 have confirmed the findings that the implementation of the UT community-engagement 2011-2013 did not adopt the CO-CD principles. The key findings that can be discussed are as follows:

[3] Sudira, Kridasakti S W, Siregar H 2015 Pengembangan Model Pengelolaan Program Abdimas Universitas Terbuka (*Development of the Open University Community Service Program Management Model*) Fundamental Research Competitive Grant Dikti 2015.

In the case of project proposal submission that which executed by the working-partners, there was a major difference between the 'Initial-Proposal' and the 'Revised-Proposal'. The difference between the two proposals were due to the wrong perception of the working-partners in which no socialization was given by PPM-UT managers prior to the proposal submission.

Quantitatively, the overall UT community-engagement programs apparently could be well organized, which 85% of the program realization was completely executed. However, when it was qualitatively measured from the perspective of CO-CD principles, the management of UT community-engagement programs was necessarily in need of managerial reformation. The reason is that the program managers and working-partners did not adopt CO-CD principles, neither is in existent in the Terms of Reference for implementation.

The absence of a written feed-back from the needs analysis results supposedly made by PPM-UT reviewers to the working-partners, it led to the uncontrollable proposal quality improvements. So it is crucial that the working-partners who submited a proposal are given a written feed back through the PPM-UT standard selection process.

In the short term there was a direct physiological and physiological impact of the intervention on the beneficiary community members. However, in the long term it was not sustainably leading to self-help. This situation was due to the absence of CO-CD Base Terms of Reference for community-engagement program implementation during 2011-2013. Therefore, the UT Terms of Reference for community-engagement program has to be developed based on the principles of sustainable self-help. The short term psychological and physiological impact of the intervention on the beneficiary community members so far have been so inefficient in donation-wise. At the same time there has always been no assurance in the achievement of sustainable self-help, and it is notified due to the absence of CO-CD base on the Terms of Reference that covers the instruments to control sustainable success rates.

In the case of CO-CD Base Management Model development, the trial had been imposed on the unfortunate community in learning access at Cilengok Village that turned out to be very effective in creating self-help. Furthermore, the formulation of CO-CD Base Management Model, the ensuing manuscript had been successfully defined.

14. Conclusion

The findings showed that the quantitative targeting was found credible as achievement (85%). However, the "qulitative targeting" of the performance management goals is indicating the far from a good-stage (≤5,2_Interval-Force 1-10). The "Gap" was due to the absent of socialization_needs-analysis_maintenance_release factors on the UT social-service grand-policy.

More specifically, the 'Program Socialization' component was not adopted in the guidelines of the implementation of the UT community-engagement programs (Abdimas-Bansos 2011-2013). The non-adoption of CO-CD principles had resulted in the absence of the Socialization component. While the components of 'Organizing', the scheduling and execution of community-engagement programs, especially Abdimas-Bansos 2012 was very weak, marked by the absence of absolute absorption of that year budget program. The programs implementation schedule was so cramped (Oct-Nop-Dec 2012), so that the outcomes were not optimal. In the 'Needs Analysis' component, the crucial issue was about *'to determine the appropriate tolerance level'* in the provision of social-aid to the prospective beneficiary communities. In the case of the 'Program Implementation' component, it was marked by a heavy workload due to the short duration of the program implementation period, and the weakness of the community-engagement program managers to perform the planning consistency with its realization. Furthermore, the 'Program Maintenance component was not adopted at all in the design of the UT Community-Engagement Terms of Reference, so that there was no assurance that the outcome of the intervention could bring about the self-help stage to the beneficiary community. Similarly in the 'Release' or disengagement component, the UT management did not adopt this component so that the results of the intervention could not achieve the maximum Self-help condition.

In the scope the trial of CO-CD Base Management Model, this model had been imposed to the beneficiary community at Cilengok Village that turned out to be very effective upto the self-help condition, and the ensuing manuscript of CO-CD Base Management Model had been successfully defined.

15. Recommendation

A CO-CD Training program is required for all Abdimas-UT program stakeholders. UT needs a CO-CD base Terms of Reference, which covers all six CO-CD components. The weak factor in the program scheduling requires the controls strengthening through sufficient time allocation including multi-year enforcement. For the weakness of the Needs Analysis, it is necessary to give timely more tolerance in the selection process to the managers and the working-partners. In the 'Implementation' component, an instrument for evaluating and monitoring is required to be implemented from the beginning of socialization until the release phase. Recommendations on the issue of non-adoption of the the 'Program Maintenance' and 'Program Release' stage in the community-engagement program intervention, it is required a university-level policy on the importance of adopting the CO-CD principle at the UT community-engagement management system, started from the strategic-plan up to the implementation guidelines and the best-practices CO-CD training programs for all UT stakeholders.

The pilot test of the CO-CD base management model had demonstrated the effective outcomes, so that this community-engagement management model should immediately be utilized as a standard terms of reference for UT community-engagement program implementation, if efficiency and effectivity are likely to be achieved.

16. Acknowledgments

First of all on behalf of the research team, we would like to thank the Research Center of UT and the Directorate of Research and Community Service of Directorate General of Higher Education for giving us opportunity to execute such this multiyear research and development (2014-2015). This crucial research and development were only possible to be completely done due to the unremitting working moral of our team mate to finish the projects optimally. Dr Sudira and Drs. Hasoloan M.Si as team mate were playing significant contribution to this project. Also, we would like to express our gratitude to all colleagues at UT Research Center, who had kindly helped providing administrative support needed to finish accountably these multiyear projects.

References

[1] Adi I R 2002 Pemikiran-pemikiran dalam Pembangunan Kesejahteraan Sosial (*Thoughts in Social Welfare Development*) Jakarta: Lembaga Penerbit Fakultas Ekonomi UI p 23

[2] Biglen S K and Bogdan R C 1998 Huberman Interactive Model of Analysis, Qualitative Research Methodology In Education Harper and Row New York

[3] Effendi K 2008 Pemberdayaan Masyarakat (*Community Empowerment*) Universitas Muhammadiyah, Cetakan 1, 2008

[4] Hadiyanti, P 2006 Kemiskinan & Upaya Pemberdayaan Masyarakat, Komunitas (*Poverty & Social Empowerment Effort, Community*) Jurnal Pengembangan Masyarakat Islam Volume 2 Nomor 1 Juni 2006

[5] Nurcholis H, Kridasakti SW, Siregar H 2014 Impact Analysis of the UT Community Service Program 2013 on the Benefeciary Community and the Target Areas, Fundamental Research Reports PP- LPPM UT and Dikti.

[6] Irawan P 2002 Logika dan Prosedur Penelitian (*Logic and Research Procedures*) STIA LAN Press. Jakarta

[7] Irawan P et al 2006 Manajemen Sumberdaya Manusia (*Human Resource Management*) STIA LAN Press Jakarta

[8] Ife J 1995 *Community Development: Creating Community Alternatives-Vision, Analysis And Practice*, Melbourne: Longman Australia, 1995 p 182

[9] Kridasakti S W 2014 Analisis Kinerja Pengelolaan Program Abdimas-Bansos Universitas Terbuka Tahun 2012 dan Dampaknya Terhadap Komunitas Masyarakat Penerima Manfaat (*Performance Analysis of The Management of Community Services through Social-Aid Program of Universitas Terbuka 2012 to the Benefiiary Community*) Thesis Graduate Study University Muhammadiyah Jakarta

[10] Midgley 1995 *Social Development: The Developmental Perspective in Social Welfare* London: SAGE Publications Ltd p 23

[11] Nugroho dan Yanuar 2007 Dilema Tanggung Jawab Korporasi (*The Dilemma of Corporate Responsibility*) Kumpulan Tulisan 10 Nopember 2007. **www.unisosdem.org** (seen on May 23rd 2013)

[12] PPM 2011 Laporan Program Abdimas-Bansos 2011 (*Program Report of Community Services 2011*) Dokumen PPM-LPPM Universitas Terbuka

[13] PPM 2012 *Jadwal Kegiatan Program Abdimas-Bansos UT 2012* (*Schedule of Program Activty of Community Services 2012*) Dokumen PPM-LPPM Universitas Terbuka
[14] Rothman J, Erlich J, & Tropman J 2007 *Strategies of Community Intervention*, 6th Edition, FE Peacock Publishers, Itasca, IL
[15] Renstra UT 2010-2015 dan RAB/RKUK PPM-LPPM-UT 2010-2013 (*UT Strategic Planning 2010-2015 and Budgeting Planning/Design Activities and Financial Proposals*) Dokumentasi PPM-UT
[16] Rudito B., dkk. 2003 Akses Peran Serta Masyarakat, lebih jauh memahami Community Development (*Access of Community Participation, Further Understanding to Community Development*) Laboratorium Antropologi "Mentawai" Fakultas Ilmu Sosial dan Ilmu Politik, Universitas Andalas
[17] Samhadi 2008 EFA 2006, Kompas 18 Maret 2008
[18] Stuffelbeam, D, L., & Shinkfield, A, J. (1985) *Systematic Evaluation*. Boston: Kluwer-Nijhoff Publishing
[19] Sudira, Kridasakti S W, Siregar H 2015 Pengembangan Model Pengelolaan Program Abdimas Universitas Terbuka (*Development of the Open University Community Service Program Management Model*) Fundamental Research Competitive Grant Dikti 2015.
[20] Tesoriero F Samuel M and Annadurai P 2006 *Building Community Strength To Address Barriers To Health And Well Being – Strength Based Strategies* Department of Social Work Madras Christian College Healthy Districts Project
[21] Wibowo A 2009 *Menumbuhkembangkan Modal Sosial Dalam Pengembangan Partisipasi Masyarakat* M POWER No.5 Vol.5 Maret 2007 Penyuluhan Pembangunan-Universitas Sebelas Maret

Table 4. Results of Observation and Interview with the Working Partner on Greening and Community Service Program Management 2011-2013.

No	Observation Object	Time / Locus	Gaps & Causes
1	Phase of Socialization of Greening / Environment Program and Community Service Programs on HDI (Hunam Development Index) dominantly on Welfare empowerment.	**2011**_ Greening the environment in Situ Gintung, South Tangerang. - Community Service Programs in Jakarta, Bogor, Depok, Tangerang, Bekasi	1) Success rate of 0% program. 2) Fundamental issues was lack of understanding of the importance of socialization stage. 3) The lack of a CO-CD perspective by all UT stakeholders, especially PPM-UT managers on Greening Programs implementation and Comdev (Community Development) progams causes the stage of socialization was not done and it is not included in the TOR of Cultivation and Reforestation Program nor Comdev Programs. . 4) No CO-CD base socialization TOR for the UT forestation and community service programs.
		2011_ Procurement of trash/plant in South Tangerang	1) 100% failure rate, pots broken & crop failure to live. 2) Ibidem (having the same case as above colomn). 3) Ibidem (having the same case as above colomn). 4) Ibidem (having the same case as above colomn).
		2011 _ Forestation Melung Village, Kedung Banteng District, Purwokerto: 12,000 trees of aren & mangit.	1) 25% failure rate plants failed to live. 2) Ibidem (having the same case as above colomn). 3) Ibidem (having the same case as above colomn). 4) Ibidem (having the same case as above colomn).
		2012 _ Forestation Muara Borneo- Pangkajene River_ Village Tekolabbua_ Pangkajene District_Makassar _Community Service Programs in Jakarta, Bogor, Depok, Tangerang, Bekasi	1) 40% failure rate failed to live. 2) Ibidem (having the same case as above colomn). 3) Ibidem (having the same case as above colomn). 4) Ibidem (having the same case as above colomn).
		2013_ Forestation Lateri Village, Teluk Dalam-Ambon. _Community Service Programs in Jakarta, Bogor, Depok, Tangerang, Bekasi	1) 40% failure rate failed to live. 2) Ibidem (having the same case as above colomn). 3) Ibidem (having the same case as above colomn). 4) Ibidem (having the same case as above colomn).
		2013_ Forestation Wonosalam Village, Rungkut-Surabaya. _Community Service Programs in Jakarta, Bogor, Depok, Tangerang, Bekasi	1) 40% failure rate failed to live. 2) Ibidem (having the same case as above colomn). 3) Ibidem (having the same case as above colomn). 4) Ibidem (having the same case as above colomn).
2	Stage of Organizing the Greening / Environment Prog and Community Service Programs on HDI (Hunam Development Index) dominantly on Welfare empowerment.	**2011**_ Greening the environment a Situ Gintung, Tangerang Selatan. - Community Service Programs in Jakarta, Bogor, Depok, Tangerang, Bekasi.	1) Success rate of organizing is 0%. 2) Fundamental issue was poor design of organizing the project that linked maintenance stage. 3) The lack of a CO-CD perspective by all UT stakeholders, especially PPM-UT managers on Greening Programs implementation causes the stage of organizing was not done properly on the TOR of Cultivation and Reforestation Program nor the Comdev programs. 4) No CO-CD base organizing in the TOR for the UT forestation and community service programs organizing. 5) Weak scheduling and execution of organizing the UT community service program.
		2011_ Procurement of trash / plant in Tangerang Selatan.	1) Failure rate is 100%, Vases/Pots were damaged & trees fail to live. 2) Ibidem (having the same case as above colomn). 3) Ibidem (having the same case as above colomn). . 4) Ibidem (having the same case as above colomn). . 5) Ibidem (having the same case as above colomn).
		2011 _Forestation Melung Village, Kedung Banteng District, Purwokerto: 12.000 trees of aren & mangit.	1) 25% failure rate failed to live. 2) Ibidem (having the same case as above colomn). 3) Ibidem (having the same case as above colomn). . 4) Ibidem (having the same case as above colomn).. 5) Ibidem (having the same case as above colomn).
		2012 _ Forestation Muara Borneo- Pangkajene River_ Tekolabbua Village_	1) 40% failure rate failed to live. 2) Ibidem (having the same case as above colomn). .

No	Observation Object	Time Locus	Gaps & Causes
		Pangkajene District_Makassar.	3) Ibidem (having the same case as above colomn). . 4) Ibidem (having the same case as above colomn).. 5) Ibidem (having the same case as above colomn). .
		2013_ Forestation Lateri Village, Teluk Dalam-Ambon.	1) 40% failure rate failed to live. 2) Ibidem (having the same case as above colomn).. 3) Ibidem (having the same case as above colomn). . 4) Ibidem (having the same case as above colomn). . 5) Ibidem (having the same case as above colomn).
		2013_ Forestation Wonosalam Village, Rungkut-Surabaya.	1) 40% failure rate failed to live. 2) Ibidem (having the same case as above colomn). . 3) Ibidem (having the same case as above colomn). . 4) Ibidem (having the same case as above colomn). . 5) Ibidem (having the same case as above colomn). .
3	Needs Analysis Phase of Greening / Environmental Program Needs and Community Service Programs on HDI (Hunam Development Index) dominantly on Welfare empowerment.	**2011**_ Greening the environment at Situ Gintung, Tangerang Selatan _ Community Service Programs in Jakarta, Bogor, Depok, Tangerang, Bekasi	1) 100% failure rate failed to live. 2) The lack of a CO-CD perspective by all UT stakeholders, especially PPM-UT managers on Greening Programs implementation and Comdev (Community Development) progams causes the stage of needs analysis was done poorly and the CO-CD principles was not included in the TOR of Cultivation and Reforestation Program nor Comdev Programs. 3) No CO-CD base needs analysis in the TOR for the UT forestation and community service programs needs analysis. 4) Weak application of firmed variables and indicators on the needs analysis instrument on the community needs for selection and decision to give social aids.
		2011_ Procurement of trash / plant in Tangerang Selatan	1) 100% failure rate Pots broken & crop failure. 2) Ibidem (having the same case as above). 3) Ibidem (having the same case as above colomn).. 4) Ibidem (having the same case as above colomn)..
		2011 _ Forestation Melung Village, Kedung Banteng District, Purwokerto: 12.000 trees of aren & mangit.	1) 25% failure rate failed to live. 2) Ibidem (having the same case as above colomn). . 3) Ibidem (having the same case as above colomn). . 4) Ibidem (having the same case as above colomn)..
		2012 _ Forestation Muara Borneo-Sungai Pangkajene_Kel Tekolabbua_Kec. Pangkajene_Makassar. _Community Service Programs in Jakarta, Bogor, Depok, Tangerang, Bekasi	1) 40% failure rate failed to live. 2) Ibidem (having the same case as above colomn). . 3) Ibidem (having the same case as above colomn). . 4) Ibidem (having the same case as above colomn). .
		2013_ Forestation Desa Lateri-Teluk Dalam Pulau Ambon. _Community Service Programs in Jakarta, Bogor, Depok, Tangerang, Bekasi	1) 40% failure rate failed to live. 2) Ibidem (having the same case as above colomn). . 3) Ibidem (having the same case as above colomn).. 4) Ibidem (having the same case as above colomn)..
		2013_ Forestation Wonosalam Village, Rungkut-Surabaya.	1) 40% failure rate failed to live. 2) Ibidem (having the same case as above). 3) Ibidem (having the same case as above colomn). . 4) Ibidem (having the same case as above colomn)..
4.	Implementation Phase of Greening / Environment Program and Community Service Programs on HDI (Hunam Development Index) dominantly on Welfare empowerment.	**2011**_Greening the environment at Situ GintungTangerang Selatan. _Community Service Programs in Jakarta, Bogor, Depok, Tangerang, Bekasi	1) Success rate of 0% program. 2) The lack of a CO-CD perspective by all UT stakeholders, especially PPM-UT managers on Greening Programs implementation and Comdev (Community Development) progams causes the stage of implementation was done poorly and the CO-CD principles was not included in the TOR of Cultivation and Reforestation Program nor Comdev Programs implementation. 3) No CO-CD base implementation in the TOR for the UT forestation and community service programs. 4) Weak monitoring of consistency between community development planning and its implementation.
		2011_Procurement of trash / plant in Tangerang	1) 100% failure rate Pots broken & crop failure.

No	Observation Object	Time Locus	Gaps & Causes
		Selatan	2) Ibidem (having the same case as above). 3) Ibidem (having the same case as above colomn). . 4) Ibidem.
		2011 _ Forestation Melung Village, Kedung Banteng District, Purwokerto: 12.000 trees of aren & mangit.	1) 25% failure rate failed to live. 2) Ibidem (having the same case as above colomn). . 3) Ibidem (having the same case as above colomn).. 4) Ibidem (having the same case as above colomn). .
		2012 _ Forestation Muara Borneo- Pangkajene River_ Tekolabbua Village_ Pangkajene District _Makassar. _Community Service Programs in Jakarta, Bogor, Depok, Tangerang, Bekasi	1) 40% failure rate failed to live. 2) Ibidem (having the same case as above colomn). . 3) Ibidem (having the same case as above colomn). . 4) Ibidem (having the same case as above colomn)..
		2013_ Forestation Lateri Village, Teluk Dalam-Ambon. _Community Service Programs in Jakarta, Bogor, Depok, Tangerang, Bekasi	1) 40% failure rate failed to live. 2) Ibidem (having the same case as above colomn). . 3) Ibidem (having the same case as above colomn).. 4) Ibidem (having the same case as above colomn). .
		2013_ Forestation Wonosalam Village, Rungkut District- Surabaya.	1) 40% failure rate failed to live. 2) Ibidem. 3) Ibidem. 4) Ibidem.
5	Maintenance Phase of Greening / Environment Program and Community Service Programs on HDI (Hunam Development Index) dominantly on Welfare empowerment.	**2011**_ Procurement of trash / plant in Tangerang Selatan _Community Service Programs in Jakarta, Bogor, Depok, Tangerang, Bekasi	1) 100% failure rate _ Pots broken & trees fail to grow. 2) The lack of a CO-CD perspective by all UT stakeholders, especially PPM-UT managers on Greening Programs implementation and Comdev (Community Development) progams causes the stage of maintenance stage was not inexistence and the CO-CD principles was not included in the TOR of Cultivation and Reforestation Program nor Comdev Programs maintenance. 3) No CO-CD base maintenance in the TOR for the UT forestation and community service programs.
		2011 _ Forestation Melung Village, Kedung Banteng District, Purwokerto: 12.000 trees of aren & mangit.	1) 25% failure rate failed to live. 2) Ibidem (having the same case as above). 3) Ibidem (having the same case as above colomn). .
		2012 _ Forestation Muara Borneo- Pangkajene River, Tekolabbua Village, Pangkajene District, Makassar. _Community Service Programs in Jakarta, Bogor, Depok, Tangerang, Bekasi	1) 40% failure rate failed to live. 2) Ibidem (having the same case as above colomn). 3) Ibidem (having the same case as above colomn)..
		2013_ Forestation Lateri Village, Teluk Dalam - Pulau Ambon. _Community Service Programs in Jakarta, Bogor, Depok, Tangerang, Bekasi	1) 40% failure rate failed to live. 2) Ibidem (having the same case as above colomn). 3) Ibidem (having the same case as above colomn). .
		2013_ Wonosalam Village, Rungkut- Surabaya.	1) 40% failure rate failed to live. 2) Ibidem (having the same case as above colomn). 3) Ibidem (having the same case as above colomn). .
6	Release Phase of Greening / Environment Program and Community Service Programs on HDI (Hunam Development Index) dominantly on Welfare empowerment.	**2011**_ Environment Greening and forestation in Situ GintungTangerang Selatan _Community Service Programs in Jakarta, Bogor, Depok, Tangerang, Bekasi	1) Success rate of 0% program. 2) The lack of a CO-CD perspective by all UT stakeholders, especially PPM-UT managers on Greening Programs implementation and Comdev (Community Development) progams causes the stage of release stage was not inexistence and the CO-CD principles was not included in the TOR of Cultivation and Reforestation Program nor Comdev Programs release. 3) No CO-CD base release in the TOR for the UT forestation and community service programs.
		2011_ Procurement of trash / plant in Tangerang Selatan	1) 100% failure rate _ Pots broken & trees failed to live. 2) Ibidem (having the same case as above). 3) Ibidem (having the same case as above colomn).

No	Observation Object	Time Locus	Gaps & Causes
		2011 _ Forestation Melung, Kec Kedung Banteng, Purwokerto: 12.000 Aren & AangitVillage trees.	1) 25% failure rate failed to live. 2) Ibidem (having the same case as above colomn). . 3) Ibidem (having the same case as above colomn). .
		2012 _ Forestation Muara Borneo- Pangkajene River, Tekolabbua Village, Pangkajene District, Makassar. _Community Service Programs in Jakarta, Bogor, Depok, Tangerang, Bekasi	1) 40% failure rate failed to live. 2) Ibidem (having the same case as above colomn).. 3) Ibidem (having the same case as above colomn). .
		2013_ Forestation Lestari Village, Teluk Dalam - Pulau Ambon. _Community Service Programs in Jakarta, Bogor, Depok, Tangerang, Bekasi	1) 40% failure rate failed to live. 2) Ibidem (having the same case as above colomn). . 3) Ibidem (having the same case as above colomn).
		2013_ Forestation Wonosalam Village, Rungkut-Surabaya.	1) 40% failure rate failed to live. 2) Ibidem (having the same case as above colomn). 3) Ibidem (having the same case as above colomn).

IOP Conf. Series: Materials Science and Engineering **332** (2018) 012045 doi:10.1088/1757-899X/332/1/012045

Optimization of territory control of the mail carrier by using Hungarian methods

S Supian[1], S Wahyuni[2], J Nahar[3], Subiyanto[4*]

[1,2,3]Department of Mathematics, Faculty of Mathematics and Natural Science, Universitas Padjadjaran, Indonesia
[4]Department of Marine Science, Faculty of Fishery and Marine Science, Universitas Padjadjaran, Indonesia

*Corresponding author: subiyanto17@gmail.com

Abstract. In this paper, the territory control of the mail carrier from the central post office Bandung in delivering the package to the destination location was optimized by using Hungarian method. Sensitivity analysis against data changes that may occur was also conducted. The sampled data in this study are the territory control of 10 mail carriers who will be assigned to deliver mail package to 10 post office delivery centers in Bandung. The result of this research is the combination of territory control optimal from 10 mail carriers as follows: mail carrier 1 to Cikutra, mail carrier 2 to Ujung Berung, mail carrier 3 to Dayeuh Kolot, mail carrier 4 to Padalarang, mail carrier 5 to Situ Saeur, mail carrier 6 to Cipedes, mail carrier 7 to Cimahi, mail carrier 8 to Soreang, mail carrier 9 to Asia-Afrika, mail carrier 10 to Cikeruh. Based on this result, manager of the central post office Bandung can make optimal decisions to assign tasks to their mail carriers.

Keyword: territory control, Hungarian method, optimal decisions, sensitivity analysis

1. Introduction

Territory control of a mail carrier is the understanding of a mail carrier in knowing the existence of the destination area. In a transportation problem, territory control variable can be considered for obtaining the optimal solution in assignment problem. The assignment problem is a special case of the transportation problem. It arises in a variety of decision-making situations [1]. A distinguishing feature of the assignment problem in this case is that one mail carrier is assigned to one and only one post office delivery centers [8].

Many papers have been discussed the assignment problem to solve decision-making problem. Sudradjat et al. [8] studied the Hungarian method to solve the assignment problem based on traveling time for a case study on the central Post Office Bandung. A new approach to one-sided assignment problems was proposed by Sasaki [6]. Maxon and Bhadury [3] discussed the assignment problem with repetitive tasks and tried to introduce a human element into the analysis. A simple random assignment problem with a unique solution was presented by Bogomolnaia and Moulin [2]. Nuass [4] presented a special purpose branch-and-bound algorithm for solving assignment problems. Sourd [7] discussed the continuous assignment problem with the aim of solving scheduling problems with irregular cost functions. Odior et al. [5] addressed a problem of the effectiveness of feasible solutions of assignment problems.

The concept of assignment problem based on territory control was applied to solve a problem for a case study on the central Post Office Bandung. This research discusses how to optimize territory control problem then solved by using Hungarian method. This optimization process is applied to a case study of the central post office Bandung in assigning employees to deliver the packet to the destination location based on several criteria owned by each employee. It is also conducted a sensitivity analysis of data changes that may occur so as not to change the optimal assignment from the initial problem.

2. Materials and Methods

The concept of assignment problem based on territory control by using Hungarian methods was applied to solve a problem for a case study on the central Post Office Bandung. The Hungarian method is a combinatorial optimization algorithm that solves the assignment problem in polynomial time for a given cost matrix [8].

Theorem 2.1: *If a number is added to or subtracted from all of the entries of any one row or column of a cost matrix, then on optimal assignment for the resulting cost matrix is also an optimal assignment for the original cost matrix.*

The Hungarian Method: The following algorithm applies the above theorem to a given $n \times n$ cost matrix to find an optimal assignment.

Step 1. Subtract the smallest number in each row from every number in the row. (This is called row reduction.) Enter the results in a new table.
Step 2. Subtract the smallest number in each column of the new table from every number in the column. (This is called column reduction.) Enter the results in another table.
Step 3. Test whether an optimal set of assignments can be made. You do this by determining the minimum number of lines needed to cover (i.e., cross out) all zeros. Since this minimum number of lines equals the maximum number of assignments that can be made to zero element positions, if the minimum number of lines equals the number of rows, an optimal set of assignments is possible. (If you find that a complete set of assignments to zero element positions is not possible, this means that you did not reduce the number of lines covering all zeros down to the minimum number.) In that case, go to step 6. Otherwise go on to step 4.
Step 4. If the number of lines is less than the number of rows, modify the table in the following way:

a. Subtract the smallest uncovered number from every uncovered number in the table.
b. Add the smallest uncovered number to the numbers at intersections of covering lines.
c. Numbers crossed out but not at the intersections of cross-out lines carry over unchanged to the next table.

Step 5. Repeat steps 3 and 4 until an optimal set of assignments is possible.
Step 6. Make the assignments one at a time in positions that have zero elements. Begin with rows or columns that have only one zero. Since each row and each column needs to receive exactly one assignment, cross out both the row and the column involved after each assignment is made. Then move on to the rows and columns that are not yet crossed out to select the next assignment, with preference again given to any such row or column that has only one zero that is not crossed out. Continue until every row and every column has exactly one assignment and so has been crossed out. The complete set of assignments made in this way is an optimal solution for the problem.

Data collected (in Likert scale) from the central Post Office Bandung is shown in Table 1 below.

Table 1. Territory control each mail carrier to each post office delivery centers (in Likert scale)

Mail Carrier	Area									
	I	II	III	IV	V	VI	VII	VIII	IX	X
1	2	3	2	2	2	3	3	3	3	2
2	3	3	2	2	2	4	2	4	3	2
3	2	2	2	2	2	4	2	4	4	3
4	4	1	2	3	4	3	3	2	2	4
5	2	1	2	3	4	3	2	2	3	2
6	4	4	3	3	2	5	2	2	2	3
7	2	3	2	2	3	4	2	1	2	3
8	3	2	3	3	2	3	4	2	3	4
9	2	3	3	3	2	4	3	4	3	3
10	3	3	3	3	2	3	2	3	4	1

Table 1 shows that the territory control of 10 mail carriers to deliver mailing packets to 10 post office delivery centers in Bandung that is Cikutra (I), Padalarang (II), Ujung Berung (III), Dayeuh Kolot (IV), Asia-Africa (V), Soreang (VI), Situ Saeur (VII), Cimahi (VIII), Cipedes (IX) and Cikeruh (X). The approach to scaling responses in survey research is used the Likert scale as follows: strongly know =1, know = 2, neither know nor don't know = 3, don't know = 4 and strongly don't know = 5.

3. Result and Discussion

The goal of the assignment problem is to minimize the territory control of 10 employees to deliver mailing packets to 10 post office delivery centers in Bandung. An important characteristic of this assignment problem is the number of workers is equal to the number of destinations. It is explained in the following way.

- Only one job is assigned to mail carrier.
- Each mail carrier is assigned to exactly one destination.

Based on the data collected, manager of the Central Post Office Bandung has ten mail carriers for ten separate destinations and the territory control of assigning each destination to each mail carrier is given. His goal is to assign one and only destination to each mail carrier in such a way that the territory control of assignment is minimized.

Balanced assignment problem: The number of mail carriers is equal to the number of destinations. The mathematical model of the assignment problem in this case can be written as follows:

$$\text{Min T} = \sum_{i=1}^{10}\sum_{j=1}^{10} c_{ij}x_{ij}$$

$$s.t \sum_{j=1}^{10} x_{ij} = 1 \quad ; i = 1,2,..,10$$

$$\sum_{i=1}^{10} x_{ij} = 1 \quad ; j = 1,2,..,10$$

$$x_{ij} = 0 \text{ or } 1, \text{ for } i \text{ and } j$$

where:

$$x_{ij} = \begin{cases} 1, \text{if mail carrier } i \text{ go to destination } j \\ 0, \text{if mail carrier } i \text{ does not go to destination } j \end{cases}, i = 1,2,...,10 \text{ and } j = 1,2,...,10.$$

Based on data in Table 1, the optimal assignments matrix was produced by applied the Hungarian method in this case as shown in Table 2 below.

Table 2. Optimal Assignment Problem Matrix

Mail Carrier	Area									
	I	**II**	**III**	**IV**	**V**	**VI**	**VII**	**VIII**	**IX**	**X**
1	0	2	0	0	0	0	1	1	1	0
2	1	2	0	0	0	1	0	2	1	0
3	0	1	0	0	0	1	0	2	2	1
4	2	0	0	1	2	0	1	0	0	2
5	0	0	0	1	2	0	0	0	1	0
6	2	3	1	1	0	2	0	0	0	1
7	1	3	1	1	2	2	1	0	1	2
8	1	1	1	1	0	0	2	0	1	2
9	0	2	1	1	0	1	1	2	1	1
10	2	3	2	2	1	1	1	2	3	0

Based on the optimal solution in Table 2, the assignment of mail carrier from Central Post Office Bandung to each post office delivery centers in Bandung can be seen in Table 3 below.

Table 3. Mail carriers assignment determination

Mail Carrier	Post office delivery centers in Bandung
1	Cikutra
2	Ujung Berung
3	Dayeuh Kolot
4	Padalarang
5	Situ Saeur
6	Cipedes
7	Cimahi
8	Soreang
9	Asia-Afrika
10	Cikeruh

In order to determine the data changes range of territory control that does not change the assigned optimal assignment of workers, the sensitivity analysis was done. Sensitivity analysis to change of objective function coefficient from territory control data was obtained as shown in Table 4.

Table 4. Range of territory control coefficient

Worker	Area									
	I	**II**	**III**	**IV**	**V**	**VI**	**VII**	**VIII**	**IX**	**X**
1	$[-\infty, 3]$	$[2, \infty]$	[1,3]	[1,3]	[1,3]	[2,4]	$[1, \infty]$	$[1, \infty]$	$[1, \infty]$	[1,3]
2	$[1, \infty]$	$[2, \infty]$	$[-\infty, 3]$	[1,3]	[1,3]	$[1, \infty]$	[1,3]	$[2, \infty]$	$[1, \infty]$	[1,3]
3	[1,3]	$[1, \infty]$	[1,3]	$[-\infty, 3]$	[1,3]	$[1, \infty]$	[1,3]	$[2, \infty]$	$[2, \infty]$	$[1, \infty]$
4	$[2, \infty]$	$[-\infty, 2]$	[1,3]	$[1, \infty]$	$[2, \infty]$	[2,4]	$[1, \infty]$	[1,3]	[1,3]	$[2, \infty]$
5	[1,3]	[0,2]	[1,3]	$[1, \infty]$	$[2, \infty]$	[2,4]	$[-\infty, 3]$	[1,3]	$[1, \infty]$	[1,3]
6	$[2, \infty]$	$[3, \infty]$	$[1, \infty]$	$[1, \infty]$	[1,3]	$[2, \infty]$	[1,3]	[1,3]	$[-\infty, 3]$	$[1, \infty]$
7	$[1, \infty]$	$[3, \infty]$	$[1, \infty]$	$[1, \infty]$	$[2, \infty]$	$[2, \infty]$	$[1, \infty]$	$[-\infty, 2]$	$[1, \infty]$	$[2, \infty]$
8	$[1, \infty]$	$[1, \infty]$	$[1, \infty]$	$[1, \infty]$	[1,3]	$[-\infty, 4]$	$[2, \infty]$	$[-\infty, 3]$	$[1, \infty]$	$[2, \infty]$
9	0	$[2, \infty]$	$[1, \infty]$	$[1, \infty]$	$[-\infty, 3]$	$[1, \infty]$	$[1, \infty]$	$[2, \infty]$	$[1, \infty]$	$[1, \infty]$
10	$[2, \infty]$	$[3, \infty]$	$[2, \infty]$	$[2, \infty]$	$[1, \infty]$	$[1, \infty]$	$[1, \infty]$	$[2, \infty]$	$[3, \infty]$	$[-\infty, 2]$

Based on Table 4, it is obtained the range of coefficients changes from the objective function allowed for each mail carrier to reach the destination location does not change the initial assignment.

4. Conclusion

In this paper, the concept of assignment problem has been applied to solve a problem for of the central post office Bandung which had a difficulty in assigning 10 mail carriers to 10 destination locations of post office delivery centers. Based on the data collected, Hungarian Method was used to solve the problem. Optimal assignments of the cases were obtained to the central post office Bandung. It was obtained that, the optimal assignment of mail carriers from Central Post Office Bandung to each post office delivery centers as follows: mail carrier 1 to Cikutra, mail carrier 2 to Ujung Berung, mail carrier 3 to Dayeuh Kolot, mail carrier 4 to Padalarang, mail carrier 5 to Situ Saeur, mail carrier 6 to Cipedes, mail carrier 7 to Cimahi, mail carrier 8 to Soreang, mail carrier 9 to Asia-Afrika, mail carrier 10 to Cikeruh. Based on this result, manager of the central post office Bandung can make optimal decisions to assign tasks to their mail carriers. The Sensitivity Analysis for assignment issues can illustrate the range of coefficient value changes for the allowed objective function so that this does not change the optimal assignment of the initial problem.

Acknowledgment

We would like to thank the financial support from Universitas Padjadjaran through the grant program of the Academic Leadership Grant (ALG) and Central Post Office Bandung for providing traveling time data.

References

[1] Boah, D.K., Adu, I. K., and Gyebil, F. J. 2015. Assignment problem of a legal firm in Kumasi, Ghana. *International Journal of Computing and Optimization*, 2(1),1-5.

[2] Bogomolnaia, A. and Moulin, H. 2002. A Simple Random Assignment Problem with a Unique Solution, *Economic Theory*, 19, 623 - 636

[3] Maxon, S. L. and Bhadury J. 2001. A Ms-excel implementation of a multi-period assignment problem with repetitive tasks, *Proceedings of the 13th Annual CSUPOM Conference, California State University San Bernardino*, 39–48.

[4] Nauss, R. M. 2003. Solving the generalized assignment problem: an optimizing and heuristic approach, *INFORMS Journal on Computing*. 15(3), 249 – 266.

[5] Odior, A.O., Owaba, C. and Oyawale, O.E. 2010. Determining Feasible Solutions of a Multicriteria Assignment Problem, *Journal of Applied Sciences and Environmental Management*, 14(1), 35-38.

[6] Sasaki, H. 1995. Consistency and Monotonicity in Assignment Problems, *International Journal of Game Theory*, 24, 373-397.

[7] Sourd F. 2004. The continuous assignment problem and its application to pre-emptive and non-pre-emptive scheduling with irregular cost functions, *Informs Journal on Computing*, 16(2), 198-208.

[8] Sudradjat Supian, Sri Wahyuni, Julita Nahar, Subiyanto. 2017. Optimization of Personnel Assignment Problem Based on Traveling Time by Using Hungarian Methods: Case Study on the Central Post Office Bandung, *4th International Conference on Operational Research (InteriOR 2017)*. Medan, Indonesia on August 21-23.

Analysis of stock investment selection based on CAPM using covariance and genetic algorithm approach

Sukono[1], D Susanti[1], M Najmia[1], E Lesmana[1], H Napitupulu[1], S Supian[1], A S Putra[2]

[1]Department of Mathematics, Faculty of Mathematics and Natuaral Sciences, Universitas Padjadjaran, INDONESIA
[2]Department of Geophysics, Faculty of Mathematics and Natuaral Sciences, Universitas Padjadjaran, INDONESIA

*Corresponding author's e-mail: sukono@unpad.ac.id

Abstract. Investment is one of the economic growth factors of countries, especially in Indonesia. Stocks is a form of investment, which is liquid. In determining the stock investment decisions which need to be considered by investors is to choose stocks that can generate maximum returns with a minimum risk level. Therefore, we need to know how to allocate the capital which may give the optimal benefit. This study discusses the issue of stock investment based on CAPM which is estimated using covariance and Genetic Algorithm approach. It is assumed that the stocks analyzed follow the CAPM model. To do the estimation of beta parameter on CAPM equation is done by two approach, first is to be represented by covariance approach, and second with genetic algorithm optimization. As a numerical illustration, in this paper analyzed ten stocks traded on the capital market in Indonesia. The results of the analysis show that estimation of beta parameters using covariance and genetic algorithm approach, give the same decision, that is, six underpriced stocks with buying decision, and four overpriced stocks with a sales decision. Based on the analysis, it can be concluded that the results can be used as a consideration for investors buying six under-priced stocks, and selling four over-priced stocks.

Keyword: CAPM, beta parameter, covariance, genetic algorithm, under-priced, over-priced.

1. Introduction

Financial investment, especially in stocks, is a liquid investment in a short term. Nevertheless, in Indonesia, real investment is a common investment done by society rather than financial investment [4]. This is assessed because there is still a lack of knowledge about stock investment instruments. In addition, the emergence of public concerns as potential investors toward risk and return in stocks as a means of investment [5; 17]. An investment, especially stock investment, certainly is not apart from the risk. Generally, investors avoid risk (*risk averse*), so for a risky investment, investors will see a large rate of return. In fact, the greater the benefits to be obtained, generally the greater the risk that must be faced by investors [14; 16]. Conversely, if you tend a lower risk, then the benefits will also be mild [13]. Therefore, thing that must be paid an attention to the investor is how the investment can generate an optimal return at a minimal level of risk. This can be done by predicting expected returns and investment risks, using the Capital Asset Pricing Model (CAPM) approach [9; 15].

There are several studies on stock investment decision analysis using the CAPM approach. Putra [13], examined the analysis of stock investment selection using CAPM and reward to variability ratio (RVAR), as a determinant of stock investment decision making. The study was conducted on corporations listed on the LQ45 index in Indonesia Stock Exchange, period February 2009 - July 2012. The results showed that from a sample of 25 stocks taken, with an analysis using the CAPM method, there are 13 stocks underpriced, with the decision to buy is above the security market line (SML). On the contrary, overpriced shares with the selling decision are below the SML line of 12 stocks. Fiarni [4], examined the investment portfolio recommendation system based on the genetic algorithm. The results showed that the optimal portfolio formed by using CAPM method and genetic algorithm, is able to give bigger profit that is equal to 32.28%, compared to stock average of individual forming of portfolio with value equal to 26.32%. The portfolio set up by the recommendation system also provides a lower risk of 1.05883, compared to the average risk of individual stocks of portfolio makers that is equal to 1.60095. Thus, the portfolio established using the recommendation system can be accepted by application users with a suitability rate of 67%. A similar study was also conducted by Zainashev [20], which is about using a genetic algorithm in the optimal investment portfolio decision-making process.

Based on the description above, in this research, the stock selection analysis using CAPM approach has been conducted. For the estimation of parameter of β in CAPM is done by two ways that is by covariance method and genetic algorithm. This is what distinguishes this research from previous research conducted by Putra [13] and Fiarni [4]. According to Mukherjee and Kumar [11], genetic algorithm has been widely applied to solving optimization problems. Through this approach, it is expected to obtain parameter estimator value β in CAPM as systematic risk. Thereafter, parameter estimator values β are used for stock selection analysis based on assumptions in CAPM. The purpose of this study is to provide information for investors in considering the selection of stock investments. As a numerical illustration, some stocks are traded on the Indonesian capital market.

2. Methodology

This methodology discusses stock returns, Capital Asset Pricing Model (CAPM), genetic algorithms, and CAPM model significance tests, used in stock selection analysis for investment.

2.1. Stock returns

In this section, the objective is to determine the stock returns. Suppose that the stock price on a day-t is P_t. To analyze financial data within horizon daily time, return R_t often given in the form of *continuously compound return* or *return* in following formula:

$$R_t = \ln\left(\frac{P_t}{P_{t-1}}\right), \tag{1}$$

with $t = 1,2,...,T$ where T is the number of observation data and it is assumed that $P_0 = 1$ [15]. Thus, we used this return for modeling in the following sections.

2.2. Capital asset pricing model

This section discusses about the Capital Asset Pricing Model (CAPM), which is used for stock selection analysis for investment. CAPM is a model that describes the expected risk and return relationship. In this model, the expected return is the risk-free return rate plus the premium that is based on systematic stock risk [1; 2]. In the equilibrium state, the required rate of return (*required return*) by the investor for a stock is affected by the risk of the stock. In this case, the only calculated risk is *systematic risk* or market risk as measured by the parameter coefficient. While the risk is not systematic (*unsystematic risk*) is assumed as no effect, because the risk can be eliminated by diversification [3; 6].

Based on CAPM theory, expected return from a stock can be calculated by using equation as follows:

$$E[R_t] = R_f + \beta(E[R_m] - R_f), \tag{2}$$

where $E[R_t]$ is expected returns of stock, $E[R_m]$ expected return of market, R_f return of risk-free assets, and β is a stock risk. In CAPM discussion, parameter β is interpreted as a systematic risk of the stock. Therefore the parameter β is considered to be representative for use in measuring systematic risk. Thus, the magnitude of a stock's risk is determined by its parameter value β [7].

Equation (2) empirically cannot be tested statistically, since equation (2) is an expectation equation, it is an unobserved value. Therefore, in order to be tested empirically, the CAPM regression equation must be changed as follows

$$R_t - R_f = \beta(R_m - R_f) + \varepsilon_t. \tag{3}$$

Since the risk-free asset return has a constant rate, it can be written as $\mu_f = E[R_f]$. Also because it is a risk-free asset, the variance is $\sigma_f^2 = Var[R_f] = 0$ [15]. Thus, equation (3) can be expressed as

$$R_t - \mu_f = \beta(R_m - \mu_f) + \varepsilon_t. \tag{4}$$

where β a *slope*, and ε_t residual of CAPM regression. Residual sequence $\{\varepsilon_t\}$ is assumed as *white noise*, which is a normal distribution with zero and variance σ_ε^2 [5].

In the covariance method approach, it is often assumed to be zero, and the magnitude of the parameter value can be estimated using the following equation:

$$\beta = \frac{Cov[R_t, R_m]}{Var[R_m]} = \frac{\sigma_{tm}}{\sigma_m^2}, \tag{5}$$

where $\sigma_{tm} = Cov[R_t, R_m]$ is a covariance of stock returns R_t and market return R_m, and $\sigma_m^2 = Var[R_m]$ is variance of market returns R_m. If $\beta > 1$, it means the increase in stock return is higher than the increase in market return, and usually the stocks are classified in *aggressive stock*. If $\beta < 1$ it means that the increase in stock return is lower than the increase in market return, and usually the stock is classified in *defensive stock* [5, 7].

2.3. Genetic algorithm

This section discusses the genetic algorithm of CAPM for parameter estimation of β, in terms of stock selection for investment.

The genetic algorithm is a search method based on the mechanisms of artificial genetics and natural selection [8]. The mechanism of artificial genetics reflect the individual's ability to engage in marriage, and produce offspring that have similar characteristics to the parent. While the mechanism of natural selection illustrates that living things can survive, if it is able to adapt to the surrounding environment [15; 19]. Therefore, it is expected that the resulting offspring's have the best combination of characteristics from the parent, and can sustain generations of offspring.

In general, the structure of the genetic algorithm can be given in the following steps [10; 18]:

a. Initial population generation, this initial population is generated randomly so as to obtain an initial solution;
b. The population is composed of a number of chromosomes representing desired solutions;
c. The formation of a new generation, in the formation of a new generation used three operators, namely reproduction / selection, cross over and mutation;

d. Evaluation of the solution, this process evaluates each population by calculating the fitness value of each chromosome, and evaluating it until the stop criteria are met. If the stop criterion has not been met, then the new generation will be re-established by repeating step b.

Referring to the general structure of the genetic algorithm, the genetic algorithm for estimating the coefficients β of CAPM as in equation (4), it can be arranged as follows [19]:

1) Generating the initial population. For each stock, the initial population generated N amounts, which is generated randomly. This initial population random number is then converted into the form of decimal values β, where $\beta \in S$;
2) Evaluation of chromosomes. Referring to the CAPM form of equation (2), the fitness value of the chromosome is the objective function values in the form of:
$$\text{Minimized } f(\beta) = \sum \varepsilon_t^2 = \sum \{R_t - \mu_f - \beta(R_m - \mu_f)\}^2 . \quad (6)$$
Based on the fitness values selected the smallest value, for the minimization program.
3) Calculation of population convergent percentage. The percentage convergent population p_c is the percentage of the number of individuals who have the same fitness value and the most. This p_c value is calculated using the following formula:
$$p_c = \frac{n}{pop} \times 100\% , \quad (7)$$
where n the number of individuals who have the same fitness and the most, and pop number of population.
4) Checking stop condition. The genetic algorithm process will stop when the generation counter has reached the specified number of generations, that is c_g = 1000, or the convergent percentage of the population reaches the defined threshold limit $\theta = 90\%$.
5) Chromosome selection. The selection process is done using roulette wheel selection. Due to the minimization program, it evaluates the fitness value $eval(v_i)$, $i = 1,...,N$, according to equation (4), with the function
$$eval(v_i) = \frac{1}{1 + f(\beta)} , \quad (8)$$
where $f(\beta)$ is the fitness value due to equation (4).
6) Crosslinking. The new population of the selection results is cross-breaded, using the Single-Point Crossover (SPC) method.
7) Mutasi. Mutations. Mutations of each generation are obtained by using $m \times pop\_size \times p_m$, with m are number of mutation, pop_size size of population, and p_m are probability of mutation (random value).
8) Decoding. It is the process of encoding the genes in a chromosome to get its value back as it was, changing the coding to decimal values.

Furthermore, this genetic algorithm is used to analyze the case of stock selection for investment as follows.

3. Result and discussion

This section discusses the results and discussion which include: data analyzed, parameter estimation using covariance method, parameter estimation using genetic algorithm method, determination of expected return, and comparison analysis.

3.1. Data analysis

This sub-section discusses about the object data being analyzed. The object of this study is the closing price of the monthly stock in period of January 1, 2016 until December 31, 2016, obtained through the internet media https://finance.yahoo.com. The data used in this research is the monthly stock closing price of AALI, ADRO, AALI, BBCA, INDF, JSMR, PGAS, SMGR, TLKM, UNTR, and Composite Stock Price Index (IHSG) in period of 2016. While the risk-free asset data used is the interest rate of Bank Indonesia. Furthermore, the price data of each stock, composite stock price index, and risk-free asset, determine the value of the return rate by using equation (1). The return data of each stock, stock price index, and risk-free asset are then used for analysis in the following sections.

3.2. Parameter estimation β using covariance method

In this sub-section, parameter β of each stock return is estimated. The parameter β of each stock is a systematic risk that is usually called market risk, which needs to be estimated based on the return data of each stock. Here the estimation is done by the covariance method which refers to equation (5). Based on (5), the parameters β of each individual stock are presented in Table 1.

Table 1. Parameter estimation β of individual stock using covariance method

Stocks	**Estimator** β
AALI	1.2808
ADRO	0.8387
ASII	1.6918
BBCA	0.9460
INDF	0.8941
JSMR	1.3414
PGAS	1.7293
SMGR	1.1153
TLKM	0.7185
UNTR	0.6186

To ensure that the estimated parameter β of each stock estimated using equation (5) is significant, a test is necessary. The significance test is done by referring equation (4), which is testing the significance of parameter estimator β of each stock. To test the significance of parameter estimators, it is performed using Wald test statistic and test statistic F . Based on the result of significance test of parameterer estimator β, it shows that parameter estimator β of each stock has significant at significance level $\alpha = 5\%$. While testing the assumption of residual normality ε_t of each stock return, is conducted by Kolmogorov-Semirnov test statistic. Based on the assumption of residual normality ε_t assumption of each stock return, it indicates that the respective residual stock return is normally distributed.

Therefore, based on the result of testing the significance of the estimator β of each stock parameter estimated using equation (5) is significant, that it can then be used for analytical purposes. Taking into account to Table 1, it shows that 5 out of 10 stocks in the research object have parameter values $\beta > 1$, so it can be said that the stocks used as samples of this study have very high average risk. In the above calculation can be seen that PGAS shares have the highest beta value, that is equal to 1.7293.

In this sub-section, the calculation of expected return from CAPM is estimated using covariance method. Using the results of the calculation of market return, risk-free return, and parameter estimators (systematic risk) β for each stock, and then we calculated the expected value of stock returns, using equation (2). The result of the calculation of expected return of 10 stock of research object is given in Table 2.

Table 2 shows that UNTR stocks have the highest expected return, which is 0.0221. Meanwhile, PGAS stock has the lowest expected return, which is equal to -0.0736. This shows that the size of the expected return rate is influenced by the size of the risk β of the stock.

Table 2. Expected return value $E[R_t]$ of CAPM using covariance method

Stocks	$E[R_t]$
AALI	-0.0350
ADRO	0.0031
ASII	-0.0704
BBCA	-0.0061
INDF	-0.0020
JSMR	-0.0402
PGAS	-0.0736
SMGR	-0.0207
TLKM	0.0135
UNTR	0.0221

So that it is obtained by investment decision of stock based on CAPM method of covariance presented in Table 3.

Table 3. Recommended decisions are based on CAPM with covariance method

Stocks	R_t	$E[R_t]$	**Evaluation**	**Recommendation**
AALI	-0.0236	-0.035	Efficient	Take or buy stock (*under-priced*)
ADRO	-0.0481	0.0031	Not efficient	Sell stock before stock price goes down (*over-priced*)
ASII	-0.0152	-0.0704	Efficient	Take or buy stock (*under-priced*)
BBCA	0.0017	-0.0061	Efficient	Take or buy stock (*under-priced*)
INDF	-0.0256	-0.002	Not efficient	Sell stock before stock price goes down (*over-priced*)
JSMR	-0.0219	-0.0402	Efficient	Take or buy stock (*under-priced*)
PGAS	-0.0398	-0.0736	Efficient	Take or buy stock (*under-priced*)
SMGR	-0.0147	-0.0207	Efficient	Take or buy stock (*under-priced*)
TLKM	0.0113	0.0135	Not efficient	Sell stock before stock price goes down (*over-priced*)
UNTR	0.0009	0.0221	Not efficient	Sell stock before stock price goes down (*over-priced*)

The recommended investment options to investors based on Table 3, are stocks of AALI, ASII, BBCA, JSMR, PGAS, TLKM, and UNTR. Because the return value of realization R_t is greater than the expected return $E[R_t]$.

3.3. Parameter estimation β using genetic algorithm

In this sub-section, the parameter estimation β of each stock is done by using genetic algorithm method. Parameter estimation using genetic algorithm method is done by referring to the general structure of genetic algorithm described in section 2.3. The estimation of each stock parameter is done with the intention of determining the parameter value that can minimize the objective function (6). The parameter estimation process is done with the help of Mat lab 2010 application program. The parameter estimation β results are given in Table 4.

As with the estimation of parameters β using the covariance method, the estimator of each stock estimated using equation (6) is significant, it is necessary to test significance. Testing is done by referring equation (4), namely testing the significance of each stock parameter estimator β. To test the significance of parameter estimators, it also performed using Wald test statistics and test statistic F.

The result of significance test of parameter estimator β for each stock has significant at significance level $\alpha = 5\%$. While testing the assumption of residual normality ε_t of each stock return, is conducted by Kolmogorov-Smirnov test statistic. The result of assumption of residual normality ε_t of each stock return shows normal distribution.

Table 4. Parameter estimation β of individual stock using genetic algorithm

Stocks	**Estimator β**
AALI	1.3366
ADRO	0.9999
ASII	1.7366
BBCA	0.8638
INDF	0.8813
JSMR	1.2366
PGAS	1.6429
SMGR	1.2068
TLKM	0.7529
UNTR	0.6907

Therefore, based on the result of testing the significance of estimator of each stock parameter estimated using equation (6) is significant, and then it can be used for further analysis purposes. Taking into account to Table 4. It appears that 5 out of 10 stocks of research object companies have parameter estimator values $\beta > 1$, so it can be said that the stocks used as samples of this study have very high average risk. In the above calculation, it can be seen that the stock PGAS has the highest beta value, that is equal to 1.7366.

In this sub-section, the expected return of each share has been calculated, based on CAPM using genetic algorithm method. Having obtained the parameter estimator value β (systematic risk), together with the expected return of the composite stock price index, and the expected return of the risk-free asset, is used to determine the expected return of each stock.

Table 5. Expected return value $E[R_t]$ of CAPM using genetic algorithm

Stocks	$E(R_i)$
AALI	-0.1769
ADRO	-0.0461
ASII	-0.0154
BBCA	0.0016
INDF	-0.0249
JSMR	-0.0221
PGAS	-0.0401
SMGR	-0.0142
TLKM	0.0111
UNTR	-1×10^{-5}

The calculation of the expected value of stock returns $E[R_t]$ is done using equation (2). The result of the calculation of expected return of the 10 stocks used as the research object is presented in Table 5.

In Table 5, it shows that TLKM shares have the highest expected return $E[R_t]$, which is 0.0113. Meanwhile, ADRO stock has the lowest expected return, which is equal to -0.0475. Based on the results in Table 5, the efficiency classification is based on expected return using the CAPM method with parameter estimators β by genetic algorithm method. The results of the efficiency classification of the stocks are presented in Table 6.

Table 6. Classification of stock's efficiency with CAPM based on parameter β which is estimated by the genetic algorithm

Stocks	R_t	$E[R_t]$	Evaluation	Recommendation
AALI	-0.0236	-0.1769	Efficient	Take or buy stock (*under-priced*)
ADRO	-0.0481	-0.0461	Not efficient	Sell stock before stock price goes down (*over-priced*)
ASII	-0.0152	-0.0154	Efficient	Take or buy stock (*under-priced*)
BBCA	0.0017	0.0016	Efficient	Take or buy stock (*under-priced*)
INDF	-0.0256	-0.0249	Not efficient	Sell stock before stock price goes down (*over-priced*)
JSMR	-0.0219	-0.0221	Efficient	Take or buy stock (*under-priced*)
PGAS	-0.0398	-0.0401	Efficient	Take or buy stock (*under-priced*)
SMGR	-0.0147	-0.0142	Not efficient	Sell stock before stock price goes down (*over-priced*)
TLKM	0.0113	0.0111	Efficient	Take or buy stock (*under-priced*)
UNTR	0.0009	-1×10^{-5}	Efficient	Take or buy stock (*under-priced*)

Investment options suggested to investors based on Table 6, are stocks of AALI, ASII, BBCA, JSMR, PGAS, TLKM, and UNTR. Because the return value of realization R_t is greater than the expected return $E[R_t]$.

3.4. Comparison analysis

In this sub-section a comparative analysis of investment decisions is recommended, based on CAPM estimated using covariance method, and CAPM estimated using genetic algorithm (GA) method. To facilitate the comparative analysis, the classification of the stock efficiency given in Table 3, and the stock efficiency classification given in Table 6, is summarized as given in Table 7.

Table 7. Comparison of stock efficiency

Saham	R_t	$E[R_t]$		Evaluation	
		Parameter β **covariance**	Parameter β **GA**	**CAPM covariance**	**CAPM GA**
AALI	-0.0236	1.2808	1.3366	Efficient	Efficient
ADRO	-0.0481	0.8387	0.9999	Not efficient	Not efficient
ASII	-0.0152	1.6918	1.7366	Efficient	Efficient
BBCA	0.0017	0.9460	0.8638	Efficient	Efficient
INDF	-0.0256	0.8941	0.8813	Not efficient	Not efficient
JSMR	-0.0219	1.3414	1.2366	Efficient	Efficient
PGAS	-0.0398	1.7293	1.6429	Efficient	Efficient
SMGR	-0.0147	1.1153	1.2068	Efficient	Efficient
TLKM	0.0113	0.7185	.0.7529	Not efficient	Not efficient
UNTR	0.0009	0.6186	0.6907	Not efficient	Not efficient

Taking into account to Table 7, it appears that stock efficiency based on CAPM estimated using covariance method, and CAPM estimated using genetic algorithm method, has no significant difference. This can be seen from the parameter estimator β value generated by both methods have no significant difference.

4. Conclussion

In this paper we have discussed the issues of stock investment selection based on CAPM using Genetic Algorithm. Here it is assumed that the stocks analyzed follow the CAPM model. Based on the analysis show that parameter estimator β by using covariance method, for stocks: AALI, ASII, JSMR, PGAS, and SMGR, these are obtained parameter estimator value β <1. As for stocks: ADRO, BBCA, INDF, TLKM, and UNTR, these are obtained parameter estimator value β <1. Meanwhile, parameter estimator using genetic algorithm method also obtained parameter estimator value β equal to those estimated using covariance method, that is for stocks: AALI, ASII, JSMR, PGAS, and SMGR, these are obtained parameter estimator value $\beta > 1$. As for stocks: ADRO, BBCA, INDF, TLKM, and UNTR, these are obtained parameter estimator value β <1. Based on the evaluation of 10 stocks of research object, it can be concluded that there are 6 stocks underpriced i.e. AALI, ASII, BBCA, JSMR, PGAS and SMGR shares with the decision to buy the stock before the price rises, and there are 4 overpriced stocks which are ADRO, INDF, TLKM, UNTR shares with the decision to sell the shares before the price falls.

Acknowledgment
Further thanks to the Rector, Director of DRPMI, and Dean of FMIPA, Universitas Padjadjaran, which provides a grant program of the Academic Leadership Grant (ALG) and a grant program of Competence Research of Lecturer of Unpad (Riset Kompetensi Dosen Unpad/RKDU).

References

[1] Abusharbeh, M.T. and Sous, S.J. 2016. Testing the Validity of Capital Assets Pricing Model: Evidence from Palestinian Exchange Market. *Journal of Accounting, Finance and Economics,* Vol. 6. No. 2. September 2016 Issue. pp. 99 – 107.

[2] Alqisie, A. and Alqurran, T. 2016. Validity of Capital Assets Pricing Model (CAPM), (Empirical Evidences from Amman Stock Exchange). Journal of Management Research. Vol. 8, No. 1 , 2016, pp. 207-223. ISSN 1941-899X. www.macrothink.org/jmr

[3] Chukwuemeka, E. 2016. Application of Capital Asset Pricing Model (CAPM) on Real Property Investment in Nigeria. *International Journal of Business & Law Research* 4(1):pp. 56-62, Jan-Mar. 2016. ISSN: 2360-8986. © Seahi Publications, 2016. www.seahipaj.org

[4] Fiarni, Cut dan Bastiyan. 2013. Sistem Rekomendasi Portofolio Investasi Berbasis Algoritma Genetika (Investment Portfolio Recommended System Based Genetic Algorithm). *Seminar Nasional Sistem Informasi Indonesia.*

[5] Hakim, S.A., Hamid, Z., and Meera, A.K.M. 2016. Capital Asset Pricing Model and Pricing of Islamic Financial Instruments. *JKAU: Islamic Econ.,* Vol. 29 No. 1, pp: 21-39 (January 2016). DOI: 10.4197 / Islec. 29-1.2

[6] Hui-Shan Lee, H.S., Cheng, F.F., and Chong, S.C. 2016. Markowitz Portfolio Theory and Capital Asset Pricing Model for Kuala Lumpur Stock Exchange: A Case Revisited. International Journal of Economics and Financial Issues, 2016, 6(S3) pp. 59-65. ISSN: 2146-4138. available at http: www.econjournals.com

[7] Irfan, I., Mubeen, M., Hussain, A., and Zubair, M. 2016. An Empirical Analysis of Higher Moment Capital Asset Pricing Model for Karachi Stock Exchange (KSE). Open Journal of Social Sciences, 2016, 4, pp. 53-60. Published Online June 2016 in SciRes. http://www.scirp.org/journal/jss. http://dx.doi.org/10.4236/jss.2016.46006

[8] Irfianti, A.D., Wardoyo, R., Hartati, S., and Sulistyoningsih, E. 2016. Determination of selection Method in Genetic Algorithm for Land Suitability. MATEC Web of Conferences 58 03002 (2016). pp. 1-5. Bisstech 2015. DOI: 10.105/matecconf/20165803002.

[9] Jovanovic, F., Milijic, N., Dimitrova, M., and Mihajlovic, I. 2016. Risk Management Impact Assessment on the Success of Strategic Investment Projects: Benchmarking Among Different Sector Companies. *Acta Polytechnica Hungarica*. Vol. 13, No. 5, 2016, pp. 221-241.

[10] Kilinc, B.K., Asikgil, B., Erar, A., and Yazici, B. 2016. Variable Selection with Genetic Algorithm and Multivariate adaptive Regression Splines in the Presence of Multicollinearity. International Journal of Advanced and Applied Sciences, 3(12) 2016, Pages: 26-31. Journal homepage: http://www.science-gate.com/IJAAS.html

[11] Mukherjee, S. and Kumar, S.M.D. 2016. Application of Genetic Algorithm and Machine Learning Techniques for Stock Market Prediction. *International Journal of Computer Trends and Technology (IJCTT)* – Volume 36 Number 1 - June 2016, pp. 59-61. ISSN: 2231-2803. http://www.ijcttjournal.org

[12] Noda, R.F., Martelanc, R., and Kayo, E.K. 2016. The Earnings/Price Risk Factor in Capital Asset Pricing Models. R. Cont. Fin. – USP, São Paulo, v. 27, n. 70, pp. 67-79, jan./fev./mar./abr. 2016. ISSN 1808-057X. DOI: 10.1590/1808-057x201412060.

[13] Putra, R. D. 2012. Analisis Pemilihan Investasi Saham dengan Menggunakan Metode Capital Asset Pricing Model (CAPM) dan Reward to Variability Ratio (RVAR) sebagai Penentu Pengambilan Keputusan Investasi Saham (Analysis of Stock Investment Selection by Using Capital Asset Pricing Model (CAPM) Method and Reward to Variability Ratio (RVAR) as Determinant of Stock Investment Decision Determination). *Penelitian.* Fakultas Ilmu Administrasi, Universitas Brawijaya Malang.

[14] Sukono, Bon, A.T.B, Mamat, M., & Supian, S. 2016. The APT with lagged, value-at-risk and asset allocations by using econometric approach. Paper presented at the *Proceedings of the International Conference on Industrial Engineering and Operations Management,* 532-542. Retrieved from www.scopus.com

[15] Sukono, Hidayat, Y., Bon· A.T.B., and Supian, S. 2017. Modelling of Capital Asset Pricing Model with Considering the lagged Effects. IOP Conf. Series: Materials Science and Engineering 166 (2017) 012001, pp. 1-7. doi:10.1088/1757-899X/166/1/012001

[16] Sukono, Sidi, P., Bon, A.T.B., & Supian, S. 2017. Modeling of mean-VaR portfolio optimization by risk tolerance when the utility function is quadratic. Paper presented at the *AIP Conference Proceedings, , 1827* doi:10.1063/1.4979451

[17] Sukono, Sidi, P., Susanti, D., & Supian, S. 2017. Quadratic investment portfolio without a risk-free asset based on value-at-risk. *Journal of Engineering and Applied Sciences, 12*(19), 4846-4850. doi:10.3923/jeasci.2017.4846.4850

[18] Tong, X., Xu, X., Plaza, A., Xie, H., Pan, H., Cao, W., and Dong, Lv. 2016. A New Genetic Method for Subpixel Mapping Using Hyperspectral Images. IEEE Journal of Selected Topics in Applied Earth Observations and Remote Sensing, Vol. 9, No. 9, September 2016, pp. 4480-4491.

[19] Zhang, B. 2016. Stock Market Risk Measurement Method Based on Improved Genetic Algorithm. Chemical Engineering Transactions, Vol. 51, 2016, pp. 631-636. ISBN 978-88-95608-43-3; ISSN 2283-9216. DOI:10.3303/CET1651106.

[20] Zainashev, T.M. 2011. Using a Genetic Algorithm in the Optimal Investment Portfolio Decision-making Process. *Proccedings of the International Multi Conference of Engineers and Computer Scientists,*2011 Vol. I, IMECS 2011, March 16-18, 2011, Hong Kong. ISBN: 978-988-18210-3-4.

IOP Conf. Series: Materials Science and Engineering **332** (2018) 012047 doi:10.1088/1757-899X/332/1/012047

Estimated value of insurance premium due to Citarum River flood by using Bayesian method

Sukono[1], I Aisah[1], Y R H Tampubolon[1], H Napitupulu[1], S Supian[1], Subiyanto[2*], P Sidi[3]

[1]Department of Mathematics, Faculty of Mathematics and Natuaral Sciences, Universitas Padjadjaran, INDONESIA

[2]Department of Marine Science, Faculty of Fishery and Marine Science, Universitas Padjadjaran, INDONESIA

[3]Department of Mathematics, Faculty of Mathematics and Natuaral Sciences, Universitas Terbuka, INDONESIA

*Corresponding author's email: subiyanto17@gmail.com

Abstract. Citarum river flood in South Bandung, West Java Indonesia, often happens every year. It causes property damage, producing economic loss. The risk of loss can be mitigated by following the flood insurance program. In this paper, we discussed about the estimated value of insurance premiums due to Citarum river flood by Bayesian method. It is assumed that the risk data for flood losses follows the Pareto distribution with the right fat-tail. The estimation of distribution model parameters is done by using Bayesian method. First, parameter estimation is done with assumption that prior comes from Gamma distribution family, while observation data follow Pareto distribution. Second, flood loss data is simulated based on the probability of damage in each flood affected area. The result of the analysis shows that the estimated premium value of insurance based on pure premium principle is as follows: for the loss value of IDR 629.65 million of premium IDR 338.63 million; for a loss of IDR 584.30 million of its premium IDR 314.24 million; and the loss value of IDR 574.53 million of its premium IDR 308.95 million. The premium value estimator can be used as neither a reference in the decision of reasonable premium determination, so as not to incriminate the insured, nor it result in loss of the insurer.

Keyword: flood insurance, right fat tail; Pareto distribution, Gamma distribution, Bayesian method.

1. Introduction

According to Bandung regency government Prevention and Preparedness, flood disaster often rush the regency of Bandung, especially Baleendah District. The intensity of flooding in Baleendah sub-district, which includes the Citarum River Basin, has resulted enormous losses in agriculture, fisheries, economic, health, and education sectors. These losses affect directly the human life, i.e. their homes, vehicles, and property flooded [12]. In anticipation of future losses, one of the most helpful solutions would be insurance protection on property, vehicles and other things vulnerable to flood damage [1]. The circumstances under which people having property insurance are environmental histories frequently affected by natural disasters, such as floods, fires, tornadoes, landslides, and so on [3]. Property insurance provides cash when customers lose insured property [4; 5; 15].

When the claim occurs, the insurer gives a sum of money to the client [7]. Insurance companies need to calculate the appropriate premium value for each claim. In this case, it is necessary to estimate the premium value [14]. Referring to Paudel et al. [9], Bayesian Inference can be applied in the estimated value of insurance premiums due to floods. The method is used to estimate the flood loss parameter that will affect the calculation of the premium. This method is suitable to be applied to studies having limited data [9]. The Bayesian approach is also used to analyse the uncertainty in the flood frequency model [2; 13].

Based on the above description, the authors conduct research on Bayesian Inference application on the estimated value of insurance premiums due to Citarum river flood (Case study: Baleendah Village). In general, the purpose of this study is to apply the author's knowledge of insurance and opportunities as well as reference materials for further researchers.

2. Methodology

In this section we will discuss: likelihood data model and function, priority distribution, posterior distribution, simulation of flood loss data and fitting curve of opportunity density, and estimated value of insurance premium.

2.1. Model Data and Likelihood Data Function

In this sub-section, the discussion of the data model and likelihood function is presented. Assuming that data loss due to flood $X_1, X_2, ..., X_8$ is a random sample of Pareto distributed with two parameters, that is as known parameter, which is the minimum value of X and θas parameter is not known. As the function of probability density X can be written as follows [13]:

$$p(x) = \begin{cases} \theta \dfrac{x_m^{\theta}}{x^{\theta+1}}; x_i \geq x_m; \theta > 1 \\ 0; \qquad \text{for } x \text{ ather} \end{cases}. \tag{1}$$

Based on (1), the likelihood function with parameter θ is not known, which is as follow:

$$L(\theta \mid X_1, ..., X_2) = \begin{cases} \theta^n x_m \prod_{i=1}^{n} x_i^{-(\theta+1)}; \ x_i \geq x_m \\ 0; \qquad \text{otherwise} \end{cases}. \tag{2}$$

Following equation (2), parameter θ is unknown, that is to be estimated, that is to determine the parameter estimator value $\hat{\theta}$ which can maximize equation (2) [6]. The estimation of parameter θ in equation (2) is done using Bayesian method.

2.2. Prior Distribution

Since the data model follows the Pareto distribution with unknown parameters θ, the conjugate prior $p(\theta)$ is proportional to the Gamma probability density function [13; 16]. The Gamma function of θ undefined with hyper parameter a and b as follows:

$$p(\theta) = \begin{cases} \theta^{a-1} \dfrac{e^{-\frac{\theta}{b}}}{b^a \Gamma(a)}; \theta \geq 0; b > 0 \\ 0; \qquad \theta < 0 \end{cases}. \tag{3}$$

This posterior distribution is further used in the discussion of posterior distribution, in the next sub-sections.

2.3. Posterior Distribution

Since the priority distribution has been determined, that is the conjugation for likelihood; the posterior distribution is a combination of likelihood and priority equations. The posterior distribution follows the same distribution as the priority, i.e. [2; 10]:

$$p(\theta \mid x) = \text{Gamma}(\alpha, \beta) = L(\theta \mid X)\, p(\theta). \tag{4}$$

By substituting equation (2) and (3) to equation (4), therefore the posterior density probability function of the flood loss is as follows:

$$p(\theta \mid x) = L(\theta \mid X)\, p(\theta)$$

$$p(\theta \mid x) = \theta^{n} x_m \prod_{i=1}^{n} x_i^{-(\theta+1)} \theta^{a-1} \frac{e^{-\frac{\theta}{b}}}{b^{a}\Gamma(a)}$$

$$p(\theta \mid x) = \frac{\theta^{a+n+1} e^{-\left(\frac{1}{b} + \ln\left(\prod_{i=1}^{n} x_i\right) - n \ln x_m\right)}}{\left(\dfrac{1}{\frac{1}{b} + \ln\left(\prod_{i=1}^{n} x_i\right) - n \ln x_m}\right)^{a+n} \Gamma(a+n)}. \tag{5}$$

Equation (5) can be rewritten as:

$$p(\theta \mid x) \propto \frac{\theta^{\alpha-1} e^{-\frac{\theta}{\beta}}}{\beta^{\alpha}\Gamma(\alpha)}, \tag{6}$$

with $\alpha = a + n$ and

$$\beta = \frac{1}{\frac{1}{b} + \ln\left(\prod_{i=1}^{n} x_i\right) - n \ln x_m} \; ; \; \beta < \frac{1}{n \ln x_m - \ln\left(\prod_{i=1}^{n} x_i\right)}. \tag{7}$$

Based on the posterior distribution function, the parameters θ can be estimated from the average posterior distribution of flood losses, i.e. [2; 6]:

$$\hat{\theta} = E[(\theta \mid x)] = \int_{-\infty}^{\infty} \theta\, p(\theta \mid x) d\theta. \tag{8}$$

This parameter estimator value $\hat{\theta}$ is then used to simulate the data loss data due to floods that are assumed to be Pareto distributed. This simulation is performed because the data loss caused by flood cannot be determined exactly.

2.4. Data simulation of flood loss and fitting curve of opportunity density

Based on the assumption that data loss caused by floods is Pareto distributed, the chance of damage for each observed area can be simulated with the following model [8]:

$$Losses = f^{-1}(\hat{\theta}, x_l, x_u). \tag{9}$$

To see the value of estimated loss based on the value of $\hat{\theta}$ obtained from Bayesian Inference, data grouping is required based on the percentile size. Suppose that the data are grouped into 10 data sets, to find the estimated value of losses per data set, the inverse of the cumulative function of Pareto distribution is as follows [3]:

$$L = Losses\ Estimator = \frac{x_m}{(1 - F(x))^{\frac{1}{\theta}}}, \tag{10}$$

with x_m the minimum value of loss derived from information II, and $F(x)$ in this case is the value of cumulative loss per percentile of data in a single RW.

Due to the data limitations, so for the chance of damage per house, it is used direct simulation data that is using random numbers by Microsoft Excel. The estimated value of losses per house is sought by using the inverse Pareto density probability function as follows [11]:

$$l = \left(\frac{\hat{\theta} x_m^{\hat{\theta}}}{f(x)} \right)^{\frac{1}{\hat{\theta}+1}} . \tag{11}$$

with l is an estimated loss per household, $f(x)$ is a chance of flood per home and $\hat{\theta}$ is a loss estimator obtained from Bayesian inference [16].

2.5. Estimated Insurance Premium Value

There are several principles used for the calculation of premiums, namely [12]:

- **Pure premium**
  Principle of pure premium calculate premium based on total loss, total premium, and annuity of premium payment, that is as follows [7; 12]

  Total Losses-Total Premium*Annuity (premium payment) = 0

  $$P = \frac{Total\ Losses}{Anuity}, \tag{12}$$

  with

  $$Annuity = \frac{1-(1+i)^{-t}}{i}$$

  i is the annual interest rate, and t is the period of payment of premium (annual).

- **Variance principle**
  This method estimates the premium value based on *Modern Actuarial Risk Theory*. The principle of variance can be written as follows [9; 12]:

  $$c = L + \rho Var(L), \tag{13}$$

  with c is the amount of premium required for property insurance, L is the loss estimator, $Var(L)$ is the variance of the estimated loss, and the ρ is risk aversion coefficient of the insurer, which is assumed to be 0.005.

- **Standard deviation principle**
  This method is also called the empirical method, which assumes the coefficient of risk aversion by the insurer depends on the standard deviation from the loss. The calculation of premiums by empirical method is as follows [12]:

  $$c = L + \rho\sqrt{Var(L)} \tag{14}$$

  with $\sqrt{Var(L)}$ is the standard deviation of loss, and ρ is coefficient of risk aversion from the insurer, which is assumed to be 0.005.

3. Result and Discussion

In this section, the discussion includes: analysed data, normality testing, estimation of loss data distribution model, and calculation of loss insurance premium due to flood.

3.1. Analysed data

This sub-section discuss about the object data for the research. Objects used in this study are data loss due to flooding, especially submerged houses, experienced by residents Baleendah Village of eight

Citizens Association (RW). The data is obtained from the Office of Baleendah district in the form of secondary data. The value of losses due to floods is obtained based on information from interviews with the Chairman of the Neighbourhood Association (RT) and residents in Baleendah Village. The data used in this study is the homes submerged by the flood of citizens of Baleendah Village. In Information I, the data of submerged housing is processed into data loss due to flood with the assumption of loss per house submerged by IDR 10,000,000. Whereas in Information II, the submerged housing data is processed with the assumption of per house perforated losses of IDR 9,500,000, because each RW is assumed to receive assistance of IDR 500,000 per incident. The loss data is used to estimate the premium value based on the chance of damage per area.

3.2. Result

Flood losses from the submerged housing data are assumed to be Pareto distributed. Since the probability density function of the Pareto distribution is known, the likelihood function can be determined. The value of θ is estimated by using Bayesian Inference. The prior value of the parameter θ needs to be determined to obtain its posterior distribution. In this study, it is used prior conjugate which means the distribution of the same prior to the posterior distribution. Posterior distribution is obtained from the multiplication of likelihood function with its priority distribution.

3.2.1. Test of data normality with Shapiro Wilk

In this sub-section, the normality test of data derived from Information II is presented. Normality test is necessary to know whether Information II is normal distribution or not. The test hypothesis is: H_0: Information II comes from the normal distribution, against alternative H_1: Information II does not come from the normal distribution. The test statistic used is Shapiro Wilk at a significance level of α_s = 0.05. The test criterion is reject H_0 if Sig $<\alpha_s$. Normality test is done using SPSS software, and the results are given in Table 1.

Table 1. Result of Normality test for Information II

	Kolmogorov-Smirnov[a]			Shapiro-Wilk		
	Statistic	df	Sig.	Statistic	df	Sig.
VAR00001	.156	8	.200*	.934	8	.557

Based on Table 1, in the Shapiro-Wilk test, the significance value Sig = 0.557 is clearly greater than α_s = 0.05. So the decision is to accept the hypothesis H_0, which means Information II comes from the normal distribution.

Furthermore, estimation of distribution data model of loss caused by Citarum River flooding, as follows.

3.2.2. Estimation of data loss distribution model

In this case, it is assumed that the data loss model $x_1,...,x_n$ derived from information II is a Pareto-distributed random sample. Random variable loss per return period of flood event with two main parameters, namely the scale parameter x_m as the minimum value of x, and the shape parameter $\theta > 1$. The first parameter value is assumed to be known, while the second is unknown. The value of the shape parameter is determined by Bayesian inference. Here is a graph of the opportunity density function of information II.

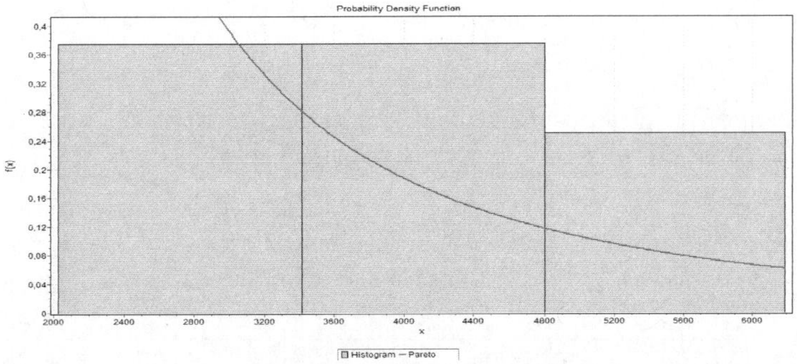

Figure 1. Graph of the density function of information opportunity II is Pareto distributed.

Whereas from Information I, that is, submerged home data which is processed with the assumption of loss per house submerged by IDR 10,000,000, used to find the prior distribution of unknown parameter. Here is a graph of the probability density function of data derived from information I.

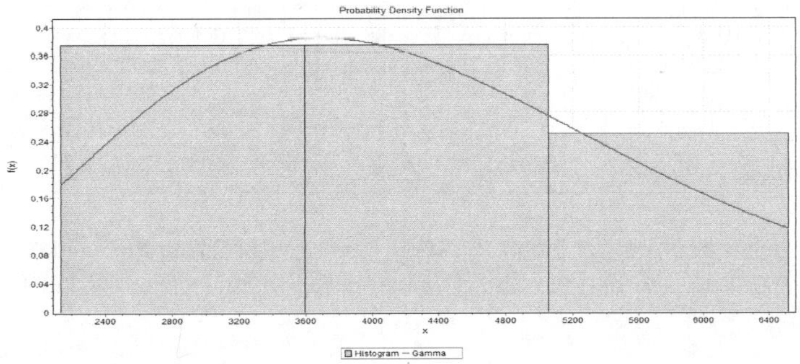

Figure 2. Graph of the density function of information opportunity I is Gamma distributed.

3.2.3. Bayesian Inference data loss model

In this sub-section, Bayesian inference analysis is performed on the data model of loss due to Citarum river flood disaster. If we substitute the parameter values x_m by 2024, which is the minimum value of x, then equation (1) can be written to:

$$p(x) = \theta \frac{2024^{\theta}}{x^{\theta+1}}\text{; } \theta > 1. \quad (15)$$

The likelihood function of the equation (14) is:

$$L(\theta \mid x_1,\ldots,x_n) = \theta^8 (2024)^{8\theta} \prod_{i=1}^{8} x_i^{-(\theta+1)}$$

In this study we assume the prior distribution for parameters θ is Gamma with hyper parameter a and b. Based on calculations with *EasyFit 5.5* software assistance, the likelihood values obtained from a and b parameters are as follows:

$$\theta \sim Gamma(7.075, 607.19)$$

Since the value a and b are known, then the posterior distribution parameters, that is α and β can be calculated. Based on equation (7), we obtained parameters estimator values α = 15.0757 and β = 0.1970494917. The x_i value is the value of loss based on information II, which is loss due to flood with the assumption of loss per house submerged by IDR 9,500,000.00.

After the parameters of the posterior distribution are estimated, the equation (4) can be written to $p(\hat{\theta} \mid x) = Gamma(15.0757, 0.1970494917)$, with the expectation of the parameter $\hat{\theta}$ is $E(\hat{\theta} \mid x)$ =

IOP Conf. Series: Materials Science and Engineering **332** (2018) 012047 doi:10.1088/1757-899X/332/1/012047

2.970659022, and the variance of the parameter $\hat{\theta}$ is $Var(\hat{\theta} \mid x) =$ 20.5853668503. So the standard deviation value is equal to $\sigma_{\hat{\theta}} = 0.765092707$.

A statistical summary of the mean, variance, and standard deviation of the estimated parameters $\hat{\theta}$ from the Citarum River flood loss data for one year is given in Table 2.

Table 2. The statistical summary of average, variance, and standard deviation of the parameters $\hat{\theta}$

Parameter	Posterior α	Posterior β	Average	Variance	Standard Deviation
$\hat{\theta}$	15.075	0.1970494	2.970659	0.5853668	0.7650927

Based on the posterior distribution function, Gamma, from equation (4) we can determine the estimated value of loss is equal to $\hat{\theta}$ = 2.970659022. In this study the loss is estimated based on the value obtained from the Bayesian inference. Parameter $\hat{\theta}$ is Pareto parameter estimator. The value of the variance and standard deviation of the loss model per RW can be determined as in Table 3.

Table 3. Summary statistics Pareto distributed loss estimates.

Parameter	Average (million IDR)	Variance (million IDR)	Standard Deviation (million IDR)
$\hat{\theta}$	3051.06	3,228,376.33	1,796.76

So that the loss value can be seen more clearly, then we draw the grouping of data into 10 groups of data based on the percentile size of the data. The estimated value of accumulated losses per RW is derived from Eq. (8) as given in Table 4.

Table 4. Estimated losses per RW with percentile size of data.

percentile		Estimated Losses (million IDR)	Information (million IDR)
0.0	0.1	2024.00	From percentile 0-10, Losses 2024
0.1	0.2	2097.09	From percentile 10-20, Losses 2097.09
0.2	0.3	2181.92	From percentile 20-30, Losses 2181.92
0.3	0.4	2282.26	From percentile 30-40, Losses 2282.26
0.4	0.5	2403.84	From percentile 40-50, Losses 2403.84
0.5	0.6	2556.03	From percentile 50-60, Losses 2556.03
0.6	0.7	2755.47	From percentile 60-70, Losses 2755.47
0.7	0.8	3035.73	From percentile 70-80, Losses 3035.73
0.8	0.9	3479.79	From percentile 80-90, Losses 3479.79
0.9	1.0	4394.51	From percentile 90-100, Losses 4394.51

Based on equation (9), the estimated value of losses per house is as given in Table 5.

Table 5. Estimated losses per house based on chance of damage caused by flood.

No	Possibility of Flood Damage	Infrastructure losses (million IDR)	No	Possibility of Flood Damage	Infrastructure losses (million IDR)
1	0.151278	629.65	31	0.359383	506.34
2	0.203541	584.30	32	0.682983	430.72
3	0.217636	574.52	33	0.282827	537.83
4	0.245526	557.34	34	0.378848	499.65
5	0.252445	553.45	35	0.106752	687.44
6	0.335881	515.04	36	0.896009	402.25
7	0.418629	487.24	37	0.819895	411.35
8	0.423930	485.70	38	0.259707	549.51
9	0.500231	465.87	39	0.261163	548.74
10	0.553040	454.24	40	0.513234	462.87
11	0.609898	443.18	41	0.056486	806.98
12	0.628365	439.86	42	0.798198	414.14
13	0.734245	422.94	43	0.586077	447.65
14	0.751399	420.49	44	0.835044	409.46
15	0.837337	409.18	45	0.395094	494.40
16	0.843138	408.46	46	0.136767	645.84
17	0.865072	405.83	47	0.581367	448.56
18	0.908786	400.82	48	0.081245	736.38
19	0.971257	394.17	49	0.189281	595.09
20	0.998131	391.47	50	0.200963	586.18
21	0.784533	415.94	51	0.475310	471.91
22	0.663157	433.93	52	0.624854	440.48
23	0.957423	395.59	53	0.297789	530.89
24	0.952795	396.08	54	0.101663	695.95
25	0.462114	475.26	55	0.842171	408.58
26	0.892646	402.64	56	0.421062	486.53
27	0.094485	708.90	57	0.476273	471.67
28	0.101737	695.82	58	0.840137	408.83
29	0.085835	726.26	59	0.967047	394.60
30	0.683979	430.57	60	0.381981	498.62

The estimated value of the loss is affected by the possibility value of each house damages. The less possibility of damage to a house due to flooding, the greater the estimation of losses in case of damage. For example, the value of loss B on the possibility of damage A indicates that if a house has a possibility of damage A, then the house has a chance of losing B.

3.3. Calculation of premiums

The amount of the premium depends on the value of the loss, the risk aversion parameters derived from the insurance company, as well as other factors. In this research, it is assumed that risk aversion parameter of insurance company ρ is equal to 0.005. Based on equations (12), (13), and (14), the value of premiums for each house with the principle of pure premiums, variance, and standard deviation, a portion of the proceeds from the calculation of premiums is presented in Table 6.

Table 6. Estimated premiums per house within 1 year based on three principles

No.	Loss per house	Variance principal (million IDR)	Standard deviation principal (million IDR)	Pure premiums (million IDR)
1	629.65	16,771.53	638.63	338.62
2	584.30	16,726.18	593.29	314.24
3	574.52	16,716.41	583.51	308.98

Based on Table 6, it can be seen that insurance premium based on the principle of variance is very much different from the principle of standard deviation and pure premium. This is due to the large value of loss variance. In determining the value of premiums, the insurance company will choose the greatest value so that the risk for the loss of the company is getting smaller. However, in many cases, there are still many factors that influence premium value besides loss, possibility of damage, and risk aversion value. If the premium is set the largest, then it will not be affordable by the community. Conversely, if the premium set too small, it will cause bankruptcy in the insurance company. Therefore, the determination of the magnitude of the insurance premium needs to consider many other factors which may affect.

4. Conclusion

In this paper, the estimated value of insurance premiums due to Citarum River flood by Bayesian method has been discussed. It is assumed that the risk data for flood losses follows the Pareto distribution with the right fat-tail. Based on the result, it can be obtained that the probability of damage to the building is Pareto distributed, with estimator parameters θ estimated using Bayesian inference method value is $\hat{\theta}$= 2.970659022. Furthermore, the insurance premium is estimated based on the probability of damage and the estimated value of the loss. The estimation results obtained based on the principle of pure premium are as follows: for the loss value of IDR 629.65 million of premium IDR 338.63 million; for a loss of IDR 584.30 million of its premium IDR 314.24 million; and the loss value of IDR 574.53 million of its premium IDR 308.95 million. Large insurance premiums based on the principle of variance are very much different high, with the principle of standard deviation and pure premium. This is due to the large value of the variance of loss. In determining the value of premiums, the insurance company will choose the greatest value so that the risk for the loss of the company is getting smaller. However, in fact, there are still many factors to consider in the determination of the premium. In addition to loss factors, building damage possibility, risk aversion values, and fairness levels also need to be considered. So this paper recommends the insurer to determine the premium value based on the principle of pure premium.

Acknowledgment

Thanks to the Rector, Director of DRPMI, and Dean of FMIPA, Universitas Padjadjaran, which provides a grant program of the Academic Leadership Grant (ALG) and a grant program of Competence Research of Lecturer of Unpad (Riset Kompetensi Dosen Unpad/RKDU).

References

[1] Abbott, E.B. 2008. Floods, Flood Insurance, Litigation, Politics – and Catastrophe: The National Flood Insurance Program. *Sea Grant Law and Policy Journal,* Vol. 1, No. 1 (June, 2008), pp. 129-155.

[2] Andraszewicz, S., Scheibehenne, B., Rieskamp, J., Grasman, R., Verhagen, J., and Wagenmakers, E.J. 2015. An Introduction to Bayesian Hypothesis Testing for Management Research. Journal of Management, Vol. 41 No. 2, February 2015, pp. 521 –543. DOI: 10.1177/0149206314560412. © The Author(s) 2014. Reprints and permissions: sagepub.com/journalsPermissions.nav

[3] Bayer, J.L. and Mechler, R. 2009. Insurance against Losses from Natural Disasters in Developing Countries. DESA Working Paper No. 85, ST/ESA/2009/DWP/85. pp. 1-37.

[4] Ekheden, E. and Hossjer, O. 2014. Pricing Catastrophe Risk in Life (Re)insurance. Scandinavian Actuarial Journal, 2014. Vol. 2014, No. 4, pp. 352-367. http://dx.doi.org/10.1080/03461238.2012.695747. http://www.tandfonline.com/loi/sact20

[5] Grace, M.F., Klein, R.W., and Kleindorfer, P.R. 2004. Homeowners Insurance with Bundled Catastrophe Coverage. *The Journal of Risk and Insurance*, 2004, Vol. 71, No. 3, pp. 351-379.

[6] Jindrová, P. and Kopecká, L. 2017. Empirical Bayes Credibility Models for Economic Catastrophic Losses by Regions. ITM Web of Conferences 9, 01006 (2017), *AMCSE 2016.* pp. 1-6. DOI: 10.1051/itmconf/20170901006

[7] Kousky, C. and Kerjan, E.M. 2015. Examining Flood Insurance Claims In The United States:Six Key Findings. © 2015 The Journal of Risk and Insurance (2015), pp. 1-32. DOI: 10.1111/jori.12106

[8] MacEachern, S.N. 2016. Nonparametric Bayesian Methods: A Gentle Introduction and Overview. Communication for Statistical Applications and Methods, 2016, Vol. 23, No. 6, pp. 445–466. http://dx.doi.org/10.5351/CSAM.2016.23.6.445. Print ISSN 2287-7843 / Online ISSN 2383-4757.

[9] Paudel, Y., Botzen,W.J.W., and Aerts, J.C.J.H. 2013. Estimation of Insurance Premiums for Coverage Against Natural Disaster Risk: An Application of Bayesian Inference. Nat. Hazards Earth Syst. Sci., 13, pp. 737–754, 2013. www.nat-hazards-earth-syst-sci.net/13/737/2013/. doi:10.5194/nhess-13-737-2013. © Author(s) 2013. CC Attribution 3.0 License

[10] Qian, S.S., Stow, C.A., and Borsuk, M.E. 2003. On Monte Carlo Methods for Bayesian Inference. Ecological Modelling 159 (2003), pp. 269-277. www.elsevier.com/locate/ecolmodel

[11] Rudari, R., Beckers, J., De Angeli, S., Rossi, L., and Trasforin, E. 2016. Impact of Modelling Scale on Probabilistic Flood Risk Assessment: The Malawi Case. E3S Web of Conferences 7, 04015 (2016), Flood R*isk* 2016 - 3rd European Conference on Flood Risk Management. pp. 1-12. DOI: 10.105/e3s conf/2016070415.

[12] Sidi, P.,Mamat, M., Sukono, and Supian, S. 2017. Evaluation Model for Risk Insurance Premiums of Building Damage Caused by Flood: Case Study in Citarum Watershed, Soutern Bandung, Indonesia. *Journal of Engineering and Applied Sciencs* 12 (17): 4420-4425, 2017. ISSN: 1816-949X. Medwell Journals, 2017.

[13] Sukono, Suyudi, M., Islamiyati, F., and Supian, S.. 2017. Estimation Model of Life Insurance Claims Risk for Cancer Patients by Using Bayesian Method. *IOP Conf. Series: Materials Science and Engineering* 166 (2017) 012022 doi:10.1088/1757-899X/166/1/012022. pp. 1-9.

[14] Sukono, Nahar, J., Putri, F. T., Subiyanto, Mamat, M., & Supian, S. 2017. Estimation of outstanding claims reserving based on inflation risk on car insurance companies by using the bootstrap method. *Far East Journal of Mathematical Sciences, 102*(4), 687-706. doi:10.17654/MS102040687

[15] Surminski, S. 2013. The Role of Insurance in Reducing Direct Risk: The Case of Flood Insurance. International Review of Environmental and Resource Economics, 2013, 7: pp. 241–278. ISSN 1932-1465; DOI 10.1561/101.00000062.

[16] Tichy, O., Smidl, V., Hofman, R., Sindelarova, K., Hiza, M., and Stihl, A. 2017. Bayesian Inverse Modeling and Source Location of an Unintended I-131 Release in Europe in the Fall of 2011. Atmos. Chem. Phys. Discuss., doi:10.5194/acp-2017-206, 2017. Manuscript under review for journal Atmos. Chem. Phys. Discussion started: 16 March 2017. Author(s) 2017. pp. 1-24.

IOP Conf. Series: Materials Science and Engineering **332** (2018) 012048 doi:10.1088/1757-899X/332/1/012048

A new chaotic attractor with two quadratic nonlinearities, its synchronization and circuit implementation

S Vaidyanathan[1*], A Sambas[2], Sukono[3], M Mamat[4], G Gundara, W S Mada Sanjaya[5], Subiyanto[6]

[1]Research and Development Centre, Vel Tech University, Avadi, Chennai, India.
[2]Department of Mechanical Engineering, Universitas Muhammadiyah Tasikmalaya, Indonesia
[3]Department of Mathematics, Faculty of Mathematics and Natural Sciences, Universitas Padjadjaran, Indonesia.
[4]Faculty of Informatics and Computing, Universiti Sultan Zainal Abidin, Kuala Terengganu, Malaysia
[5]Department of Physics, Universitas Islam Negeri Sunan Gunung Djati, Bandung, Indonesia
[6]Department of Marine Science, Faculty of Fishery and Marine Science, Universitas Padjadjaran, Indonesia

[*]Email: sundarvtu@gmail.com

Abstract. A 3-D new chaotic attractor with two quadratic nonlinearities is proposed in this paper. The dynamical properties of the new chaotic system are described in terms of phase portraits, equilibrium points, Lyapunov exponents, Kaplan-Yorke dimension, dissipativity, etc. We show that the new chaotic system has three unstable equilibrium points. The new chaotic attractor is dissipative in nature. As an engineering application, adaptive synchronization of identical new chaotic attractors is designed via nonlinear control and Lyapunov stability theory. Furthermore, an electronic circuit realization of the new chaotic attractor is presented in detail to confirm the feasibility of the theoretical chaotic attractor model.

1. Introduction

Chaos theory deals with nonlinear dynamical systems that are highly sensitive to initial conditions, which is also characterized by the existence of a positive Lyapunov chaos exponent [1-2]. In the last few decades, chaotic and hyperchaotic systems have attracted the interest of the engineering and science community because of their wide applications in many scientific and engineering fields [3-18].

Motivated by the research on chaotic systems with dissipative chaotic attractors with quadratic nonlinearities [19-24], we prove a new 3-D new chaotic system with two quadratic nonlinearities in this paper. The new chaotic system is dissipative and it has three unstable equilibrium points.

Section 2 describes the new chaotic system and details the properties such as Lyapunov exponents and Kaplan-Yorke dimension. Section 3 describes the adaptive synchronization of the new chaotic system with unknown parameters. Furthermore, an electronic circuit realization of the new chaotic

IOP Conf. Series: Materials Science and Engineering **332** (2018) 012048 doi:10.1088/1757-899X/332/1/012048

system is presented in detail in Section 4. The circuit experimental results of the new chaotic attractor show agreement with the numerical simulations. Section 5 contains the conclusions of this work.

2. A new chaotic system with two quadratic nonlinearities

In this paper, we announce a new 3-D chaotic system with two quadratic nonlinearities given by

$$\begin{cases} \dot{x}_1 = a(x_2 - x_1) - x_3 \\ \dot{x}_2 = x_1 x_3 \\ \dot{x}_3 = 50 - b x_1^2 - x_3 \end{cases} \quad (1)$$

where x_1, x_2, x_3 are state variables and a, b are positive constants.

In this paper, we show that the system (1) is *chaotic* for the parameter values

$$a = 3, \quad b = 1 \quad (2)$$

For numerical simulations, we take the initial values of the system (1) as

$$x_1(0) = 0.2, \quad x_2(0) = 0.2, \quad x_3(0) = 0.2 \quad (3)$$

Figure 1 shows the phase portraits strange attractor of the new chaotic system (1) for the parameter values (2) and initial conditions (3). Figure 1 (a) shows the 3-D phase portrait of the new chaotic system (1). Figures 1 (b)-(c) show the projections of the new chaotic system (1) in (x_1, x_2), (x_2, x_3) and (x_1, x_3) coordinate planes, respectively.

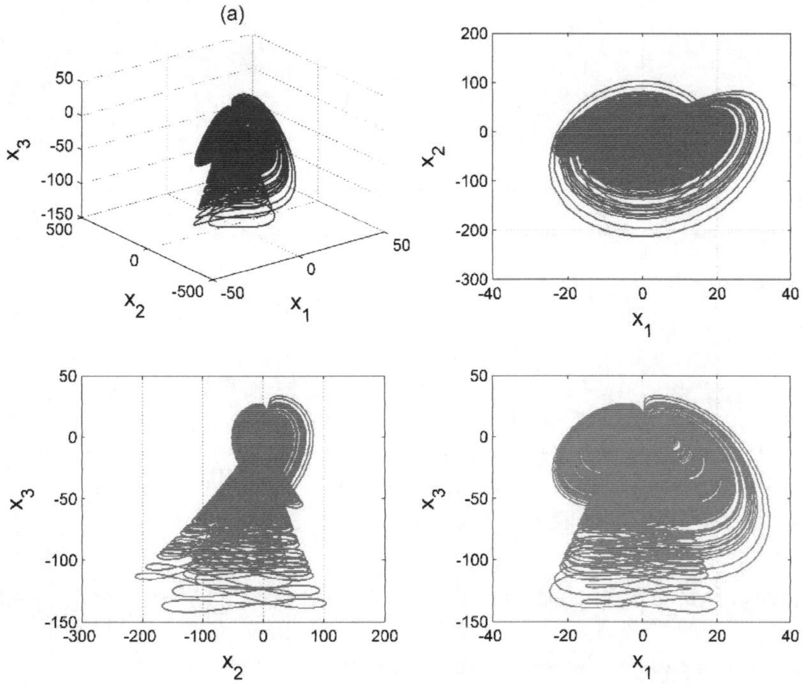

Figure 1. Phase portraits of the new chaotic system (1) for $a = 3, \ b = 1$

For the rest of this section, we take the parameter values as in the chaotic case (2). The equilibrium points of the new chaotic system (1) are obtained by solving the system of equations

$$a(x_2 - x_1) - x_3 = 0, \quad x_1 x_3 = 0, \quad 50 - b x_1^2 - x_3 = 0 \quad (4)$$

Solving the equations in (4) we obtain the equilibrium points of the system (1) as

$$E_1 = \begin{bmatrix} 0 \\ 50/3 \\ 50 \end{bmatrix}, \quad E_2 = \begin{bmatrix} \sqrt{50} \\ \sqrt{50} \\ 0 \end{bmatrix}, \quad E_3 = \begin{bmatrix} -\sqrt{50} \\ -\sqrt{50} \\ 0 \end{bmatrix} \tag{5}$$

It is easy to show that E_1 is a saddle point, while E_2 and E_3 are saddle-focus points.

For the parameter values as in the chaotic case (2) and the initial state as in (3), the Lyapunov exponents of the new 3-D system (2) are determined using Wolf's algorithm as

$$L_1 = 1.0676, \; L_2 = 0, \; L_3 = -5.0676 \tag{6}$$

Since $L_1 > 0$, the new 3-D system (1) is chaotic. Thus, the system (1) exhibits a chaotic attractor. Also, we note that the sum of the Lyapunov exponents in (6) is negative. This shows that the new 3-D chaotic system (1) is dissipative.

The Kaplan-Yorke dimension of the new 3-D system (1) is determined as

$$D_{KY} = 2 + \frac{L_1 + L_2}{|L_3|} = 2.2107, \tag{7}$$

which indicates the complexity of the new chaotic system (1).

Figure 2 shows the Lyapunov exponents of the new chaotic system (1) with a strange attractor.

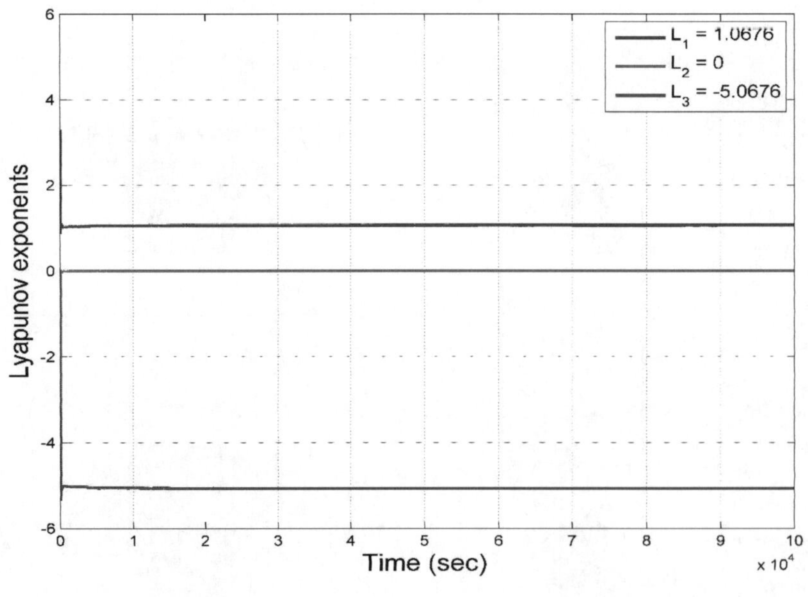

Figure 2. Lyapunov exponents of the new chaotic system (1) for $a = 3, \; b = 1$

3. Adaptive synchronization of the new chaotic systems

In this section, we devise adaptive controller so as to synchronize the respective states of identical new chaotic systems with unknown parameters considered as *master* and *slave* systems respectively. As the master system, we consider the new chaotic system given by

$$\begin{cases} \dot{x}_1 = a(x_2 - x_1) - x_3 \\ \dot{x}_2 = x_1 x_3 \\ \dot{x}_3 = 50 - b x_1^2 - x_3 \end{cases} \tag{8}$$

where x_1, x_2, x_3 are the states and a, b are unknown parameters.

As the slave system, we consider the new chaotic system given by

$$\begin{cases} \dot{y}_1 = a(y_2 - y_1) - y_3 + u_1 \\ \dot{y}_2 = y_1 y_3 + u_2 \\ \dot{y}_3 = 50 - b y_1^2 - y_3 + u_3 \end{cases} \tag{9}$$

where y_1, y_2, y_3 are the states and u_1, u_2, u_3 are adaptive controls to be designed.

The synchronization error between the systems (8) and (9) is defined as

$$\begin{cases} e_1 = y_1 - x_1 \\ e_2 = y_2 - x_2 \\ e_3 = y_3 - x_3 \end{cases} \tag{10}$$

The error dynamics is obtained as

$$\begin{cases} \dot{e}_1 = a(e_2 - e_1) - e_3 + u_1 \\ \dot{e}_2 = y_1 y_3 - x_1 x_3 + u_2 \\ \dot{e}_3 = -b(y_1^2 - x_1^2) - e_3 + u_3 \end{cases} \tag{11}$$

We consider the adaptive control defined by

$$\begin{cases} u_1 = -\hat{a}(t)(e_2 - e_1) + e_3 - k_1 e_1 \\ u_2 = -y_1 y_3 + x_1 x_3 - k_2 e_2 \\ u_3 = \hat{b}(t)(y_1^2 - x_1^2) + e_3 - k_3 e_3 \end{cases} \tag{12}$$

where k_1, k_2, k_3 are positive gain constants.

Substituting (12) into (11), we obtain the closed-loop system

$$\begin{cases} \dot{e}_1 = [a - \hat{a}(t)](e_2 - e_1) - k_1 e_1 \\ \dot{e}_2 = -k_2 e_2 \\ \dot{e}_3 = -[b - \hat{b}(t)](y_1^2 - x_1^2) - k_3 e_3 \end{cases} \tag{13}$$

We define the parameter estimation errors as

$$\begin{cases} e_a(t) = a - \hat{a}(t) \\ e_b(t) = b - \hat{b}(t) \end{cases} \tag{14}$$

Using (14), we can simplify (13) as

$$\begin{cases} \dot{e}_1 = e_a(e_2 - e_1) - k_1 e_1 \\ \dot{e}_2 = -k_2 e_2 \\ \dot{e}_3 = -e_b(y_1^2 - x_1^2) - k_3 e_3 \end{cases} \tag{15}$$

Differentiating (14) with respect to t, we obtain

$$\begin{cases} \dot{e}_a(t) = -\dot{\hat{a}}(t) \\ \dot{e}_b(t) = -\dot{\hat{b}}(t) \end{cases} \tag{16}$$

Next, we consider the Lyapunov function defined by

$$V(e_1, e_2, e_3, e_a, e_b) = \frac{1}{2}\left(e_1^2 + e_2^2 + e_3^2 + e_a^2 + e_b^2\right) \tag{17}$$

which is positive definite on $\boldsymbol{R}^5$.

Differentiating V along the trajectories of (15) and (16), we obtain

$$\dot{V} = -k_1e_1^2 - k_2e_2^2 - k_3e_3^2 + e_a\left[e_1(e_2 - e_1) - \dot{\hat{a}}\right] + e_b\left[-e_3(y_1^2 - x_1^2) - \dot{\hat{b}}\right] \tag{18}$$

In view of (18), we take the parameter update law as

$$\begin{cases} \dot{\hat{a}} = e_1(e_2 - e_1) \\ \dot{\hat{b}} = -e_3\left(y_1^2 - x_1^2\right) \end{cases} \tag{19}$$

Next, we prove the main theorem of this section.

Theorem 2. The new chaotic systems (8) and (9) with unknown parameters are globally and asymptotically stabilized by the adaptive control law (12) and the parameter update law (19), where k_1, k_2, k_3 are positive constants.

Proof. The Lyapunov function V defined by (17) is quadratic and positive definite on R^5.

By substituting the parameter update law (19) into (18), we obtain the time-derivative of V as

$$\dot{V} = -k_1e_1^2 - k_2e_2^2 - k_3e_3^2 \tag{20}$$

which is negative semi-definite on R^5.

Thus, by Barbalat's lemma [25], it follows that the closed-loop system (15) is globally asymptotically stable for all initial conditions $e(0) \in R^3$.

Hence, we conclude that the new chaotic systems (8) and (9) with unknown parameters are globally and asymptotically stabilized by the adaptive control law (12) and the parameter update law (19), where k_1, k_2, k_3 are positive constants.

This completes the proof. ■

For numerical simulations, we take the gain constants as

$$k_1 = 10, \quad k_2 = 10, \quad k_3 = 10 \tag{21}$$

We take the parameter values as in the chaotic case (2), i.e.

$$a = 3, \quad b = 1 \tag{22}$$

We take the initial conditions of the states of the master system (8) as

$$x_1(0) = 5.2, \quad x_2(0) = 8.9, \quad x_3(0) = 3.1 \tag{23}$$

We take the initial conditions of the states of the slave system (9) as

$$y_1(0) = 12.5, \quad y_2(0) = 2.1, \quad y_3(0) = 14.7 \tag{24}$$

We take the initial conditions of the parameter estimates as

$$\hat{a}(0) = 7.3, \quad \hat{b}(0) = 10.6 \tag{25}$$

Figure 3 shows the synchronization of the states of the new chaotic systems (8) and (9). Figure 4 shows the time-history of the synchronization errors e_1, e_2, e_3.

IOP Conf. Series: Materials Science and Engineering **332** (2018) 012048 doi:10.1088/1757-899X/332/1/012048

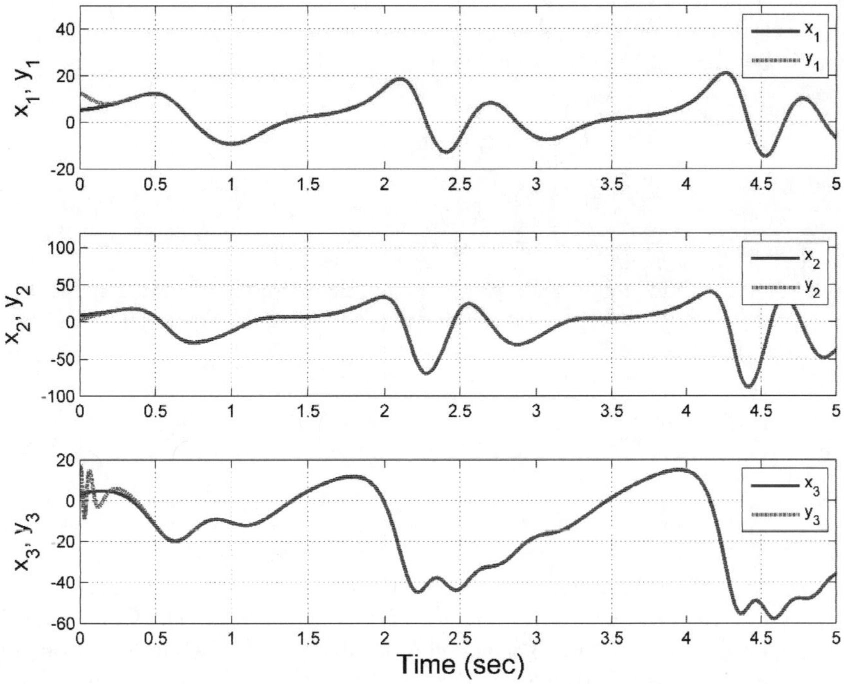

Figure 3. Complete synchronization of the new chaotic systems

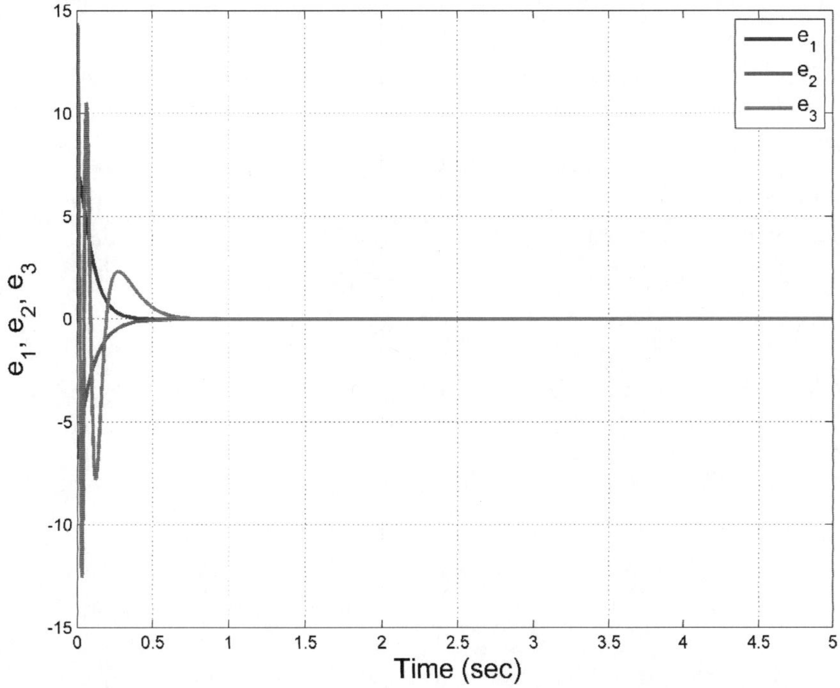

Figure 4. Time-history of the synchronization errors for the new chaotic systems

4. Circuit implementation of the new chaotic system

The electronic circuit modeling the new chaotic system (1) is realized by using off-the-shelf components such as resistors, capacitors, operational amplifiers, and multipliers. The circuit electronic of a new chaotic system (1) by MultiSIM is shown in **Figure 5**. In this work, a linear scaling is considered as follows:

$$\left.\begin{aligned} \dot{x}_1 &= a(2x_2 - x_1) - x_3 \\ \dot{x}_2 &= 2.5x_1x_3 \\ \dot{x}_3 &= 10 - 5x_1^{\ 2} - x_3 \end{aligned}\right\} \tag{26}$$

The circuit equation corresponding to each state of the scaled system (26) using Kirchhoff's laws can be obtained as:

$$\left.\begin{aligned} \dot{x}_1 &= \frac{1}{C_1R_1}x_2 - \frac{1}{C_1R_2}x_1 - \frac{1}{C_1R_3}x_3 \\ \dot{x}_2 &= \frac{1}{C_2R_4}x_1x_3 \\ \dot{x}_3 &= \frac{1}{C_3R_7}V_1 - \frac{1}{C_3R_5}x_1^{\ 2} - \frac{1}{C_3R_6}x_3 \end{aligned}\right\} \tag{27}$$

where the variables x_1, x_2, and x_3 are the outcomes of the integrators *U1A*, *U3A*, *U5A*.

We choose $R_2 = R_8 = R_9 = R_{10} = R_{11} = 100$ kΩ, $R_1 = 50$ KΩ, $R_3 = R_6 = 300$ KΩ, $R_4 = 120$ KΩ, $R_5 = 60$ KΩ, $R_7 = 30$ KΩ, $C_1 = C_2 = C_3 = 3.2$ nF. The supplies of all active devices are ±15 Volt. The designed electronic circuit is implemented in MultiSIM. The obtained results are presented in **Figures 6** (a) - (c), which show the phase portraits of the chaotic attractor in x_1-x_2, x_2-x_3 and x_1-x_3 planes, respectively. It is easy to see a good agreement between the circuital attractor and theoretical attractor.

5. Conclusions

This work announced a new chaotic system with two quadratic nonlinearities. First, the qualitative properties of the new chaotic system are reported. Dynamical behaviors of the new chaotic system with two quadratic nonlinearities are investigated through equilibrium points, projections of chaotic attractors, Lyapunov exponents, and Kaplan–Yorke dimension. In addition, adaptive synchronization scheme of new chaotic systems is shown via adaptive control approach. Furthermore, an electronic circuit realization of the new chaotic system using the electronic simulation package MultiSIM confirmed the feasibility of the theoretical model.

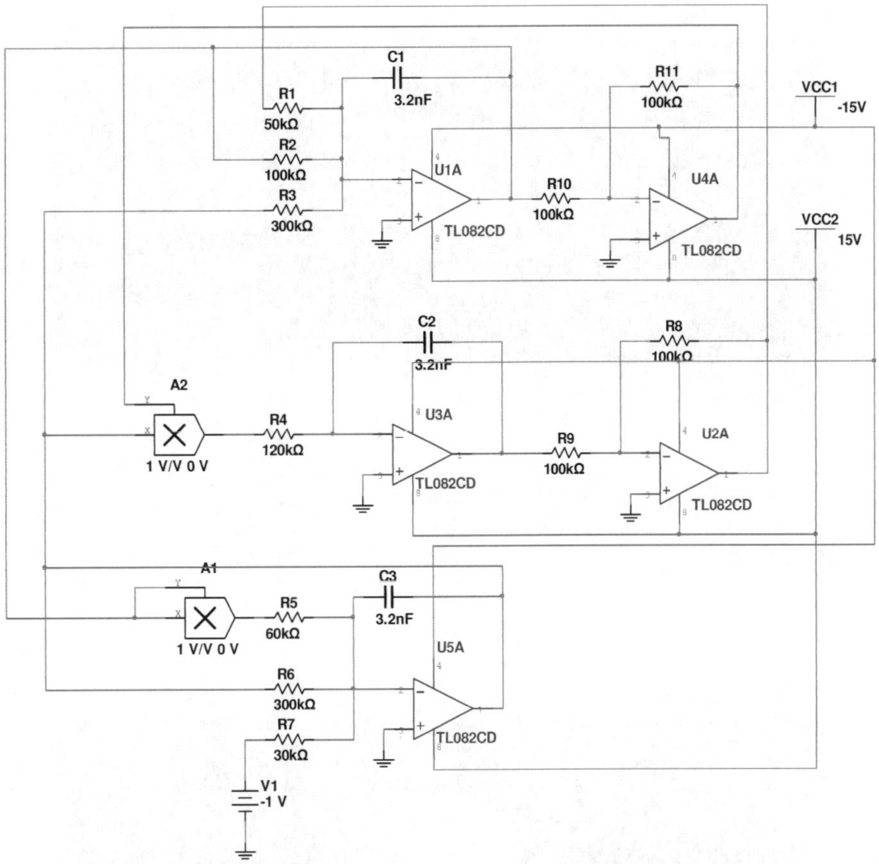

Figure 5 Circuit design for new chaotic system (1) by MultiSIM

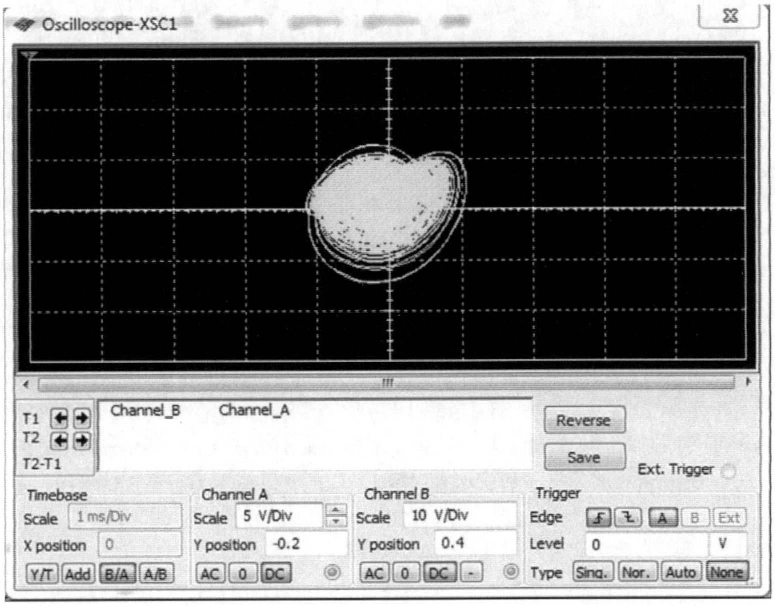

(a)

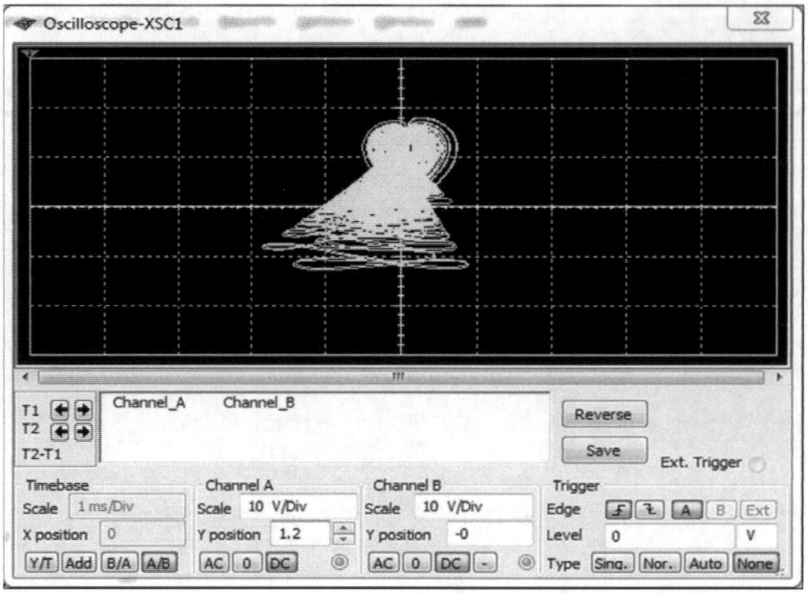

(b)

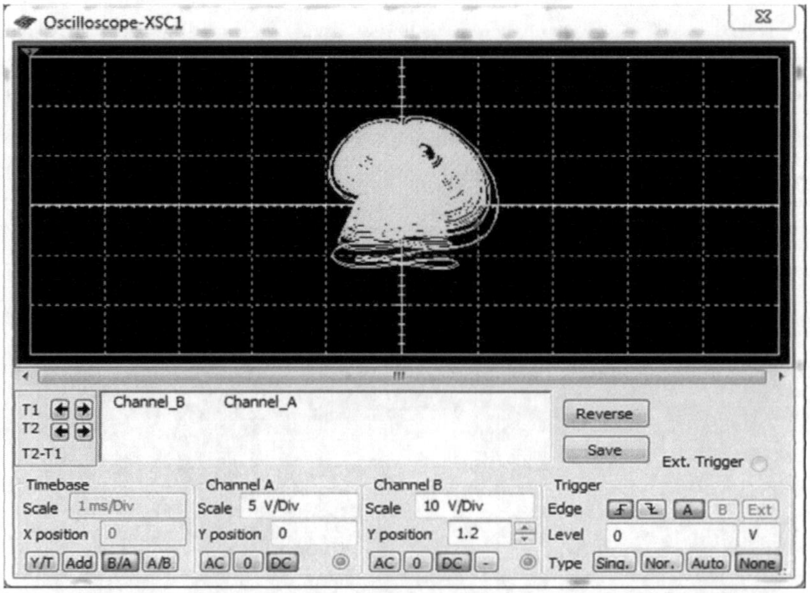

(c)

Figure 6 The phase portraits of new chaotic system (1) observed on the oscilloscope in different planes (a) *x*-*y* plane, (b) *y*-*z* plane and (c) *x*-*z* plane by MultiSIM

References

[1] Azar A T and Vaidyanathan S 2015 *Chaos Modeling and Control Systems Design* (Berlin: Springer)
[2] Azar A T and Vaidyanathan, S 2016 *Advances in Chaos Theory and Intelligent Control* (Berlin: Springer)
[3] Kengne J, Chedjou J C, Kenne G and Kyamakya K 2012 *Communications in Nonlinear Science and Numerical Simulation* **17** 2914-2923
[4] Vaidyanathan S 2015 *International Journal of PharmTech Research* **8** 106-116
[5] Vaidyanathan S 2016 *International Journal of ChemTech Research* **9** 297-304
[6] Xu C and Wu Y 2015 *Applied Mathematical Modelling* **39** 2295-2310
[7] Bodale I and Oancea V A 2015 *Chaos Solitons and Fractals* **78** 1-9
[8] Vaidyanathan S 2015 *International Journal of ChemTech Research* **8** 73-85
[9] Punjabi A and Boozer A 2014 *Physics Letters A* **378** 2410-2416
[10] Vaidyanathan S 2015 *International Journal of ChemTech Research* **8** 818-827
[11] Voorsluijs V and Decker Y D 2016 *Physica D: Nonlinear Phenomena* **335** 1-9
[12] Chattopadhyay J, Pal N, Samanta S, Venturino E and Khan Q J A 2015 *Biosystems* **138** 18-24
[13] Sambas, A., Vaidyanathan, S., Mamat, M., Sanjaya, W. M., and Rahayu, D. S. 2016 *Advances and Applications in Chaotic Systems* (Berlin: Springer)
[14] Gritli H, Khraief N and Belghith S 2012 *Communications in Nonlinear Science and Numerical Simulation* **17** 4356-4372
[15] Sambas, A., WS, M. S., Mamat, M., and Prastio, R. P 2016 *Advances in Chaos Theory and Intelligent Control* (Berlin: Springer)
[16] Sambas, A., Mamat, M., Mada Sanjaya, W. S., Salleh, Z., and Mohamad, F. S 2015 *Advanced Studies in Theoretical Physics* **9** 379-394
[17] Yang J, Wang L, Wang Y and Guo T 2017 *Neurocomputing* **227** 142-148
[18] Bao B C, Bao H, Wang N, Chen M and Xu Q 2017 *Chaos Solitons and Fractals* **94** 102-111
[19] Vaidyanathan S, Rajagopal K, Volos C K, Kyprianidis I M and Stouboulos I N 2015 *Journal of Engineering Science and Technology Review* **8** 130-141
[20] Vaidyanathan S, Volos C K and Pham V T 2015 *Journal of Engineering Science and Technology Review* **8** 232-244
[21] Vaidyanathan S 2015 *Journal of Engineering Science and Technology Review* **8** 106-115
[22] Vaidyanathan S and Rhif A 2017 *International Journal of Modelling, Identification and Control* **28** 28-39
[23] Li C, Wu L, Li H and Tong Y 2013 *Nonlinear Analysis: Modelling and Control* **18** 66-77
[24] Vaidyanathan S and Sampath S 2017 *International Journal of Modelling, Identification and Control* **27** 3-13
[25] Khalil H K 2002 *Nonlinear Systems* (New York: Prentice Hall)

Spatial weighting approach in numerical method for disaggregation of MDGs indicators

S D Permai[1*], U Mukhaiyar[2], N L P Satyaning P P[3], M Soleh[4], Q Aini[5]

[1]Statistics Department, School of Computer Science, Bina Nusantara University, Jakarta 11480, Indonesia
[2]Statistics Research Division, Faculty of Mathematics and Natural Sciences, Institut Teknologi Bandung, Bandung 40132, Indonesia
[3]Department of Statistics, Institut Teknologi Sepuluh Nopember, Surabaya 60111
[4]Department of Statistics, Bogor Agricultural University, Bogor 16680, Indonesia
[5]Regional Balance and Statistical Analysis Sub-Division, Central Statistics Agency, West Nusa Tenggara 84455, Indonesia

[*]Corresponding author: syarifah.permai@binus.ac.id

Abstract. Disaggregation use to separate and classify the data based on certain characteristics or on administrative level. Disaggregated data is very important because some indicators not measured on all characteristics. Detailed disaggregation for development indicators is important to ensure that everyone benefits from development and support better development-related policymaking. This paper aims to explore different methods to disaggregate national employment-to-population ratio indicator to province- and city-level. Numerical approach applied to overcome the problem of disaggregation unavailability by constructing several spatial weight matrices based on the neighbourhood, Euclidean distance and correlation. These methods can potentially be used and further developed to disaggregate development indicators into lower spatial level even by several demographic characteristics.

1. Introduction

The 17 Sustainable Development Goals (SDGs) build on the successes of the previous eight Millennium Development Goals (MDGs), while including new areas such as climate change, economic inequality, innovation, sustainable consumption, peace and justice, among other priorities. Despite substantial progress has been made on many of MDGs, the progress has been uneven across regions and countries [1]. Millions of people are being left behind, especially the poorest and the vulnerable groups because of their gender, age, disability, ethnicity or geographic location.

Learning from MDGs, one of the highlights of SDGs is "leaving no one behind". It can be seen that the SDGs targets itself requires more disaggregated data by several demographic characteristics as mentioned above. Since disaggregation are not available for MDGs indicator, there will be a limitation to analyze both SDGs and MDGs data together for monitoring and research purpose. It is indeed important to have disaggregation as detail as possible for development indicators in order to (i) ensure that the benefit of the development reach everyone and (ii) assist the formulation of better policy to achieve the goals and targets. However, the important point is that disaggregation according to these dimensions would not be relevant for all indicators [2].

This study focus on estimating development indicator at the local level. The local level is the geographical level at which data are requested with a view to planning sub-regional policies or evaluating the results of policy [3]. Spatial disaggregation methods are based on area interpolation techniques. The spatial relationship imposed in disaggregation data process. This way used in all such techniques to decrease error value [4]. Several methods are proposed and piloted to spatially disaggregate one of important indicators in development goals, which is employment-to-population ratio. Employment-to-population ratio is one of indicators for the second target of Goal 1 Eradicate poverty and hunger: achieve full and productive employment and decent work for all, including women and young people. The national-to-province and province-to-city disaggregation has been done using 2011 data.

2. Simple Proportion

One of disaggregation method is weighted method. Weighting method using proportion is the simplest approach for disaggregating data. This method assumes that target variable (Y_i) is uniformly distributed in each area. The target variable can be estimated as [5],

$$Y_i = \frac{A_i}{\bar{A}} Y \tag{1}$$

Where,
$i = 1, 2, \ldots, n$,
Y_i : value of indicator for unit i
Y : value of MDGs indicator in higher level
A_i : value of non-MDGs indicator for unit i
$\bar{A}$: average value of non MDGs indicator in higher level
Note that A_i is a variable that highly correlated or has similar pattern with respective MDGs indicator. For this study, proportion of working population to the total population is used as A_i.

3. Numerical Method Approach

In this approach, numerical method principle is applied. There are two categories in numerical methods, direct methods and iterative methods. Direct methods give exact solution of problem without rounding error. Iterative methods find solution from a sequence of approximation solutions. This method using starting point $Y^{(0)}$ and generate sequence of approximate solutions $Y^{(k)}$. The latest approximations to the components of Y are used in the update of subsequent components [6]. The simple proportion defined in previous subsection can be considered as the initial value $Y^{(0)}$.

In this paper, numerical method used is iterative method. Iterative methods generate a sequence of approximations to the desired solution, often referred as successive approximation or trial and error method. This method is start with a function, which maps one approximation into another better. In this way a sequence of possible solutions to the problem is generated. The approximation obtained acceptably accurate when the solution is convergent. [7] The sequence is said to converge to the limit if $|Y - Y^{(m)}| < \varepsilon$. Iterative methods to find a sequence of approximation solutions following

$$\vec{Y}^{(k+1)} = \boldsymbol{W} * \vec{Y}^{(k)} \tag{2}$$

$$\vec{Y}^{(k)} = \left(Y_1{}^{(k)} \quad Y_2{}^{(k)} \quad \cdots \quad Y_n{}^{(k)}\right)' \tag{3}$$

$$Y_i{}^{(0)} = \frac{A_i}{\bar{A}} Y \tag{4}$$

$$\bar{A} = \frac{\sum_{i=1}^{n} A_i}{n} \tag{5}$$

W is a spatial weight matrix $\{w_{ij}\}$ and $\vec{Y}^{(k)}$ is k-th iteration value of indicator in i-th area.

Stopping rule is defined as if $\left|\vec{Y}^{(k)} - \vec{Y}^{(actual)}\right| < \varepsilon$, for any small real values $\varepsilon > 0$. This approach is used by knowing the real values so that how good this approach can be identified. From this result then data disaggregation to smaller areas can be executed using the same approaches.

The most important thing in numerical method approach for data disaggregation is determining the spatial weight matrix. The spatial weights matrix is an integral part of spatial modeling and defined as the formal expression of spatial dependence between observations [8]. There are several methods can be used to construct the spatial weight matrix. Based on Tobler's first law said that everything is related to everything else, but near things are more related than distant things [9]. Therefore, in this paper several methods of constructing spatial weight matrix using geographical proximity between areas are experimented.

3.1. Neighbourhood Based

Nearest neighbor method uses the simplest way to determine the weight. This method uses the determination of spatial unit share a boundary or not. The next step is to create a matrix ***M*** which contains the coding between the units that have shared a boundary or not. This method also called as rook contiguity [10].

$$\boldsymbol{M} = \{m_{ij}\} = \begin{bmatrix} m_{11} & \cdots & m_{1n} \\ \vdots & \ddots & \vdots \\ m_{n1} & \cdots & m_{nn} \end{bmatrix} \tag{6}$$

$$m_{ij} = \begin{cases} 1 & \text{location } i^{th} \text{ and } j^{th} \text{ share the same borderline} \\ 0 & \text{otherwise} \end{cases} \tag{7}$$

In most cases it is convenient to normalize spatial weights to remove dependence on extraneous scale factors. This produces row normalization matrix called matrix ***W***.

$$T = diag\left(\frac{1}{\sum_{j=1}^{n} m_{1j}}, \frac{1}{\sum_{j=1}^{n} m_{2j}}, \ldots, \frac{1}{\sum_{j=1}^{n} m_{nj}}\right) \tag{8}$$

$$\boldsymbol{W} = T * \boldsymbol{M} \tag{9}$$

3.2. Euclidean Based

Geographical proximity can be measured using distance. The most common distance, Euclidean distance, is applied in this paper. Given x and y is longitude and latitude coordinate, respectively, below is the formula for calculating the distance between the two units [11].

$$d_{ij} = \sqrt{(x_i - x_j)^2 + (y_i - y_j)^2} \tag{10}$$

The problem is there is no maximum limit value of the distance, so that the distance values must be normalized to obtain a spatial weight matrix as follow.

$$w_{ij} = \begin{cases} 0 & if\ i = j \\ \dfrac{\left(\dfrac{1}{1 + d_{ij}}\right)}{\sum_{i=1}^{n}\left(\dfrac{1}{1 + d_{ij}}\right)} & if\ i \neq j \end{cases} \tag{11}$$

3.3. Correlation Based

Methods based on correlation are desirable if the relationships among original distances do not follow a mathematically predictable pattern or are thought to be non-linear. The correlations do not change when distances are transformed [12]. Define correlation matrix of intended units based on data history and construct a distance matrix D as follow:

$$d_{ij} = \sqrt{2 * (1 - r_{ij})^2} \tag{12}$$

IOP Conf. Series: Materials Science and Engineering **332** (2018) 012049 doi:10.1088/1757-899X/332/1/012049

Where r_{ij} is sample correlation between location *i*-th and *j*-th. For obtaining this correlation, the history data (previous observations) of respective MDG's indicator are needed. Two units which have higher correlation means the distance between two units are nearer. So, spatial weight matrix is improved by the correlation among neighbors who shared a boundary.

$$w_{ij} = \frac{m_{ij}\, {d_{ij}}^{-1}}{\sum_{\substack{k=1 \\ k \neq i}}^{n} m_{ik}\, {d_{ik}}^{-1}} \tag{13}$$

4. Data and Methods

In this research, the MDGs indicator data used are the employment to population ratio index in 2011 for the national level and province level. The provincial level used are DKI Jakarta and West Java. The data will be disaggregated from national to province and from province to city. Non-MDGs indicator data that used are proportion of working population to the total population in 2011 and administrative map of Indonesia, DKI Jakarta and West Java. The data disaggregation is performed using simple proportion and spatial weighting approach.

5. Results and Discussion

The focus of this section is discussing the results and evaluating the method to conclude the best method so far. Aggregation from national to province level has firstly been done using simple proportion and numerical approach with three methods of weight matrix construction explained above. The disaggregation models developed show different results for each province, as shown in Figure 1. There are little differences of estimated pattern among big islands in Indonesia.

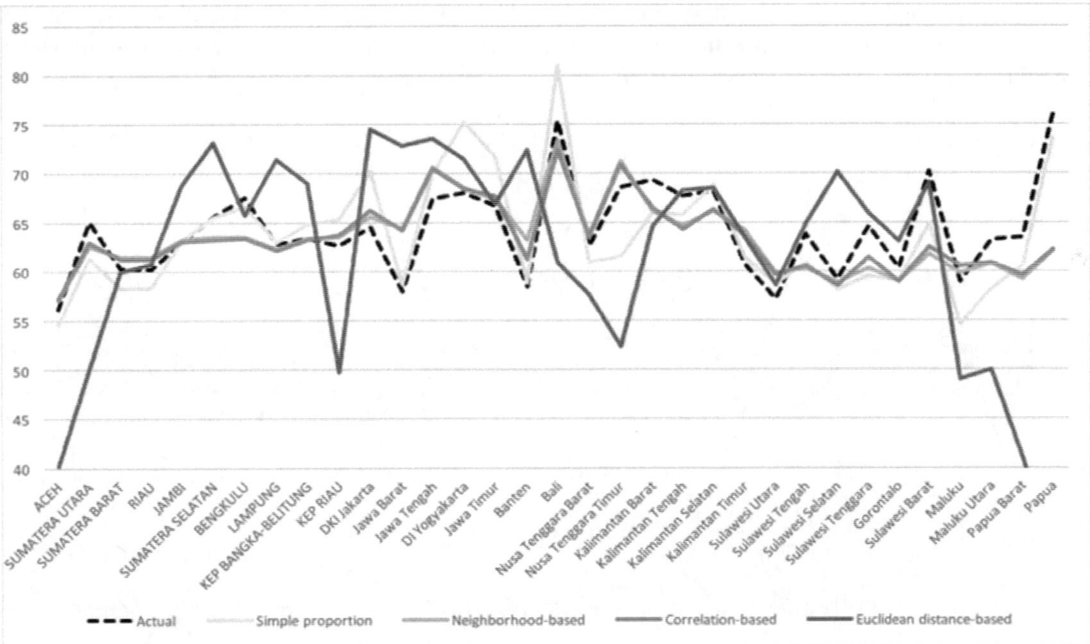

Figure 1. National to province disaggregation results and the actual data.

Based on Figure 1, most of the models underestimate the employment to population ratios for Aceh, Papua and Papua Barat, and provinces situated in Sulawesi island. The highest deviation found in estimating the employment to population ratio of Papua Barat. Besides the underestimation, overestimation can be found in provinces situated in Java islands. Employment to population ratio of provinces in Sumatera island, Bali, and Nusa Tenggara are closely estimated, where in Bali and Nusa

IOP Conf. Series: Materials Science and Engineering **332** (2018) 012049 doi:10.1088/1757-899X/332/1/012049

Tenggara the variance is smaller while in Sumatera island it is higher. One of the reasons of getting either overestimation or underestimation is that the non-MDGs official statistics used for initial proportion weighting of the respective provinces have different pattern with the employment-to-population ratio. Some provinces are not really affected by their neighborhood, while some are influenced a lot by them. This can also lead to the higher deviation in estimating an indicator in lower spatial level.

Table 1. National-to-province model evaluation

Method	MAE	MAPE	MSE
Simple proportion	2.689	4.1	11.323
Numerical Method Approach:			
Neighborhood-based	2.726	4.2	14.710
Euclidean distance-based	8.347	12.9	147.018
Correlation-based	2.615	4.0	13.319

In order to evaluate the models, several statistics to measure the goodness of models have been calculated as shown in Table 1. The lower the value of mean average error (MAE), mean average percentage error (MAPE), and mean squared error (MSE), the better the model. From these three criterion, correlation-based numerical approach have better estimated the employment-to-population ratio of province-level with the lowest MAE and MAPE, 2.615 and 4 respectively. It is also noted that Euclidean distance-based method gives the worst estimation (highest value for the three criterion) results among proposed methods, which indicates that the closer distance does not lead to the higher dependency among locations. That neighborhood-based model is better that the Euclidean distance one indicates that the locations which share same administrative borderline have a bigger chance to influence each other's.

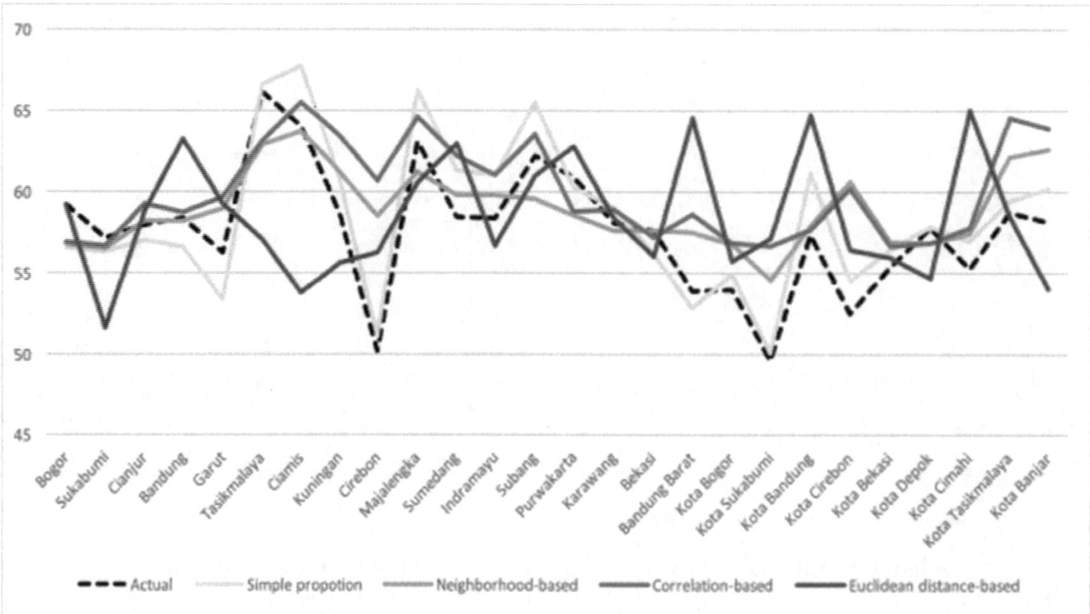

Figure 2. Province to city disaggregation results and the actual data for West Java

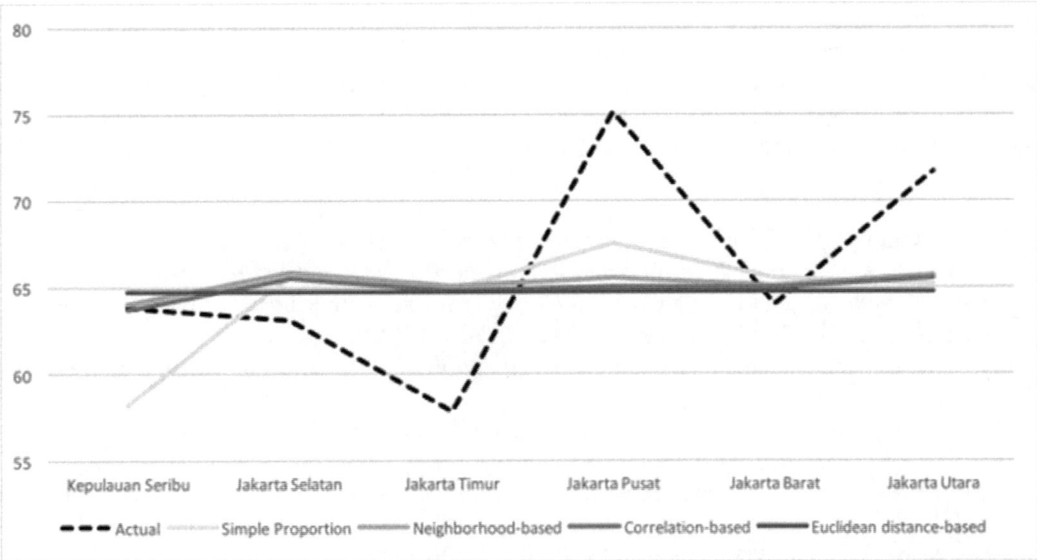

Figure 3. Province to city disaggregation results and the actual data for DKI Jakarta

Province to city disaggregation are also done for DKI Jakarta and West Java using the same methods, showed in Figure 2 and Figure 3. Although there are some underestimations (e.g. Ciamis, Tasikmalaya) and a lot of overestimations, the models for West Java disaggregation can closely estimate several cities, for instance Majalengka, Sumedang, Indramayu, Subang, Purwakarta and Karawang. Based on the evaluation criterion in Table 2, simple proportion with the lowest MAE, MAPE and MSE (1.739, 3 and 4.174 respectively) is the better method to disaggregate province level data into city level data.

Table 2. Province to city model evaluation

Province	Method	MAE	MAPE	MSE
West Java	Simple proportion	1.739	3	4.174
	Numerical Method Approach:			
	Neighborhood-based	2.445	4.4	10.591
	Euclidean distance-based	4.061	7.1	26.938
	Correlation-based	3.062	5.5	15.834
DKI Jakarta	Simple proportion	5.137	7.8	31.606
	Numerical Method Approach:			
	Neighborhood-based	4.490	6.7	31.592
	Euclidean distance-based	4.604	6.8	34.711
	Correlation-based	4.468	6.6	32.653

However, the models for DKI Jakarta are not well estimating the employment to population ratio of its cities and the value for all cities are almost the same towards one number. One crucial aspects that affecting this result is that the weight spatial matrix developed does not suit DKI Jakarta. It can be that the characteristic of five cities are very similar as well as the cities share almost the same borderlines and almost all cities become the neighbor of others. It is obvious that the best model is the simple proportion one, since the weight matrix does not work well for DKI Jakarta.

6. Conclusion

Spatial weighting using numerical approach can be potentially used as to estimate development indicators at lower spatial level. Further improvement needed in order to get the most suitable spatial weight matrix, since it is indeed the most crucial part in numerical method disaggregation. Another important thing is finding the non-MDGs official statistics that highly correlated or have similar pattern with the respective MDGs indicator to construct initial proportion weight. This paper contributes well in proposing the methodologies of data disaggregation to monitor the achievement of development indicators at local level, and therefore, to make sure that no one left behind.

References

[1] United Nations 2015 *The millennium development goals report*
[2] Espey J and Karoubi E M 2015 *A Global Initiative For The United Nations* **26**
[3] Pratesi M, Petrucci A and Salvati N 2015 *Global Strategy* Technical Report Series GO-07-2015
[4] Global Strategy 2015 *Technical Report Series GO-07-2015*
[5] Flowerdew R and Green M 1994 *Areal Interpolation and Types of Data* (London: Taylor and Francis)
[6] Kubicek M, Janovska D and Dubcova M 2005 *Numerical Methods and Algorithm* (Praha: Vysokáškolachemicko-technologická v Praze)
[7] McDonough J M 2007 *Lectures in Basic Computational Numerical Analysis* (USA : University of Kentucky)
[8] Getis A and Aldstadt J 2003 *Geographical Analysis* **36** No 2 p 90 – 104
[9] Drukker D M, Peng H, Prucha I R and Raciborski R 2013 *The Stata Journal* **13** No 2 p 242 – 286
[10] Tang B and He H 2015 *IEEE Computational Intelligence Magazine* 1556-603x/15©2015IEEE p 52 – 60
[11] Krislock N and Wolkowicz H 2011 *Handbook of Semidefinite, Cone and Polynomial Optimization: Theory, Algorithms, Software and Applications*
[12] Amerise I L and Tarsitano A 2012 *Working Paper* vol 09 (Italy: Dipartimento di Economia e Statistica, Universita Della Calabria)

IOP Conf. Series: Materials Science and Engineering **332** (2018) 012050 doi:10.1088/1757-899X/332/1/012050

Economic impact due to Cimanuk river flood disaster in Garut district using Cobb-Douglas analysis with least square method

T A S Bestari[1]*, S Supian[2], S Purwani[2]

[1]Master Student of Mathematics Study Program, Faculty of Mathematics and Natural Sciences, Universitas Padjadjaran, Indonesia
[2]Departement of Mathematics, Faculty of Mathematics and Natural Sciences, Universitas Padjadjaran, Indonesia.

*Email: tegarbestari@gmail.com

Abstract. Cimanuk River, Garut District, West Java which have upper course in Papandayan Mountain have an important purpose in dialy living of Garut people as a water source. But in 2016 flash flood in this river was hitted and there was 26 peple dead and 23 peole gone. Flash flood which hitted last year make the settlement almost align with the ground, soaking school and hospital. BPLHD Jawa Barat saw this condition as a disaster which coused by distroyed upper course of Cimanuk River. Flash Flood which happened on the 2016 had ever made economic sector paralized.
Least square method selected to analyze economic condition in residents affected post disaster, after the mathematical equations was determined by Cobb Douglas Method. By searching proportion value of the damage, and the result expected became a view to the stakeholder to know which sector that become a worse and be able to make a priority in development

Keyword : Cobb Douglas, Least Square Method, Proportion value

1. Introduction

Since long time ago, people have settled near the river and today this condition still continued. River becomes one choice for transportation, provides a water for industry and agriculture also be able to increase the quality of life because living in the in the middle of a beautiful nature. But, living near a river have a risk of flood, one of the most dangerous disaster on earth [1]. Cimanuk River location in Garut Resident, West Java, Indonesia which the upstream in Papandayan Mountain have an important role in daily life of Garut citizen such as a source of water. In 2016, was happened a big flash flood in this river and as a consequence there was 26 deaths dan 23 gone also the the economic activity was paralyzed. BPLH (Environmental Management Agency) Jawa Barat saw this condition caused by the damaged land in upstream of Cimanuk River. The flash flood in Cimanuk River in 2016 was the most damage flood ever in Garut. Data from BPLHD (Disaster Management Agency) Garut Resident showed that economic lost in that disaster touch 671 billion rupiahs [2]. In this case, the civilization activity still

IOP Conf. Series: Materials Science and Engineering **332** (2018) 012050 doi:10.1088/1757-899X/332/1/012050

near river like hospital, school and housing that means just only upstream and river revitalization can make this disaster stopped. In the other side, upstream revitalization need a time so this this paper aimed to model the economic lost if the condition still the same.

Some research related to this paper are Develop a dynamic optimization model, where inter-temporal decision of an economic agent are interacts with the hydrological system [3]. Input-Output Tables used to study the analysis of the economic impact of a Natural Disaster Tokai Flood in 2000 in Japan [4]. They proposed analysis methodology to estimate returns to scale in case of Cobb-Douglas production function [5]. Partial least square used to identify the information about the correlation of multivariate cognitive abilities and local brain structure in children and adolescents [6]

This paper showed the mathematical model which formed by Cobb-Douglas Method and the index and constant gotten by Partial Least Square Method. The main data was generated by normal distribution. This paper aims to derive a mathematical model of Cimanuk river flash flood, that can be used to predict the total of economic lost

2. Preliminaries

This section describe the main theoretical tools that used to formed mathematical model for flash flood in Cimanuk River.

2.1. Cobb-Douglas Method.

Cobb-Douglas production function is a function or a model that involves a dependent variable and two or more independent variables. The model can be written as follow:

$$Y=aX_1^bX_2^c \tag{1}$$

Where:
Y is an output
a is an efficiency index
X_1,X_2 is a kind of input
b,c is a production elasticity from the used input

In order for data to be obtained can be analysed by Cobb-Douglas Method, so the data must be transformed into a linier form by using natural Logarithm (Ln). The model showed:

$$\ln Y = \ln a+ b \ln X_1 + c \ln X_2 . \tag{2}$$

and transform back to (1) [7]

2.2. Partial Least Square Method

Partial Least Square (PLS) Method designed to be an alternative from Structural Equation Modelling (SEM) which used to solved the condition between variable really complex but the sample size is small [8]. This method was introduced by Herman O.A Wold to forming a prediction model. (Talbot) PLS used to predict the effect of X variable to Y variable and explained the theoretical correlation between the variable. PLS is a regression method that can be used to identify the factor which combination X variable as an explanation and Y variable as a response.

PLS regression is a method to find a component from X variable which have correlation with Y variable. The concept of PLS regression is describing Y variable as a response and X variable as a predictor by equation:

$$y=Tq^T+f \tag{3}$$

$$x=TP^T+E \tag{4}$$

Where :
y is a vector variable response
x is a predictor variable matrix
T is a PLS component score
q is a vector loading from y

P is a factor loading matrix from x

Vector Q obtained from the regression the indicator variable with the PLS component score using Ordinary Least Square (OLS) [9]

2.3 Significance Test

2.3.1 Significance Test (F statistics test)

The F test aims to determine the significance of the correlation of independent variable to dependent variable. The F test is done by comparing the value of F-test and F-table with a level of significant 95%. The way to test the hypothesis are:

1. Determine the hypothesis
   H_0 = the lost of housing, lost of infrastructure, lost of social criteria, lost of economic source and lost of cross sector simultaneous do not have a correlation into the economic impact.
   H_1 = the lost of housing, lost of infrastructure, lost of social criteria, lost of economic source and lost of cross sector simultaneous have a correlation into the economic impact.
2. Determine the level of significant (α)
   In this case, the level of significant is 5% with the level of confidence 95%
3. Determine the Testing Criteria
   H_0 is not accepted if F-test > F-table
   H_1 is accepted if F-test $\leq$ F-table
4. Conclusion
   If H_0 not accepted so the lost of housing, lost of infrastructure, lost of social criteria, lost of economic source and lost of cross sector simultaneous have a correlation into the economic impact.
   If H_1 accepted so the lost of housing, lost of infrastructure, lost of social criteria, lost of economic source and lost of cross sector simultaneous have a correlation into the economic impact.

2.3.2 Wald Test (t test)

Test t aims to determine the differences of each other data by means. The t test also showing how significant the difference are. The t test is done by comparing the value of t-test and t-table with a level of significant 95%. The way to test the hypostasis are:

1. Determine the hypothesis
   a) $H_{01}:\beta_1=0$, lost of housing as a partial have no correlation to economic impact.
      $H_{11}:\beta_1{}^1=0$, lost of housing as a partial have correlation to economic impact.
   b) $H_{02}:\beta_2=0$, lost of infrastructure as a partial have no correlation to economic impact.
      $H_{12}:\beta_2{}^1=0$, lost of infrastructure as a partial have correlation to economic impact.
   c) $H_{03}:\beta_1=0$, lost of social criteria as a partial have no correlation to economic impact.
      $H_{13}:\beta_3{}^1=0$, lost of social criteria as a partial have correlation to economic impact.
   d) $H_{04}:\beta_4=0$, lost of economic as a partial have no correlation to economic impact.
      $H_{14}:\beta_4{}^1=0$, lost of economic as a partial have correlation to economic impact.
   e) $H_{05}:\beta_5=0$, lost of cross sector as a partial have no correlation to economic impact.
      $H_{15}:\beta_5{}^1=0$, lost of cross sector as a partial have no correlation to economic impact.
2. Determine the level of significant (α)
   In this case, the level of significant is 5% with the level of confidence 95%.
3. Determine the testing criteria
   H_{01},H_{02},H_{03},H_{04},H_{05} are not accepted if : t-test > t -table or -t-test < -t-table.

IOP Conf. Series: Materials Science and Engineering **332** (2018) 012050 doi:10.1088/1757-899X/332/1/012050

$H_{01}, H_{02}, H_{03}, H_{04}, H_{05}$ are accepted if : -t-table $\leq$ t-test $\leq$ t-table

4. Conclusion
   a) If H_{01} not accepted so the lost of housing as partial not have correlation to economic impact.
      If H_{01} accepted so the lost of housing have correlation to economic impact.
   b) If H_{02} not accepted so the lost of infrastructure not have correlation to economic impact.
      If H_{02} accepted so the lost of infrastructure have correlation to economic impact.
   c) If H_{03} not accepted so the social criteria not have correlation to economic impact.
      If H_{03} accepted so the social criteria have correlation to economic impact.
   d) If H_{04} not accepted so the lost of economic not have correlation to economic impact.
      If H_{04} accepted so the lost of economic have correlation to economic impact.
   e) If H_{05} not accepted so cross sector not have correlation to economic impact.
      If H_{05} accepted so cross sector have correlation to economic impact.

2.3.3 Normality assumption test

Normality test aims to measure the data having the normal distribution so be able to use on parametric statistic. In this case, Kolmogorov-Smirnov test used to determine the normal distribution from data. If the probability > 0.05 so the data is normal distribution and if probability $\leq$ 0.05 the data is not normal distribution instead.

3. Results and Discussion

To know the economic impact, the data using the total economic loss (Y) which consist in 5 variable there are 1. The loss of housing 39.318% (X_1), 2. Loss of infrastructure 15,477% (X_2), 3. Social 4,779% (X_3), 4. Economic source 7,021% (X_4), 5. Cross sector such as banking and living environment 33,405% (X_5). Those data will generated into 50 data using Normal Distribution with 1% standard deviation from original data.

3.1 Significance Test (F statistics test)

After the data convert to logarithm natural and processed with the F test, the result is

Table 1 ANOVA table which show the F-test value

	df	*SS*	*MS*	*F*	*Significance F*
Regression	5	0,000836	0,000167	2,767859	0,029363274
Residual	44	0,002659	6,04E-05		
Total	49	0,003495			

Based on the table of ANOVA showed that F-test 2.767859 > F-table 2.42 and the value of significance > 0.05 (α), so the conclusion is H_0 not accepted that means the lost of housing, lost of infrastructure, lost of social criteria, lost of economic source and lost of cross sector simultaneous have a correlation into the economic impact.

3.2 Wald Test (t test)

After the data convert to logarithm natural and processed with the t test, the result is

Table 2 Table which show the value of t-test

	Coefficients	*Standard Error*	*t Stat*
Intercept	26,96087815	5,463150823	4,93504189
Lost of Housing	-0,02969591	0,098459932	-0,301604
Lost of Infrastructure	-0,230823501	0,124361944	-1,85606219
Lost in Social	-0,15389522	0,113426691	-1,35678136
Lost of Economic	0,244718139	0,110096817	2,22275398
Lost in Cross Sector	0,176663318	0,107127912	1,64908766

a) First hypothesis
The value of t-test -0.30160 > t-table -1.68 so other lost of housing have correlation to economic impact
b) Second hypothesis
The value of t-test -1,85606219 < t-table -1.68 so the lost of infrastructure not have correlation to economic impact
c) Third hypothesis
The value of t-test -1,35678136 > t-table -1.68 so the social criteria have correlation to economic impact.
d) Fourth hypothesis
The value of t-test 2,22275398 > t-table 1.68 so the lost of economic have correlation to economic impact
e) Fifth hypothesis
The value of t-test 1,64908766 < t-table 1.68 so cross sector have no correlation to economic impact.

3.3 Normality assumption test
After the data convert to logarithm natural and processed with the normality assumption test, the result is

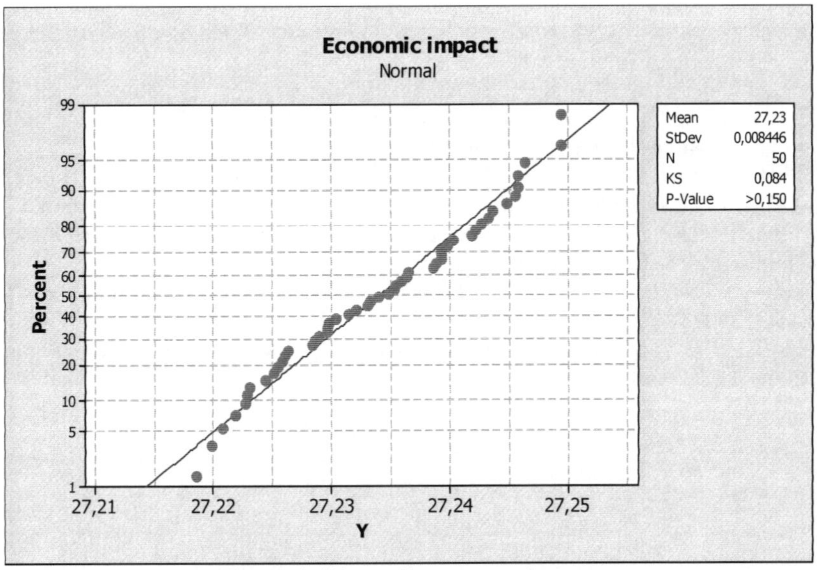

Figure 1. Normality assumption test

The value of KS 0.084 and the P-Value > 0.150 which mean bigger than 0.05 (α) so the variable determined as a normal distribution.

3.4 Partial Least Square Method

The 50 data changed into the Ln data to follow Cobb-Douglas function. After all data processed by PLS the result is

Constant = 26.9609, X_1 = -0.0297, X_2 = -0.2308, X_3 =-0.1539, X_4 = 0.2447, X_5 = 0.1767

3.5 Cobb-Douglas Method

so the regression result followed Cobb-Douglas Function (2)

$$\ln Y = \ln 26.9609 - 0.0297 \ln X_1 - 0.2308 \ln X_2 - 0.1539 \ln X_3 + 0.2447 \ln X_4 + 0.1767 \ln X_5$$

and can be written as (1)

$$Y=511.647.000.000\ X_1^{-0.0297} X_2^{-0.2308} X_3^{-0.1539} X_4^{0.2447} X_5^{0.1767}$$

4. Conclusion

The model solution in the economic impact due to Cimanuk river flood impact was created by Cobb-Douglas Method after the constant got by Partial Least Square Method. From the result of the discussions, it can be conclude that The Cobb-Douglas Method be able to forming the mathematical model of the flood depend on the data of the economic lost. And the mathematical model can be used by government to make the regulation to anticipate the flood.

References

[1] Ohl, C., Tapsell, S., 2000. Flooding and human health: the dangers posed are not always obvious. Br. Med. J. 321 (7270). 1167e1168

[2] BPBD reports. 2016. Kegiatan Penanggulangan Bencana Banjir Bandang Sungai Cimanuk Kab. Garut. Penerbit BPBD Kabupaten Garut

[3] Grames Johanna, et al. 2016. Modeling the interaction between flooding events and economic growth. Ecological Economics 129 (2016)193-209

[4] Ishikawa Yoshifumi, Katada Toshitaka. 2006. Analysis of the Economic Impacts of a Natural Disaster Using Interregional Input-Output Tables for the Affected Region: A Case Study of the Tokai Flood of 2000 in Japan. Intermediate Input-Output meeting

[5] Pavelescu Florin Marius. 2014. Methodological consideration regarding the estimated returns to scale in case of Cobb-Douglas production function. Elsevier. Procedia Economic and Finance 8 (2014) 535-542

[6] ziegler G, et al. Partial least square correlation of multivariate cognitive abilities and local brain structure in children and adolescents. Elsevier. Neurolmage volume 82 pages 284-294

[7] Soekartawi. 1990. Teori Ekonomi Produksi Dengan Pokok Bahasan Analisis fungsi Cobb-Douglas. Jakarta : Radar Jaya Offset.

[8] Hair, Black, Babin, & Anderson. 2010. Multivariate Data Analysis 7th. New Jersey: Prentice Hall.

[9] Abdi, H. 2003. Partial Least Squares (PLS) Regression. The University of Texas at Dallas.

IOP Conf. Series: Materials Science and Engineering **332** (2018) 012051 doi:10.1088/1757-899X/332/1/012051

Numerical solution of a logistic growth model for a population with Allee effect considering fuzzy initial values and fuzzy parameters

Z Amarti[1*], N S Nurkholipah[1], N Anggriani[2], A K Supriatna[2]

[1]Master Student of Mathematics Study Program, Faculty of Mathematics and Natural Sciences, Universitas Padjadjaran, INDONESIA
[2]Department of Mathematics, Faculty of Mathematics and Natural Sciences, Universitas Padjadjaran, INDONESIA

*E-mail: zeniamarti@gmail.com

Abstract. Predicting the future of population number is among the important factors that affect the consideration in preparing a good management for the population. This has been done by various known method, one among them is by developing a mathematical model describing the growth of the population. The model usually takes form in a differential equation or a system of differential equations, depending on the complexity of the underlying properties of the population. The most widely used growth models currently are those having a sigmoid solution of time series, including the Verhulst logistic equation and the Gompertz equation. In this paper we consider the Allee effect of the Verhulst's logistic population model. The Allee effect is a phenomenon in biology showing a high correlation between population size or density and the mean individual fitness of the population. The method used to derive the solution is the Runge-Kutta numerical scheme, since it is in general regarded as one among the good numerical scheme which is relatively easy to implement. Further exploration is done via the fuzzy theoretical approach to accommodate the impreciseness of the initial values and parameters in the model.

1. Introduction

Analysis a model that describes the behavior of the population dynamics is important to understanding and predicting the growth of a population. In this paper, we discuss the population model in the form of differential equations that include the possibility of growth and extinction. In this case population growth influenced by the Allee effect. The Allee effect was first described by Warder Clyde Allee in 1930. Allee observed that extinctions may occur to populations of low population densities. Low population density makes it difficult for individuals to interact, which increases the likelihood of their extinction [1].

Moreover, many researchers had assumed that the parameters in their models are constant, in the sense that it is crisp, hence the model is crisp deterministic. But in fact, in real life the parameters are not always certain, sometime they are uncertain and imprecise. This is because the dynamics of the population can be affected by environmental changes such as food availability, birth and death rates, and so on. This situation may cause the parameter of the population model to vary from time, and difficult to estimate. In this case an imprecise estimation is likely to occur. This may cause inaccuracies in making mathematical models, even in classical deterministic approaches. This can be due to the nature of the state variable involved, such as the impreciseness in the model parameters and in the initial conditions. Some authors have used the fuzzy set theory in their papers in exploring the impreciseness, which among them is in the case of population dynamics [2] [3] and epidemiology problems [4-7]. The fuzzy set theory, introduced by Zadeh in 1965, intended to illustrate some parts of the real world in the mathematical concepts of subjective topics in many cases. It becomes an important tool for a better understanding of some real situations, especially in the dynamics of population [2].

The model given in this paper is the Verhulst model with the Allee effect. We use fuzzy set theory to accommodate the impreciseness in parameters and the initial value. Model parameters and the initial value are assumed to be fuzzy numbers and represented by the triangular membership function. Model parameters and the initial values are given hypothetically and are not based on real data, just to illustrate the idea of the model. The fuzziness of the initial population size and model parameters propagates to the output in the numerical method used to obtain the solution to the model. In this case we use the fourth order Runge-Kutta method.

2. Preliminaries

In this section, we briefly describe the main theoretical tools that will be used in the analysis of the proposing mathematical model for a population with Allee effect considering fuzzy initial values and fuzzy parameters.

2.1. Population growth model with Allee effect

Generally, a population has a slow growth in the beginning, but a rapid acceleration begins at a certain time and finally a phase of stationary growth is achieved, such as in the Verhulst's logistic model. Later, Warder Clyde Allee observed a fact that in a small population such growth may exist but there may be also the extinction of the species. This observation related to the initial density of the population. Allee concluded that there is a critical point in which if the initial density of the population is below this critical point then the population vanishes and if it is above this critical point then the population has a positive growth phase. In other words, a population with the Allee effect may have a threshold population size, below which the population goes extinct deterministically. This threshold is often called the Allee threshold [8]. The graph of the solution pattern of the population growth with the Allee effect is illustrated with figure 1.

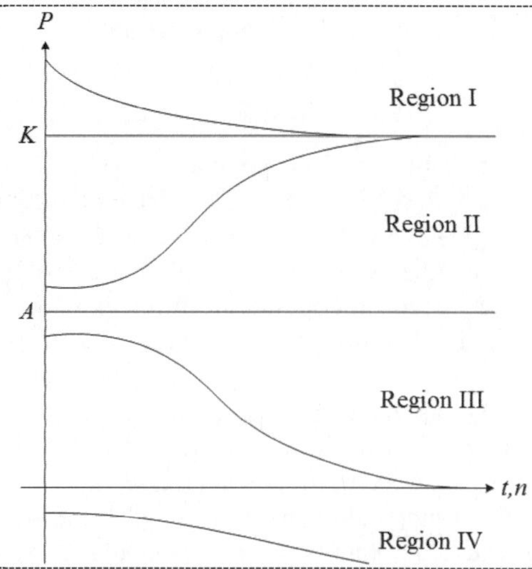

Figure 1. Illustration of the solution pattern of the population growth model with Allee effect, which is divided into four regions.

Figure 1 illustrates the solution pattern of the population growth model with Allee effect, which is divided into four regions. Let *P* denotes the number of population densities, *K* denotes the carrying capacity, and *A* denotes the Allee threshold. Among the simplest equations of the population growth model with Allee effect is formalized by generalizing the known Verhultst's logistic differential equation, given in the form:

$$\frac{dP}{dt} = rP\left(1-\frac{P}{K}\right)\left(\frac{P}{A}-1\right) \tag{2.1}$$

where *t* represent time and the constant *r* defines the growth rate. In this model, the per capita growth rate $\left(\frac{dP}{dt}\right)$ is negative if the population size is above the carrying capacity (*K*) and positive otherwise. However, in the presence of the Allee effect, it also decreases below a given population size, and even might be negative below the critical population threshold (*A*) which eventually might lead to extinction [1].

From the model, there are three equilibrium solution to the population growth model with the Allee effect. First, is the equilibrium level where the population is zero. The second equilibrium corresponding to the maximum sustainable population *K* and the third one corresponding to the minimum sustainable population *A*, as illustrated in figure 1. These three equilibrium solution divide the region into four distinct subregions. In Region I, where $P > K$, we should expect that the solutions decay toward *K*, as we have seen in the original logistic model. In Region II, where $A < P < K$, we should expect that the solutions rise toward *K* eventually, in an asymptotic manner, much as they behave with the logistic model. In Region III, where $0 < P < A$, we should expect that the solutions decay toward zero. Finally, in Region IV, where $P < 0$, we should expect that the solutions decay toward $-\infty$. In this case, region IV does not have a biological interpretation. From figure 1, the upper one at the height of *K* is a *stable equilibrium* since any solution that starts near that level converges to that equilibrium; the upper one at the height *A* is an *unstable equilibrium* because any solution that starts near it diverges from that level; the equilibrium at $P = 0$ is a *semi-stable equilibrium* because a solution that starts near it is converges and some are diverges from that level. This stability behavior of the equilibrium points can be analyzed easily through the standard stability analysis via the investigation of the eigen values of the jacobian of the respective equilibrium [9].

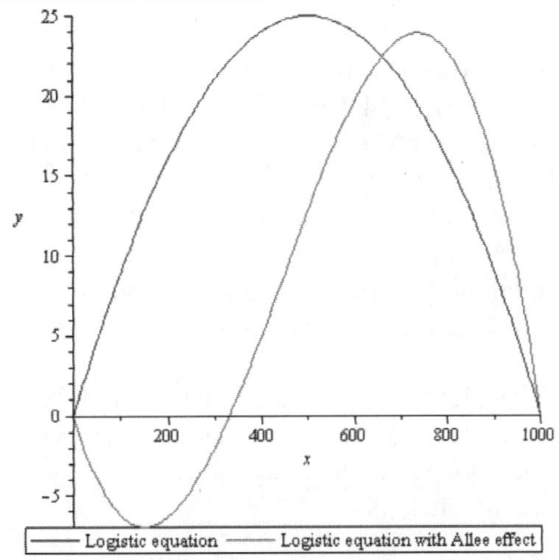

Figure 2. Illustration of logistic equation and logistic equation with Allee effect.

The per capita growth of the original logistic equation is always positive while the logistic equation with Allee effect is negative if the population is below the Allee threshold (330 in figure 2). The Allee effect has strong evidence for ecological systems and sometime related to phenomena like conspecific interactions, rarity and animal sociality. In many situations, environmental conditions which influence the growth of population is subject to imprecise factors. To obtain a more appropriate strategy in managing a population, this impreciseness should be incorporated in the model. One way to do this is by modeling the parameters in the model in a fuzzy form. The aim at the present paper is to investigate the effect of fuzzy initial value and fuzzy parameters in the solution of the Verhulst logistic population model. The solution that we will consider is the numerical solution.

2.2. *The fourth-order Runge-Kutta method*

The classical Runge-Kutta method of order four for solving differential equation $\frac{dy}{dt} = y' = f(y)$ with $y(t_0) = y_0$ is as follows [10]:

$$y_{n+1} = y_n + \frac{1}{6}h\left[k_1 + 2k_2 + 2k_3 + k_4\right] \tag{2.2}$$

with

$$k_1 = f\left(y_n\right),$$
$$k_2 = f\left(y_n + k_1\frac{h}{2}\right),$$
$$k_3 = f\left(y_n + k_2\frac{h}{2}\right),$$
$$k_4 = f\left(y_n + k_3 h\right).$$

All k values are recursively related. It means k_1 appears in the equation for k_2, which appears again in equation k_3 and so on, with y_0 is the initial value. This method can be easily to apply to equation (2.1) with $f(P) = rP\left(1 - \frac{P}{K}\right)\left(\frac{P}{A} - 1\right)$ and $P(t_0) = P_0$.

2.3. *Fuzzy theory*

The basic idea of fuzzy set theory is that real-world phenomena cannot be clearly divided into black and white divisions. Fuzzy set theory provides a mathematical framework that represents impreciseness, unclear and inaccuracies of the abundance of information available. The fuzzy set theory is an extension of the classical set theory. In a classical set theory, the set of membership functions has only two possible memberships, i.e. $B = \{0,1\}$. In the fuzzy set theory, the limits set have ambiguity and are not clearly defined. The fuzzy set declares elements to be members of multiple sets with different degrees of membership (μ). The set of membership functions is expanded from the set $B = \{0,1\}$ which contains only two alternatives, at unit interval $U = [0,1]$ which has a number of alternatives. The degree of membership of each element in the fuzzy set can be obtained by the membership function approach [11].

2.3.1. Membership function. There are several functional approaches that can be used to define membership value. The most commonly used are the triangular membership function, the trapezoidal membership function and the bell shape membership function. The following shows the various of membership function of the fuzzy numbers [11]:

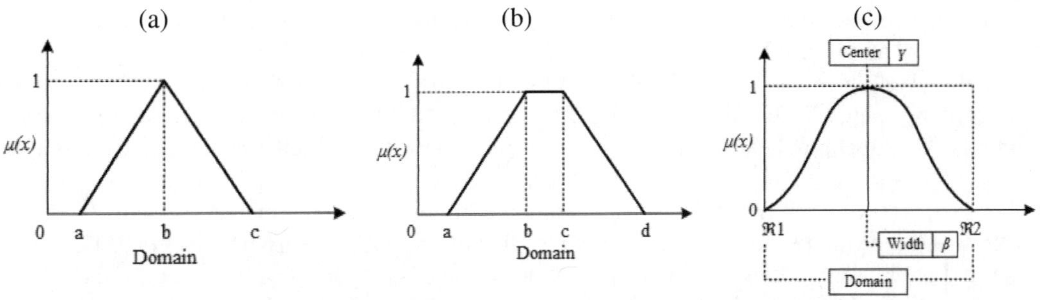

Figure 3. Graph of membership function of: (a) representation of the triangular curve; (b) representation of the trapezoidal curve; and (c) representation of the bell shape curve.

The equations for the fuzzy membership numbers presented in the graph of figure 3 are given as follows:

- Representation of the triangular curve

$$\mu(x)=\begin{cases}0 & ;x\le a\\ \frac{x-a}{b-a} & ;a<x\le b\\ \frac{c-x}{c-b} & ;b<x\le c\\ 0 & ;x\ge c\end{cases}, \tag{2.3}$$

- Representation of the trapezoidal curve

$$\mu(x)=\begin{cases}0 & ;x\le a\\ \frac{x-a}{b-a} & ;a<x\le b\\ 1 & ;b<x\le c,\\ \frac{d-x}{d-c} & ;c<x\le d\\ 0 & ;x\ge d\end{cases} \tag{2.4}$$

- Representation of the bell shape curve

$$\mu(x)=\begin{cases}e^{-\frac{(x-\gamma)^2}{2\beta^2}} & ;\Re 1\le x\le \Re 2.\\ 0 & ;\text{otherwise}\end{cases} \tag{2.5}$$

3. Results and discussion

The population growth model with the Allee effect used in this paper is given in equation (2.1). Let P be the initial value of an amount greater than the Allee threshold (A) and Z is the initial value for a smaller amount than the Allee threshold (A). We explore the solution to equation (2.1) for various initial values and parameters. The solution to the model is obtained by using the fourth-order Runge-Kutta method. Assumed that $P(0)$ is around 3040; $Z(0)$ is around 2960; A is around 3000; r is around 0.0407; and K is around 8000. To model impreciseness, such as "around" some value number, an initial value and parameters is fuzzified into a fuzzy form with triangular membership function. For the case of triangular fuzzy number, let $\bar{P}(0) = (P(0)_1/P(0)_2/P(0)_3)$ is above A and $\bar{Z}(0) = (Z(0)_1/Z(0)_2/Z(0)_3)$ is below A. Note that we follow the notation in [12] to write the fuzzy numbers. To solve equation (2.1) we do numerically by using Maple program. Technically, in triangular fuzzy number, since we assume that the fuzziness of the input propagates to the output, then we assume that initially there are three initial values as the input to the Runge-Kutta scheme, i.e. $P(0)_1$, $P(0)_2$, and $P(0)_3$ correspond to the triangular fuzzy number $\bar{P}(0) = (P(0)_1/P(0)_2/P(0)_3)$. The results of the triangular fuzzy number are presented in figures 4 to 7. We stack the graph for consecutive times t, so that the height of the triangular is one. The length of the curve (blue line) for a fixed t in figure 4 to 7 part b and c is the radius or the length of the domain of the fuzzy number at time t, while the red line is the membership function for those fuzzy numbers.

Figure 4(a) shows the two cases simultaneous graph of solution to $\bar{P}(0)$ and $\bar{Z}(0)$. If it is given a greater initial value than A, i.e. $P_1(0) = 3020$; $P_2(0) = 3040$; and $P_3(0) = 3060$, then the graph of the solution increases and asymptotically towards carrying capacity. It means, if the initial population is greater than the Allee threshold, the population number will grow approaching the carrying capacity. It is eventually stable, no increase or decrease in population numbers. On the other hand, if it is given a smaller initial value than A, i.e. $Z_1(0) = 2940$; $Z_2(0) = 2960$; and $Z_3(0) = 2980$, then the graph of solution decreases and asymptotically toward the zero point. It means, if the initial population is smaller than the Allee threshold, the population number will decline to the point where there is no population and the population is potentially extinct. Figure 4(b) shows the graph of triangular membership function of the model solution to 100 iterations. The right graph shows the model solution to the triangular membership function for an initial value greater than A, while the left graph shows the solution to the triangular membership function of a smaller initial value than A. From this figure, it appears that the domain in the triangular membership function is getting larger until the 100th iteration. It means, the possible solution to this model is wider until the 100th iterations. Figure 4(c) shows the graph of triangular membership function of the model solution to 3000 iterations. From this figure, it appears that the domain in the triangular membership function is getting larger until certain iteration, but eventually it closes to a crisp membership function. It shows that the initial values greatly affect the solution obtained from the model with the Allee effect. The figure shows a highly significant difference growth pattern between the populations, depending on their initial values.

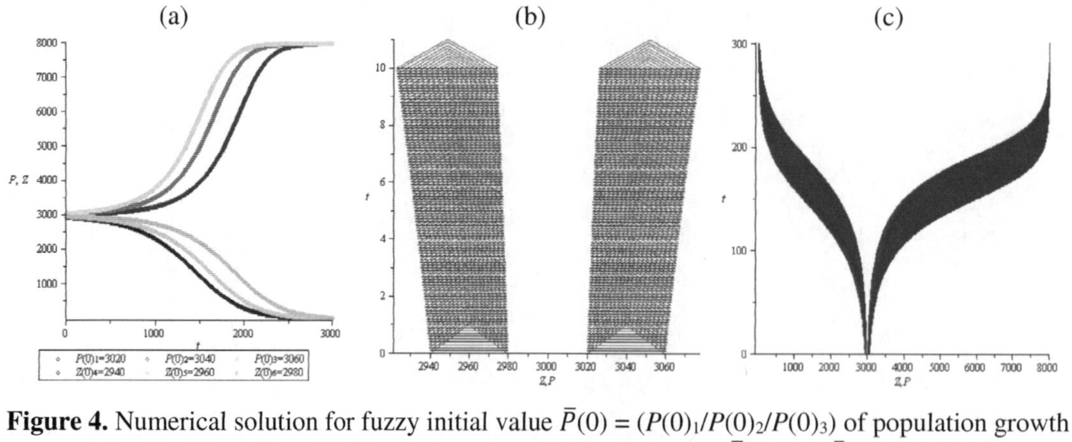

Figure 4. Numerical solution for fuzzy initial value $\bar{P}(0) = (P(0)_1/P(0)_2/P(0)_3)$ of population growth model with Allee effect. (a) Simultaneous graph of solution for $\bar{P}(0)$ and $\bar{Z}(0)$; (b) Graph of triangular membership function for the model solution with $T = 100$; and (c) Graph of triangular membership function for the model solution with $T = 3000$.

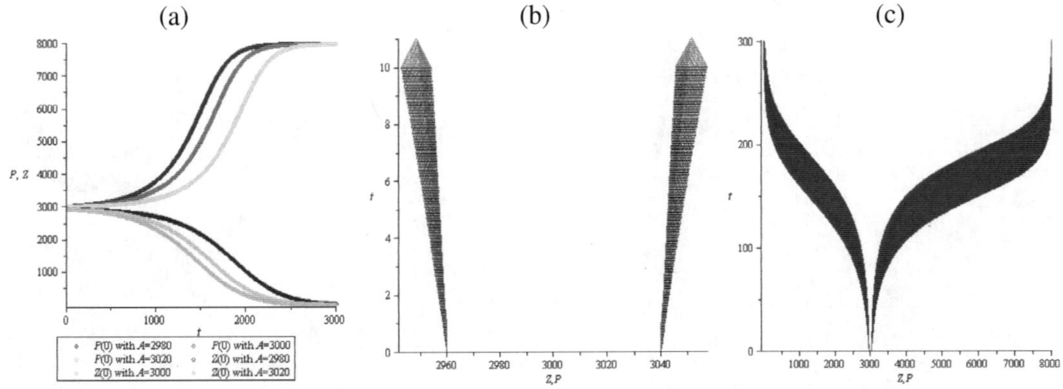

Figure 5. Numerical solution to fuzzy parameter $\bar{A} = (A_1/A_2/A_3)$ of population growth model with Allee effect. (a) Simultaneous graph of solution to $P(0)$ and $Z(0)$; (b) Graph of triangular membership function for the model solution to $T = 100$; and (c) Graph of triangular membership function for the model solution to $T = 3000$.

Figure 5(a) shows the two cases simultaneous graph of solution for $P(0)$ and $Z(0)$ with triangular fuzzy number for the Allee thresholds ($\bar{A}$), i.e. A_1= 2980, A_2 = 3000 and A_3 = 3020. Figure 5(b) shows the graph of triangular membership function of the model solution to 100 iterations. The right graph shows the model solution to the triangular membership function for an initial value greater than A, while the left graph shows the solution to the triangular membership function of a smaller initial value than A. Figure 5(c) shows a graph of triangular membership function of the model solution to 3000 iterations. From this figure, it appears that the solution to the model with the triangular membership function obtained is has the same pattern as figure 4. It means, if the Allee threshold is represented in a fuzzy number, then the probability of a change in the number of populations still depends on the number of the initial population. These behaviour is qualitatively the same as figure 4 in the case of triangular fuzzy number for initial population numbers.

The similar pattern as figure 4 and 5 are also shown in figure 6 and 7. Figure 6 shows the two cases simultaneous graph of solution to $P(0)$ and $Z(0)$ with triangular fuzzy number for the carrying capacity ($\bar{K}$), i.e. K_1= 7500, K_2 = 8000 and K_3 = 8500. While figure 7 shows the two cases simultaneous graph of solution to $P(0)$ and $Z(0)$ with triangular fuzzy number for the growth rate ($\bar{r}$), i.e. r_1= 0.0357, r_2 = 0.0407 and r_3 = 0.0457.

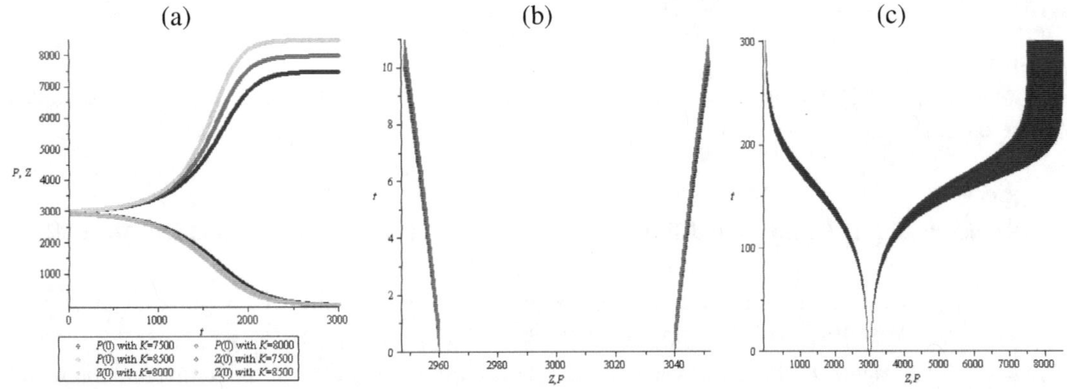

Figure 6. Numerical solution to fuzzy parameter $\bar{K} = (K_1/K_2/K_3)$ of population growth model with Allee effect. (a) Simultaneous graph of solution to $P(0)$ and $Z(0)$; (b) Graph of triangular membership function for the model solution to $T = 100$; and (c) Graph of triangular membership function for the model solution to $T = 3000$.

IOP Conf. Series: Materials Science and Engineering **332** (2018) 012051 doi:10.1088/1757-899X/332/1/012051

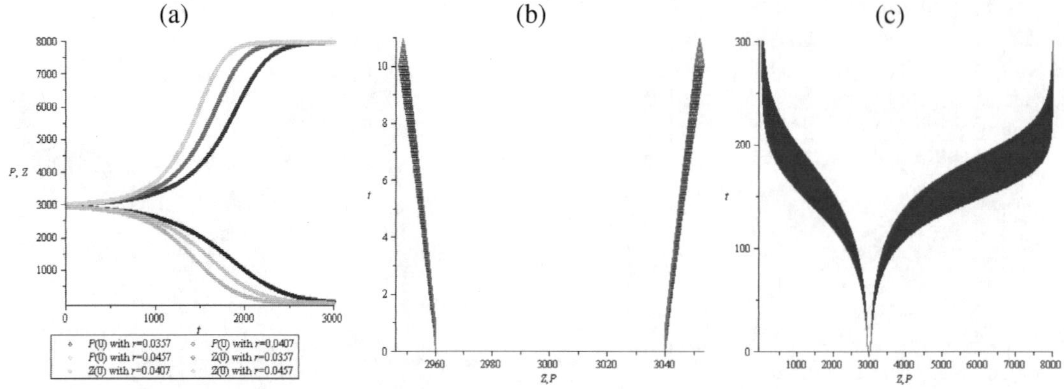

Figure 7. Numerical solution of fuzzy parameter $\bar{r} = (r_1/r_2/r_3)$ of population growth model with Allee effect. (a) Simultaneous graph of solution to $P(0)$ and $Z(0)$; (b) Graph of triangular membership function for the model solution to $T = 100$; and (c) Graph of triangular membership function for the model solution to $T = 3000$.

4. Conclusion

Population growth model used in this paper is the Verhulst logistic model with Allee effect. Model parameters and initial population number are determined hypothetically and not based on real data. The initial value of the population number is divided into 2 parts, first is for the initial value greater than the Allee threshold ($P(0) > A$) and second is the initial value smaller than the Allee threshold ($\bar{Z}(0) < A$). To accommodate the impreciseness in the model, fuzzy set theory approach is used. In this case, the initial population number and model parameters are assumed to be a triangular fuzzy number. $\bar{P}(0) = (P(0)_1/P(0)_2/P(0)_3)$ is triangular fuzzy number for the initial population number, $\bar{A} = (A_1/A_2/A_3)$ is triangular fuzzy number for the Allee threshold, $\bar{K} = (K_1/K_2/K_3)$ is triangular fuzzy number for the carrying capacity, and $\bar{r} = (r_1/r_2/r_3)$ is triangular fuzzy number for the growth rate. Fuzzy number of initial values and model parameters give the same qualitative behaviour of the solution.

Generally, the initial value greatly affects the solution to the population model with Allee effect. Modeling imprecise initial value around the critical value A with a crisp initial value could cause the extinction of population to a certain possibility degree, depending on the supposedly membership function of the input. From the discussion, it can be concluded that the mathematical model with the fuzzy set theory approach is considered appropriate in accommodating the inadequacy of solving the numerical solution to the existing model.

Acknowledgments
The authors thank to the Universitas Padjadjaran, due to the funding of this work through the scheme of Academic Leadership Grant (ALG) with contract number 855/UN6.3.1/PL/2017.

References

[1] F. Courchamp, T. C. Brock and B. Grenfell, "Inverse density dependence and the Allee effect," *Trends in Ecology & Evolution,* vol. 14, no. 10, pp. 405-410, 1999.

[2] L. C. Barros, R. C. Bassanezi and P. A. Tonelli, "Fuzzy modelling in population dynamics," *Ecological Modelling,* vol. 128, pp. 27-33, 2000.

[3] M. J. P. Castanho, R. P. Mateus and K. D. Hein, "Fuzzy model of Drosophila mediopunctata population dynamics," *Ecological modelling,* vol. 287, pp. 9-15, 2014.

[4] E. Massad, N. R. S. Ortega, C. J. Struchiner and M. N. Burattini, "Fuzzy epidemics," *Artificial Intelligence in Medicine,* vol. 29, pp. 241-259, 2003.

IOP Conf. Series: Materials Science and Engineering **332** (2018) 012051 doi:10.1088/1757-899X/332/1/012051

[5] R. M. Jafelice, L. C. d. Barros, R. C. Bassanezi and F. Gomide, "Fuzzy modeling in symptomatic HIV virus infected population," *Bulletin of Mathematical Biology,* vol. 66, pp. 1597-1620, 2004.

[6] B. K. Mishra and S. K. Pandey, "Fuzzy epidemic model for the transmission of worms in computer network," *Nonlinear Analysis,* vol. 11, pp. 4335-41, 2010.

[7] A. Plerou, E. Vlamou and B. Papadopoulos, "Fuzzy logic models in epidemic control," *Precision Medicine,* vol. 1, pp. 1-6, 2016.

[8] R. V. d. Santos, F. L. Ribeiro and A. S. Martinez, "Models for Allee effect based on physical principles," *Journal of Theoretical Biology,* pp. 143-152, 2015.

[9] S. P. Gordon, "Modeling population growth and extinction," *Problems, Resources, and Issues in Mathematics Undergraduate,* vol. 19, no. 6, pp. 548-560, 2009.

[10] K. Atkinson, W. Han and D. Stewart, Numerical solution of ordinary differential equations, New Jersey: Wiley-Interscience, 2009.

[11] S. Kusumadewi and H. Purnomo, Aplikasi Logika Fuzzy untuk Pendukung Keputusan, 2 ed., Yogyakarta: Graha Ilmu, 2010.

[12] J. J. Buckley, E. Eslami and T. Feuring, Fuzzy Mathematics in Economics and Engineering, 1 ed., vol. 91, Physica-Verlag Heidelberg, 2002.

IOP Conf. Series: Materials Science and Engineering **332** (2018) 012052 doi:10.1088/1757-899X/332/1/012052

On the multiple depots vehicle routing problem with heterogeneous fleet capacity and velocity

F Hanum, A P Hartono, T Bakhtiar*

Division of Operations Research, Department of Mathematics, Bogor Agricultural University, Jl. Meranti, Kampus IPB Dramaga, Bogor 16680, Indonesia

*Corresponding author: tbakhtiar@ipb.ac.id

Abstract. This current manuscript concerns with the optimization problem arising in a route determination of products distribution. The problem is formulated in the form of multiple depots and time windowed vehicle routing problem with heterogeneous capacity and velocity of fleet. Model includes a number of constraints such as route continuity, multiple depots availability and serving time in addition to generic constraints. In dealing with the unique feature of heterogeneous velocity, we generate a number of velocity profiles along the road segments, which then converted into traveling-time tables. An illustrative example of rice distribution among villages by bureau of logistics is provided. Exact approach is utilized to determine the optimal solution in term of vehicle routes and starting time of service.

1. Introduction

In the academic literature, the vehicle routing problem (VRP) is one of the most challenging combinatorial optimization tasks and has become the topic of research for decades. VRP is one of the most studied among the combinatorial optimization problems, due both to its practical relevance and to its considerable difficulty [13]. An enormous number of researches have been devoted to studying the VRP since 1959, as it has been described by Dantzig & Ramser as a generalization of Travelling Salesman Problem, where the use of several vehicles are permitted in delivering products and all vehicles must return to the depot after visiting some customers [2]. In this effort, much attentions were mainly emphasized on the development of the problem features and assumptions leading to a huge number of problem variant formulations as well as various heuristic and metaheuristic algorithms advancement to address the problem. Since then VRP has been the heart of the transportation, distribution, and logistics management [10].

The capacitated vehicle routing problem (C-VRP) is known as the most basic and classical variant of VRP. In this case, the optimal routes from a depot to a set of customers with known finite demand should be sought in such a way they consist of the delivery of products conducted by vehicles with limited carrying capacity [6], [9], [14]. This variant is then extended to include the case where multiple depots are available to serve customers in which it is known as multiple depots vehicle routing problem (MD-VRP), see e.g., [11] and [15]; and one where each customer should be served in a specified time window, i.e., vehicle routing problem with time windows (VRP-TW), see for instance, [7], [8].

This present paper discusses MD-VRP-TW with heterogeneous fleet capacity and velocity. The unique feature of heterogeneous velocity is undertaken by generating a number of velocity profiles along the road segments, which then it is converted into several tables of traveling-time. In fact, in some part of modelling, we motivated by the work of [1].

2. Model of VRP

In general, VRP refers to a problem of seeking delivery routes for a fleet of vehicles of known capacities to supply a number of nodes of known locations and demands for a certain commodity, given a set of constraints. Constraints can include any combination of the following: each vehicle is originating and terminating at a depot upon completion of its route, each customer must be served by only one vehicle and each dispatched vehicle should visit at least one customer [5], [8], [12]. Routes for the vehicles should be generated to minimize some cost functions such as the total distance of trips and the cost of delivery.

The volume by [4] summarizes the most significant results for the VRP and its variants since 2000. They present significant methodological developments or unique approaches for addressing existing VRPs as well as novel problems that have introduced in the vehicle routing domain and underline new challenges for the field. In this section we present a brief overview regarding the VRP and its variants under consideration.

2.1. C-VRP

C-VRP is a VRP in which a fleet of identical vehicles with limited known capacity must service a number of customers with known in advance and may not be split demands for a single commodity from a depot at minimum traveling cost or distance needed to serve all the customers. That is, CVRP is VRP with the additional constraint that every vehicles must have uniform capacity of a single commodity, i.e., only the capacity restrictions for the vehicles are imposed.

Suppose there are n customers to be served and $\mathbb{F}$ is the set of all available vehicles. If x_{ij}^k denotes the decision variables, i.e., $x_{ij}^k = 0$ if there is no arc from customer i to customer j by vehicle k, and $x_{ij}^k = 1$ otherwise; q_i is the demand of customer i; and C is the capacity of vehicle, then the capacity restriction in a route is written as

$$\sum_{i=1}^{n} q_i \sum_{j=1}^{n} x_{ij}^k \leq C, \quad \forall k \in \mathbb{F}.$$

It is conducted by [8] a literature review on the recent developments and publications involving the vehicle routing problem and its variants, including C-VRP and VRP-TW.

2.2. VRP-TW

VRP-TW arises when a natural constraint for VRP is added, that is a customer can only receive a delivery or service in a specified window of time. By this constraint, a vehicle now has to visit a customer within a certain time frame. It means that we need a way to calculate the time a vehicle has to spend serving a customer and the travel time to the next customer on its route. More specifically, we must calculate the service time, the travel time and the total time.

If the time window of customer i is $[a_i, b_i]$ then vehicle may arrive before the time window opens but the customer cannot be served until the time windows open:

$$a_i \leq y_i^k, \quad \forall k \in \mathbb{F},$$

where y_i^k is a decision variable denotes the starting time to serve customer i by vehicle k. It is not allowed to arrive after the time window has closed, which means that

$$y_i^k + s_i \leq b_i, \quad \forall k \in \mathbb{F},$$

where s_i is the known service time for customer i.

VRP with time windows can be approached either by exacts, heuristics or metaheuristics methods. Readers may consult the survey paper of [3] for the comprehensive overview of the method.

2.3. MD-VRP

The case of MD-VRP is posed if a company may have more than one depots from which it can serve its customers. If the customers are clustered around depots, then a number of independent VRPs can be performed to serve customers. However, if the customers and the depots are intermingled then a MD-

VRP should be solved. In fact, an MD-VRP necessitates the assignment of customers to depots and a fleet of vehicles is organized at each depot basis. It means that each vehicle originates from one depot must service the customers assigned to that depot and return to the same depot at the end of their assigned tours as loosely can be expressed by

$$\sum_{j=1}^{n} x_{ij}^{k} \leq 1, \quad \forall k \in \mathbb{F}_i, \forall i \in \mathbb{W},$$

$$\sum_{i=1}^{n} x_{ij}^{k} \leq 1, \quad \forall k \in \mathbb{F}_j, \forall j \in \mathbb{W},$$

where $\mathbb{F}_i$ denotes the set of all available vehicles at depot i and $\mathbb{W}$ is the set of all depots. Complete formulation of such a problem can be found, for instance, in [11].

3. Model of Interest

In this section we present our model of interest, namely time windowed multiple depots vehicle routing problem with heterogeneous capacity and velocity of fleet (MD-VRP-TW-HCV). To facilitate our model formulation, we define a number of sets, parameters and variables.

We denote by $\mathbb{W}, \mathbb{V}, \mathbb{N}$, respectively the set of all depots with cardinality m, that of all customers with cardinality n and that of all nodes, i.e., $\mathbb{N} = \mathbb{W} \cup \mathbb{V}$. For clarity we define $\mathbb{W} = \{1,2,\dots,m\}$ and $\mathbb{V} = \{m+1, m+2, \dots, m+n\}$. We define by $\mathbb{F}_l$ the set of all available vehicles at depot l and by $\mathbb{F}$ the set of all available vehicles, i.e., $\mathbb{F} = \cup_{l=1}^{m} \mathbb{F}_l$, with cardinality p. We also define the following parameters:

q_i	:	demand of customer i (in unit),
s_i	:	duration of service time for customer i (in hour),
$[a_i, b_i]$	:	time window of customer i,
C_k	:	capacity of vehicle k (in unit),
$\bar{c}_k$	:	fixed cost of vehicle k (in currency unit),
v_k	:	variable cost of vehicle k (in currency unit per kilometer),
d_{ij}	:	distance between node (depot or customer) i and node j (in kilometer),
c_{ij}^{k}	:	traveling cost between node i and node j by using vehicle k, i.e., $c_{ij}^{k} = d_{ij} v_k$ (in currency unit),
t_{ij}	:	travelling time between node i and node j (in hour),
M	:	big positive constant (big-M),
Z	:	the value of objective function.

For decision variables we have the following binary variable

$$x_{ij}^{k} = \begin{cases} 1; \text{if vehicle } k \text{ travel from node } i \text{ to node } j \\ 0; \text{otherwise,} \end{cases}$$

and $y_i^k \in \mathbb{R}^+$ which defined by the starting time to serve customer i by vehicle k, thus $y_i^k \in [a_i, b_i]$. We search the routes such that minimize the total cost which includes both the vehicle fixed cost and the traveling cost

$$\min Z = \sum_{i=1}^{m+n} \sum_{j=1}^{m+n} \sum_{k=1}^{p} c_{ij}^{k} x_{ij}^{k} + \sum_{i=1}^{m} \sum_{j=m+1}^{m+n} \sum_{k=1}^{p} \bar{c}_k x_{ij}^{k},$$

subject to a set of constraints:

(a). each customer is serviced exactly once by a vehicle and after visiting a customer it should be ensured that vehicle will visit the next customer:

$$\sum_{i=1}^{m+n} \sum_{k=1}^{p} x_{ij}^{k} = 1, \quad \forall j \in \mathbb{V},$$

$$\sum_{j=1}^{m+n} \sum_{k=1}^{p} x_{ij}^k = 1, \quad \forall i \in \mathbb{V},$$

(b). an operated vehicle must start from a depot and must return to the same depot after completing the tour; it is not necessary to utilize all available vehicles:

$$\sum_{j=m+1}^{m+n} x_{ij}^k \leq 1, \quad \forall k \in \mathbb{F}_i, \forall i \in \mathbb{W},$$

$$\sum_{i=m+1}^{m+n} x_{ij}^k \leq 1, \quad \forall k \in \mathbb{F}_j, \forall j \in \mathbb{W},$$

(c). there is no trip conducted by a vehicle not belong to a depot:

$$\sum_{j=m+1}^{m+n} x_{ij}^k = 0, \quad \forall k \notin \mathbb{F}_i, \forall i \in \mathbb{W},$$

$$\sum_{i=m+1}^{m+n} x_{ij}^k = 0, \quad \forall k \notin \mathbb{F}_j\, ; \forall j \in \mathbb{W},$$

(d). route continuity:

$$\sum_{i=1}^{m+n} x_{ir}^k = \sum_{j=1}^{m+n} x_{rj}^k, \quad \forall k \in \mathbb{F}, \forall r \in \mathbb{V},$$

(e). time window:

$$a_i \leq y_i^k, \quad \forall k \in \mathbb{F}, \forall i \in \mathbb{N},$$

$$y_i^k + s_i \leq b_i, \quad \forall k \in \mathbb{F}, \forall i \in \mathbb{N},$$

(f). sub-tour elimination:

$$y_i^k + s_i + t_{ij}(y_i^k + s_i) - y_j^k \leq M(1 - x_{ij}^k), \quad \forall k \in \mathbb{F}, \forall i \in \mathbb{N}, \forall i \in \mathbb{V},$$

(g). the capacity of any vehicle is not violated:

$$\sum_{i=m+1}^{m+n} q_i \sum_{j=1}^{m+n} x_{ij}^k \leq C_k, \quad \forall k \in \mathbb{F},$$

(h). there is no inter-depot route:

$$x_{ij}^k = 0, \quad \forall k \in \mathbb{F}; \ \forall i, j \in \mathbb{W}.$$

4. Model Implementation

4.1. Problem statement

As an illustrative example we consider a distribution problem of rice for the poor (RASKIN), where the bureau of logistics operates two depots ($m = 2$) to service eighteen villages as customers ($n = 18$). To show their indices, we denote $\mathbb{W} = \{1,2\}$ and $\mathbb{V} = \{3,4,\dots,20\}$. Each village may have different demand of rice and time window as depicted by Table 3 in the appendix. The duration of service time is known and is provided in the same table.

In the distribution process, the bureau operates a fleet consisting of six vehicle ($p = 6$). The first three vehicles belong to Depot 1 and the rest belongs to Depot 2. Thus, $\mathbb{F}_1 = \{1,2,3\}$ and $\mathbb{F}_2 = \{4,5,6\}$. Capacity, fixed cost and variable cost of each vehicle are given by Table 4 in the appendix. Distance between nodes is assumed to be symmetric and the data is provided in Table 5 in the appendix.

By having twenty nodes, in fact we have $C(20,2) = 190$ road segments with different condition and thus different expected velocity. However, for the sake of simplicity, we assume that there are only six different profiles of velocity applicable to all road segments as illustrated by Figure 1. In this figure, horizontal axis indicate the time span (in minute) and the vertical axis measures the velocity (in

IOP Conf. Series: Materials Science and Engineering **332** (2018) 012052 doi:10.1088/1757-899X/332/1/012052

kilometer per minute). In the first sub-figure, we can see that in the first five minutes, due to the road condition, vehicle can only travel 10 km/min and enhanced to 20 km/min during the next five minutes. Once distance between two nodes of certain profile is provided (see Table 4 in the appendix), then we can calculate the traveling time between them. Subsequently, a table of traveling time among nodes can be constructed (see Table 5 in the appendix).

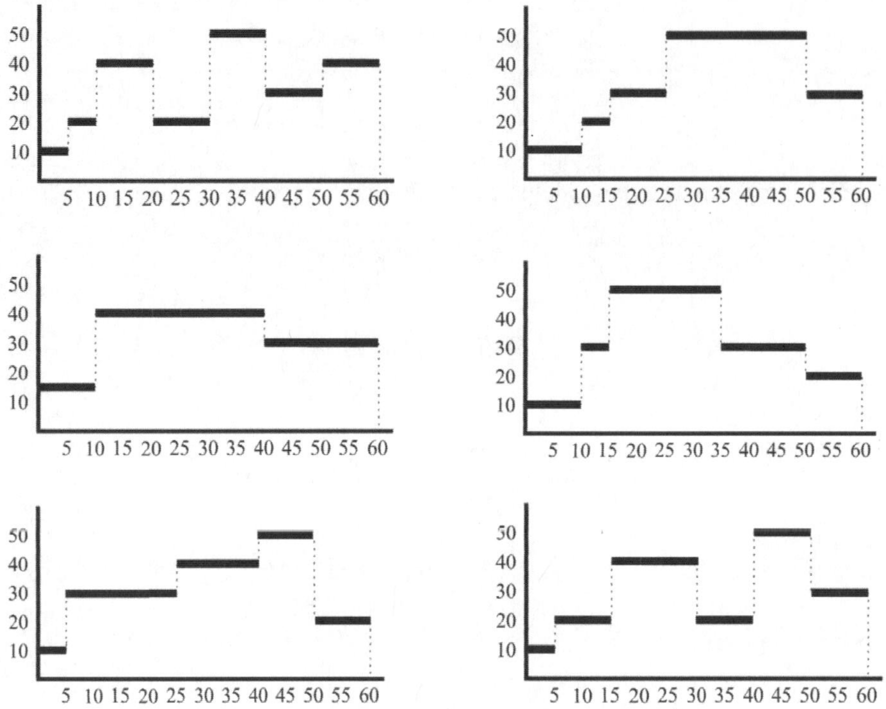

Figure 1. Velocity profiles

4.2. The optimal route

We implement the MD-VRP-TW-HCV model formulated in the previous section in Lingo 11 and run by using PC desktop of CPU 2.30 GHz and RAM 4 GB. Execution time was 1 minute and 3 seconds, providing the (global) optimal cost of 767,500 currency unit. Optimal distribution route of each vehicle is presented in Table 1 or Figure 2. It can be concluded from the table that all the vehicles should be operated to service all retailers. More particular, Vehicle 2 travels the most (39 km) to provide service to four retailers and Vehicle 6 the least for one retailer.

Table 1. The optimal route

Vehicle	Distribution Route	Amount of distributed product	Distance Traveled
1	1 → 5 → 6 → 14 → 12 → 1	338	35
2	1 → 4 → 3 → 10 → 13 → 1	343	39
3	1 → 16 → 7 → 8 → 19 → 1	285	26
4	2 → 15 → 18 → 20 → 2	167	43
5	2 → 17 → 9 → 2	150	26
6	2 → 11 → 2	90	4

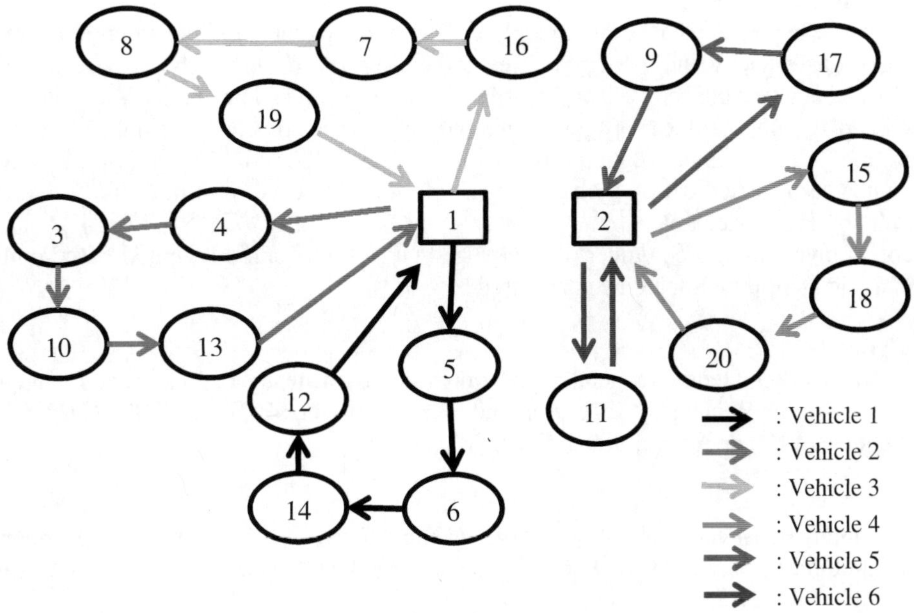

Figure 2. Distribution route

To inspect the fulfilment of demand and time window, let consider the distribution process undertaken by Vehicle 1. This vehicle delivers products to villages 5, 6, 14 and 12, respectively. The total demand of rice requested by these four villages is $99 + 100 + 59 + 80 = 338$, as shown in Table 3 in the appendix and confirmed by optimal solution in Table 1. Even though Depot 1 is open since 6 am, Vehicle 1 can leave depot at 08.54 at the latest as it requires 6 minutes to reach village 5 and starting service at 9 am for one hour. The driver can have a waiting time or reserve time until 12.10 before proceeding to village 6, and so forth. Distribution task is completed by Vehicle 1 at 21.00 before returning to Depot 1. The time flow for Vehicle 1 may be depicted by Figure 3. Tables 8-12 in the appendix present the fulfilment of demand and time window for other vehicles.

Table 2. Service time by Vehicle 1

Node (i)	Time Windows $[a_i, b_i]$	Duration of service (s_i)	Start time (y_i^1)	End time	Time limit for leaving	Time to next node (t_{ij})
1	06.00 - 22.00	0	06.00.00	06.00.00	08.54.00	6 min
5	09.00 - 14.00	1	09.00.00	10.00.00	12.10.30	14 min 30 sec
6	10.00 - 15.00	1	12.25.00	13.25.00	16.28.04	10 min 45 sec
14	16.00 - 21.00	1	16.38.49	17.38.49	19.52.00	8 min
12	16.00 - 21.00	1	20.00.00	21.00.00	21.11.40	48 min 20 sec
1	06.00 - 22.00	0	-	-		

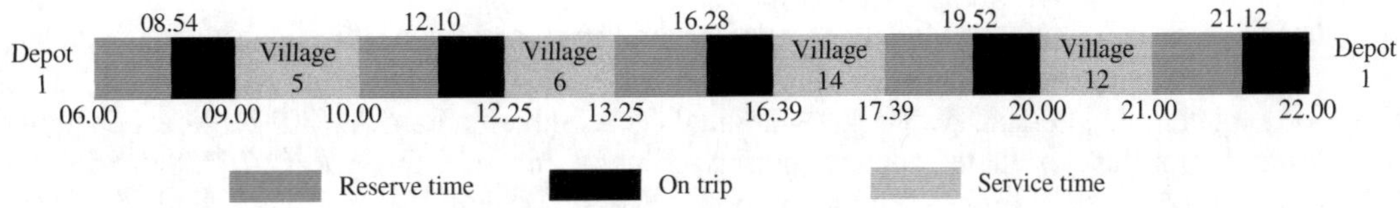

Figure 3. Time flow of Vehicle 1

5. Concluding Remark

In this paper we have presented a variant model of VRP, namely time windowed multiple depots VRP with heterogeneous capacity and velocity of fleet, where the last feature of homogeneous velocity is approached by characterizing the velocity profile of road segments. The model is formulated in the framework of mixed integer linear programming and an exact approached is employed to obtain the optimal solution. A model implementation of rice for the poor (RASKIN) distribution between depot and villages has been carried-out. A global optimal solution is reached in term of transportation route for vehicles including service time.

For problem involving more number of depots and customers, solution method based on heuristic and metaheuristic algorithms should be considered.

Acknowledgments
Authors gratefully acknowledge financial support provided by Ministry of Research, Technology and Higher Education of the Republic of Indonesia under grant No. 011/SP2H/LT/DRPM/IV/2017 through the PUPT Scheme of Bogor Agricultural University.

References

[1] Afshar-Nadjafi B, Afshar-Nadjafi A. (2014). A constructive heuristic for time-dependent multi-depot vehicle routing problem with time-windows and heterogeneous fleet. *J. King Saud Univ. – Eng. Sci.*, **29**, 29-34.
[2] Dantzig, G. B. & Ramser, J. H. (1959). The truck dispatching problem, *Management Science*, **6**, 80-91.
[3] El-Sherbeny, N. A. (2010). Vehicle routing with time windows: An overview of exact, heuristic and metaheuristic methods. *J. King Saud Univ. – Sci.*, **22,** 123-131.
[4] Golden, B., Raghavan, S., & Wasil, E. (2008). *The vehicle routing problem: latest advances and new challenges*. New York: Springer.
[5] Kaur, H. (2013). Review paper on multi-depot vehicle routing problem. *Int. J. Sci. & Eng. Res.*, **4**, 410-413.
[6] Kir, S., Yazgan, H. R. & Tüncel, E. (2017). A novel heuristic algorithm for capacitated vehicle routing problem. *J. Industr. Eng. Int.*, **13**, 323-330.
[7] Kirci, P. (2016). An optimization algorithm for a capacitated vehicle routing problem with time windows. *Sadhana*, **41**, 519-530.
[8] Kumar, S. N. & Panneerselvam, R. (2012). A survey on the vehicle routing problem and its variants. *Intell. Inform. Manag.*, **4**, 66-74.
[9] Lysgaard, J., Letchford, A. & Eglese, R. (2004). A new branch-and-cut algorithm for the capacitated vehicle routing problem. *Math. Prog. A*, **100**, 423-445.
[10] Moons, S., Ramaekers, K., Caris, A., & Arda, Y. (2017). Integrating production scheduling and vehicle routing decisions at the operational decision level: a review and discussion. *Comp. & Industr. Eng.*, *104*, 224–245.
[11] Salhi, S., Imran, A., & Wassan, N. A. (2014). The multi-depot vehicle routing problem with heterogeneous vehicle fleet: Formulation and a variable neighborhood search implementation. *Comp. & Oper. Res.*, **52-B**, 315-325.
[12] Sen, A. & Bulbul, K. (2008). A survey on multi trip vehicle routing problem. *International Logistics and Supply Chain Congress' 2008*.
[13] Toth, P. and Vigo, D. (2002). Models, relaxations and exact approaches for the capacitated vehicle routing problem. *Disc. App. Math.*, **123**, 487-512.
[14] Uchoa E., Pecin, D., Pessoa, A., Poggi, M., Vidal, T. & Subramanian, A. (2017). New benchmark instances for the capacitated vehicle routing problem. *Europ. J. Oper. Res.*, **257**, 845-858.
[15] Yalian, T. (2016). An improved ant colony optimization for multi-depot vehicle routing problem. *Int. J. Eng. Tech.,* **8**, 385-388.

IOP Conf. Series: Materials Science and Engineering **332** (2018) 012052 doi:10.1088/1757-899X/332/1/012052

Appendix

Table 3. Demand, duration of service time and time windows

Node (i)	Demand (q_i)	Service time (s_i)	Time Windows $[a_i, b_i]$
1 (depot)	0	0	06.00-22.00
2 (depot)	0	0	06.00-22.00
3	90	1	07.00-12.00
4	96	1	08.00-13.00
5	99	1	09.00-14.00
6	100	1	10.00-15.00
7	57	1	11.00-16.00
8	68	1	12.00-17.00
9	66	1	13.00-18.00
10	95	1	14.00-19.00
11	90	1	15.00-20.00
12	80	1	16.00-21.00
13	62	1	17.00-21.00
14	59	1	16.00-21.00
15	58	1	07.00-11.00
16	82	1	09.00-13.00
17	84	1	11.00-15.00
18	52	1	13.00-17.00
19	78	1	15.00-19.00
20	57	1	17.00-21.00

Table 4. Capacity, fixed cost and variable cost

Vehicle (k)	Capacity (C_k)	Fixed Cost ($\bar{c}_k$)	Variable Cost (v_k)
1	550	120,000	1,000
2	450	100,000	1,000
3	300	97,000	1,000
4	275	95,000	1,000
5	195	92,500	1,000
6	100	90,000	1,000

Table 5. Distance between nodes

Node	1	2	3	4	5	6	7	8	9	10	11	12	13	14	15	16	17	18	19	20
1	0	14	20	10	1	3	20	2	8	10	3	25	16	15	15	15	17	2	4	17
2		0	23	25	22	25	9	18	11	7	2	14	12	21	13	6	12	19	24	15
3			0	1	21	12	18	24	15	8	3	23	18	4	20	11	25	18	14	5
4				0	16	10	3	17	5	4	11	7	16	5	22	6	22	21	10	11
5					0	4	6	20	22	5	19	5	3	6	25	16	13	19	1	5
6						0	12	16	6	19	16	10	8	3	15	16	8	21	7	25
7							0	4	19	19	13	1	11	3	15	2	13	12	23	24
8								0	23	2	4	10	25	16	11	3	16	21	1	23
9									0	20	12	13	12	22	19	11	3	16	18	24
10										0	18	13	4	25	6	4	24	21	16	6
11											0	18	4	16	7	22	7	18	23	21
12												0	19	2	7	9	22	4	10	12
13													0	11	17	25	19	1	14	17
14														0	2	3	18	8	23	8
15															0	18	10	10	22	24

Node	1	2	3	4	5	6	7	8	9	10	11	12	13	14	15	16	17	18	19	20
16																0	6	25	20	4
17																	0	11	15	4
18																		0	13	5
19																			0	23
20																				0

Table 6. Condition of road segment between nodes

Node	1	2	3	4	5	6	7	8	9	10	11	12	13	14	15	16	17	18	19	20
1	-	3	2	3	3	1	3	4	2	2	2	1	5	6	4	2	5	2	6	5
2		-	1	6	3	4	4	5	3	6	5	5	5	6	2	5	3	6	6	1
3			-	5	3	6	4	6	1	2	2	1	5	4	3	2	2	4	1	4
4				-	4	2	5	5	4	5	1	1	2	3	3	3	1	5	2	2
5					-	6	6	4	2	6	3	4	1	6	6	2	6	3	2	4
6						-	5	5	5	5	6	6	1	1	1	4	2	2	6	2
7							-	4	4	1	3	2	5	5	2	4	2	6	6	6
8								-	3	5	3	3	5	4	3	6	2	5	4	3
9									-	2	6	5	5	5	6	6	3	2	3	1
10										-	1	1	5	1	4	1	6	6	4	6
11											-	1	6	2	2	4	5	1	1	2
12												-	2	4	3	6	4	1	4	4
13													-	4	2	2	2	3	3	6
14														-	6	2	3	2	6	2
15															-	3	2	4	1	6
16																-	3	3	3	5
17																	-	6	4	1
18																		-	3	2
19																			-	1
20																				-

Table 7. Traveling time between nodes (10^{-2} hour)

Node	1	2	3	4	5	6	7	8	9	10	11	12	13	14	15	16	17	18	19	20
1	00	45	65	37	10	18	57	13	41	45	23	81	55	54	48	55	57	18	24	57
2		00	74	82	62	75	33	60	39	32	12	50	45	74	51	26	41	70	80	55
3			00	09	59	45	55	80	55	41	23	74	60	20	57	47	75	55	53	23
4				00	50	45	16	57	23	19	43	28	57	27	62	29	71	67	45	47
5					00	24	30	60	69	27	55	23	18	30	82	57	47	55	10	23
6						00	45	55	26	62	59	40	30	18	55	50	41	67	32	75
7							00	20	58	63	43	10	42	16	55	13	51	45	78	80
8								00	66	12	24	37	75	50	39	19	57	67	07	66
9									00	65	45	47	45	69	70	42	21	57	53	77
10										00	61	51	19	81	25	20	80	74	50	30
11											00	61	24	57	37	65	29	61	74	67
12												00	63	13	31	37	65	20	35	40
13													00	38	59	75	63	10	45	64
14														00	14	23	53	41	78	41
15															00	33	45	35	71	80
16																00	29	72	57	19
17																	00	42	48	20
18																		00	43	31
19																			00	74
20																				00

Table 8. Service time by Vehicle 2

Node (i)	Time Windows $[a_i, b_i]$	Duration of service (s_i)	Start time (y_i^2)	End time	Time limit for leaving	Time to next node (t_{ij})
1	06.00 - 22.00	0	06.00.00	06.00.00	07.50.00	22 min
4	08.00 - 13.00	1	08.12.00	09.12.00	10.10.05	5 min 20 sec
3	07.00 - 12.00	1	10.15.25	11.15.25	15.25.01	24 min 20 sec
10	14.00 - 19.00	1	15.49.21	16.49.21	19.48.40	11 min 20 sec

Node (i)	Time Windows $[a_i, b_i]$	Duration of service (s_i)	Start time (y_i^2)	End time	Time limit for leaving	Time to next node (t_{ij})
13	17.00 - 21.00	1	20.00.00	21.00.00	21.27.15	32 min 45 sec
1	06.00 - 22.00	0	-	-		

Table 9. Service time by Vehicle 3

Node (i)	Time Windows $[a_i, b_i]$	Duration of service (s_i)	Start time (y_i^3)	End time	Time limit for leaving	Time to next node (t_{ij})
1	06.00 - 22.00	0	06.00.00	06.00.00	08.45.00	33 min
16	09.00 - 13.00	1	09.18.00	10.18.00	11.32.24	8 min
7	11.00 - 16.00	1	11.40.24	12.40.24	15.03.24	12 min 15 sec
8	12.00 - 17.00	1	15.15.39	16.15.39	17.56.00	4 min
19	15.00 - 19.00	1	18.00.00	19.00.00	21.45.30	14 min 30 sec
1	06.00 - 22.00	0	-	-		

Table 10. Service time by Vehicle 4

Node (i)	Time Windows $[a_i, b_i]$	Duration of service (s_i)	Start time (y_i^4)	End time	Time limit for leaving	Time to next node (t_{ij})
2	06.00 - 22.00	0	06.00.00	06.00.00	08.33.00	30 min 36 sec
15	07.00 - 11.00	1	09.03.36	10.03.36	13.16.07	21 min 15 sec
18	13.00 - 17.00	1	13.37.22	14.37.22	19.41.40	18 min 20 sec
20	17.00 - 21.00	1	20.00.00	21.00.00	21.27.00	33 min
2	06.00 - 22.00	0	-	-	-	
2	06.00 - 22.00	0	06.00.00	06.00.00	08.33.00	30 min 36 sec

Table 11. Service time by Vehicle 5

Node (i)	Time Windows $[a_i, b_i]$	Duration of service (s_i)	Start time (y_i^5)	End time	Time limit for leaving	Time to next node (t_{ij})
2	06.00 - 22.00	0	06.00.00	06.00.00	10.35.36	24 min 24 sec
17	11.00 - 15.00	1	11.00.00	12.00.00	16.47.20	12 min 40 sec
9	13.00 - 18.00	1	17.00.00	18.00.00	21.36.48	23 min 12 sec
2	06.00 - 22.00	0	-	-	-	

Table 12. Service time by Vehicle 6

Node (i)	Time Windows $[a_i, b_i]$	Duration of service (s_i)	Start time (y_i^6)	End time	Time limit for leaving	Time to next node (t_{ij})
2	06.00 - 22.00	0	06.00.00	06.00.00	14.52.40	7 min 20 sec
11	15.00 - 20.00	1	15.00.00	16.00.00	21.52.40	7 min 20 sec
2	06.00 - 22.00	0	-	-	-	

Demand analysis of flood insurance by using logistic regression model and genetic algorithm

P Sidi[1*], M B Mamat[2], Sukono[3], S Supian[3], A S Putra[4]

[1]Department of Mathematics, Faculty of Mathematics and Natural Sciences, Universitas Terbuka, INDONESIA
[2]Faculty of Informatics and Computing, Universiti Sultan Zainal Abidin, MALAYSIA
[3]Department of Mathematics, Faculty of Mathematics and Natural Sciences, Universitas Padjadjaran, INDONESIA
[4]Department of Geophysiscs, Faculty of Mathematics and Natural Sciences, Universitas Padjadjaran, INDONESIA

[*]Corresponding author: pram@ecampus.ut.ac.id

Abstract. Citarum River floods in the area of South Bandung Indonesia, often resulting damage to some buildings belonging to the people living in the vicinity. One effort to alleviate the risk of building damage is to have flood insurance. The main obstacle is not all people in the Citarum basin decide to buy flood insurance. In this paper, we intend to analyse the decision to buy flood insurance. It is assumed that there are eight variables that influence the decision of purchasing flood assurance, include: income level, education level, house distance with river, building election with road, flood frequency experience, flood prediction, perception on insurance company, and perception towards government effort in handling flood. The analysis was done by using logistic regression model, and to estimate model parameters, it is done with genetic algorithm. The results of the analysis shows that eight variables analysed significantly influence the demand of flood insurance. These results are expected to be considered for insurance companies, to influence the decision of the community to be willing to buy flood insurance.

Keywords: Citarum River, building damage, floods insurance, logistic regression, and genetic algorithm.

1. Introduction

Flood disasters often occur in the Citarum River Basin in the area of South Bandung, Indonesia, and often cause a lot of damage to buildings. The government and several community agencies provide compensation in the form of assistance such as building equipment and building repair costs, for houses damaged [11]. But often the amount of compensation does not fully compensate, so there is a need to find ways to reduce the losses [13; 14]. The community participation to reduce the risk of building disasters is important, by raising awareness and capacity of the community. Because the community is a party who has direct experience in the event of a disaster, so that the understanding has become the capital for the reduction of risk of loss due to flood disaster [4]. In the context of reducing the risk of damage to buildings, insurance is seen as the right solution, because insurance can minimize losses due to the disaster [2]. Because insurance is a financial protection system (or financial compensation) for life, property, health and so on, getting reimbursed from unforeseen events, which involves regular premium payments over a period of time in exchange policy to guarantee such protection [16; 17]. Indeed, people living in areas affected by Citarum river flood have realized, how important the role of insurance to reduce the risk of damage to buildings due to flood disaster. But until now, not all the people living in areas affected by the flood of the Citarum River are willing to insure their buildings. Therefore, analysis of purchasing decision of this flood insurance is very important to do.

Based on studies from several literatures, there are some researchers who have conducted an analysis of the purchase decisions of flood insurance. Among them, Esau [9], examines the factors that influence consumer purchasing decisions on insurance products in Manado Indonesia. This research uses Classic Assumption test model and multiple regression analysis. This model is used to verify and prove the research hypothesis. The analysis results show that product, price, promotion, people, and process together influence consumer purchase decision. Atreya et al. [1], conducted an analysis of what drives households to buy flood insurance in the Georgian region of the United States. The analysis was done using econometric model. The results show that more educated individuals, older, and African-Americans, all the same, are more likely to buy flood insurance. Ulbinaite et al. [19], examines the factors affecting insurance purchase decision making in Lithuania. In this research, factor analysis and multiple regression analysis are used to determine how factors are formed and how much their relative weights are. In general, it can be concluded that insurance purchasing decisions are affected by age, income, education level, household status, and number of family members. Similar studies have also been conducted by previous researchers, among others by Desrosiers [7], Mathur & Tripathi [10], Shaw [12], and Ulbinaite & Kucinskiene [18]. They all do research on the factors that influence the purchase decisions of insurance products.

Based on the above description, this paper intends to analyse the decision to buy flood insurance by using logistic regression model and genetic algorithm. The objective is to obtain a significant logistic regression model estimator, so it can be used to determine what variables significantly influence the purchase decision of flood insurance. This analysis is very useful in order to determining the socialization strategy about flood insurance products, which can reduce the risk of damage to buildings due to flooding. Especially for people living in the Citarum River Basin, South Bandung, Indonesia.

2. Materials and methods

This section discusses the materials and methods used in this study. Discussion of the methods used, including: logistic regression model, log likelihood function, genetic algorithm, and model significance test statistic.

2.1. Materials

This sub-section discuss about the material used in the analysis. The data used in this analysis is obtained from interviews with community representatives in the Citarum River flood disaster area in South Bandung area in 2016. The data consists of 100 samples, of which $n_1 = 18$ belong to category 1, which suffered damage to the building and decided to buy flood insurance, and $n_0 = 82$ belonging to category 0, which suffered damage to the building and decided not to buy flood insurance. Factors contributing to the decision to purchase flood insurance, is assumed that there are 8 factors, including: Z_1 family income; Z_2 level of education; Z_3 the distance of the house with the source of the flood; Z_4 building elevation by road; Z_5 the frequency of floods in a year; Z_6 prediction of future floods; Z_7 perceptions of insurance company services, and Z_8 perceptions of government efforts in tackling floods.

Against observation data, normality tests need to be done, because the non-free variable is the data whose values fluctuate between high and low. This is to avoid the estimation of biased model parameters. Normality test of observational data is done using SPSS statistical software.

2.2. *Methods*
This subsection discuss about the methodology used for the analysis. The discussion includes: logistic regression model, log likelihood function, genetic algorithm, and statistical test of model parameter significance.

2.2.1. *Logistic Regression Model*
Binary logistic regression model, is a regression model that can be used to estimate the effect of some independent variables Z_i ($i=1,...,F$ where F many independent variables), to non-free variables D, which are binary values 0 and 1 [3; 15]. The binary regression model used in this analysis is shaped as follows [5]:

$$\pi(Z_i)=\frac{e^{\sum_{j=0}^{J}\beta_k Z_{ij}}}{1+e^{\sum_{j=0}^{J}\beta_k Z_{ij}}},\ i=1,...,F. \tag{1}$$

When the left and right segments of equation (1), are drawn natural logarithms, then we obtain a new equation of the following form [15; 8]:

$$g(Z_i)=\ln\left(\frac{\pi(Z_i)}{1-\pi(Z_i)}\right)=\sum_{j=0}^{J}\beta_j Z_{ij},\ i=1,...,F. \tag{2}$$

To estimate the parameters of equation (2), it can be done using the likelihood method, as described below.

2.2.2. *The Likelihood Function of the Logistic Regression Model*
Suppose that the vector parameter model is known as $\boldsymbol{\beta}'=(\beta_0,\beta_1,...,\beta_J)$. According to Sukono et al. [15], the purpose of estimating the regression model parameters is to determine the values β_j ($j=0,1,...,J$) that contribute to equation (2). Suppose there are J independent variables, $Z_1,...,Z_J$ the conditional density function of the dependent variable D with the parameter vector $\boldsymbol{\beta}$, is to follow the distribution of Bernoulli in the form of the following:

$$f(D\,|\,\boldsymbol{\beta})=\prod_{i=1}^{F}\pi_i^{D_i}(1-\pi_i)^{1-D_i}\ ;\ D_i=0 \text{ or } 1. \tag{3}$$

Maximum Likelihood Estimator (MLE), is estimating the values of the parameter vectors $\boldsymbol{\beta}$ that can maximize likelihood function in (3).

The dependent variable D_i is given a value of 0 or 1, for each pair (Z_i, D_i). If given D_i = 1, then contributing to the likelihood function is $\pi(Z_i)$, and if given value Z_i = 0, then contributing to the likelihood function is $1-\pi(Z_i)$. Therefore, the contribution of the pair of points (Z_i, D_i) to the likelihood function can be written as follows [15; 3]:

$$L(\boldsymbol{\beta})=\prod_{i=1}^{F}\pi_i^{D_i}(1-\pi_i)^{1-D_i}\ ;\ D_i=0 \text{ or } 1. \tag{4}$$

If equation (1) is substituted into (4), then the equation is given as follows:

$$L(\boldsymbol{\beta})=\prod_{i=1}^{F}\left(e^{\sum_{j=0}^{J}\beta_j Z_{ij}}\right)^{D_i}\left(1+e^{\sum_{j=0}^{J}\beta_j Z_{ij}}\right)^{-1}. \tag{5}$$

If the left and right sides of the equation (5) take the natural logarithm, then the following equations are obtained [15]:

$$\ell(\boldsymbol{\beta})=\sum_{i=1}^{F}\left\{D_i\sum_{j=0}^{J}\beta_j Z_{ij}-\ln\left(1+e^{\sum_{j=0}^{J}\beta_j Z_{ij}}\right)\right\}. \tag{6}$$

The values of the elements of the parameter vector $\boldsymbol{\beta}$ in equation (6) are estimated using the genetic algorithm, as follows.

2.2.3. Estimates Using Genetic Algorithms

Genetic algorithm is a genetic mechanism-based search method, and natural generation selection [6]. This genetic mechanism represents the individual's ability to engage in marriage, and produce offspring that have characteristics similar to that of the mother. While the generation selection naturally represents that living things can survive, when able to adapt to the surrounding environment [20]. Therefore, future generations are expected to have a combination of the best characteristics of the parent, and can sustain the next generations [6].

In general, the structure of the genetic algorithm is the following stages [15]:

a) Early population generation, generated randomly to obtain the initial solution of a dummy;
b) The resurrected population includes a number of chromosomes representing the solution obtained;
c) The establishment of a new generation, a new generation formed involving three operators, namely reproduction / selection, cross over, and mutation;
d) The evaluation process, each population raised is determined by the fitness value for each chromosome, and the evaluation is performed until the stopping criterion is reached. If the stopping criterion has not been reached, then a new generation needs to be established in a way like stage b).

Based on the structure of the genetic algorithm, to maximize the log likelihood function in equation (6), a generalized genetic algorithm can be prepared as follows [6; 20]:

1) The initial population determination, determined by the number of J, is generated randomly. This random number of the initial population generated is converted into decimal values β_j, where $j = 1,..., J$;
2) Chromosome evaluation, the fitness value of chromosome is the value of log likelihood function as equation (6). The resulting fitness values are the largest, for the maximization program.
3) Calculation of the convergent population percentage, convergent population c_p, is the percentage of the number of individuals who can produce the same fitness value and the most. The c_p value is calculated using the following formula:

$$c_p = \frac{n}{s_p} \times 100\% \quad (7)$$

where n are the number of fitness-producing individuals which is the same and the largest, and s_p is the total population.

4) Evaluation of the stopping condition, the genetic algorithm process is dismissed if the generation counter has reached the number of generations s_g, which is in the research is set at 1000, or the percentage of conventional population s_p has reached the threshold limit which in this study is determined by $\tau = 90\%$.
5) Chromosome selection, the process is based on the principle of roulette wheel selection. Since the maximization program, the evaluation of the fitness value $eval(v_i)$, $i = 1,..., N$ is done referring to equation (6), i.e. by using the equation:

$$eval(v_i) = f(\boldsymbol{\beta}) \quad (8)$$

where $f(\boldsymbol{\beta})$ is the fitness value referring to equation (6), and $\boldsymbol{\beta}' = (\beta_0, \beta_1, ..., \beta_J)$.

6) Cross-breeding, the new population of the selection is done by cross-breeding using the Single-Point Crossover (SPC) method.
7) Mutations, each generation is done by counting $m \times s_p\_size \times p_m$, which is m the number of mutations, the size of the population s_p_size, and p_m the probability of mutations whose value is determined randomly.

8) Decoding, represents the process of encoding genes in a chromosome to recover its original value, i.e. converting coding into decimal values.
   Furthermore, this genetic algorithm is used to analyse observation data about purchasing decision of damage insurance for buildings caused by Citarum river flood.

2.2.4. Test of Significance of Estimator of Logistic Regression Model

This section discusses the significance test of estimator logistic regression model. Testing is done which includes: verification test and validation test.

Verification Test. For verification test of parameter estimator of logistic regression model, this research is done by using Wald test statistic. According to [15], Hosmer and Lameshow in 1989, parameter verifier estimation β_j ($j = 0, 1, ..., J$) testing can be performed using Wald's individual test statistics. The Wald test statistic used is following the standard normal distribution, as follows:

$$\hat{t}_{Stat} = \frac{\hat{\beta}_j}{SE(\hat{\beta}_j)}; \; j = 0, 1, ..., J, \tag{9}$$

where $\hat{\beta}_j$ is an estimator of parameter β_j ($j = 0, 1, ..., J$), and $SE(\hat{\beta}_j)$ is standard deviation of parameter estimator $\hat{\beta}_j$. Hypothesis used for this Wald test is H_0: $\hat{\beta}_j = 0$, with alternative H_1: $\hat{\beta}_j \neq 0$ ($j = 0, 1, ..., J$). The test criterion is to reject H_0 if $\hat{t}_{Stat} < t_{\frac{1}{2}(1-\alpha)}$ or $\hat{t}_{Stat} > t_{\frac{1}{2}(\alpha)}$, in reverse, to accept H_0 if $t_{\frac{1}{2}(1-\alpha)} \leq \hat{t}_{Stat} \leq t_{\frac{1}{2}(\alpha)}$, where $t_{\frac{1}{2}(\alpha)}$ is the percentile of standard normal distribution with significance level of $(1-\alpha)\%$.

Validation Test. In this research, validation test is done by using Likelihood Ratio Test, Hosmer & Lameshow Test, and Test R^2 Determination.

2.2.5. Likelihood Ratio Test

For the validation of estimator estimation of logistic regression model parameters, this study was conducted using likelihood ratio test statistic $\hat{G}$, as follows [6]:

$$\hat{G} = 2\left\{ \sum_{i=1}^{F} D_i \ln \hat{\pi}_i + \sum_{i=1}^{F} (1 - D_i) \ln(1 - \hat{\pi}_i) - n_1 \ln n_1 - n_0 \ln n_0 + F \ln F \right\}. \tag{10}$$

Hypothesis used for likelihood ratio test is H_0: $\hat{\beta}_0 = \hat{\beta}_1 = ... = \hat{\beta}_J = 0$ ($j = 0, 1, ..., J$), with alternative H_1: $\exists \hat{\beta}_0 \neq \hat{\beta}_1 \neq ... \neq \ddot{\beta}_J \neq 0$ ($j = 0, 1, ..., J$). Since statistic $\hat{G}$ is distributed asymptotically with Chi-Square $\chi^2_{(\alpha,df)}$, therefore the criterion test used is to reject H_0, if $\hat{G} > \chi^2_{(\alpha,df)}$, in reverse to accept H_0 if $\hat{G} \leq \chi^2_{(\alpha,df)}$, where α is significance level, and $df = J - 1$ with J is the number of logistic regression model parameters.

2.2.6. Hosmer & Lemeshow Test

The validation test of parameter estimator of logistic regression model can be done by using Hosmer & Lemeshow test statistic $\hat{H}$, as follows [15; 8]:

$$\hat{H} = \sum_{j=1}^{g} \frac{(y_j - n_j x_j)}{n_j x_j (1 - x_j)} \text{ atau } P_{Value} = \Pr(\hat{H}), \tag{11}$$

where $y_j = \sum_{j=1}^{n_j} D_j$ dan $x_j = \sum_{j=1}^{n_j} (m_j \bar{\pi}_j / n_j)$. Hypothesis for Hosmer & Lemeshow test is H_0: There is no difference between the observations and the model used; with alternatives H_0: There is a difference between the observations and the model used. Statistic Hosmer & Lemeshow is distributed asymptotically with Chi-Square $\chi^2_{(\alpha,df)}$, therefore the criterion test used is to reject H_0 if $\hat{H} > \chi^2_{(\alpha,df)}$ or $P_{Value} < (1-\alpha)$, in reverse to accept H_0 if $\hat{H} \leq \chi^2_{(\alpha,df)}$ or $P_{Value} \geq (1-\alpha)$, where α is significance level, and $df = g-2$ with generally $g = 10$.

2.2.7. *Determination* R^2

According to [15], Hosmer & Lemeshow use deterministic values in logistic regression models, to show the correlation strength between independent variables and non-free variables. Statistics of deterministic R^2 coefficients are calculated using the following equation:

$$R^2 = 1 - \exp\left\{-\left(\frac{\hat{\ell}(\boldsymbol{\beta})}{N}\right)\right\}, \tag{12}$$

where $\hat{\ell}(\boldsymbol{\beta})$ is the maximum likelihood log value estimator in equation (6), and N the number of observation data. When the value $R^2 \to 1$, then the correlation between the independent variable and the dependent variable is strong. Conversely, when value $R^2 \to 0$, then the correlation between the independent variable and the dependent variable is not strong.

The Methodology mentioned above, and then used to analyse observation data, as follows.

3. Materials and methods

In this section, we discussed about the data conducted analysis, analysis results, and discussion of the results of analysis, based on the data processed. The data analysed are as described in section 2.1 on the material.

The first step in multivariate analysis, before the data is done the analysis needs to be tested normality. The normality test is intended to ensure that the data is normally distributed. Since the data are good for multivariate analysis, the data is normally distributed. Testing of data normality, can be done using Anderson-Darling statistic with the help of software of Minitab 16. If data is normal distribution, then data can be used to process parameter estimation from binary logistic regression model.

3.1. *Results*

In this section we will discuss the results of parameter estimation from binary logistic regression model. The result of normality test to the data conducted by the analysis shows that the data follows the normal distribution. Therefore, the data can then be used for the process of parameter estimation, which in this study is expressed as a vector parameter $\boldsymbol{\beta}' = (\beta_0, \beta_1, ..., \beta_8)$. The parameter estimate to be obtained is parameters that maximize the log likelihood function value in equation (6). In the process of parameter estimation, in this paper is done by using genetic algorithm. The estimation process using genetic algorithm is done with Matlab 7.0 software. The result of the parameter estimation process is given in Table 1.

Towards the estimators of the coefficient parameters given in Table 1, it is necessary to verify and validate. Verification and validation testing is conducted to ensure that the parameter estimators are significant, either individually or in whole against the logistic regression model estimator.

Partial verification test of parameter estimator of logistic regression model is done by using Wald statistic, referring to equation (9). The hypothesis used in the Wald test is $H_0 : \hat{\beta}_j = 0$, with an alternative $H_1 : \hat{\beta}_j \neq 0$ ($j = 0, 1, ..., 8$). If the significance level is determined by α = 0.95, then from the standard normal distribution table, the percentile values $t_{\frac{1}{2}(1-0.95)}$ = -0.27 and $t_{\frac{1}{2}(0.95)}$ = 0.27 are obtained.

Looking at the results given in Table 1, column (d), because the values of $\hat{t}_j$ ($j = 0,1,...,8$), $\hat{t}_j < t_{\frac{1}{2}(1-0.95)}$ or $\hat{t}_j > t_{\frac{1}{2}(0.95)}$, then reject the hypothesis H_0. This means that all parameter estimators of logistic regression models are significant.

Table 1. Parameter estimator and standard error

Factors of (Z_j)	Parameters Estimator ($\hat{\beta}_j$)	Standard Error $SE(\hat{\beta}_j)$	Ratio $\hat{t}_j = \frac{\hat{\beta}_j}{SE(\hat{\beta}_j)}$	Prob($\hat{t}_j$)	Significances
(a)	**(b)**	**(c)**	**(d)**	**(e)**	**(f)**
Z_1	-0.202521	0.084525	-2.395989	0.000002	Significance
Z_2	3.914214	1.319371	2.966727	0.000031	Significance
Z_3	0.045176	0.013214	3.418798	0.000000	Significance
Z_4	0.065321	0.013523	4.830363	0.000000	Significance
Z_5	-0.018932	0.003261	-5.805581	0.000003	Significance
Z_6	-2.982538	1.058321	-2.818179	0.000001	Significance
Z_7	1.716352	0.83479	2.056028	0.000000	Significance
Z_8	-2.964273	1.161934	-2.551154	0.000051	Significance
Constant	-1.813250	0.503184	-3.603553	0.000024	Significance

Log likelihood statistic $\hat{G}$ = 24.753526

Furthermore, the validation test is performed on the logistic regression model parameter estimators, it is intended to test that the parameters of the logistic regression model parameters together are significant influencing $\pi(Z)$. Validation test of parameter estimators of logistic regression model, in this paper is done by using Likelihood Ratio statistic referring to equation (10). The hypothesis used in this verification test is $H_0 : \hat{\beta}_0 = \hat{\beta}_1 = ... = \hat{\beta}_8 = 0$, with an alternative hypothesis $H_1 : \hat{\beta}_0 = \hat{\beta}_1 = ... = \hat{\beta}_8 = 0$. If determined significance level of α = 0.95 and *df* = 8, then from the Chi-Square distribution table obtained percentile value of $\chi^2_{(1-0.95,8)}$ = 2.7326. While statistics $\hat{G}$ = 24.753526, it is clear that $\hat{G} > \chi^2_{(1-0.95,8)}$. Therefore, the decision rejects the hypothesis H_0. This means that the parameter estimators of the logistic regression model, are of significant influence $\pi(Z)$ together.

The next validation test is performed using the Hosmer & Lemeshow test statistic, with reference to (11). The hypotheses for Hosmer & Lemeshow test is H_0 : There is a difference between the observation data and the estimator model, with the alternative hypothesis H_1: There is no difference between the observation data and the estimator model. Hosmer & Lemeshow statistical tests can also be performed using statistics P_{Value}, with reference to (11). In this test, obtained probability $P_{Value} = 0.2561325$. When the significance level is set to α = 0.95, it is clear that $P_{Value} > (1-\alpha)$. Therefore, the decision rejects the hypothesis H_0, which means that there is no difference between the observational data and the logistic regression model estimator.

Furthermore, is to measure the strength of the correlation between independent and non-free variables, it can be done by using the coefficient of determination R^2, with reference to (12). Results of data processing obtained that the value of determination $R^2 = 0.784392$. This shows that there is a strong correlation between independent variables $Z_1,..., Z_8$ and non-free variable $\pi(Z)$.

3.2. Discussion

Taking into account the logistic regression model estimators given in Table 1 shows that:

(i) The value of estimated coefficient of variables for educational level, the distance of the house with the source of flood, the election of the house with the road, and the public perception of the insurance company, are positive and significant at the significance level of α = 0.95. This illustrates that the higher level of education of people living in areas affected by the Citarum river floods, is very influential on the awareness to buy flood insurance. The closer the house, the closer to the source of the flood, and the position of the house building that is lower or closer to the same level with the high road in front of it, the greater awareness that the residential building has the risk of damaged by the flood. Similarly, the better the public perception of insurance companies, the more likely they will not mind buying flood insurance products.

(ii) The value of estimated variable coefficient for income level, the frequency of floods in a year, the predicted number of floods, and the perceptions of government efforts in tackling the flood are negative and significant at the significance level α = 0.95. This shows that for people with low income levels, frequent flood frequencies within a year, predicted floods are frequent and negative perceptions of government payments to cope with floods, choose not to buy flood insurance products. Interview results explain those who have large funds, prefer to move the location or do the arrangement of the location of the house by raising the building.

Furthermore, when parameter estimators are given on equation (1), we obtain a logistic regression model which can be used to estimate the probability of damage to the building due to Citarum river flood. The equation estimator of the logistic regression model, with the value of variable coefficient estimators rounded up to three decimal places, is as follows:

$$\pi(Z) = \frac{e^{-1.813-0.202Z_1+3.914Z_2+0.045Z_3+0.065Z_4-0.019Z_5-2.983Z_6+1.716Z_7-2.964Z_8}}{1+e^{-1.813-0.202Z_1+3.914Z_2+0.045Z_3+0.065Z_4-0.019Z_5-2.983Z_6+1.716Z_7-2.964Z_8}}. \quad (13)$$

For example, a person living in an area affected by Citarum river floods has the following categories: income levels that are large or in category $Z_1 = 1$; education level is high or in category $Z_2 = 1$; the distance of the house with the source of flood is relatively close or in category $Z_3 = 1$; the building position is classified under the road in front of the house or in category $Z_4 = 1$; the frequency of floods in a year is classified frequently or in category $Z_5 = 1$; predicted floods in the future will be classified as often $Z_6 = 1$; public perception of insurance company is good or in category $Z_7 = 1$; and perceptions of the government's efforts in dealing with floods are low or in category $Z_8 = 0$. The values of this category factor when substituted into equation (13) together can be calculated the probability of damage to buildings due to floods, as follows:

$$\pi(Z) = \frac{e^{-1.813-0.202(1)+3.914(1)+0.045(1)+0.065(1)-0.019(1)-2.983(1)+1.716(1)-2.964(0)}}{1+e^{-1.813-0.202(1)+3.914(1)+0.045(1)+0.065(1)-0.019(1)-2.983(1)+1.716(1)-2.964(0)}}$$

$$= 0.91157$$

Thus, a person living in a flooded area of the Citarum River has a category as mentioned above; having a probability of buying flood insurance products is 0.91157. The example, together with nine other peoples, based on their respective characteristics, the probability prediction of the purchase decision of the flood insurance product is given as in Table 2.

Based on the predicted probability of purchasing decision of the flood insurance product, the insurance company should be able to optimally use, to increase the participation of the people living in the affected area of Citarum river flood, to be willing to buy flood insurance products. Because for people living in areas affected by Citarum river floods, buying flood insurance products can reduce the risk of damage to buildings due to the Citarum river flood.

Table 2. Predicted Probability of purchasing decision of flood insurance products based on their respective characteristics

Buildings	Z_1	Z_2	Z_3	Z_4	Z_5	Z_6	Z_7	Z_8	Probabilty
1	1	1	1	1	1	1	1	0	0.91157
2	1	1	1	1	0	0	1	0	0.99520
3	0	1	0	1	0	1	0	1	0.10239
4	0	1	0	0	1	0	1	0	0.99554
5	1	0	1	0	1	0	1	0	0.79200
6	1	0	1	1	0	1	0	1	0.00194
7	1	1	1	0	1	0	0	1	0.63899
8	1	0	1	0	0	1	1	0	0.16424
9	0	0	0	1	0	0	0	0	0.46555
10	0	0	0	0	0	0	0	0	0.44942

4. Conclusion

In this paper, we have been analysis of purchasing decision of flood insurance by using logistic regression model and genetic algorithm. Based on the analysis result, it can be concluded that the analysis of purchasing decision of flood insurance can be done by using logistic regression model. For estimation of logistic regression model parameters, it can be done using genetic algorithm. The estimation of logistic regression model parameters shows that eight factors $Z_1,...,Z_8$ significantly influence the probability function $\pi(Z)$ in equation (1). Using the equation estimator (1), for a person living in the area affected by Citarum river floods that have the following categories: income levels are large; the level of education is high; the distance of houses with the source of flood is quite close; the building position belongs to the bottom of the road in front of the house; the frequency of floods in a year is frequent; predictions of future floods are often classified; public perception of insurance companies is good; and perceptions of government efforts in the prevention of floods are low, have the probability of decision to buy flood insurance products amounted to 0.91157. Predicted probability of purchasing decision of flood insurance products, by the insurance company can be used optimally to increase public participation in purchasing flood insurance products.

Acknowledgments

Authors thank Rector, Director of DRPMI, and Dean of FMIPA, Universitas Padjadjaran, which provides a grant program of the Academic Leadership Grant (ALG) and a grant program of Competence Research of Lecturer of Unpad (*Riset Kompetensi Dosen Unpad*/RKDU).

References

[1] Atreya, A., S. Ferreira, and E. Michel-Kerjan, 2014, What Drives Households to Buy Flood Insurance Evidence from Georgia, Working Paper # 2014-08, Risk Management and Decision Processes Center, The Wharton School, University of Pennsylvania.

[2] Aliagha, A.G., T. E. Jin, W. W. Choong, M. Nadzri Jaafar, and H. M. Ali, 2014, Factors affecting flood insurance purchase in residential properties in Johor, Malaysia, Nat. Hazards Earth Syst. Sci., 14, 3297–3310, 2014, http: www.nat-hazards-earth-syst-sci.net/14/3297/2014/, doi:10.5194/nhess-14-3297-2014.

[3] Bayaga, A., 2010, Multinomial Logistic Regression: Usage and Application in Risk Analysis. Journal of Applied Quantitative Methods. Vol. 5, No. 2, Summer 2010. pp. 288-297.

[4] Ciumas, C. and R.A. Coca, 2015, An Analysis of the Factors Influencing the Demand for Catastrophe Insurance, Journal of Public Administration, Finance and Law, Special Issue 2/2015, pp. 69-78.

[5] Cramer, J.S., 2002, The Origins of Logistic Regression. Tinbergen Institute Discussion Paper, T1.2002-119/4. Faculty of Economics and Econometrics University of Amsterdam and Tinbergen

Institute, pp. 1-15.

[6] Demir, E. And O. Akkus, 2015, An Introductory Study on "How the Genetic Algorithm Works in the Parameter Estimation of Binary Logit Model?". International Journal of Sciences: Basic and Applied Research (IJSBAR), Volume 19, No 2, pp 162-180.

[7] Desrosiers, M.A., 2012, How Individuals Purchase Insurance: Going Beyond Expected Utility Theory, *Casualty Actuarial Society E-Forum*, Winter 2012-Volume 2, pp. 1-18.

[8] Eshima, N., M. Tabata, and G. Zhi, 2001, Path Analysis with Logistic Regression Models Effect Analysis of Fully Recursive Causal Systems of Categorical Variables. *J. Japan Statist. Soc.* Vol 31, No. 1, pp. 1-14.

[9] Esau, E.Y.S, 2015, Factors Affecting Consumer Purchase Decision on Insurance Product in PT. Prudential Life Assurance Manado, Jurnal EMBA, Vol.3 No.3 Sept. 2015, pp.491-501, ISSN 2303-11.

[10] Mathur, A. and A. Tripathi, 2014, Factors Influencing Customer's Choice for Insurance Companies-A Study of Ajmer City, IOSR Journal of Business and Management (IOSR-JBM), e-ISSN: 2278-487X, p-ISSN: 2319-7668. Volume 16, Issue 2. Ver. I (Feb. 2014), pp. 35-43.

[11] Setiadi, H.A. 2013. Identification of Building Damage and Infrastructure Function due toCitarum Flood in Bandung Regency, Working Paper, *Pusat Litbang Permukiman, Balitbang, Kementerian Pekerjaan Umum Jalan Panyawangan – Cileunyi Wetan, Kabupaten Bandung.*

[12] Shaw, M.M., 2004, Group Flood Insurance Program and Flood Insurance Purchase Decisions, International Journal of Mass Emergencies and Disasters, August 2004, Vol. 22, No. 2, pp. 59–75.

[13] Sidi, P., M. Mamat, Sukono, and S. Supian, 2017a, Supply and demand analysis for flood insurance by using logistic regression model: case study at Citarum watershed in South Bandung, West Java, Indonesia, IOP Conf. Series: Materials Science and Engineering 166 (2017) 012002 doi:10.1088/1757-899X/166/1/012002

[14] Sidi, P., M. Mamat, Sukono, and S. Supian, 2017b, Insurance Premium Risk Evaluation Model for Building Damage Caused by Flood of the River: Case Study on Citarum Watershed in South Bandung Indonesia, Journal of Engineering and Applied Sciences 12 (17): 4420-4425.

[15] Sukono, A. Sholahuddin, M. Mamat, and K. Prafidya, 2014, Credit Scoring for *Cooperative of Financial Services* Using Logistic Regression Estimated by Genetic Algorithm. Applied Mathematical Sciences, Vol. 8, 2014, no. 1, pp. 45-57. http://dx.doi.org/10.12988/ams.2014.310600

[16] Sukono, Nahar, J., Putri, F. T., Subiyanto, Mamat, M., & Supian, S. 2017a. Estimation of outstanding claims reserving based on inflation risk on car insurance companies by using the bootstrap method. *Far East Journal of Mathematical Sciences, 102*(4), 687-706. doi:10.17654/MS102040687.

[17] Sukono, Suyudi, M., Islamiyati, F., & Supian, S. 2017b. Estimation model of life insurance claims risk for cancer patients by using bayesian method. Paper presented at the *IOP Conference Series: Materials Science and Engineering, 166*(1) doi:10.1088/1757-899X/166/1/012022

[18] Ulbinaite, A. and M. Kucinskiene, 2013a, Insurance Service Purchase Decision-Making Rationale: Expert-Based Evidence From Lithuania, Journal of EKONOMIKA 2013 Vol. 92(2), pp. 137-155.

[19] Ulbinaite, A., M. Kucinskiene, and Y. L. Moullec, 2013b, Determinants of Insurance Purchase Decision Making in Lithuania, Inzinerine Ekonomika-Engineering Economics, 2013, 24(2), 144-159. http://dx.doi.org/10.5755/j01.ee.24.2.3439

[20] Yan, X., W. Luo, W.Li, W.Chen, C. Zhang, and H. Liu, 2013, An Improved Genetic Algorithm and Its Application in Classification. IJCSI International Journal of Computer Science Issues, Vol. 10, Issue 1, No 1, January 2013, pp. 337-346.

Fuzzy logic and A* algorithm implementation on goat foraging games

P Harsani*, I Mulyana, D Zakaria

Computer Science Study Program, Pakuan University, Jl. Pakuan PO Box 452 Bogor 16143, Jawa Barat, Indonesia

*Email: prihastuti.harsani@unpak.ac.id

Abstract. *Goat foraging* is one of the games that apply the search techniques within the scope of artificial intelligence. This game involves several actors including players and enemies. The method used in this research is fuzzy logic and Algorithm A*. Fuzzy logic is used to determine enemy behaviour. The A* algorithm is used to search for the shortest path. There are two input variables: the distance between the player and the enemy and the anger level of the goat. The output variable that has been defined is the enemy behaviour. The A* algorithm is used to determine the closest path between the player and the enemy and define the enemy's escape path to avoid the player. There are 4 types of enemies namely farmers, planters, farmers and sellers of plants. Players are goats that aims to find a meal that is a plant. In this game goats aim to spend grass in the garden in the form of a maze while avoiding the enemy. The game provides an application of artificial intelligence and is made in four difficulty levels.

1. Introduction

Artificial intelligence is often applied in a variety of games that aim to make the game more interesting, one of which is on enemy behaviour aimed at defeating players. In the AI game there is a goal to calculate the optimization of actors' behaviour against players. The result must be trustworthy and fun [1]. In the AI game there are two main actions that must be done, namely pathfinding and steering. Pathfinding is how actors can find routes by searching methods in AI. While steering is the method of directing the action based on the environment encountered. One of the games that implement search techniques is Goat Foraging. Game Goat Foraging can be said as a *pacman* game with the labyrinth arena where the labyrinth is a garden or rice fields and players as a goat that can walk in the labyrinth path. Players in this game have a goal to eat all the grass that is in every lane and also besides it must avoid enemies who try to chase the player, where the enemy in this game is human.

The fuzzy method in the game Goat Foraging is a determinant of the behaviour of the enemy whether to chase players, wait or get away from the players. There are 3 characters to be determined fuzzy variables. The characters are breeders, planters, farmer and seller. The breeder will determine the shortest distance with the goat. Variables used are distance, anger and behaviour. There are four rules to describe the relationship between input and output. The specified variables are distance and level of anger. The method used for calculating the distance between enemies and farmers is the Manhattan method.

There are previous studies related to the use of fuzzy logic and the shortest route search algorithm in the game. Based on previous study [2,3,4] the using of fuzzy logic in the game allows game developers to express actor behaviour in a more natural form and ease of programming because it does not require elaborate math formulation. Application of the A* algorithm in game applications is also applied in artificial intelligence games. One of them is the game for pathfinding and steering. Other studies had constructed games with behavioural-reading approach and A* algorithm implementation are performed by [5,6]. The game was built using action-RPG. The behaviour was determined by the fuzzy method. The actor whose behaviour was determined is the attacker and the archer [3]. The next game was a maze game. The A* algorithm was applied to determine the shortest path the player must pass. The next research is to make a maze game. The A* algorithm is applied to determine the shortest path. The player must go through the path to find the solution. The current study is to combine fuzzy and A* methods to produce a goat foraging game.

2. Methods

This research builds a game that implements fuzzy logic and A* algorithm. Games are built using a methodology approach to multimedia design. The stages are as follows (Figure 1):

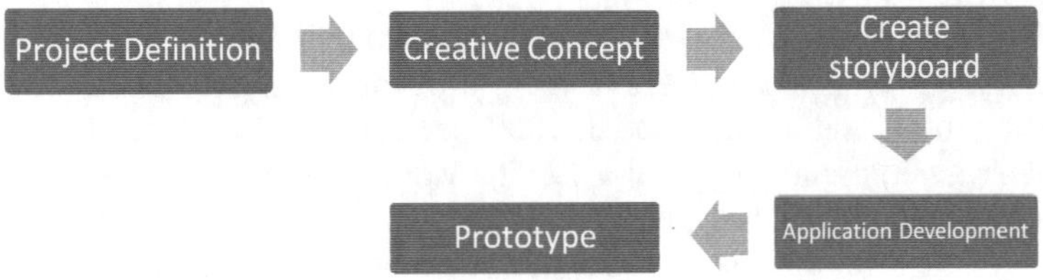

Figure 1. Steps Games Development

Project definition: In this stage determined the basic elements in the games, namely users, content and planning. The game to be built is a game with the type of game platform. The game platform is a unique game and has a simple game system. Platform games usually require gamers to pass through some obstacles to achieve the goal.

Creative concept: This step define the following stages: (i) determination of actors in the game. The actor in the game is divided into 2, namely the player and the enemy. Players are goats that aim to find food, while the enemy consists of farmers, planters, farmers and sellers of plants, (ii) game scenario determination. The goat foraging game scenario involves two actions to be performed in the AI game, namely steering and pathfinding. Steering is how enemy characters behave based on actions performed by players. The behaviours performed are presented in Table 1.

Table 1. Behaviour of enemy character

Enemy Character	Behaviour	Methods
Breeder	the shortest route of search and the level of anger	Fuzzy rule, A*
Planter	Predict the goat route with the level of anger	Fuzzy rule, A*
Farmer	the shortest route of search and the level of anger	Fuzzy rule, A*
Seller	Behaviours such as breeders and planters	A*, Randomize

2.1. Fuzzy Logic

The three enemy behaviours in this game are catch up, leave and waiting. The scale used is 0-200. The behaviour will be determined by two inputs. The inputs are distance and anger level. Distance has a scale of 0 to 50 and the level of anger has a scale of 0 to 100. Fuzzy rule is different for all enemies. This is based on the behaviour that has been determined on the scenario. Fuzzy rule for breeders is as follows:

[R1] IF And anger CALM THEN behaviour CATCH UP;
[R2] IF distance AWAY And anger CALM THEN behaviour CATCH UP;

[R3] IF distance CLOSE And anger ANGRY THEN behaviour LEAVE;
[R4] IF distance AWAY And anger ANGRY THEN behaviour WAIT.

Fuzzy rules like these are basically formed with attention to human behaviour. Humans who want to catch a goat will pay attention goat condition. If the goat is angry then people will tend to avoid goats so as not to get hurt, then if the goats are calm then man will try to catch the goat, because that is formed fuzzy rule.

2.1.1. Fuzzyfication. Fuzzyfication is the process of mapping the input values (crisp input) derived from a controlled system (non-fuzzy quantity) into the fuzzy set according to its membership function. The fuzzy set is a fuzzy input that will be processed fuzzy on the next process. To convert input crisp to fuzzy input, first must specify membership function for each input crisp, then fuzzyfication process will take crisp input and compare with existing membership function to generate fuzzy input. Input variable (distance and anger) and output variable (behaviour) are depicted in the fuzzy scale and the membership function is obtained as shown in (1), (2), (3) and (4).

There are 2 input variables, namely distance and anger level. The distance variable consists of 2 fuzzy sets, near and far. For the closest distance is 0, while the farthest distance is 50 meters (1000 pixels). The value of 1000 is the conversion value of many grids to 50 meters if 1 meter is represented by 20 pixels. The value 1000 is the maximum value of the grid size. Distance value 0 means the enemy and the goat are at the same place and farthest is 50 (1000 pixels). The level of anger consists of 2 fuzzy sets, which are angry and calm. The anger level has an interval of 0-100. The value of 100 is a multiple of 2 out of 50.

$$\mu\, CLOSE[x] = \begin{cases} 1, & x \leq 0 \\ \frac{50-x}{50-0}, & 0 \leq x \leq 50 \\ 0, & x \geq 50 \end{cases}$$

$$\mu\, AWAY[x] = \begin{cases} 1, & x \geq 50 \\ \frac{x-0}{50-0}, & 0 \leq x \leq 50 \\ 0, & x \leq 0 \end{cases} \tag{1}$$

$$\mu\, CALM[y] = \begin{cases} 1, & x \leq 0 \\ \frac{100-y}{100-0}, & 0 \leq y \leq 100 \\ 0, & x \geq 100 \end{cases}$$

$$\mu\, ANGRY[y] = \begin{cases} 1, & y \geq 100 \\ \frac{y-0}{100-0}, & 0 \leq y \leq 100 \\ 0, & y \leq 0 \end{cases} \tag{2}$$

The output variable that will be determined membership function is behaviour. This variable consists of three fuzzy sets, leave, wait and catch-up. Three domains are obtained by observing human behaviour in pursuit. There are 3 values, namely 0 for the set of blur, 100 for waiting and 200 to pursue. This value is derived based on human behaviour that will wait before pursuit.

$$\mu\, LEAVE[z] = \begin{cases} 1, & z \leq 0 \\ \frac{100-z}{100-0}, & 0 < z \leq 100 \\ 0, & z \geq 0 \end{cases}$$

$$\mu\, WAIT[z] = \begin{cases} 1, & z = 100 \\ \frac{z-0}{100-0}, & 0 < z < 100 \\ \frac{200-z}{200-100}, & 100 < z \leq 200 \\ 0, & z \leq 0, z \geq 200 \end{cases} \tag{3}$$

$$\mu\, CATCHUP[z] = \begin{cases} 1, & z \geq 200 \\ \frac{z-100}{200-100}, & 100 \leq z \leq 200 \\ 0, & z \leq 100 \end{cases} \tag{4}$$

2.1.2. Fuzzy Implication. Operator used for fuzzy implication is operator MIN. Every fuzzy rule that has been defined will produce implications. For R1, fuzzy implications are presented in Figure 2. This will apply to all rules.

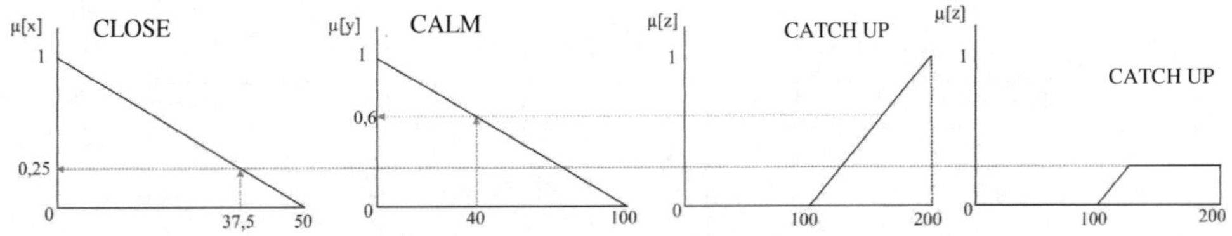

Figure 2. Fuzzy implication

2.1.3. Rule Composition. The composition of all rules will be generated by applying the MAX method to the implication function (Figure 3)

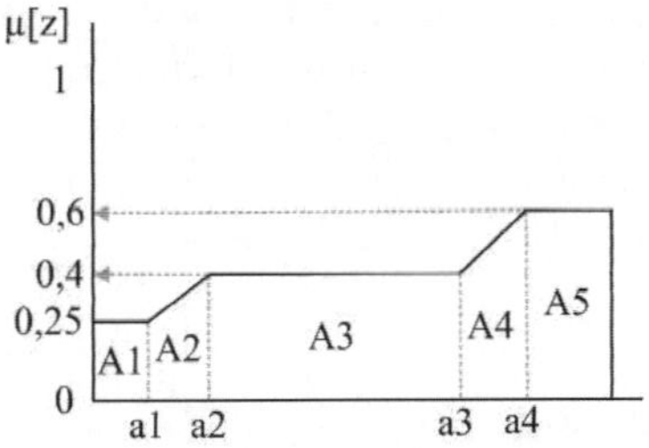

Figure 3. Rule Composition

2.1.4. Defuzzyfication. The defuzzification used is the centroid method. The results obtained from fuzzyfication can determine the speed of the enemy chasing the goats for each of its frames [3].

2.2. A Algorithm*

The A* algorithm is used to determine the shortest distance between enemies and goats. With the A* algorithm the enemy will run effectively to get to the goats making it difficult for the goats to avoid the enemies who chase them. The A* algorithm is in the process of using nodes that form dots or boxes as determinants of places or paths that can be passed. The A* algorithm is one of the Branch & Bound algorithms, also called an algorithm to search for solutions using additional information (heuristics) in generating the optimal solution [6].

2.3. Manhattan Heuristic

Manhattan heuristics are used to determine the distance between enemy and player, the process of calculating the distance using Manhattan is done by summing the horizontal distance of the enemy and the goat with the vertical distance of the enemy and the goat. If the enemy P1 is at the point (x_1, y_1) and the goat P2 is at the point (x_2, y_2), then to calculate the Manhattan distance d_M between the enemy and the goat is:

$$d_M = |x_1 - x_2| + |y_1 - y_2| \tag{5}$$

Create Storyboard: In this stage is designed things as follows, design a navigation map, define screen layouts, contents and actions, design user interaction controls. This game consist of nine main screen that is opening, menu, menu selection, game, game complete, game ends, help, about and exit. The implementation of this game is a hierarchy.

Application Development: This stage is transforming design into games using Adobe FlashCS6. There are three parts in this stage that is define and implement visual layout of the interface, produce/edit graphics, videos, sound and create controls and insert elements.

3. Result and Discussion

This game has three levels that have a maze and setting different places, the degree of difficulty is distinguished by the types of labyrinth such as shown in Table 2:

Table 2. Types of Difficulty Levels in the Maze

Level	Branches	Number of Flower	Dead ends
Level 1	59 Branches	10 flowers	-
Level 2	40 Branches	8 flowers	-
Level 3	34 Branches	8 flowers	8 Dead ends

This game has a place in the garden gardens with a fence as a barrier and a light brown path that can be passed by goats and enemies. The player will control the movement of the goat. The enemy will move to pursue the player with the shortest path. Determination of the shortest path using A* algorithm. If players eat flowers then the enemy will run away but this is only temporary because the level of anger will decrease slowly and the enemy will again catch up if the level of anger is up. Players will be deemed victorious if the grass and the existing interest have been exhausted and the player will be deemed lost or game over if the player comes in contact with the enemy. In general, the game structure can be seen in Figure 4.

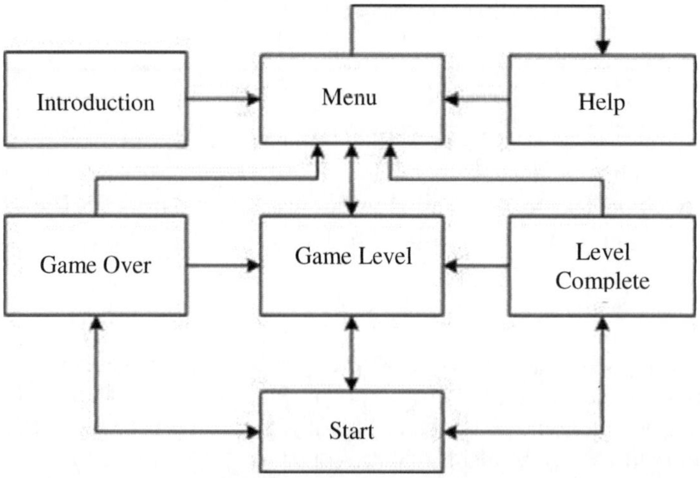

Figure 4. Navigation Structure

Level 1 in this game has a place setting in the garden maze with a fence as a barrier and a light brown path that can be passed by goats and enemies (Figure 5) The game starts when the player presses the direction button on the keyboard, and at that moment the enemy will start to chase the player with the shortest path. If players eat flowers then the enemy will run away but this is only temporary because the level of anger will decrease slowly. The enemy will again catch up if the level of anger is up. Players will be deemed victorious if the grass and flowers are exhausted and the player will be considered defeated. If the player comes into contact with the enemy then the game will be finished. The other levels only have differences in the difficulty levels mentioned in Table 2. Each level will have different settings.

Manhattan distance calculations and A* algorithm implementation use the coordinates as location reference and will use the frame as a character movement. Images and pictures show examples of distance calculations (Figure 6). In the path determination and distance calculation, A* implementation is performed using the grid. (Figure 6). The use of grids in A* implementation is one of the strategies to reduce state space [8]. The larger the size of the grid fed the size of the state space will be greater and further increase the time complexity.

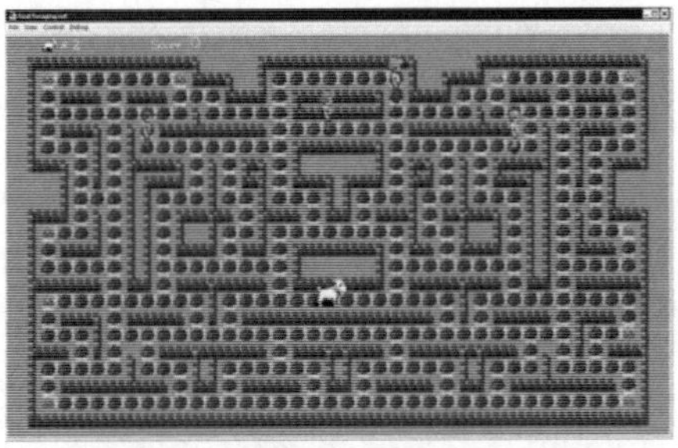

Figure 5. Game Layout

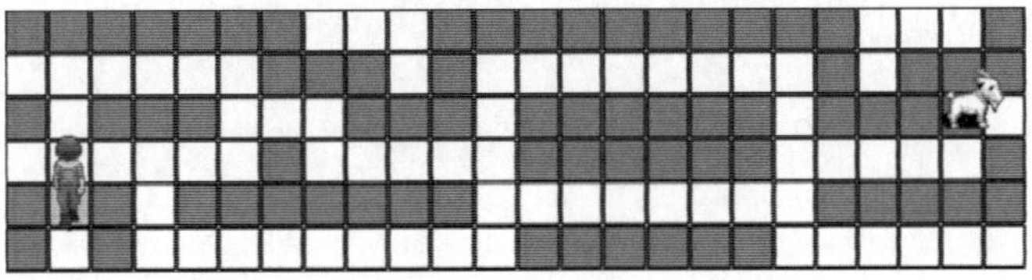

Figure 6. Manhattan

The known coordinates of the enemy are 208,183 and the coordinates of the goats are 887,112. Then the distance can be known as follows:

$$h(breeder) = |\, x(breeder) - x(goat)\,| + |\, y(breeder) - y(goat)|$$
$$= |\, 208 - 887\,| + |\, 183 - 112|$$
$$= 679 + 71 = 750\, pixel$$

Then the distance is 750 pixels, then converted into meters,

$$distance\ (meter) = \frac{750}{20} = 37{,}5\ meter$$

The 1 meter size is assumed to require 20 pixels. This is taking into account the distance on a fuzzy that is limited to 0 to 50 or if in pixels equals 0 to 1000 pixels (50 multiplied by 20). 1000 pixels adjust to the size of the labyrinth of 1152x640 and a maximum resolution of 1280x750.

4. Conclusion

This research builds a platform games. Game rules implemented using fuzzy logic and A* algorithm. Fuzzy logic is used to determine the level of enemy characters and actions performed while the A* algorithm is used to determine the escape path for the enemy and its target. This game is made using Adobe Flash CS6 and action script. This study shows that the games can be made using the concept of artificial intelligence and able to apply the degree of difficulty based on the shape of the path and speed of pursuit by the enemy. To increase game difficulty, fuzzy rule can be added in the determination of enemy characters and target action.

References

[1] Narayek, A. 2004. *AI in Computer Games* Volume 1 Issue 10 Game Development. New York: ACM.

[2] Ha JW, Pyo H and Kim J. 2004. *Fuzzy Logic in Agent-Based Game Design*. IEEE Explore Volume 2. IEEE Annual Meeting of the Fuzzy Information, Processing NAFIPS '04. Canada: IEEE.

[3] Pirovano, M. 2012. The use of Fuzzy Logic for Artificial Intelligence in Games.

[4] Hui YC, Prakash EC, and Chaudhari NS. 2004. *Game AI: Artificial Intelligence for 3D Path Finding*, TENCON 2004. IEEE Region 10 Conference, Volume B, 306- 309, Thailand: IEEE.

[5] Purba KR, Hasanah, RN, and Muslim, MA. 2013. *Implementasi Logika Fuzzy untuk*

Mengatur Perilaku Musuh dalam Game Bertipe Action-RPG. Jurnal EECCIS, Universitas Brawijaya Vol 7, No 1. Malang: Universitas Brawijaya.

[6] Rengga, DP, Djuriatno W., and Aswin M. 2013. *Pencarian Rute Terdekat pada Labirin Menggunakan Metode A**. Jurnal Mahasiswa TEUB Vol. 1, No. 2. Malang: Universitas Brawijaya.

[7] Simanjuntak, H. and Sigiro, M. 2006. *Penerapan Algoritma A* sebagai Salah Satu Algoritma Branch & Bound pada Aplikasi Global Positioning System (GPS)*. Teknik Informatika STEI ITB. Bandung: Institut Teknologi Bandung.

[8] Hu J, Wan, WG and Yu, X. 2012 *Pathfinding Algorithm in Real-Time Strategy Game Based on Unity3D*. International Conference on Audio Language and Image Processing 16-18 July 2012 Shanghai. China: IEEE.